D1162287

Communications and Control Engineering

Springer
London
Berlin
Heidelberg
New York
Barcelona
Hong Kong
Milan
Paris
Singapore
Tokyo

Published titles include:

Markov Chains and Stochastic Stability
S.P. Meyn and R.L. Tweedie

Robust Control: Systems with Uncertain Physical Parameters
J. Ackermann in co-operation with A. Bartlett, D. Kaesbauer, W. Sienel and R. Steinhauser

Optimal Sampled-Data Control Systems
Tongwen Chen and Bruce Francis

Nonlinear Control Systems (3rd edition)
Alberto Isidori

Theory of Robot Control
C. Canudas de Wit, B. Siciliano and G. Bastin (Eds)

Fundamental Limitations in Filtering and Control
María M. Seron, Julio Braslavsky and Graham C. Goodwin

Constructive Nonlinear Control
R. Sepulchre, M. Jankovic and P.V. Kokotović

A Theory of Learning and Generalization
M. Vidyasagar

Adaptive Control
I.D. Landau, R. Lozano and M.M'Saad

Stabilization of Nonlinear Uncertain Systems
Miroslav Krstić and Hua Deng

Passivity-based Control of Euler-Lagrange Systems
Romeo Ortega, Antonio Loría, Per Johan Nicklasson and Hebertt Sira-Ramírez

Stability and Stabilization of Infinite Dimensional Systems with Applications
Zheng-Hua Luo, Bao-Zhu Guo and Omer Morgul

Nonsmooth Mechanics (2nd edition)
Bernard Brogliato

Nonlinear Control Systems II
Alberto Isidori

L_2-Gain and Passivity Techniques in Nonlinear Control
Arjan van der Schaft

Control of Linear Systems with Regulation and Input Constraints
Ali Saberi, Anton A. Stoorvogel and Peddapullaiah Sannuti

Robust and $H\infty$ Control
Ben M. Chen

Efim N. Rossenwasser and Bernhard P. Lampe

Computer Controlled Systems

Analysis and Design with Process-orientated Models

With 111 Figures

Springer

Efim N. Rosenwasser, Prof Dr
Department of Automatic Control, St Petersburg State Marine Technical University, St Petersburg, Russia

Bernhard P. Lampe, Prof Dr
University of Rostock, Institute of Automation, D-18051 Rostock, Germany

Series Editors

E.D. Sontag • M. Thoma

ISSN 0178-5354

ISBN 1-85233-307-3 Springer-Verlag London Berlin Heidelberg

British Library Cataloguing in Publication Data
Rosenwasser, Efim N.
Computer controlled systems : analysis and design with process-orientated models. - (Communications and control engineering)
1.Automatic control
I.Title II.Lampe, Bernhard P.
629.8'95
ISBN 185233307-3

Library of Congress Cataloging-in-Publication Data
A catalog record for this book is available from the Library of Congress

Apart from any fair dealing for the purposes of research or private study, or criticism or review, as permitted under the Copyright, Designs and Patents Act 1988, this publication may only be reproduced, stored or transmitted, in any form or by any means, with the prior permission in writing of the publishers, or in the case of reprographic reproduction in accordance with the terms of licences issued by the Copyright Licensing Agency. Enquiries concerning reproduction outside those terms should be sent to the publishers.

© Springer-Verlag London Limited 2000
Printed in Great Britain

"MATLAB® and is the registered trademark of The MathWorks, Inc., http://www.mathworks.com"

The use of registered names, trademarks, etc. in this publication does not imply, even in the absence of a specific statement, that such names are exempt from the relevant laws and regulations and therefore free for general use.

The publisher makes no representation, express or implied, with regard to the accuracy of the information contained in this book and cannot accept any legal responsibility or liability for any errors or omissions that may be made.

Typesetting: Camera ready by authors
Printed and bound by Athenæum Press Ltd., Gateshead, Tyne & Wear
69/3830-543210 Printed on acid-free paper SPIN 10761438

To Jelena and Bärbel

Preface

Digital elements such as signal processors, microcontrollers or industrial PCs are widely used for filtering and control of continuous-time processes or plants. Because of the serial operation mode of these digital elements the continuous-time processes cannot be served continuously. That is why continuous-time and discrete-time signals occur simultaneously in systems of this kind. The continuous (real) time also contains the discrete time instants – therefore, a complete and accurate description of such systems in continuous time is possible. However, this description proves to be more complicated, because the hybrid system becomes time-variable, also in cases, where all continuous-time processes and all digital elements are time invariant. Often digital elements operate with a fixed time period T, that leads to a periodically time-variable system.
Besides the time sampling the use of analogue to digital and digital to analogue converters together with the finite numerical precision of the processors also results in amplitude quantization of the measuring and control signals. Taking into account these effects would lead to nonlinear models. Because these effects are often of minor influence but difficult in handling, in theory and practice it is generally accepted to consider mainly the pure time-quantization effect. Systems containing both continuous-time and discrete-time signals are usually called sampled-data systems (SD systems).

In the well known textbook of Åström and Wittenmark (1984) two principally different approaches for modelling computer controlled systems were formulated. The first possibility is to look at a system from the point of view of a computer. Then all measuring and control signals are considered at the sampling instants only. This results in quite simple discrete-time, time-invariant models of the system. This method is called *Computer Oriented Approach.* The second possibility is to observe the system from the continuous-time process. This needs a more detailed description. The models required are necessarily more complicated. This method is called *Process Oriented Approach.*

The computer-oriented methods are comparatively simple, and can be effectively used for the solution of a lot of problems in computer controlled systems. For these reasons, many textbooks and monographs in this field use only computer-oriented models.

On the other hand, it was shown in numerous research projects over the last decade that many important problems in analysis and design of computer controlled systems can only be solved on the basis of process-oriented models. Recently, it has became evident that the computer-oriented approach has fundamental limitations for the following reasons. First, a rigorous discrete-time model of the continuous-time plant is not always available. Second and more important, digital control design techniques based on computer-oriented plant models cannot, in many cases, ensure the acceptable performance of the system to be designed, see Feuer and Goodwin (1996).

Such situations often occur if the performance of sampled-data systems is investigated under disturbances affecting continuous-time networks directly. Hence, problems of this kind are among the most important in modern control theory, because of the mass introduction of computers in control systems for continuous-time plants and processes, see Anderson (1993).

A rigorous solution of these problems involves theoretical difficulties, because the use of routine investigation methods for time-invariant systems (both continuous and discrete time) is not effective in the above situations. This results from the fact that SD systems considered in continuous time are time-variable systems with periodically varying parameters. For this reason, the investigation of process-oriented models has recently attracted a great deal of attention in the literature. A summery of the results in this field and further bibliography can be found in Ackermann (1988); Chen and Francis (1995); Dullerud (1996); and Sågfors (1998).

The present book for the first time gives a detailed, step-by-step presentation of a theory of linear computer controlled systems on the basis of process-oriented models. The main idea consists in regarding SD systems as a special class of linear systems with periodically varying parameters. The approach used in the book is based exclusively on Input-/Output representations and methods in the frequency domain and does not involve a transformation to the time domain. This opens up new important possibilities from both theoretical and computational viewpoints, especially in application of modern polynomial methods, EUROPOLY. Various aspects of the present book have been presented in the monographs of Rosenwasser (1973, 1994c), and in the published results of the joint investigations in the field of direct SD synthesis performed in 1989–1999 at St. Petersburg State Marine Technical University and the University of Rostock.

The approach developed in the book is based on a mathematical description of sampled-data systems by means of the bilateral Laplace transformation in continuous time and the parametric transfer function (PTF) concept. When applied to SD systems, this approach is constructive and gives a universal description of SD systems over the whole time axis $-\infty < t < \infty$. This description includes the ordinary linear time-invariant (LTI) models (in continuous or discrete time) as special cases.

This technique provides general analytical and design methods for SD systems, which in principle can be extended to systems with distributed parameters. Such systems include, for instance, those with time delay which are also discussed in the book.
The theory presented below is conceptually based on the work of Ya. Z. Tsypkin, who proposed a general frequency-domain description of sampled-data systems in the complex plane, and L. A. Zadeh, who introduced the concept of PTF into automatic control theory. Other significant results in this field are due to J. R. Raggazini, J. T. Tou, and S. S. L. Chang.

The book consists of five parts and two appendices.
In the first part (Chapters 1–4) a theory of some functional transforms is presented which generalizes the concept of the modified discrete Laplace transform. These transforms have proved to be an effective tool for the mathematical description of SD systems in continuous time.

The second part (Chapters 5–8) deals with operator description of linear systems with periodically varying parameters. In this context the concepts of parametric transfer function and frequency response characteristic of a linear periodic operator and linear periodic system are introduced and a rigorous mathematical justification is given for the application of the Laplace transform and the PTF technique to the analysis of linear periodic systems under deterministic and stochastic disturbances.

In Part III (Chapters 9–12) various classes of linear periodic SD systems are investigated as special cases of general linear periodic systems. This idea makes it possible, on the basis of results obtained in Part II, to develop a general frequency-domain description for various kinds of SD systems. Open-loop and closed-loop systems with one or more sampling units (including those with phase shift) are considered step by step. Closed expressions for parametric transfer functions and output operational transforms are given for various types of SD systems.
The results developed in Part III clearly demonstrate the wide possibilities of the adopted approach. For instance, the concept of transfer function appears to be extensible to a general class of sampled-data systems for which transfer functions in the classical sense do not exist. Such systems include for example, SD systems in which exogenous disturbances act upon continuous-time elements, and periodic SD systems with several sampling units.

Part IV (Chapters 13–15) presents frequency-domain analysis methods for SD systems under deterministic and stochastic disturbances. Methods for the analysis of transients and quasi-stationary modes are given. The stability and stabilization of periodic SD systems of various classes are considered in detail. Investigation of process-oriented models makes it possible to find some remarkable features in the behaviour of computer controlled systems which cannot be described by computer-oriented models. For example, the concept of steady-state error has to be modified for SD systems, because, in general, the steady-state process at a constant input is periodic. The steady-state response of a stable system to a stationary stochastic input signal is a periodically non-stationary stochastic process. Nevertheless, some concepts of LTI, continuous-time system theory, for instance the concept of frequency response, can be effectively generalized for sampled-data systems.

The fifth, and last part of the book (Chapters 16–18) presents frequency-domain and polynomial methods for direct SD system design. We consider the problems of quadratic optimization of SD systems in continuous time in the intervals $0 < t < \infty$ (LQ–problem) and $-\infty < t < \infty$ ($\mathcal{H}_2$–problem), and also the associated $\mathcal{H}_\infty$–problem. In addition, statements of some robust and $\mathcal{L}_2$–optimization problems are presented, which can be reduced to an $\mathcal{H}_\infty$–optimisation problem for an equivalent discrete-time system. Chapter 18 presents a polynomial technique for the realization of the proposed design methods. It has been written by Dr. K. Polyakov, who developed these computational procedures and the relevant MATLAB toolbox ***DirectSD***, which was used to compute the numerical examples given in Part V.
The results of Part V form the basis for the conclusion that a process-oriented approach is of great practical importance for optimal design of computer controlled systems. For instance, a discrete-time control design based on computer-oriented models of continuous-time plants sometimes gives closed-loop systems with small stability margin. In such cases the error signal can be very large between the sampling instants, and this effect can become more significant as the sampling period decreases.

Appendix A describes general properties of rational periodic functions, which constitute the basis for the computational procedures given in the book.
Appendix B is a description of the MATLAB toolbox ***DirectSD***. A reduced release of this toolbox is free available as ***SDshort*** at the homepage of the second author

http://www-at.e-technik.uni-rostock.de/~bl/Matlab-toolbox.html

With the help of Appendix B and the toolbox ***SDshort*** the reader gains ability to calculate numerous examples of the book, and to see by himself the practical relevance of the theory.

In principle, the book does not call for preliminary knowledge in the field of computer-oriented system theory. Nevertheless, a reader familiar with the standard methods for computer-oriented models, e.g. with the already mentioned book of Åström and Wittenmark (1984), can better appreciate the new possibilities offered by the process-oriented approach.
To make the material more comprehensible, new ideas and concepts are mainly discussed for single-input/single-output (SISO) cases, though most of the results, without major modifications, can be extended to multivariable systems. Examples of such generalizations are given in the relevant parts of the book.
The bulk of the book is illustrated by theoretical and numerical examples, which are an integral part of the text. The results of some examples are of

independent significance. Most of them are chosen in such a way that their solution by computer-oriented methods would involve great difficulties, or even are unsolvable.

The book is addressed to engineers and scientific workers involved in the investigation and design of computer controlled systems. It also can be used as a complementary textbook for process-oriented methods in computer controlled systems by students engaged in control theory, communication engineering, and related fields. It also offers interesting insights for practically oriented mathematicians and engineers in system theory.

The mathematical tools used in the book in general are included in basic math syllabuses for engineers at technical universities. Necessary additional material is given directly in the text.

The bibliography is by no means complete, for it includes only works directly used by the authors.

The authors gratefully acknowledge financial supports from the Deutsche Forschungsgemeinschaft and the Deutsche Akademische Austauschdienst. We thank colleagues in Saint Petersburg and Rostock for their help during the preparation of the print pattern, especially, Dr. Konstantin Polyakov for his effective cooperation and the first English translation, and Patrick Plant for his conscientious corrections. We are especially indebted to the Professors J. Ackermann, B.D.O. Anderson, K.J. Åström, M. Grimble, B. Francis, P.M. Frank, B. Friedland, G.C. Goodwin, T. Kaczorek, and A. Weinmann for many helpful discussions and valuable hints.

Efim Rosenwasser and Bernhard Lampe

Rostock, February 2000

Contents

Part I

Operational transformations of functions of continuous and discrete arguments

Introduction As is known, operational methods play a major part in the theory of linear stationary continuous and discrete-time systems. These methods are based on the Laplace transformation for continuous systems or on the z-transformation, discrete Laplace transformation and other similar techniques for discrete-time systems. Such approaches appear to be unacceptable for problems considered in the present book, because sampled-data systems are hybrid in nature, so that both continuous and discrete-time processes take place, simultaneously. Therefore, to extend operational calculus to sampled-data systems, we have to employ more general operational transformations and their combinations. The first part of the book contains a theory of such transformations and constitutes the basis for the investigation methods developed subsequently.

Chapter 1

Bilateral Laplace transforms

1.1 Exponentially bounded functions

In this section we introduce a special class of functions which will be of prime importance in what follows.

Definition. A function $f(t)$ of the real argument t defined over $-\infty < t < \infty$ will be called *exponentially bounded* in the interval $[\alpha, \beta]$ if there exists a real constant $M > 0$ such that

$$|f(t)| < \begin{cases} Me^{\alpha t} & t \geq 0 \\ Me^{\beta t} & t \leq 0 \end{cases} \tag{1.1}$$

where α and β are real constants and $\alpha < \beta$. In what follows the set of functions $f(t)$ satisfying (1.1) will be denoted by $\mathrm{H}[\alpha, \beta]$.
If, in particular,

$$f(t) = 0 \quad \text{for} \quad t \leq 0 \tag{1.2}$$

then, assuming $\beta = \infty$, instead of (1.1) we obtain

$$|f(t)| < Me^{\alpha t}, \quad t \geq 0\,. \tag{1.3}$$

The set of functions $f(t)$ satisfying (1.2) and (1.3) will be henceforth denoted by $\mathrm{H}_+[\alpha, \infty]$. If

$$\begin{aligned} f(t) &= 0 \quad \text{for} \quad t \geq 0 \\ |f(t)| &< Me^{\beta t}, \quad t \leq 0 \end{aligned} \tag{1.4}$$

then we shall write $f(t) \in \mathrm{H}_-[-\infty, \beta]$.

Let us formulate a general property indicating that a function $f(t)$ belongs to $\mathrm{H}[\alpha, \beta]$.

Theorem 1.1 *A function $f(t)$ satisfies $f(t) \in \mathrm{H}[\alpha, \beta]$ if and only if there exists a constant $M > 0$ such that*

$$\left|f(t)\mathrm{e}^{-st}\right| < M, \quad -\infty < t < \infty \tag{1.5}$$

for all complex s in $\alpha \leq \operatorname{Re} s \leq \beta$.

Proof *Sufficiency:* Suppose that for all $\alpha < \operatorname{Re} s < \beta$ the estimate (1.5) holds. Then, for $s = \alpha$, $t \geq 0$ we have

$$\left|f(t)\mathrm{e}^{-\alpha t}\right| < M, \quad t \geq 0$$

i.e., $|f(t)| < M\mathrm{e}^{\alpha t}$ for $t \geq 0$. Similarly we obtain $|f(t)| < M\mathrm{e}^{\beta t}$ for $t \leq 0$, therefore (1.1) holds and $f(t) \in \mathrm{H}[\alpha, \beta]$.
Necessity: Let q be a real value. Using (1.1) we have

$$\left|f(t)\mathrm{e}^{-qt}\right| < \begin{cases} M\mathrm{e}^{(\alpha-q)t} & t \geq 0 \\ M\mathrm{e}^{(\beta-q)t} & t \leq 0 . \end{cases} \tag{1.6}$$

Hence, for all $\alpha \leq q \leq \beta$,

$$\left|f(t)\mathrm{e}^{-qt}\right| < M, \quad -\infty < t < \infty . \tag{1.7}$$

Taking into account that for any complex s we have $|\mathrm{e}^{-st}| = \mathrm{e}^{-\operatorname{Re} st}$, we conclude that (1.5) follows from (1.7). ■

Corollary From Theorem 1.1 it follows that for $f(t) \in \mathrm{H}[\alpha, \beta]$ we also have $f(t) \in \mathrm{H}[\alpha', \beta']$ for any $\alpha < \alpha' < \beta' < \beta$.

In the following we shall write $f(t) \in \mathrm{H}(\alpha, \beta)$, if $f(t) \in \mathrm{H}[\alpha', \beta']$ for any $\alpha' > \alpha$, $\beta' < \beta$, $\alpha' < \beta'$. The notations $\mathrm{H}_+(\alpha, \infty)$ and $\mathrm{H}_-(-\infty, \beta)$ are defined similarly.

Theorem 1.2 *Let $f(t) \in \mathrm{H}(\alpha, \beta)$ and $\alpha < \operatorname{Re} s < \beta$. Then there exist constants $\gamma > 0$ and $M > 0$ such that*

$$\left|f(t)\mathrm{e}^{-st}\right| < M\mathrm{e}^{-\gamma|t|}, \quad -\infty < t < \infty . \tag{1.8}$$

Proof Let $f(t) \in \mathrm{H}(\alpha, \beta)$ and $\alpha < \operatorname{Re} s < \beta$. Then there exists a positive constant $\gamma > 0$ such that

$$\alpha < \operatorname{Re} s \pm \gamma < \beta . \tag{1.9}$$

Hence, using (1.5) we have

$$\left|f(t)\mathrm{e}^{(-s\pm\gamma)t}\right| < M, \quad -\infty < t < \infty , \tag{1.10}$$

and further

$$\left|f(t)\mathrm{e}^{-st}\right| < \begin{cases} M\mathrm{e}^{-\gamma t} & t \geq 0 \\ M\mathrm{e}^{\gamma t} & t \leq 0 \end{cases} \tag{1.11}$$

which is equivalent to (1.8). ■

Corollary 1 For $f(t) \in \mathrm{H}_+(\alpha, \infty)$ and $\alpha < \alpha'$ we have

$$\left|f(t)\mathrm{e}^{-st}\right| < M\mathrm{e}^{-\gamma t}, \quad t \geq 0, \quad \operatorname{Re} s \geq \alpha'. \tag{1.12}$$

Corollary 2 For $f(t) \in \mathrm{H}(-\lambda, \lambda)$, where λ is a positive number, we have

$$|f(t)| < M\mathrm{e}^{-(\lambda-\epsilon)|t|}, \quad -\infty < t < \infty \tag{1.13}$$

where ϵ is an arbitrarily small positive value.

1.2 Bilateral Laplace transforms

For a function $f(t) \in \mathrm{H}(\alpha, \beta)$ the integral

$$F(s) = \int_{-\infty}^{\infty} f(t)\mathrm{e}^{-st}\,\mathrm{d}t \tag{1.14}$$

converges absolutely for any s in the stripe $\alpha < \operatorname{Re} s < \beta$. Indeed, for $\alpha < \operatorname{Re} s < \beta$, by (1.8) we have

$$|F(s)| \leq \int_{-\infty}^{\infty} \left|f(t)\mathrm{e}^{-st}\right|\mathrm{d}t < M\int_{-\infty}^{\infty} \mathrm{e}^{-\gamma|t|}\,\mathrm{d}t = \frac{2M}{\gamma}. \tag{1.15}$$

The function $F(s)$ is called the *bilateral Laplace transform* for the function $f(t)$. If, as a special case, $f(t) \in \mathrm{H}_+(\alpha, \infty)$, instead of (1.14) we have

$$F(s) = \int_{0}^{\infty} f(t)\mathrm{e}^{-st}\,\mathrm{d}t \tag{1.16}$$

and, for $f(t) \in \mathrm{H}_-(-\infty, \beta)$, (1.14) reduces to

$$F(s) = \int_{-\infty}^{0} f(t)\mathrm{e}^{-st}\,\mathrm{d}t. \tag{1.17}$$

In the interest of brevity bilateral Laplace transforms will henceforth simply be called Laplace transforms. The functions $f(t)$ and $F(s)$ are called original and image, respectively. Sometimes instead of (1.14), (1.16) and (1.17) we shall use the notations

$$\begin{aligned} f(t) &\longrightarrow F(s) \\ f(t) &\stackrel{\cdot}{\longrightarrow} F(s) \\ f(t) &\underset{\cdot}{\longrightarrow} F(s)\,. \end{aligned} \tag{1.18}$$

1.3 Inverse transforms

In many cases it is possible to reconstruct the original $f(t)$ by the image $F(s)$. In order to formulate the appropriate theorem we introduce a new class of functions. From now on we shall write $f(t) \in \Lambda(\alpha,\beta)$ if $f(t) \in \mathrm{H}(\alpha,\beta)$ and $f(t)$ has bounded variation over any finite interval of argument variation. The sets $\Lambda_+(\alpha,\infty)$ and $\Lambda_-(-\infty,\beta)$ are defined in a similar way. The properties of functions of bounded variation are considered, for example, in Titchmarsh (1932). Readers who are not acquainted with this concept may simply suppose that the function $f(t)$ is piece-wise smooth in any finite interval of argument variation. As is known, for a function of bounded variation for any fixed $t=\tilde{t}$ there exist finite limits

$$f(\tilde{t}+0) = \lim_{\epsilon\to+0} f(\tilde{t}+\epsilon), \quad f(\tilde{t}-0) = \lim_{\epsilon\to-0} f(\tilde{t}+\epsilon). \tag{1.19}$$

Besides, at any point where the function is continuous we have

$$f(\tilde{t}+0) = f(\tilde{t}-0) = f(\tilde{t}). \tag{1.20}$$

Theorem 1.3 (van der Pol and Bremmer (1959)) *Let $f(t) \in \Lambda(\alpha,\beta)$ and $F(s)$ be its image. Then, for all $-\infty < t < \infty$ there exists the limit*

$$\hat{f}(t) = \frac{1}{2\pi \mathrm{j}} \lim_{a\to\infty} \int_{c-\mathrm{j}a}^{c+\mathrm{j}a} F(s)\mathrm{e}^{st}\,\mathrm{d}s, \quad \alpha < c < \beta \tag{1.21}$$

independent of c. Then, for any t,

$$\hat{f}(t) = \frac{f(t+0)+f(t-0)}{2} \tag{1.22}$$

and, at the points of continuity,

$$\hat{f}(t) = f(t). \tag{1.23}$$

■

Formula (1.21) is named the *inverse Laplace transformation.* In applications (1.21)–(1.23) are often written in the form

$$f(t) = \frac{1}{2\pi \mathrm{j}} \int_{c-\mathrm{j}\infty}^{c+\mathrm{j}\infty} F(s)\mathrm{e}^{st}\,\mathrm{d}s, \quad \alpha < c < \beta. \tag{1.24}$$

Nevertheless, the true meaning of the last equation is given by (1.21) and (1.22). If the integral (1.24) converges absolutely, i.e., the integral

$$f_a(t) = \frac{1}{2\pi \mathrm{j}} \int_{c-\mathrm{j}\infty}^{c+\mathrm{j}\infty} |F(s)|\,\mathrm{d}s \tag{1.25}$$

converges, the integral (1.24) converges in the ordinary sense. Further, it can be proved that the related original $f(t)$ is continuous in this case.

1.4 Discretization of inversion integrals

Let $\omega > 0$ be a real number and $T = \frac{2\pi}{\omega}$. Then, the integral (1.24) can be represented in the form

$$f(t) = \frac{1}{2\pi \mathrm{j}} \sum_{k=-\infty}^{\infty} \int_{c+k\mathrm{j}\omega}^{c+(k+1)\mathrm{j}\omega} F(s)\mathrm{e}^{st}\,\mathrm{d}s\,. \tag{1.26}$$

Since

$$\int_{c+k\mathrm{j}\omega}^{c+(k+1)\mathrm{j}\omega} F(s)\mathrm{e}^{st}\,\mathrm{d}s = \int_{c}^{c+\mathrm{j}\omega} F(q+k\mathrm{j}\omega)\mathrm{e}^{(q+k\mathrm{j}\omega)t}\,\mathrm{d}q \tag{1.27}$$

it follows from (1.24) that

$$f(t) = \frac{1}{2\pi \mathrm{j}} \sum_{k=-\infty}^{\infty} \int_{c}^{c+\mathrm{j}\omega} F(q+k\mathrm{j}\omega)\mathrm{e}^{(q+k\mathrm{j}\omega)t}\,\mathrm{d}q\,. \tag{1.28}$$

Changing the order of summation and integration and substituting s for q, we obtain

$$f(t) = \frac{T}{2\pi \mathrm{j}} \int_{c}^{c+\mathrm{j}\omega} \mathcal{D}_F(T,s,t)\,\mathrm{d}s \tag{1.29}$$

with the notation

$$\mathcal{D}_F(T,s,t) \triangleq \frac{1}{T} \sum_{k=-\infty}^{\infty} F(s+k\mathrm{j}\omega)\mathrm{e}^{(s+k\mathrm{j}\omega)t}\,. \tag{1.30}$$

Of course, the passage from (1.24) to (1.29) must be justified in each particular case. It is admissible if the integral (1.25) converges.
The function $\mathcal{D}_F(T,s,t)$ will be called the *discrete Laplace transform (DLT)* for the image $F(s)$ for reasons explained below. From (1.30) we immediately obtain

$$\mathcal{D}_F(T,s,t) = \mathcal{D}_F(T,s+\mathrm{j}\omega,t)\,. \tag{1.31}$$

Therefore, formula (1.29) can be written in the form of an integral with symmetric limits.

$$f(t) = \frac{T}{2\pi \mathrm{j}} \int_{c-\mathrm{j}\omega/2}^{c+\mathrm{j}\omega/2} \mathcal{D}_F(T,s,t)\,\mathrm{d}s\,. \tag{1.32}$$

In what follows the passage from (1.24) to (1.32) will be called the *discretization of the inversion integral* with step $\mathrm{j}\omega$ (or with period T).

1.5 Theorems about image functions

In many cases it is necessary to find whether a function $F(s)$ is the Laplace transform of a function $f(t)$. The following statement gives some sufficient conditions which can be easily employed.

Theorem 1.4 (Jevgrafov (1965)) *Let a function $F(s)$ be analytic in the stripe $\alpha \le \mathrm{Re}\, s \le \beta$. Suppose that for $|s| \to \infty$ and $\alpha \le \mathrm{Re}\, s \le \beta$ the following estimate holds:*

$$|F(s)| < A|s|^{-1-\lambda} \tag{1.33}$$

where $A > 0$ and $\lambda > 0$ are constants. Then the integral

$$f(t) = \frac{1}{2\pi \mathrm{j}} \int_{c-\mathrm{j}\infty}^{c+\mathrm{j}\infty} F(s)\mathrm{e}^{st}\, \mathrm{d}s, \quad \alpha \le c \le \beta \tag{1.34}$$

converges absolutely, the function $f(t)$ is independent of c and continuous with respect to t. Moreover, $f(t) \in \Lambda(\alpha, \beta)$ and $f(t) \longrightarrow F(s)$ in $\alpha < \mathrm{Re}\, s < \beta$. In the special case, when $\beta = \infty$, we obtain $f(t) = 0$ for $t < 0$ and $f(t) \in \Lambda_+(\alpha, \infty)$. ■

Let us note that the conditions of Theorem 1.4 are far from necessary. In particular, the associated original could also be continuous, if the above conditions are not fulfilled, and on the other hand discontinuous functions have great practical importance. Nevertheless, Theorem 1.4 can be quite useful in many practical cases. In more complex cases the following theorem can be used.

Theorem 1.5 *Let the function $F(s)$ be analytic in the stripe $\alpha \le \mathrm{Re}\, s \le \beta$ and let the following estimate hold for $|s| \to \infty$, $\alpha \le \mathrm{Re}\, s \le \beta$*

$$|F(s)| < A|s|^{-1}\,. \tag{1.35}$$

Let γ be a constant such that the function

$$F_\gamma(s) \triangleq \frac{F(s)}{s-\gamma} \tag{1.36}$$

is analytic in $\alpha \le \mathrm{Re}\, s \le \beta$. Due to Theorem 1.4 the original

$$f_\gamma(t) = \frac{1}{2\pi \mathrm{j}} \int_{c-\mathrm{j}\infty}^{c+\mathrm{j}\infty} F_\gamma(s)\mathrm{e}^{st}\, \mathrm{d}s, \quad \alpha \le c \le \beta \tag{1.37}$$

does exist. Assume that there exists the derivative $\frac{\mathrm{d}f_\gamma(t)}{\mathrm{d}t} \in \Lambda(\alpha, \beta)$. Then, integral (1.34) converges as defined in (1.21), for the associated original we have $f(t) \in \Lambda(\alpha, \beta)$ and

$$f(t) = \frac{\mathrm{d}f_\gamma(t)}{\mathrm{d}t} - \gamma f_\gamma(t)\,. \tag{1.38}$$

In this case $f(t) \longrightarrow F(s)$. If we have $\beta = \infty$ under the conditions of the theorem, then $f(t) \in \Lambda_+(\alpha, \infty)$. ■

Example 1.1 Find by means of Theorem 1.5 the inverse Laplace transform for

$$F(s) = \frac{1}{s}, \quad \operatorname{Re} s \ge \alpha > 0. \tag{1.39}$$

In this case Theorem 1.4 is not applicable. Consider the function

$$F_\gamma(s) = \frac{1}{s(s-\gamma)}, \quad \gamma < 0. \tag{1.40}$$

As is known, Doetsch (1967), the original for the image (1.40) for $\operatorname{Re} s > 0$ is

$$f_\gamma(t) = \frac{1}{\gamma}\left(\mathrm{e}^{\gamma t} - 1\right) \mathbb{1}(t) \tag{1.41}$$

where

$$\mathbb{1}(t) \triangleq \begin{cases} 1 & \text{for } t > 0 \\ 0 & \text{for } t < 0 . \end{cases} \tag{1.42}$$

The function $f_\gamma(t)$ is continuous and has the derivative

$$\frac{\mathrm{d}f_\gamma(t)}{\mathrm{d}t} = \begin{cases} \mathrm{e}^{\gamma t} & t > 0 \\ 0 & t < 0 . \end{cases} \tag{1.43}$$

Moreover, $\frac{\mathrm{d}f_\gamma(t)}{\mathrm{d}t} \in \Lambda_+(0,\infty)$, because $\gamma < 0$. Because of (1.38) the original for the image (1.39) appears as

$$f(t) = \frac{\mathrm{d}f_\gamma(t)}{\mathrm{d}t} - \gamma f_\gamma(t) = \mathbb{1}(t) . \tag{1.44}$$

□

1.6 General properties of images

For $f(t) \in \Lambda(\alpha,\beta)$ the image (1.14) is an analytic function in s in the stripe $\alpha < \operatorname{Re} s < \beta$. Since $f(t)$ has finite variation in any internal stripe $\alpha < \alpha' \le \operatorname{Re} s \le \beta' < \beta$ and for $|s| \to \infty$, due to Bochner (1959) we have

$$|F(s)| < A|s|^{-1}, \quad A = \text{const.} \tag{1.45}$$

In the general case $F(s)$ can be analytically continued outside of this stripe, which gives an analytical function defined in its natural range.

1.7 Addition of images

Let

$$f_i(t) \longrightarrow F_i(s), \quad \alpha_i < \operatorname{Re} s < \beta_i, \quad i = 1,2 . \tag{1.46}$$

Then, if the stripes of convergence for $F_i(s)$ overlap,

$$f_1(t) + f_2(t) \longrightarrow F_1(s) + F_2(s), \quad \max(\alpha_1, \alpha_2) < \operatorname{Re} s < \min(\beta_1, \beta_2). \quad (1.47)$$

If the stripes of convergence for $F_1(s)$ and $F_2(s)$ have no common part, then (1.47) is meaningless.
For ordinary Laplace transforms addition of images is always possible. In this case

$$f_i(t) \quad \dot{\longrightarrow} \quad F_i(s), \qquad \operatorname{Re} s > \alpha_i, \quad i = 1, 2 \qquad (1.48)$$

$$f_1(t) + f_2(t) \quad \dot{\longrightarrow} \quad F_1(s) + F_2(s), \quad \operatorname{Re} s > \max(\alpha_1, \alpha_2). \qquad (1.49)$$

1.8 Time shift for originals

Let $f(t) \in \Lambda(\alpha, \beta)$ and $f(t) \longrightarrow F(s)$. The function

$$f_\tau(t) \triangleq f(t + \tau) \qquad (1.50)$$

where τ is a constant, will be called *the time shift* of the function $f(t)$ by τ (to the right for $\tau < 0$ and to the left for $\tau > 0$). Next we show that $f_\tau(t) \in \Lambda(\alpha, \beta)$. Indeed, from $f(t) \in \Lambda(\alpha, \beta)$ it follows that

$$\left| f(t+\tau)\mathrm{e}^{-s(t+\tau)} \right| < M = \text{const.}, \quad \alpha < \operatorname{Re} s < \beta. \qquad (1.51)$$

Hence,

$$\left| f(t+\tau)\mathrm{e}^{-st} \right| < M\mathrm{e}^{\operatorname{Re} s\tau}, \quad \alpha < \operatorname{Re} s < \beta. \qquad (1.52)$$

Therefore, there exists the Laplace transform

$$F_\tau(s) = \int_{-\infty}^{\infty} f(t+\tau)\mathrm{e}^{-st}\,\mathrm{d}t = \mathrm{e}^{s\tau}F(s), \quad \alpha < \operatorname{Re} s < \beta. \qquad (1.53)$$

It is noteworthy that these formulae have a different form for the ordinary Laplace transform, van der Pol and Bremmer (1959). If

$$f(t) \dot{\longrightarrow} F(s), \quad \operatorname{Re} s > \alpha \qquad (1.54)$$

for $\tau < 0$ we have an equation similar to (1.53):

$$f_\tau(t) \dot{\longrightarrow} \mathrm{e}^{s\tau}F(s), \quad \operatorname{Re} s > \alpha \quad \tau < 0 \qquad (1.55)$$

but for $\tau > 0$ we obtain

$$f_\tau(t) \dot{\longrightarrow} \mathrm{e}^{s\tau}\left[F(s) - \int_0^\tau f(t)\mathrm{e}^{-st}\,\mathrm{d}t\right], \quad \operatorname{Re} s > \alpha, \quad \tau > 0. \qquad (1.56)$$

The difference between (1.55) and (1.56) is due to the fact that the theory of the ordinary Laplace transformation only considers originals that are zero for $t < 0$.

1.9 Image shift in the complex domain

Let $f(t) \in \Lambda(\alpha,\beta)$ and $f(t) \longrightarrow F(s)$. Then, for a complex λ,

$$e^{-\lambda t} f(t) \longrightarrow F(s+\lambda)\,, \quad \alpha - \operatorname{Re}\lambda < \operatorname{Re} s < \beta - \operatorname{Re}\lambda\,. \tag{1.57}$$

Indeed, by the above conditions

$$\left| f(t) e^{-\lambda t} e^{-st} \right| < M\,, \quad \alpha < \operatorname{Re}\lambda + \operatorname{Re} s < \beta \tag{1.58}$$

i.e., $e^{-\lambda t} f(t) \in \Lambda(\alpha_1, \beta_1)$ with $\alpha_1 = \alpha - \operatorname{Re}\lambda$ and $\beta_1 = \beta - \operatorname{Re}\lambda$. In the stripe $\alpha_1 < \operatorname{Re} s < \beta_1$ we have

$$e^{-\lambda t} f(t) \longrightarrow \int_{-\infty}^{\infty} f(t) e^{-(\lambda+s)t}\, dt = F(s+\lambda)\,. \tag{1.59}$$

For the ordinary Laplace transform we have a similar formula

$$e^{-\lambda t} f(t) \stackrel{\cdot}{\longrightarrow} F(s+\lambda)\,, \quad \operatorname{Re} s > \alpha - \operatorname{Re}\lambda\,. \tag{1.60}$$

1.10 Convolution of originals

Let $f_i(t) \in \Lambda(\alpha,\beta)$ $(i = 1,2)$. By *the convolution* of the functions $f_1(t)$ and $f_2(t)$ we mean the following expression

$$f_1 * f_2(t) \triangleq \int_{-\infty}^{\infty} f_1(t-\tau) f_2(\tau)\, d\tau\,. \tag{1.61}$$

It is known that

$$f_1 * f_2(t) = f_2 * f_1(t) = \int_{-\infty}^{\infty} f_2(t-\tau) f_1(\tau)\, d\tau\,. \tag{1.62}$$

Theorem 1.6 *Let $f_i(t) \in \Lambda(\alpha,\beta)$ and $f_i(t) \longrightarrow F_i(s)$. Then*

$$f_1 * f_2(t) \longrightarrow F_1(s) F_2(s)\,, \quad \alpha < \operatorname{Re} s < \beta \tag{1.63}$$

and $f_1 * f_2(t) \in \Lambda(\alpha,\beta)$. In this case $f_1 * f_2(t)$ is continuous.

Proof Equation (1.63) follows from the assumption $f_i(t) \in \Lambda(\alpha,\beta)$, van der Pol and Bremmer (1959). In this case for any $F_i(s)$ the estimate (1.45) holds, therefore the function $F(s) = F_1(s)F_2(s)$ decreases for $|s| \to \infty$ as $|s|^{-2}$. Then, as a result of Theorem 1.4, the function $f_1 * f_2(t) \in \Lambda(\alpha,\beta)$ is continuous. ■

It is noteworthy that for $f_i(t) \in \Lambda_+(\alpha,\infty)$ from (1.63) and Theorem 1.4 it follows that $f_1 * f_2(t) \in \Lambda(\alpha,\infty)$, and the corresponding formulae for the convolution are the form

$$f_1 * f_2(t) = \begin{cases} \int\limits_0^t f_1(t-\tau) f_2(\tau)\, d\tau = \int\limits_0^t f_2(t-\tau) f_1(\tau)\, d\tau & t \geq 0 \\ 0 & t \leq 0\,. \end{cases} \tag{1.64}$$

1.11 Parseval's formula

Let the complex valued functions $f(t)$ and $g(t)$ be in $\Lambda(\alpha,\beta)$ with $\alpha < 0$, $\beta > 0$, i.e., the stripe of convergence contains the imaginary axis. In this case the following Parseval's formula holds:

$$\frac{1}{2\pi \mathrm{j}} \int_{-\mathrm{j}\infty}^{\mathrm{j}\infty} F(s)\overline{G(s)}\,\mathrm{d}s = \int_{-\infty}^{\infty} f(t)\overline{g(t)}\,\mathrm{d}t \tag{1.65}$$

where $f(t) \longrightarrow F(s)$, $g(t) \longrightarrow G(s)$ and the bar denotes the complex conjugate value. If $f(t)$ and $g(t)$ are real, Parseval's formula can be written in a different form. In this case $\overline{G(s)} = G(\overline{s})$, and on the imaginary axis we have $s = \mathrm{j}\nu$ and $\overline{s} = -\mathrm{j}\nu = -s$, i.e. $\overline{G(s)} = G(-s)$. Therefore

$$\frac{1}{2\pi \mathrm{j}} \int_{-\mathrm{j}\infty}^{\mathrm{j}\infty} F(s)G(-s)\,\mathrm{d}s = \int_{-\infty}^{\infty} f(t)g(t)\,\mathrm{d}t \tag{1.66}$$

Taking $f(t) = g(t)$ in (1.66), we obtain

$$\frac{1}{2\pi \mathrm{j}} \int_{-\mathrm{j}\infty}^{\mathrm{j}\infty} F(s)F(-s)\,\mathrm{d}s = \int_{-\infty}^{\infty} f^2(t)\,\mathrm{d}t\,. \tag{1.67}$$

If, in addition, $f(t) = 0$ for $t < 0$, Eq. (1.67) yields

$$\frac{1}{2\pi \mathrm{j}} \int_{-\mathrm{j}\infty}^{\mathrm{j}\infty} F(s)F(-s)\,\mathrm{d}s = \int_{0}^{\infty} f^2(t)\,\mathrm{d}t\,. \tag{1.68}$$

1.12 Differentiation of originals

Let $f(t) \in \Lambda(\alpha,\beta)$ and $f(t) \longrightarrow F(s)$. The function $f(t)$ is assumed to have a derivative $\frac{\mathrm{d}f(t)}{\mathrm{d}t} \in \Lambda(\alpha,\beta)$. Then

$$\frac{\mathrm{d}f(t)}{\mathrm{d}t} \longrightarrow sF(s)\,, \quad \alpha < \operatorname{Re} s < \beta\,. \tag{1.69}$$

Let us note that for the ordinary Laplace transform the formula for the image of the derivative has another form. It is known that if $f(t) \in \Lambda_+(\alpha,\infty)$ and $\frac{\mathrm{d}f(t)}{\mathrm{d}t} \in \Lambda_+(\alpha,\infty)$, we have

$$\frac{\mathrm{d}f(t)}{\mathrm{d}t} \stackrel{\cdot}{\longrightarrow} sF(s) - f(+0), \quad \operatorname{Re} s > \alpha\,. \tag{1.70}$$

Equation (1.70) differs from (1.69) and turns into the latter only for $f(+0) = 0$. This is due to the fact that the function

$$\hat{f}(t) = \begin{cases} f(t) & t > 0 \\ 0 & t < 0 \end{cases} \tag{1.71}$$

is not differentiable for $f(+0) \neq 0$ and (1.69) is inapplicable, because the associated conditions are not satisfied. Using (1.69) several times, we obtain for any natural n

$$\frac{\mathrm{d}^n f(t)}{\mathrm{d}t^n} \longrightarrow s^n F(s)\,, \quad \alpha < \operatorname{Re} s < \beta\,. \tag{1.72}$$

This formula holds if all the derivatives up to the n-th order exist and are in $\Lambda(\alpha,\beta)$. For the ordinary Laplace transformation instead of (1.72) we have

$$\frac{\mathrm{d}^n f(t)}{\mathrm{d}t^n} \stackrel{\cdot}{\longrightarrow} s^n F(s) - f(+0)s^{n-1} - f'(+0)s^{n-2} - \ldots - f^{(n-1)}(+0) \tag{1.73}$$

with $\operatorname{Re} s > \alpha$.

1.13 Differentiation of images

Let $f(t) \in \Lambda(\alpha,\beta)$ and

$$F(s) = \int_{-\infty}^{\infty} f(t)\mathrm{e}^{-st}\,\mathrm{d}t\,, \quad \alpha < \operatorname{Re} s < \beta\,. \tag{1.74}$$

Under the given assumptions the function $F(s)$ is analytic in the stripe of convergence $\alpha < \operatorname{Re} s < \beta$ and there has derivatives of all orders with respect to s. These derivatives can be obtained by differentiating the integrand in (1.74). Thus, we obtain

$$\frac{\mathrm{d}^n F(s)}{\mathrm{d}s^n} = (-1)^n \int_{-\infty}^{\infty} f(t)t^n \mathrm{e}^{-st}\,\mathrm{d}t\,, \quad \alpha < \operatorname{Re} s < \beta \tag{1.75}$$

and the integrals on the right-hand side of (1.75) converge, because of $f(t) \in \Lambda(\alpha,\beta)$. Equation (1.75) can be rewritten in the equivalent form

$$(-1)^n t^n f(t) \longrightarrow \frac{\mathrm{d}^n F(s)}{\mathrm{d}s^n}\,. \tag{1.76}$$

1.14 Inversion formulae for real rational images

In this section we consider real rational images of the form

$$F(s) = \frac{m(s)}{d(s)} \tag{1.77}$$

where $m(s)$ and $d(s)$ are polynomials in s. If $\deg m(s) \leq \deg d(s)$, where 'deg' denotes the degree of a polynomial, the ratio $F(s)$ is called *proper*. If $\deg m(s) < \deg d(s)$, then $F(s)$ is called *strictly proper*. In this section we obtain general inversion formulae for bilateral strictly proper images. In this case

$$\begin{aligned} m(s) &= m_1 s^{p-1} + m_2 s^{p-2} + \ldots + m_p \\ d(s) &= s^p + d_1 s^{p-1} + \ldots + d_p\,. \end{aligned} \tag{1.78}$$

The following inversion formulae are based on two simple relations.

1. Let

$$f(t) = f_+(t,a) \triangleq \begin{cases} e^{at} & t > 0 \\ 0 & t < 0 \end{cases} \tag{1.79}$$

where a is a complex constant. Then, the assigned image is given by

$$F_+(s,a) = \int_0^\infty f_+(t,a)e^{-st}\,dt = \frac{1}{s-a}, \quad \operatorname{Re} s > \operatorname{Re} a. \tag{1.80}$$

2. If

$$f(t) = f_-(t,a) \triangleq \begin{cases} 0 & t > 0 \\ -e^{at} & t < 0 \end{cases} \tag{1.81}$$

takes place the Laplace image appears as

$$F_-(s,a) = \int_{-\infty}^0 f_-(t,a)e^{-st}\,dt = \frac{1}{s-a}, \quad \operatorname{Re} s < \operatorname{Re} a. \tag{1.82}$$

Obviously, $f_-(t,a) \in \Lambda_-(-\infty, \operatorname{Re} a)$ and $f_+(t) \in \Lambda_+(\operatorname{Re} a, \infty)$. Therefore, the inversion formula (1.21) is valid. Using the simplified form (1.24), from (1.80) and (1.82) we have

$$\frac{1}{2\pi j}\int_{c-j\infty}^{c+j\infty} \frac{1}{s-a} e^{st}\,ds = \begin{cases} e^{at} & t > 0 \\ 0 & t < 0 \end{cases} \quad c > \operatorname{Re} a \tag{1.83}$$

$$\frac{1}{2\pi j}\int_{c-j\infty}^{c+j\infty} \frac{1}{s-a} e^{st}\,ds = \begin{cases} 0 & t > 0 \\ -e^{at} & t < 0 \end{cases} \quad c < \operatorname{Re} a. \tag{1.84}$$

From (1.83) and (1.84), using (1.76) for any natural n we obtain

$$\frac{1}{2\pi j}\int_{c-j\infty}^{c+j\infty} \frac{1}{(s-a)^n} e^{st}\,ds = \begin{cases} \frac{1}{(n-1)!}t^{n-1}e^{at} & t > 0 \\ 0 & t < 0 \end{cases} \quad c > \operatorname{Re} a \tag{1.85}$$

$$\frac{1}{2\pi j}\int_{c-j\infty}^{c+j\infty} \frac{1}{(s-a)^n} e^{st}\,ds = \begin{cases} 0 & t > 0 \\ -\frac{1}{(n-1)!}t^{n-1}e^{at} & t < 0 \end{cases} \quad c < \operatorname{Re} a. \tag{1.86}$$

Using (1.83)–(1.86), we can derive the general inversion formulae for images of the form (1.77), where $F(s)$ is irreducible and strictly proper.
The poles of the function $F(s)$ are the roots of the equation

$$d(s) = s^p + d_1 s^{p-1} + d_2^{p-2} + \dots + d_p = 0. \tag{1.87}$$

Let s_i $(i = 1, \dots, \ell)$ be the roots of equation (1.87) with multiplicity ν_i, so that $\nu_1 + \nu_2 + \dots + \nu_\ell = p$. Then, the function $F(s)$ can be developed into partial fractions

$$F(s) = \sum_{i=1}^{\ell}\sum_{k=1}^{\nu_i} \frac{f_{ik}}{(s-s_i)^k} \tag{1.88}$$

where the constants f_{ik} can be calculated by known techniques. Henceforth, the values s_i are assumed to be ordered so that $\operatorname{Re} s_i \le \operatorname{Re} s_{i+1}$. Denote by

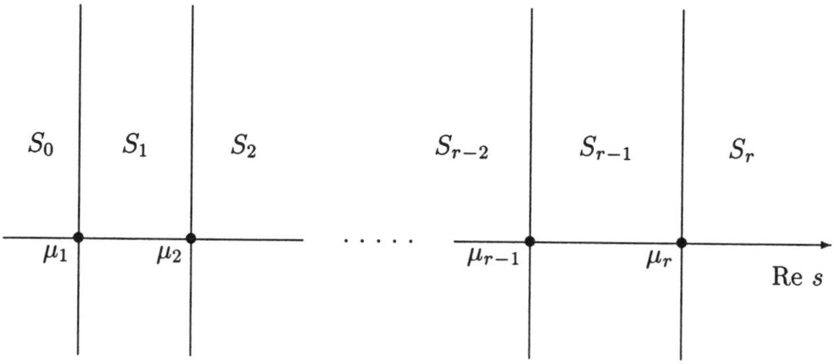

Figure 1.1: Dissection of the complex plane by $F(s)$

$\mu_1, ..., \mu_r$ the different real parts of the roots s_i numbered so that $\mu_i < \mu_{i+1}$. Let also $\mu_0 = -\infty$ and $\mu_{r+1} = \infty$. Consider the open intervals

$$S_- \triangleq S_0 : (-\infty, \mu_1),\ S_1 : (\mu_1, \mu_2),\ \dots\ S_{r-1} : (\mu_{r-1}, \mu_r),\ S_+ \triangleq S_r : (\mu_r, \infty)$$

on the real axis (Fig. 1.1), which will be called *regularity intervals* of the image $F(s)$. The intervals S_- and S_+ will be called outer, while the remaining ones will be called inner intervals, Rosenwasser (1977) . Consider the integral

$$f_m(t) = \frac{1}{2\pi \mathrm{j}} \int_{c-\mathrm{j}\infty}^{c+\mathrm{j}\infty} F(s)\mathrm{e}^{st}\, \mathrm{d}s\,, \quad c \in S_m\,. \tag{1.89}$$

For a given m all the roots s_i can be divided into two groups, where the first group includes the roots with $\mathrm{Re}\, s_i \le \mu_m$, and the second one consists of the roots with $\mathrm{Re}\, s_i \ge \mu_{m+1}$. Let the roots $s_1, ..., s_\rho$ belong to the first group and $s_{\rho+1}, ..., s_\ell$ belong to the second one. Then, the partial fraction expansion (1.88) can be represented in the form

$$F(s) = F_{m+}(s) + F_{m-}(s) \tag{1.90}$$

where

$$F_{m+}(s) = \sum_{i=1}^{\rho} \sum_{k=1}^{\nu_i} \frac{f_{ik}}{(s - s_i)^k} \tag{1.91}$$

$$F_{m-}(s) = \sum_{i=\rho+1}^{\ell} \sum_{k=1}^{\nu_i} \frac{f_{ik}}{(s - s_i)^k}\,. \tag{1.92}$$

Substituting (1.90) into (1.89), we obtain

$$f_m(t) = f_{m+}(t) + f_{m-}(t) \tag{1.93}$$

where

$$f_{m+}(t) = \frac{1}{2\pi j}\int_{c-j\infty}^{c+j\infty} F_{m+}(s)e^{st}\,ds\,, \quad \mu_m < c < \mu_{m+1} \qquad (1.94)$$

$$f_{m-}(t) = \frac{1}{2\pi j}\int_{c-j\infty}^{c+j\infty} F_{m-}(s)e^{st}\,ds\,, \quad \mu_m < c < \mu_{m+1}\,. \qquad (1.95)$$

Since $\operatorname{Re} s_i < c$ in the integral (1.94), using (1.85) we have

$$f_{m+}(t) = \begin{cases} \displaystyle\sum_{i=1}^{\rho}\sum_{k=1}^{\nu_i} \frac{f_{ik}}{(k-1)!}t^{k-1}e^{s_i t} & t > 0 \\ 0 & t < 0 \end{cases} \qquad (1.96)$$

Similarly, using (1.95) and (1.86), we obtain

$$f_{m-}(t) = \begin{cases} 0 & t > 0 \\ \displaystyle -\sum_{i=\rho+1}^{\ell}\sum_{k=1}^{\nu_i} \frac{f_{ik}}{(k-1)!}t^{k-1}e^{s_i t} & t < 0\,. \end{cases} \qquad (1.97)$$

Adding (1.96) and (1.97) and using (1.93), we find

$$f_{m}(t) = \begin{cases} \displaystyle\sum_{i=1}^{\rho}\sum_{k=1}^{\nu_i} \frac{f_{ik}}{(k-1)!}t^{k-1}e^{s_i t} & t > 0 \\ \displaystyle -\sum_{i=\rho+1}^{\ell}\sum_{k=1}^{\nu_i} \frac{f_{ik}}{(k-1)!}t^{k-1}e^{s_i t} & t < 0\,. \end{cases} \qquad (1.98)$$

As a special case, for the interval S_+, i.e., for $c > \mu_r$, from (1.98) we have

$$f_r(t) \triangleq f_+(t) = \begin{cases} \displaystyle\sum_{i=1}^{\ell}\sum_{k=1}^{\nu_i} \frac{f_{ik}}{(k-1)!}t^{k-1}e^{s_i t} & t > 0 \\ 0 & t < 0\,. \end{cases} \qquad (1.99)$$

Analogously, for the interval S_-, i.e., $c < \mu_1$, we have

$$f_0(t) \triangleq f_-(t) = \begin{cases} 0 & t > 0 \\ \displaystyle -\sum_{i=1}^{\ell}\sum_{k=1}^{\nu_i} \frac{f_{ik}}{(k-1)!}t^{k-1}e^{s_i t} & t < 0\,. \end{cases} \qquad (1.100)$$

Next we formulate some general results that follow from the above formulae.

1. In the general case the image (1.88) is associated with $r+1$ various originals, the number of which is equal to the number of the corresponding regularity intervals S_m. Hereinafter the function $f_m(t)$ will be called *the original* for the image $F(s)$ *in the regularity interval* S_m. The functions $f_m(t)$ (1.98) taken in the aggregate will be called *the family of originals* associated with the image $F(s)$ and denoted by $\{f_m(t)\}$.

2. Let $[\mu'_m, \mu'_{m+1}]$ be an arbitrary closed interval inside a regularity interval S_m. Using (1.98), we can easily verify that

$$\left|f_m(t)e^{-st}\right| < M_m = \text{const.}, \quad \mu'_m \leq \operatorname{Re} s \leq \mu'_{m+1}$$

whence, according to Theorem 1.1, follows

$$f_m(t) \in \Lambda(\mu_m, \mu_{m+1}) . \tag{1.101}$$

3. Let us note that any of the functions $f_m(t)$ defines all of the functions in the family $\{f_m(t)\}$ uniquely. Indeed, if we know $f_k(t)$ for any k, we can find the associated image in the stripe $\mu_k < \operatorname{Re} s < \mu_{k+1}$, which will be called the k-th regularity stripe. Then, using analytical continuation of $F(s)$ over the whole complex plane, and using (1.89), we can construct the whole family $\{f_m(t)\}$.

4. From (1.98) we immediately obtain

$$f_m(+0) = \sum_{i=1}^{\rho} f_{i1} , \qquad f_m(-0) = - \sum_{i=\rho+1}^{\ell} f_{i1} . \tag{1.102}$$

Hence,

$$\Delta f_m(0) \triangleq f_m(+0) - f_m(-0) = \sum_{i=1}^{\ell} f_{i1} . \tag{1.103}$$

The left side of (1.103) is the break altitude of the function $f_m(t)$ at the point $t = 0$. Equation (1.88) yields

$$\sum_{i=1}^{\ell} f_{i1} = m_1 \tag{1.104}$$

where m_1 is the corresponding coefficient in (1.78). From (1.104) and (1.103) we obtain

$$\Delta f_m(0) = m_1 . \tag{1.105}$$

5. As follows from (1.105), for $m_1 \neq 0$ all functions $\{f_m(t)\}$ have breaks at $t = 0$ of the altitude m_1. It can be easily shown that for $m_1 = 0$ and $m_2 \neq 0$ the functions $f_m(t)$ are continuous and its derivatives are piece-wise continuous. For $m_1 = m_2 = 0$ and $m_3 \neq 0$ the functions $f_m(t)$ have continuous first derivatives and piece-wise continuous second derivatives, and so on.

Chapter 2

Operational transformations for discrete-time functions

2.1 Power series

As is known, a power series (Taylor series) is a series of the form

$$A(\zeta) \simeq a_0 + a_1\zeta + a_2\zeta^2 + \ \dots \tag{2.1}$$

where $\{a_i\}$ $(i = 0, 1, 2, \dots)$ is a sequence of complex numbers.
For any series of the form (2.1) there exists a number $R \geq 0$ such that the series converges for $|\zeta| < R$ and diverges for $|\zeta| > R$. The value of R is called *the radius of convergence* of the power series (2.1) and can be calculated by the following Cauchy-Hadamard formula

$$R^{-1} = \limsup_{n\to\infty} \sqrt[n]{|a_n|}\,. \tag{2.2}$$

For some series $R = 0$ is valid, i.e. the series (2.1) converges only for $\zeta = 0$. For some series we have $R = \infty$. Functions defined by such series are called *entire functions.* The open disc $|\zeta| < R$ is called *the disc of convergence* of the series (2.1). The sum of the series (2.1) exists inside the disc of convergence and is an analytical function in ζ. Therefore, with $R \neq 0$ we can write, instead of (2.1),

$$A(\zeta) = a_0 + a_1\zeta + a_2\zeta^2 + \ \dots\ , \qquad |\zeta| < R\,. \tag{2.3}$$

Conversely, a function $A(\zeta)$ that is analytic inside a circle with radius R can be uniquely represented as sum of a power series (2.3). The coefficients of this series can be expressed by the differential formulae

$$a_n = \frac{1}{n!}\left.\frac{\mathrm{d}^n A(\zeta)}{\mathrm{d}\zeta^n}\right|_{\zeta=0} \tag{2.4}$$

or by the integral formulae

$$a_n = \frac{1}{2\pi j} \oint \frac{A(\zeta)}{\zeta^{n+1}} \, d\zeta \tag{2.5}$$

where the contour integral is taken anti-clockwise along an arbitrary circle $|\zeta| = \rho$ lying inside the convergence disc (i.e., $\rho < R$). Moreover, the following Cauchy inequality holds

$$|a_n| < \frac{M_\rho}{\rho^n} \tag{2.6}$$

where

$$M_\rho \triangleq \max_{|\zeta|=\rho} |A(\zeta)| \,. \tag{2.7}$$

2.2 Power series as transformations

Let us consider a numerical sequence $\{a_i\}$, $(i = 0, 1, 2, \ldots)$. The sequence $\{a_i\}$ can be associated with a series of the form (2.1) that can be considered as a transform of the initial sequence. If the obtained series has non-zero radius of convergence, there is an analytic function $A(\zeta)$ defined inside the disc of converges (2.3), which can be viewed as a transform of the sequence $\{a_i\}$. In principle, the function $A(\zeta)$ can admit an analytic continuation outside the disc of convergence.

Example 2.1 Find the transform of the sequence $\{a_i\}$ with $a_i = 1$, $(i = 0, 1, 2, \ldots)$. The corresponding formal series (2.1) takes the form

$$A(\zeta) = A_+(\zeta) \triangleq 1 + \zeta + \zeta^2 + \ \ldots \ . \tag{2.8}$$

The radius of convergence of this series is $R = 1$, and its sum is given by

$$A_+(\zeta) = \frac{1}{1-\zeta}, \qquad |\zeta| < 1 \,. \tag{2.9}$$

The function (2.9) can be viewed as a transform of the initial sequence. It is noteworthy that the function (2.9) admits the analytical continuation over the whole complex plane except for the point $\zeta = 1$. □

It should be noted that the correspondence between the sequence $\{a_i\}$ and the image $A(\zeta)$ is bi-unique, because, on the one hand, sequence $\{a_i\}$ determines the sum of series (2.3) uniquely, and, on the other hand, the function $A(\zeta)$ determines the sequence $\{a_i\}$ by (2.4) and (2.5) uniquely.

2.3 Laurent series as transformations

A *Laurent series* (with the center at $\zeta = 0$) is a series of the form

$$A(\zeta) \simeq \sum_{n=-\infty}^{\infty} a_n \zeta^n \,, \tag{2.10}$$

which contains positive as well as negative powers of ζ. Relation (2.10) can be represented in the form

$$A(\zeta) = A_-(\zeta) + A_+(\zeta) \tag{2.11}$$

where

$$A_-(\zeta) \triangleq \sum_{n=-\infty}^{-1} a_n \zeta^n \,, \qquad A_+(\zeta) \triangleq \sum_{n=0}^{\infty} a_n \zeta^n \,. \tag{2.12}$$

The series $A_+(\zeta)$ is called the ***proper part*** of the Laurent series, and the series $A_-(\zeta)$ is called its ***main part.*** The proper part is a power series that converges on a disc $|\zeta| < R_+$. The series $A_-(\zeta)$ is a power series in $z = \zeta^{-1}$ that converges in an area $|\zeta| > R_-$. If $R_+ > R_-$, both series (2.12) converge in the ring $R_- < |\zeta| < R_+$ called the ring of convergence. Hence, the Laurent series (2.10) converges also. Moreover,

$$A(\zeta) = \sum_{n=-\infty}^{\infty} a_n \zeta^n \,, \quad R_- < |\zeta| < R_+ \,. \tag{2.13}$$

If $R_+ < R_-$, the series (2.10) does not converge anywhere. If there exists a ring of convergence, the function $f(\zeta)$ is analytic in that ring, and the coefficients a_n are determined uniquely by the formulae

$$a_n = \frac{1}{2\pi \mathrm{j}} \oint \frac{A(\zeta)}{\zeta^{n+1}} \, \mathrm{d}\zeta \tag{2.14}$$

where the integral is taken along any circle $|\zeta| = \rho$ lying inside the ring of convergence, i.e., $R_- < \rho < R_+$. With

$$M_\rho \triangleq \max_{|\zeta|=\rho} |A(\zeta)| \tag{2.15}$$

the following Cauchy inequality holds:

$$|a_n| < \frac{M_\rho}{\rho^n} \,. \tag{2.16}$$

Given an infinite bilateral sequence $\{a_n\}$, $(n = 0, \pm 1, \pm 2, \dots)$, the sum of the associated Laurent series (2.10), in the case of its convergence, can be considered as a transform of this sequence. As for the Taylor series, the relation between the set of sequences $\{a_n\}$, for which the series (2.10) converges, and the set of functions, which are analytical in a round ring, is biunique. Henceforth, the transform of a sequence $\{a_n\}$, $(n = 0, \pm 1, \pm 2, \dots)$ by means of the Laurent series (2.13) will be called ***bilateral.*** From this point of view the transformation by means of construction of the Fourier series can be considered as

a special case of the bilateral transformation, if the unilateral sequence $\{a_i\}$, $(i=0,1,2,\ldots)$ is completed by $a_n = 0$, $(n=-1,-2,\ldots)$.
A characteristic feature of the bilateral transformation is the fact that the same function of a complex variable is the image of different numerical sequences in various rings of convergence. This situation is similar to that for the bilateral Laplace transformation.

Example 2.2 Let us consider the properties of of the transform (2.9) in the region $|\zeta| > 1$. As follows from Example 2.1, for a bilateral numerical sequence $\{a_n\}_+$, where $a_n = 1$ for $n \geq 0$ and $a_n = 0$ for $n < 0$, the corresponding bilateral image in $|\zeta| < 1$ has the form (2.9). On the other hand, for a numerical sequence $\{a_n\}_-$, in which $a_n = 0$ for $n \geq 0$ and $a_n = -1$ for $n < 0$, for $|\zeta| > 1$, $z = \zeta^{-1}$ we have

$$A(\zeta) \triangleq A_-(\zeta) = -\sum_{n=-\infty}^{-1} \zeta^n = -\sum_{n=1}^{\infty} z^n = -\frac{z}{1-z} = \frac{1}{1-\zeta}\,. \tag{2.17}$$

Thus, the function $A(\zeta) = (1-\zeta)^{-1}$ is the image of different bilateral sequences in $|\zeta| > 1$ and $|\zeta| < 1$. □

2.4 Discrete Laplace transforms

Using the substitution

$$\zeta = \mathrm{e}^{-sT} \tag{2.18}$$

in (2.13), where s is a complex variable and $T > 0$ is a real constant, we obtain the series

$$A^*(s) \triangleq A(\zeta)\,|_{\zeta=\mathrm{e}^{-sT}} = \sum_{n=-\infty}^{\infty} a_n \mathrm{e}^{-nsT}\,. \tag{2.19}$$

The function $A^*(s)$ is called the *discrete Laplace transform* (bilateral) for the sequence $\{a_n\}$. Comparing (2.13) with (2.18), we find that the transform (2.19) converges, with respect to s, in the vertical stripe

$$\alpha < \operatorname{Re} s < \beta \tag{2.20}$$

where α and β are real constants such that

$$\mathrm{e}^{-\beta T} = R_-, \qquad \mathrm{e}^{-\alpha T} = R_+\,. \tag{2.21}$$

The stripe (2.20) henceforth will be called the stripe of convergence of the discrete Laplace transform. We note that for a unilateral sequence $\{a_n\}$ instead of (2.19) we have

$$A^*(s) = \sum_{n=0}^{\infty} a_n \mathrm{e}^{-nsT}\,. \tag{2.22}$$

In this case the region of convergence is the half-plane $\operatorname{Re} s > \alpha$.
If the discrete Laplace transform is known, the elements of the sequence $\{a_n\}$ can be found by the following formulae:

$$a_n = \frac{T}{2\pi \mathrm{j}} \int_{c-\mathrm{j}\omega/2}^{c+\mathrm{j}\omega/2} A^*(s) \mathrm{e}^{nsT} \, \mathrm{d}s \tag{2.23}$$

where $\omega = 2\pi/T$ and $\alpha < c < \beta$. Formulae (2.23) can be obtained from (2.14) with the substitution (2.18).

Example 2.3 Find the discrete Laplace transforms for (2.9) and (2.17). As follows from the results obtained in examples 2.1 and 2.2, the discrete Laplace transforms of the sequences $\{a_n\}_+$ and $\{a_n\}_-$ take the form

$$A_+^*(s) = \frac{1}{1-\mathrm{e}^{-sT}}, \quad \operatorname{Re} s > 0, \qquad A_-^*(s) = \frac{1}{1-\mathrm{e}^{-sT}}, \quad \operatorname{Re} s < 0. \tag{2.24}$$

Hence, the function $A^*(s) = 1/(1-\mathrm{e}^{-sT})$ is a common discrete Laplace transform of different sequences in the half-planes $\operatorname{Re} s > 0$ and $\operatorname{Re} s < 0$, respectively. □

Let us formulate convergence conditions for discrete Laplace transforms. A bilateral sequence $\{a_n\}$ will be called *exponentially bounded* in the interval (α, β), if there exists a constant $M_d > 0$ such that

$$\left|a_n \mathrm{e}^{-nsT}\right| < M_d, \qquad \alpha < \operatorname{Re} s < \beta, \; (n = 0, \pm 1, \pm 2, \dots). \tag{2.25}$$

In what follows, the set of such sequences is denoted by $\Lambda_d(\alpha, \beta)$. If a sequence $\{a_n\}$ is right unilateral, we use $\beta = \infty$ in (2.25). The set of the corresponding unilateral sequences is henceforth denoted by $\Lambda_{d+}(\alpha, \infty)$.

Theorem 2.1 *The discrete Laplace transform (2.19) for a bilateral sequence* $\{a_n\}$ *converges in the stripe (2.20) iff* $\{a_n\} \in \Lambda_d(\alpha, \beta)$.

Proof *Necessity:* Let R'_- and R'_+ be constants such that $R_- < R'_- < R'_+ < R_+$. As follows from the Cauchy inequality (2.16), for $R'_- \le |\zeta| = \rho \le R'_+$ we have

$$|a_n \rho^n| < M_d \tag{2.26}$$

where M_d is a positive constant. Relations (2.26) are equivalent to

$$\left|a_n \mathrm{e}^{-nsT}\right| < M_d, \quad \alpha < \alpha' \le \operatorname{Re} s \le \beta' < \beta. \tag{2.27}$$

Comparing (2.25) with (2.27), we prove the necessity of the conditions of the theorem.

Sufficiency: Let us have (2.25) and let $[\alpha', \beta']$ be an arbitrary closed interval in (α, β). For any $\operatorname{Re} s$ from the interval $[\alpha', \beta']$ there is a constant $\gamma > 0$ such that $\alpha < \operatorname{Re} s \pm \gamma < \beta$. Then, from (2.25) it follows that

$$\left|a_n e^{-(s\pm\gamma)nT}\right| < M_d, \quad (n = 0, \pm 1, \pm 2, \ldots). \tag{2.28}$$

From (2.23) we obtain

$$\left|a_n e^{-snT}\right| < M_d e^{-\gamma nT}, \quad n \geq 0, \qquad \left|a_n e^{-snT}\right| < M_d e^{\gamma nT}, \quad n < 0 \tag{2.29}$$

that is equivalent to

$$\left|a_n e^{-snT}\right| < M_d e^{-\gamma |n| T}, \quad \alpha' \leq \operatorname{Re} s \leq \beta'. \tag{2.30}$$

Using (2.30) and (2.19), for $\alpha' \leq \operatorname{Re} s \leq \beta'$ we obtain

$$|A^*(s)| \leq \sum_{n=-\infty}^{\infty} \left|a_n e^{-nsT}\right| < M_d \sum_{n=-\infty}^{\infty} e^{-\gamma|n|T} = M_d \frac{1+e^{-\gamma T}}{1-e^{-\gamma T}} \tag{2.31}$$

i.e., the series (2.13) converges absolutely and uniformly in any closed stripe $\alpha < \alpha' \leq \operatorname{Re} s \leq \beta' < \beta$. ■

2.5 Discrete Laplace transforms for sampled functions

Let us have a continuous function in a continuous argument $f(t)$ defined over the axis $-\infty < t < \infty$. Consider, for a real number $T > 0$, the *sampling period*, a sequence of equidistant time instants $t_k = kT$, $(k = 0, \pm 1, \pm 2, \ldots)$. The sequence t_k is associated with the following sequence of function values

$$\{f_k\} \triangleq \{f(kT)\} \tag{2.32}$$

which will be called the *sampling* of the function $f(t)$ associated with the sequence $t_k = kT$. The initial function $f(t)$ is called the *envelope* of the sampled function $\{f_k\}$. Each sampled function (2.32) can be associated with its discrete Laplace transform

$$F^*(s) \triangleq \sum_{k=-\infty}^{\infty} f(kT) e^{-ksT}. \tag{2.33}$$

Suffcient conditions for the convergence of the transform (2.33) are given by the following theorem.

Theorem 2.2 *Let $f(t) \in \Lambda(\alpha, \beta)$ and be continuous. Then, the discrete Laplace transform (2.33) converges absolutely in any stripe $\alpha < \alpha' \leq \operatorname{Re} s \leq \beta' < \beta$.*

Proof From the condition $f(t) \in \Lambda(\alpha, \beta)$ it follows that

$$\left|f(t)\mathrm{e}^{-st}\right| < M\,, \quad \alpha' \leq \operatorname{Re} s \leq \beta'\,. \tag{2.34}$$

Hence, for $t = kT$ we obtain

$$\left|f(kT)\mathrm{e}^{-ksT}\right| < M\,, \quad \alpha' \leq \operatorname{Re} s \leq \beta' \tag{2.35}$$

i.e., the sequence $\{f(kT)\}$ satisfies the conditions of Theorem 2.2. ■

In many practical cases the function under sampling $f(t)$ has discontinuities at some points. Henceforth, it is always assumed that the function $f(t)$ has finite limits from the left and from the right for any $t = \tilde{t}$

$$f(\tilde{t}+0) = \lim_{\epsilon \to +0} f(\tilde{t}+\epsilon)\,, \quad f(\tilde{t}-0) = \lim_{\epsilon \to -0} f(\tilde{t}+\epsilon)\,. \tag{2.36}$$

If the limits (2.36) coincide, the function $f(t)$ is continuous at the point $t = \tilde{t}$. If the limits are different, the function has a finite discontinuity for the given $\tilde{t}$

$$\Delta f(\tilde{t}) \triangleq f(\tilde{t}+0) - f(\tilde{t}-0)\,. \tag{2.37}$$

The value $f(\tilde{t}+0)$, (or $f(\tilde{t}-0)$) will be henceforth called *right (or left) value* of the function $f(t)$ at the point $\tilde{t}$.
Assume that the envelope under investigation is discontinuous for $t = kT$. Then, under the above assumptions, the function $f(t)$ can be associated with two samplings, constructed from its right and left values, respectively

$$\{f_k^-\} \triangleq \{f(kT-0)\}\,, \quad \{f_k^+\} \triangleq \{f(kT+0)\}\,. \tag{2.38}$$

Each of the sequences $\{f_k^-\}$ und $\{f_k^+\}$ is associated with its own discrete Laplace transform, which will be denoted by $F_-^*(s)$ and $F_+^*(s)$, respectively. In this case Theorem 2.2 holds as well.

We often have to deal with discontinuous functions $f(t)$ considering transforms of unilateral functions. Indeed, let $f(t) = 0$ for $t < 0$, then $f(-0) = 0$. If, in addition, $f(+0) \neq 0$, the function $f(t)$ is discontinuous at $t = 0$.

Example 2.4 Calculate the discrete Laplace transforms for the left resp. right values (2.38) of the function

$$f(t) = \begin{cases} \mathrm{e}^{\alpha t} & t > 0 \\ 0 & t < 0\,. \end{cases} \tag{2.39}$$

In this case $f(-0) = 0$ and $f(+0) = 1$, and therefore the function $f(t)$ is discontinuous at $t = 0$. The sequence $\{f_k^-\}$ takes the form $f_k^- = 0$ $(k \leq 0)$, $f_k^- = \mathrm{e}^{k\alpha T}$, $(k > 0)$, therefore,

$$F_{-}^{*}(s)=\sum_{k=1}^{\infty}\mathrm{e}^{k\alpha T}\mathrm{e}^{-ksT}=\frac{\mathrm{e}^{(\alpha-s)T}}{1-\mathrm{e}^{(\alpha-s)T}}\,,\quad \operatorname{Re}s>\operatorname{Re}\alpha\,. \tag{2.40}$$

Similarly, the sequence $\{f_k^+\}$ takes the form $f_k^+=0$ $(k<0)$ and $f_k^+=\mathrm{e}^{k\alpha T}$, $(k\geq 0)$, so that

$$F_{+}^{*}(s)=\sum_{k=0}^{\infty}\mathrm{e}^{k\alpha T}\mathrm{e}^{-ksT}=\frac{1}{1-\mathrm{e}^{(\alpha-s)T}}\,,\quad \operatorname{Re}s>\operatorname{Re}\alpha\,. \tag{2.41}$$

□

2.6 Modified discrete Laplace transforms

For a given T the function of continuous argument $f(t)$ can be represented as an aggregate of sampled functions depending on a parameter. With this aim in view, we introduce a parameter ε, $(0\leq\varepsilon<T)$ and consider the following number sequence

$$\{f_k^{\pm}(\varepsilon)\}\triangleq\{f(kT+\varepsilon\pm 0)\}. \tag{2.42}$$

Obviously, $\{f_k^{\pm}(\varepsilon)\}$ is the sequence of values of the function $f(t)$ at the points $t_k=kT+\varepsilon\pm 0$. For each fixed ε from (2.42) we can derive sampled functions that can be associated with the bilateral discrete Laplace transform

$$F_{\pm}^{*}(s,\varepsilon)\triangleq\sum_{k=-\infty}^{\infty}f(kT+\varepsilon\pm 0)\mathrm{e}^{-ksT}\,. \tag{2.43}$$

If the function $f(t)$ is continuous, equation (2.43) defines a unique function

$$F^{*}(s,\varepsilon)\triangleq\sum_{k=-\infty}^{\infty}f(kT+\varepsilon)\mathrm{e}^{-ksT} \tag{2.44}$$

that will be called the *modified discrete Laplace transform* of the function $f(t)$. Henceforth by the modified discrete Laplace transform we shall mean the relations (2.43). In the general case, the convergence of the series (2.43) is dependent on the choice of ε. It is possible that the series (2.43) converges for some ε and diverges for others. Nevertheless, this is not the case if $f(t)\in\Lambda(\alpha,\beta)$.

Theorem 2.3 *If $f(t)\in\Lambda(\alpha,\beta)$, the series (2.43) converges absolutely for any $0\leq\varepsilon<T$ and $\alpha<\alpha'\leq\operatorname{Re}s\leq\beta'<\beta$.*

Proof Substituting $t+\varepsilon$ for t in (1.1), we have

$$|f(t+\varepsilon)|<\begin{cases} M\mathrm{e}^{\alpha'\varepsilon}\mathrm{e}^{\alpha' t} & t\geq 0\\ M\mathrm{e}^{\beta'\varepsilon}\mathrm{e}^{\beta' t} & t\leq 0\,.\end{cases} \tag{2.45}$$

Hence, for $0\leq\varepsilon<T$ it follows that

$$|f(t+\varepsilon)| < \begin{cases} M_1 e^{\alpha' t} & t \geq 0 \\ M_1 e^{\beta' t} & t \leq 0 \end{cases} \tag{2.46}$$

where M_1 is a new positive constant. As follows from (2.46), for a fixed ε we have $f(t+\varepsilon) \in \Lambda(\alpha,\beta)$. Therefore, the proposition to be proved follows from Theorem 2.2. ■

Example 2.5 Let us find the modified discrete Laplace transform for the discontinuous function (2.39). For $0 < \varepsilon < T$ we have

$$f(kT+\varepsilon) = \begin{cases} e^{\alpha\varepsilon} e^{\alpha k T} & k \geq 0 \\ 0 & k \leq 0\,. \end{cases} \tag{2.47}$$

Therefore, for $0 < \varepsilon < T$ and $\operatorname{Re} s > \operatorname{Re} \alpha$ we obtain

$$F^*(s,\varepsilon) = e^{\alpha\varepsilon} \sum_{k=0}^{\infty} e^{(\alpha-s)kT} = \frac{e^{\alpha\varepsilon}}{1 - e^{(\alpha-s)T}}\,. \tag{2.48}$$

For $\varepsilon = \pm 0$ we have the samplings $\{f_k^-\}$ and $\{f_k^+\}$ considered in example 2.4. Their discrete Laplace transforms are defined by Eqs. (2.40) and (2.41). Therefore, the transform $F_+^*(s,\varepsilon)$ is given by (2.48) and (2.41), while $F_-^*(s,\varepsilon)$ is defined by (2.48) and (2.40). □

Chapter 3

Displaced pulse frequency response

3.1 Basic definitions

Definition Let $f(t) \in \Lambda(\alpha, \beta)$ be continuous. Construct the series

$$\varphi_f(T, s, t) \triangleq \sum_{k=-\infty}^{\infty} f(t + kT)\mathrm{e}^{-s(t+kT)}, \quad \alpha < \operatorname{Re} s < \beta \tag{3.1}$$

where $T > 0$ is a constant and s is a complex variable. The sum of the series (3.1) will be called the *displaced pulse frequency response* (DPFR) of the function $f(t)$.

The physical reasons for this choice of terminology are explained in Rosenwasser (1973). Sufficient conditions for the convergence of the series (3.1) are given by

Theorem 3.1 *Let $f(t) \in \Lambda(\alpha, \beta)$. Then, the series (3.1) converges absolutely and uniformly with respect to s in any stripe $\alpha < \alpha' \leq \operatorname{Re} s \leq \beta' < \beta$ and with respect to t on any finite interval.*

Proof Using (1.8), for any k we have

$$\left| f(t + kT)\mathrm{e}^{-s(t+kT)} \right| < M\mathrm{e}^{-\gamma|t+kT|}, \quad \alpha' \leq \operatorname{Re} s \leq \beta' \tag{3.2}$$

where $\gamma > 0$. In this case

$$\left| f(t + kT)\mathrm{e}^{-s(t+kT)} \right| < M \quad \alpha' \leq \operatorname{Re} s \leq \beta', \quad -\infty < t < \infty. \tag{3.3}$$

Let $-qT < t < qT$, where q is an arbitrary natural number. The series (3.1) can be represented in the form

$$\varphi_f(T,s,t) = \sigma_+ + \sigma_- + \sigma_0 \tag{3.4}$$

where

$$\begin{aligned}
\sigma_+ &\triangleq \sum_{k=q}^{\infty} f(t+kT)\mathrm{e}^{-s(t+kT)} \\
\sigma_- &\triangleq \sum_{k=-\infty}^{-q} f(t+kT)\mathrm{e}^{-s(t+kT)} \\
\sigma_0 &\triangleq \sum_{k=-q+1}^{q-1} f(t+kT)\mathrm{e}^{-s(t+kT)}\,.
\end{aligned} \tag{3.5}$$

Consider the series σ_-. Since it is assumed that $t-qT<0$, we have $t+kT<0$ for $k \le -q$. Therefore, for all terms of the series σ_- we have $|t+kT| = -t-kT$ and for $k \le q$ Eq. (3.2) gives the estimate

$$\left|f(t+kT)\mathrm{e}^{-s(t+kT)}\right| < M\mathrm{e}^{-\gamma|t+kT|} \le M\mathrm{e}^{\gamma qT}\mathrm{e}^{\gamma kT}\,. \tag{3.6}$$

Hence,

$$|\sigma_-| \le \left|\sum_{k=-\infty}^{-q} f(t+kT)\mathrm{e}^{-s(t+kT)}\right| \le M\mathrm{e}^{\gamma qT}\sum_{k=-\infty}^{-q}\mathrm{e}^{\gamma kT}\,. \tag{3.7}$$

Obviously,

$$\sum_{k=-\infty}^{-q}\mathrm{e}^{\gamma kT} = \sum_{q}^{\infty}\mathrm{e}^{-\gamma kT} = \frac{\mathrm{e}^{-\gamma qT}}{1-\mathrm{e}^{-\gamma T}}\,. \tag{3.8}$$

Therefore, from (3.7) we have

$$|\sigma_-| < \frac{M}{1-\mathrm{e}^{-\gamma T}}\,. \tag{3.9}$$

Moreover, the terms of the series σ_- are majorated, for $\alpha' \le \operatorname{Re} s \le \beta'$ and $-qT < t < qT$, by the terms of a convergent geometric progression. Hence, the series σ_- converges absolutely and uniformly with respect to s and t in the specified regions.
In a similar way we can prove the absolute and uniform convergence of the series σ_+. Taking due account of (3.3), for the sum σ_0 we have $|\sigma_0| < M(2q-1)$. Thus, the terms of the series (3.1), except for a limited number of summands, are majorated by terms of a convergent series. Hence, as has been established, Titchmarsh (1932), the series (3.1) converges absolutely and uniformly in any region of variation of the arguments s and t. ■

Corollary 1 From Definition (3.1) it follows that

$$\varphi_f(T,s,t) = \varphi_f(T,s,t+T). \tag{3.10}$$

Therefore, it suffices to investigate the properties of the DPFR $\varphi_f(T,s,t)$ for $0 \le t \le T$. Since the sum of the series σ_+ satisfies an estimate similar to (3.9), with due account for (3.10) we can state that the DPFR $\varphi_f(T,s,t)$ is uniformly bounded in any stripe $\alpha' \le \operatorname{Re} s \le \beta'$, i.e.,

$$|\varphi_f(T,s,t)| < d = \text{const.}, \quad \alpha' \le \operatorname{Re} s \le \beta'. \tag{3.11}$$

Corollary 2 Each term of the series (3.1) is analytic with respect to s in the stripe $\alpha' \le \operatorname{Re} s \le \beta'$. Therefore, $\varphi_f(T,s,t)$ is analytic with respect to s for $\alpha' \le \operatorname{Re} s \le \beta'$ as the sum of a uniformly convergent series of analytic functions.

Corollary 3 Since $f(t)$ is continuous and the series (3.1) converges, $\varphi_f(T,s,t)$ is continuous with respect to t.

We note that if the function $f(t)$ is right unilateral, i.e., $f(t) = 0$ for $t < 0$, for $0 \le t < T$ the terms of the series (3.1) corresponding to negative k are zero, so that

$$\varphi_f(T,s,t) = \sum_{k=0}^{\infty} f(t+kT)\mathrm{e}^{-s(t+kT)}, \quad 0 \le t < T, \quad \operatorname{Re} s > \alpha. \tag{3.12}$$

3.2 Properties of DPFR for discontinuous functions

Equations (3.1) and (3.12) are based on the assumption that the function $f(t)$ is continuous. Nevertheless, in many applications, including those considered in this book, the function $f(t)$ has limited discontinuities for some t. In this section we investigate the properties of the DPFR for such functions. Henceforth, as in Chapter 2, it is assumed that the function $f(t)$ has finite limits

$$f(\tilde{t}+0) = \lim_{\epsilon \to +0} f(\tilde{t}+\epsilon), \qquad f(\tilde{t}-0) = \lim_{\epsilon \to -0} f(\tilde{t}+\epsilon) \tag{3.13}$$

for any $t = \tilde{t}$. As is known, the limits (3.13) exist if the function $f(t)$ has bounded variation. In what follows this is taken to be true. Then, using (3.13), we can define the following two transformations

$$\varphi_f(T,s,t\pm 0) \triangleq \sum_{k=-\infty}^{\infty} f(t+kT\pm 0)\mathrm{e}^{-s(t+kT)}, \quad \alpha < \operatorname{Re} s < \beta. \tag{3.14}$$

If the function $f(t)$ is continuous, both transforms (3.14) coincide with (3.1). The following theorem can be proved similarly to Theorem 3.1.

Theorem 3.2 *Let $f(t) \in \Lambda(\alpha,\beta)$. Then, the series (3.14) are defined and converge absolutely and uniformly for $\alpha < \alpha' \le \operatorname{Re} s \le \beta' < \beta$ on any finite interval of t.*

We note that Corollaries 1 and 2 of Theorem 3.1 remain valid, while Corollary 3 does not hold.
In order to investigate the properties of the DPFR $\varphi_f(T,s,t)$ for a discontinuous function $f(t)$ as a function in t in detail, we introduce the following definitions.

Definition The value $t = \tilde{t}$ will be called *regular* if the function $f(t)$ is continuous for $t = \tilde{t} + kT$ with any k. If $f(t)$ is discontinuous at $t = \tilde{t} + kT$ at least for a single k, such argument value $\tilde{t}$ will be called *singular*.

As follows from (3.14), for any regular $\tilde{t}$ the functions $\varphi_f(T,s,\tilde{t}+0)$ and $\varphi_f(T,s,\tilde{t}-0)$ coincide. Therefore, for any regular t we can denote

$$\varphi_f(T,s,t) = \varphi_f(T,s,t+0) = \varphi_f(T,s,t-0)\,. \tag{3.15}$$

As follows from (3.15), for a discontinuous function at regular points we can define a single DPFR $\varphi_f(T,s,t)$ given by (3.15) and (3.14). Therefore, the transforms $\varphi_f(T,s,t+0)$ and $\varphi_f(T,s,t-0)$ differ only for singular t.
To investigate the behaviour of the function $\varphi_f(T,s,t)$ at singular points we shall use the following Theorem, Titchmarsh (1932).

Theorem *The limit of the sum of an uniformly convergent functional series, in which each function tends to some limit, equals the sum of the limits of these member-functions.* ∎

Let $t = t_0$ be a singular point. Assume that there exists a sufficiently small number $\epsilon > 0$ such that the function $f(t)$ is continuous in the closed intervals $\mathcal{P}_{k-} = [t_0 + kT - \epsilon, t_0 + kT - 0]$ and $\mathcal{P}_{k+} = [t_0 + kT + 0, t_0 + kT + \epsilon]$ for an arbitrary k. Consider the set of functions

$$u_k(t) \triangleq f(t+kT)\mathrm{e}^{-s(t+kT)}, \quad \alpha < \alpha' \le \operatorname{Re} s \le \beta' < \beta. \tag{3.16}$$

On any interval $\mathcal{P}_{k-}$, $\mathcal{P}_{k+}$ the series for $\varphi_f(T,s,t)$ converges uniformly. In this case, since $f(t) \in \Lambda(\alpha,\beta)$, there exist the following limits:

$$\begin{aligned} u_{k0}^{+} &\triangleq \lim_{\tau\to+0} f(t_0+kT+\tau)\mathrm{e}^{-s(t_0+kT)} = f(t_0+kT+0)\mathrm{e}^{-s(t_0+kT)} \\ u_{k0}^{-} &\triangleq \lim_{\tau\to-0} f(t_0+kT+\tau)\mathrm{e}^{-s(t_0+kT)} = f(t_0+kT-0)\mathrm{e}^{-s(t_0+kT)}\,. \end{aligned} \tag{3.17}$$

Hence, using the above theorem, we can state that

$$\begin{aligned} \sum_{k=-\infty}^{\infty} f(t_0+kT+0)\mathrm{e}^{-s(t_0+kT)} &= \lim_{\tau\to+0} \varphi_f(T,s,t_0+\tau) = \varphi_f(T,s,t_0+0) \\ \sum_{k=-\infty}^{\infty} f(t_0+kT-0)\mathrm{e}^{-s(t_0+kT)} &= \lim_{\tau\to-0} \varphi_f(T,s,t_0+\tau) = \varphi_f(T,s,t_0-0)\,. \end{aligned} \tag{3.18}$$

Thus, if we construct the DPFR $\varphi_f(T,s,t)$ for regular points, its limits (3.18) define the related transforms (3.14) for all singular points. It is noteworthy that, due to (3.10), we can always assume that $0 \le t_0 < T$ without loss of generality.

Example 3.1 Apply relation (3.18) to study the properties of the DPFR of the discontinuous function

$$f(t) = \begin{cases} e^{at} & t > 0 \\ 0 & t < 0 \,. \end{cases} \tag{3.19}$$

Its singular points are $t_{0k} = kT, \quad (k = 0, \pm 1, \pm 2, \dots)$, all other points are regular. Obviously, $f(t) \in \Lambda_+(a, \infty)$, therefore, for $\operatorname{Re} s > \operatorname{Re} a$ and $0 < t < T$ we have

$$\varphi_f(T,s,t) = \sum_{k=0}^{\infty} e^{a(t+kT)} e^{-s(t+kT)} = \frac{e^{(a-s)t}}{1 - e^{(a-s)T}} \,. \tag{3.20}$$

In this case from (3.20) it follows that

$$\begin{aligned} \varphi_f(T,s,+0) = \lim_{t \to +0} \varphi_f(T,s,t) &= \frac{1}{1 - e^{(a-s)T}} \\ \varphi_f(T,s,-0) = \lim_{t \to -0} \varphi_f(T,s,t) &= \frac{e^{(a-s)T}}{1 - e^{(a-s)T}} \end{aligned} \tag{3.21}$$

which corresponds to Eqs. (2.40) and (2.41) obtained in Example 2.4. □

3.3 Representation of DPFR as Fourier series

Henceforth we shall assume that the DPFR $\varphi_f(T,s,t)$ is a function of bounded variation with respect to t. In applications considered in the present book this always holds. Taking (3.10) into account, we can find the expansion of $\varphi_f(T,s,t)$ into a Fourier series with respect to t. Let this series have the form

$$\varphi_F(T,s,t) \triangleq \sum_{m=-\infty}^{\infty} \varphi_m(s) e^{mj\omega t} \,, \quad \omega = \frac{2\pi}{T} \tag{3.22}$$

where the Fourier coefficients $\varphi_m(s)$ are calculated by the known formulae

$$\varphi_m(s) = \frac{1}{T} \int_0^T \varphi_f(T,s,t) e^{-mj\omega t} \, dt \tag{3.23}$$

It is assumed that $f(t) \in \Lambda(\alpha, \beta)$. To calculate the coefficients, we substitute (3.1) into (3.23). In this case the discontinuities of the function $f(t)$ do not matter, because they do not affect the value of the integral. Then, we have

$$\varphi_m(s) = \frac{1}{T} \int_0^T \left[\sum_{k=-\infty}^{\infty} f(t+kT) e^{-s(t+kT)} \right] e^{-mj\omega t} \, dt \,. \tag{3.24}$$

Since the series (3.1) converges uniformly, the order of summation and integration in (3.24) can be interchanged, so that

$$\varphi_m(s) = \frac{1}{T} \sum_{k=-\infty}^{\infty} \int_0^T f(t+kT)\mathrm{e}^{-s(t+kT)}\mathrm{e}^{-mj\omega t}\,\mathrm{d}t\,. \tag{3.25}$$

It can be easily verified that

$$\int_0^T f(t+kT)\mathrm{e}^{-s(t+kT)}\mathrm{e}^{-mj\omega t}\,\mathrm{d}t = \int_{kT}^{(k+1)T} f(t)\mathrm{e}^{-(s+mj\omega)t}\,\mathrm{d}t\,. \tag{3.26}$$

Substituting (3.26) into (3.25), we obtain

$$\varphi_m(s) = \frac{1}{T} \sum_{k=-\infty}^{\infty} \int_{kT}^{(k+1)T} f(t)\mathrm{e}^{-(s+mj\omega)t}\,\mathrm{d}t = \frac{1}{T}\int_{-\infty}^{\infty} f(t)\mathrm{e}^{-(s+mj\omega)t}\,\mathrm{d}t\,. \tag{3.27}$$

Comparing this equation with (1.14), we have

$$\varphi_m(s) = \frac{1}{T}F(s+mj\omega) \tag{3.28}$$

where $F(s)$ is the Laplace transform of the function $f(t)$, which does exist, because $f(t) \in \Lambda(\alpha,\beta)$. From (3.22) and (3.28) we obtain

$$\varphi_F(T,s,t) = \frac{1}{T} \sum_{k=-\infty}^{\infty} F(s+kj\omega)\mathrm{e}^{kj\omega t}\,, \quad \omega = \frac{2\pi}{T}\,. \tag{3.29}$$

The series (3.29) is uniquely defined by the Laplace transform $F(s)$. Therefore, the sum of the series (3.29) will be called the *displaced pulse frequency response* (DPFR) for the image $F(s)$.

Since $\varphi_f(T,s,t)$ is a function of bounded variation by assumption, we have at regular points of $\varphi_f(T,s,t)$

$$\varphi_F(T,s,t) = \varphi_f(T,s,t) \tag{3.30}$$

while at any points of discontinuity $t = \tilde{t}$

$$\varphi_F(T,s,\tilde{t}) = \frac{1}{2}\left[\varphi_f(T,s,\tilde{t}+0) + \varphi_f(T,s,\tilde{t}-0)\right]\,. \tag{3.31}$$

Therefore, for a comparison of the properties of $\varphi_f(T,s,t)$ and $\varphi_F(T,s,t)$ it is necessary to use additional information on the properties of $F(s)$, $f(t)$ and $\varphi_f(T,s,t)$. Henceforth, we shall employ the following statement.

Theorem 3.3 *Let a function $F(s)$ satisfy the conditions of Theorem 1.4, i.e., it is analytic in the stripe $\alpha \le \operatorname{Re} s \le \beta$ and there it satisfies the estimate $|F(s)| < A|s|^{-1-\lambda}$. Then, the following statements hold:*

1. *The series (3.29) converges in the stripe $\alpha \leq \operatorname{Re} s \leq \beta$ absolutely and uniformly with respect to s and t. Its sum is analytic with respect to s and continuous with respect to t.*
2. *For all t and $\alpha < \operatorname{Re} s < \beta$ the equalities*

$$\varphi_F(T,s,t) = \varphi_f(T,s,t) = \sum_{k=-\infty}^{\infty} f(t+kT)\mathrm{e}^{-s(t+kT)} \tag{3.32}$$

hold, where

$$f(t) = \frac{1}{2\pi \mathrm{j}} \int_{c-\mathrm{j}\infty}^{c+\mathrm{j}\infty} F(s)\mathrm{e}^{st}\,\mathrm{d}s \tag{3.33}$$

is the associated original for the image $F(s)$.
3. *If $\beta = \infty$, i.e., the estimate (1.33) holds for all $\operatorname{Re} s > \alpha$, we have $f(t) = 0$ for $t < 0$ and*

$$\varphi_F(T,s,t) = \varphi_f(T,s,t) = \sum_{k=0}^{\infty} f(t+kT)\mathrm{e}^{-s(t+kT)}\,, \quad 0 \leq t < T\,. \tag{3.34}$$

Proof 1. As follows from the estimate (1.33), for $\alpha \leq \operatorname{Re} s \leq \beta$ we have

$$|F(s+k\mathrm{j}\omega)| \leq B|k|^{-1-\lambda} \tag{3.35}$$

where B and λ are positive constants. Since the numerical series

$$I_\lambda \triangleq \sum_{k=-\infty}^{\infty} |k|^{-1-\lambda} \tag{3.36}$$

converges for $\lambda > 0$, from (3.35) it follows that all terms of the series (3.29) are majorated by positive numbers that form a convergent series. Therefore, the series (3.29) converges for $\alpha \leq \operatorname{Re} s \leq \beta$ uniformly with respect to s and t. Hence, the trigonometric series (3.29) is the Fourier series of its sum, which is continuous. Moreover, the function $\varphi_f(T,s,t)$ is analytic with respect to s, because the series (3.29) converges uniformly.

2. Let $f(t)$ be the original for $F(s)$ given by (3.33). In accordance with Theorem 1.4, this function belongs to $\Lambda(\alpha,\beta)$ and is continuous. As follows from Theorem 3.1, the series for $\varphi_f(T,s,t)$ converges absolutely and uniformly, and its sum is continuous with respect to t. Therefore, due to uniform convergence of the series (3.29), we obtain (3.32).

3. If $\beta = \infty$, by Theorem 1.4 we have $f(t) = 0$ for $t < 0$, and from (3.32) follows (3.34) for $0 \leq t < T$. ∎

If the estimate (1.34) does not hold for $F(s)$, the DPFR $\varphi_f(T,s,t)$ is, generally speaking, discontinuous with respect to t. Next we consider the relations between the functions $\varphi_f(T,s,t)$ and $\varphi_F(T,s,t)$ at singular points of $f(t)$. Let

$t = t_0$ be a singular point. Without loss of generality, we assume $0 \leq t_0 < T$. Denote

$$f_k^+(t_0) \triangleq f(kT + t_0 + 0), \quad f_k^-(t_0) \triangleq f(kT + t_0 - 0). \tag{3.37}$$

Then, the altitude of discontinuity of the function $f(t)$ at the point $t = t_0 + kT$ is

$$\Delta f_k(t_0) \triangleq f_k^+(t_0) - f_k^-(t_0). \tag{3.38}$$

Using (3.14) and (3.18), we obtain

$$\begin{aligned} \varphi_f(T, s, t_0 + 0) &= \sum_{k=-\infty}^{\infty} f_k^+(t_0) e^{-s(t_0 + kT)} \\ \varphi_f(T, s, t_0 - 0) &= \sum_{k=-\infty}^{\infty} f_k^-(t_0) e^{-s(t_0 + kT)}. \end{aligned} \tag{3.39}$$

Let $\Delta\varphi(t_0)$ be the value of discontinuity of $\varphi_f(T, s, t)$ at the point t_0

$$\Delta\varphi(t_0) \triangleq \varphi_f(T, s, t_0 + 0) - \varphi_f(T, s, t_0 - 0). \tag{3.40}$$

Substituting (3.39) into (3.40) and taking (3.38) into account, we obtain

$$\Delta\varphi(t_0) = e^{-st_0} \Delta_f^*(s, t_0) \tag{3.41}$$

where

$$\Delta_f^*(s, t_0) \triangleq \sum_{k=-\infty}^{\infty} \Delta f_k(t_0) e^{-ksT} \tag{3.42}$$

is the discrete Laplace transform for the sequence of discontinuities $\Delta f_k(t_0)$. Thus, for a singular point t_0 we have

$$\varphi_f(T, s, t_0 + 0) - \varphi_f(T, s, t_0 - 0) = e^{-st_0} \Delta_f^*(s, t_0). \tag{3.43}$$

Considering (3.43) and (3.31) together, we obtain

$$\varphi_f(T, s, t_0 + 0) = \varphi_F(T, s, t_0) + \frac{1}{2} e^{-st_0} \Delta_f^*(s, t_0). \tag{3.44}$$

Taking into account that

$$\varphi_F(T, s, t_0) = \frac{1}{T} \sum_{k=-\infty}^{\infty} F(s + kj\omega) e^{kj\omega t_0} \tag{3.45}$$

we can represent (3.44) in the form

$$\varphi_f(T, s, t_0 + 0) = \frac{1}{T} \sum_{k=-\infty}^{\infty} F(s + kj\omega) e^{kj\omega t_0} + \frac{1}{2} e^{-st_0} \Delta_f^*(s, t_0). \tag{3.46}$$

As a special case, for $t_0 = 0$ we obtain

$$\varphi_f(T,s,+0) = \frac{1}{T}\sum_{k=-\infty}^{\infty} F(s+kj\omega) + \frac{1}{2}\Delta_f^*(s,0)\,. \tag{3.47}$$

We note that if only a finite number of the points $t_{0k} = t_0 + kT$ are points of discontinuity of the function $f(t)$, the series (3.42) reduces to an ordinary sum. As a special case, let the function $f(t)$ have a discontinuity at $t = 0$ and be continuous for $t = kT$, $k \neq 0$. In this case, if $f(-0) = 0$, then $\Delta_f^*(s,+0) = f(+0)$ and (3.47) takes the form

$$\varphi_f(T,s,+0) = \frac{1}{T}\sum_{k=-\infty}^{\infty} F(s+kj\omega) + \frac{1}{2}f(+0)\,. \tag{3.48}$$

If the specified restrictions are not satisfied, Eq. (3.48) becomes inapplicable and we have to use the general relations (3.46) and (3.47).

Example 3.2 Examine the correctness of formula (3.48) for a discontinuous function. Consider, as in Example 3.1, the function

$$f(t) = \begin{cases} e^{at} & t > 0 \\ 0 & t < 0\,. \end{cases} \tag{3.49}$$

This function has a single singular point $t_0 = 0$. Moreover, $f(-0) = 0$ and the applicability conditions of (3.48) hold. We show that (3.48) gives the correct result in this case. Using (3.20) and (3.21), for $\operatorname{Re} s > \operatorname{Re} a$ we have

$$\varphi_f(T,s,t) = \frac{e^{(a-s)t}}{1 - e^{(a-s)T}}\,, \qquad 0 < t < T \tag{3.50}$$

so that

$$\varphi_f(T,s,+0) = \frac{1}{1 - e^{(a-s)T}}\,, \quad \varphi_f(T,s,-0) = \frac{e^{(a-s)T}}{1 - e^{(a-s)T}}\,. \tag{3.51}$$

The Laplace image of the function $f(t)$ equals

$$F(s) = \frac{1}{s-a}\,, \qquad \operatorname{Re} s > a \tag{3.52}$$

i.e., the Fourier series (3.45) for the function (3.49) takes the form

$$\varphi_F(T,s,t) = \frac{1}{T}\sum_{k=-\infty}^{\infty} \frac{1}{s+kj\omega-a} e^{kj\omega t}\,. \tag{3.53}$$

In this case, Eq. (3.31) gives, taking due account of (3.51),

$$\varphi_F(T,s,0) = \frac{1}{T}\sum_{k=-\infty}^{\infty} \frac{1}{s+kj\omega-a} = \frac{1}{2}\cdot\frac{1+e^{(a-s)T}}{1-e^{(a-s)T}}\,. \tag{3.54}$$

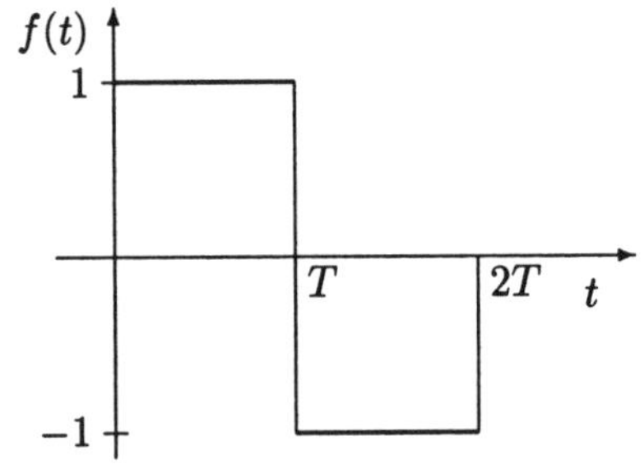

Figure 3.1: Function from Example 3.3

Then, taking into account that $f(+0) = 1$, we obtain

$$\varphi_F(T,s,0) + \frac{1}{2}f(+0) = \frac{1}{1-e^{(a-s)T}} = \varphi_f(T,s,+0) \tag{3.55}$$

and Eq. (3.48) holds. □

Example 3.3 An example is given, where formula (3.48) is not applicable, and instead (3.47) must be used. Consider the function shown in Fig. 3.1

$$f(t) = \begin{cases} 0 & t<0 \\ 1 & 0<t<T \\ -1 & T<t<2T \\ 0 & t>2T\,. \end{cases} \tag{3.56}$$

Its Laplace image has the form

$$F(s) = \frac{\left(1-e^{-sT}\right)^2}{s}\,. \tag{3.57}$$

In this case

$$\varphi_F(T,s,t) = \left(1-e^{-sT}\right)^2 \frac{1}{T}\sum_{k=-\infty}^{\infty} \frac{1}{s+kj\omega}\, e^{kj\omega t}\,. \tag{3.58}$$

It can immediately be verified that

$$\frac{1}{T}\sum_{k=-\infty}^{\infty} \frac{1}{s+kj\omega}\, e^{kj\omega t} = \frac{e^{-st}}{1-e^{-sT}}\,, \quad 0<t<T\,. \tag{3.59}$$

Therefore, using (3.58), we obtain

$$\varphi_f(T,s,t) = \left(1-e^{-sT}\right)e^{-st}\,, \quad 0<t<T\,. \tag{3.60}$$

It can be verified that the series (3.58) is indeed the Fourier series for the function defined by (3.60) in the interval $0<t<T$. From (3.60) we have

$$\varphi_f(T,s,+0) = 1 - \mathrm{e}^{-sT}, \quad \varphi_f(T,s,-0) = \mathrm{e}^{-sT}\left(1-\mathrm{e}^{-sT}\right). \tag{3.61}$$

Hence, using (3.58), from (3.31) with $t = 0$ we obtain

$$\varphi_F(T,s,0) = \frac{1}{2}\left[\varphi_f(T,s,+0) + \varphi_f(T,s,-0)\right] = \frac{1-\mathrm{e}^{-2sT}}{2}. \tag{3.62}$$

Moreover, as follows from Fig. 3.1, the function $f(t)$ has discontinuities at $t = 0$, $t = T$, $t = 2T$, and its altitudes are 1, -2 and 1, respectively. Therefore, the function $\Delta^*\varphi(s,0)$, obtained from (3.41) for $t = 0$, appears as

$$\Delta^*\varphi(s,0) = 1 - 2\mathrm{e}^{-sT} + \mathrm{e}^{-2sT} = \left(1-\mathrm{e}^{-sT}\right)^2. \tag{3.63}$$

From (3.62) and (3.63) we find

$$\begin{aligned}\varphi_F(T,s,0) + \frac{1}{2}\Delta^*\varphi(s,0) &= \frac{1}{2}(1-\mathrm{e}^{-2sT}) + \frac{1}{2}\left(1-2\mathrm{e}^{-sT}+\mathrm{e}^{-2sT}\right)\\ &= 1-\mathrm{e}^{-sT} = \varphi_f(T,s,+0)\end{aligned} \tag{3.64}$$

i.e., Eq. (3.44) holds. □

3.4 Inversion formulae

The series (3.1) and (3.29) can be considered as functional transforms of the functions $f(t)$ and $F(s)$. Next, we show that conversely, under some restrictions, we can determine $f(t)$ and $F(s)$, knowing the DPFR $\varphi_f(T,s,t)$ and $\varphi_F(T,s,t)$.

Theorem 3.4 *Let $f(t) \in \Lambda(\alpha,\beta)$. Then,*

$$f(t) = \frac{T}{2\pi \mathrm{j}}\int_{c-\mathrm{j}\omega/2}^{c+\mathrm{j}\omega/2} \varphi_f(T,s,t)\mathrm{e}^{st}\,\mathrm{d}s, \quad \alpha < c < \beta, \tag{3.65}$$

$$F(s) = \int_0^T \varphi_f(T,s,t)\,\mathrm{d}t, \quad \alpha < \operatorname{Re} s < \beta. \tag{3.66}$$

Proof As follows from (3.1),

$$\varphi_f(T,s,t)\mathrm{e}^{st} = \sum_{k=-\infty}^{\infty} f(t+kT)\mathrm{e}^{-ksT}. \tag{3.67}$$

The function e^{st} for $\alpha < \alpha' \le \operatorname{Re} s \le \beta' < \beta$ is uniformly bounded on any finite interval of t. Therefore, the series on the right side of (3.67) converges uniformly for $\alpha < \alpha' \le \operatorname{Re} s \le \beta' < \beta$, because it is obtained by multiplying a uniformly convergent series by a bounded function, Titchmarsh (1932). Hence, for $\alpha' \le c \le \beta'$ we have

$$\frac{T}{2\pi \mathrm{j}}\int_{c-\mathrm{j}\omega/2}^{c+\mathrm{j}\omega/2}\varphi_f(T,s,t)\mathrm{e}^{st}\,\mathrm{d}s=\frac{T}{2\pi \mathrm{j}}\sum_{k=-\infty}^{\infty}f(t+kT)\int_{c-\mathrm{j}\omega/2}^{c+\mathrm{j}\omega/2}\mathrm{e}^{-ksT}\,\mathrm{d}s\,. \quad (3.68)$$

Taking into account that

$$\frac{T}{2\pi \mathrm{j}}\int_{c-\mathrm{j}\omega/2}^{c+\mathrm{j}\omega/2}\mathrm{e}^{-ksT}\,\mathrm{d}s=\begin{cases}1 & k=0\\ 0 & k\neq 0\end{cases} \quad (3.69)$$

from (3.68) we obtain (3.65). To prove the equality (3.66) we use Eqs.(3.23) and (3.28), which yield

$$\int_0^T \varphi_f(T,s,t)\,\mathrm{d}t=T\varphi_0(s)=F(s)\,. \quad (3.70)$$

■

Theorem 3.5 *Let $f(t)\in\Lambda(\alpha,\beta)$. Then,*

$$F(s)=\int_0^T \varphi_F(T,s,t)\,\mathrm{d}t\,. \quad (3.71)$$

If, moreover, the series (3.29) $\varphi_F(T,s,t)$ converges absolutely and uniformly for $\alpha<\alpha'\leq \operatorname{Re}s\leq\beta'<\beta$, the function $f(t)$ is continuous and

$$f(t)=\frac{T}{2\pi \mathrm{j}}\int_{c-\mathrm{j}\omega/2}^{c+\mathrm{j}\omega/2}\varphi_F(T,s,t)\mathrm{e}^{st}\,\mathrm{d}s\,,\quad \alpha<c<\beta\,. \quad (3.72)$$

Proof For $f(t)\in\Lambda(\alpha,\beta)$ the series $\varphi_f(T,s,t)$ converges and its sum is integrable as the sum of a series of integrable functions. In this case the series (3.29) is term-wise integrable as a Fourier series of an integrable function, Titchmarsh (1932). Consequently,

$$\int_0^T \varphi_F(T,s,t)\,\mathrm{d}t=\frac{1}{T}\sum_{k=-\infty}^{\infty}F(s+k\mathrm{j}\omega)\int_0^T \mathrm{e}^{k\mathrm{j}\omega t}\,\mathrm{d}t\,. \quad (3.73)$$

Taking into account that

$$\int_0^T \mathrm{e}^{k\mathrm{j}\omega t}\,\mathrm{d}t=\begin{cases}T & k=0\\ 0 & k\neq 0\end{cases} \quad (3.74)$$

from (3.73) we obtain (3.71). To prove (3.72), we note that by assumption the series

$$I=\sum_{k=-\infty}^{\infty}|F(s+k\mathrm{j}\omega)| \quad (3.75)$$

converges for $\alpha<\operatorname{Re}s<\beta$. Hence, the integral

$$f(t)=\frac{T}{2\pi \mathrm{j}}\int_{c-\mathrm{j}\infty}^{c+\mathrm{j}\infty}F(s)\mathrm{e}^{st}\,\mathrm{d}s \quad (3.76)$$

converges, and the function $f(t)$ is continuous. Since the integral (3.76) converges, equalities (1.29) and (1.32) hold, which is equivalent to (3.72). ■

3.5 Properties of DPFR of real rational images

If the Laplace image of a function $f(t)$ has the form (1.77), the properties of the DPFR $\varphi_f(T,s,t)$ and $\varphi_F(T,s,t)$ can be investigated in detail. In the given case

$$\varphi_F(T,s,t) = \frac{1}{T}\sum_{k=-\infty}^{\infty} \frac{m(s+kj\omega)}{d(s+kj\omega)}\mathrm{e}^{kj\omega} \tag{3.77}$$

where $m(s)$ and $d(s)$ are the polynomials appearing in (1.78).

Theorem 3.6 *Let us assume the partial fraction expansion*

$$F(s) = \sum_{i=1}^{\ell}\sum_{k=1}^{\nu_i} \frac{f_{ik}}{(s-s_i)^k}\,. \tag{3.78}$$

Then, the series (3.77) is the Fourier series of the periodic function

$$\begin{aligned} \tilde{\varphi}_F(T,s,t) &= \hat{\varphi}_F(T,s,t)\,, \qquad 0<t<T \\ \tilde{\varphi}_F(T,s,t) &= \tilde{\varphi}_F(T,s,t+T) \end{aligned} \tag{3.79}$$

where

$$\hat{\varphi}_F(T,s,t) \triangleq \sum_{i=1}^{\ell}\sum_{k=1}^{\nu_i} \frac{f_{ik}}{(k-1)!}\frac{\partial^{k-1}}{\partial s_i^{k-1}}\frac{\mathrm{e}^{(s_i-s)t}}{1-\mathrm{e}^{(s_i-s)T}}\,. \tag{3.80}$$

Proof Let the Fourier series of the function (3.79) take the form

$$\tilde{\varphi}_F(T,s,t) = \sum_{\mu=-\infty}^{\infty} \varphi_\mu(s)\mathrm{e}^{\mu j\omega t} \tag{3.81}$$

where

$$\varphi_\mu(s) = \frac{1}{T}\int_0^T \hat{\varphi}_F(T,s,t)\mathrm{e}^{-\mu j\omega t}\,\mathrm{d}t\,. \tag{3.82}$$

Using (3.80), we have

$$\varphi_\mu(s) = \frac{1}{T}\sum_{i=1}^{\ell}\sum_{k=1}^{\nu_i} \frac{f_{ik}}{(k-1)!}\frac{\partial^{k-1}}{\partial s_i^{k-1}}\int_0^T \frac{\mathrm{e}^{(s_i-s)t}\mathrm{e}^{-\mu j\omega t}}{1-\mathrm{e}^{(s_i-s)T}}\,\mathrm{d}t\,. \tag{3.83}$$

Calculating the integral on the right-hand side of the latter equation gives

$$\int_0^T \frac{\mathrm{e}^{(s_i-s)t}\mathrm{e}^{-\mu j\omega t}}{1-\mathrm{e}^{(s_i-s)T}}\,\mathrm{d}t = \frac{1}{s-s_i+\mu j\omega}\,. \tag{3.84}$$

From the obvious equality

$$\frac{1}{(s-s_i+\mu j\omega)^k} = \frac{1}{(k-1)!}\frac{\partial^{k-1}}{\partial s_i^{k-1}}\frac{1}{s-s_i+\mu j\omega} \tag{3.85}$$

follows that

$$\frac{1}{(k-1)!}\frac{\partial^{k-1}}{\partial s_i^{k-1}}\int_0^T \frac{e^{(s_i-s)t}e^{-\mu j\omega t}}{1-e^{(s_i-s)T}}\,dt = \frac{1}{(s-s_i+\mu j\omega)^k}\,. \tag{3.86}$$

Equations (3.83), (3.86) and (3.78) yield

$$\varphi_\mu(s) = \frac{1}{T}\sum_{i=1}^{\ell}\sum_{k=1}^{\nu_i}\frac{f_{ik}}{(s-s_i+\mu j\omega)^k}\,. \tag{3.87}$$

Hence, the required Fourier series has the form (3.77). ■

Let $S_m : (\mu_m, \mu_{m+1})$ be a regularity intervals of the function $F(s)$. Each interval S_m is associated with an original $f_m(t)$ defined by (1.89). The originals $f_m(t)$, taken as an aggregate, define a family $\{f_m(t)\}$. Since, due to (1.98) $f_m(t) \in \Lambda(\mu_m, \mu_{m+1})$, the following DPFRs are defined for the stripes $\mu_m < \operatorname{Re} s < \mu_{m+1}$:

$$\varphi_{f_m}(T, s, t) = \sum_{k=-\infty}^{\infty} f_m(t+kT)e^{-s(t+kT)}\,. \tag{3.88}$$

Theorem 3.7 *For* $\mu_m < \operatorname{Re} s < \mu_{m+1}$ *and* $t \neq kT$ *the equality*

$$\varphi_{f_m}(T, s, t) = \tilde{\varphi}_F(T, s, t) \tag{3.89}$$

holds, where $\tilde{\varphi}_F(T, s, t)$ *is the function defined by (3.79) and (3.80).*

Proof To prove the assertion it suffices to show that the sum of each series (3.88) in the interval $0 < t < T$ is given by the following expression, which is independent of m

$$\varphi_{f_m}(T, s, t) = \sum_{i=1}^{\ell}\sum_{k=1}^{\nu_i}\frac{f_{ik}}{(k-1)!}\frac{\partial^{k-1}}{\partial s_i^{k-1}}\frac{e^{(s_i-s)t}}{1-e^{(s_i-s)T}}\,. \tag{3.90}$$

Introduce the functions

$$f_+(t,a) \triangleq \begin{cases} e^{at} & t>0 \\ 0 & t<0 \end{cases}, \quad f_-(t,a) \triangleq \begin{cases} 0 & t>0 \\ -e^{at} & t<0 \end{cases} \tag{3.91}$$

where a is a complex value. Then, for $\operatorname{Re} s > \operatorname{Re} a$ and $0 < t < T$ we have

$$\varphi_{f_+}(T, s, t) = \sum_{k=0}^{\infty} e^{a(t+kT)}e^{-s(t+kT)} = \frac{e^{(a-s)t}}{1-e^{(a-s)T}}\,. \tag{3.92}$$

Similarly, for $\operatorname{Re} s < \operatorname{Re} a$, $0 < t < T$ we obtain

$$\varphi_{f_-}(T, s, t) = -\sum_{k=-\infty}^{-1} e^{a(t+kT)}e^{-s(t+kT)} = \frac{e^{(a-s)t}}{1-e^{(a-s)T}}\,. \tag{3.93}$$

Introduce the functions

$$f_{n+}(t,a) \triangleq t^n f_+(t,a)\,, \quad f_{n-}(t,a) \triangleq t^n f_-(t,a)\,. \tag{3.94}$$

Henceforth, we consider the terms of the series in (3.92) and (3.93) as analytic functions in the complex variable s. The series in (3.92) and (3.93) converge uniformly, therefore they can be differentiated by a in the corresponding regions of convergence, Titchmarsh (1932). Performing differentiation, from (3.92) and (3.93) for $0 < t < T$ we obtain

$$\sum_{k=0}^{\infty}(t+kT)^n \mathrm{e}^{a(t+kT)}\mathrm{e}^{-s(t+kT)} = \frac{\partial^n}{\partial a^n}\frac{\mathrm{e}^{(a-s)t}}{1-\mathrm{e}^{(a-s)T}}\,, \quad \operatorname{Re} s > \operatorname{Re} a \tag{3.95}$$

$$-\sum_{k=-\infty}^{-1}(t+kT)^n \mathrm{e}^{a(t+kT)}\mathrm{e}^{-s(t+kT)} = \frac{\partial^n}{\partial a^n}\frac{\mathrm{e}^{(a-s)t}}{1-\mathrm{e}^{(a-s)T}}\,, \quad \operatorname{Re} s < \operatorname{Re} a\,. \tag{3.96}$$

Taking (3.94) into account, we can represent these equalities in the form

$$\varphi_{f_{n+}}(T,s,t) = \frac{\partial^n}{\partial a^n}\frac{\mathrm{e}^{(a-s)t}}{1-\mathrm{e}^{(a-s)T}}\,, \quad \operatorname{Re} s > \operatorname{Re} a \tag{3.97}$$

$$\varphi_{f_{n-}}(T,s,t) = \frac{\partial^n}{\partial a^n}\frac{\mathrm{e}^{(a-s)t}}{1-\mathrm{e}^{(a-s)T}}\,, \quad \operatorname{Re} s < \operatorname{Re} a\,. \tag{3.98}$$

Next, we consider an arbitrary regular interval $S_m : (\mu_m, \mu_{m+1})$ and the related original $f_m(t)$ defined by (1.98). Using (3.94), we can write (1.98) in the form

$$f_m(t) = \begin{cases} \displaystyle\sum_{i=1}^{\rho}\sum_{k=1}^{\nu_i}\frac{f_{ik}}{(k-1)!}f_{(k-1)+}(t,s_i) & t>0 \\ \displaystyle\sum_{i=\rho+1}^{\ell}\sum_{k=1}^{\nu_i}\frac{f_{ik}}{(k-1)!}f_{(k-1)-}(t,s_i) & t<0\,. \end{cases} \tag{3.99}$$

If $\mu_m < \operatorname{Re} s < \mu_{m+1}$, calculating the DPFR (3.88) with due account for (3.97) and (3.98), we obtain (3.90). ∎

Using (3.89), closed expressions for the functions $\tilde{\varphi}_F(T,s,t)$ and $\varphi_{f_m}(T,s,t)$ can be derived for any t. Let $\lambda T < t < (\lambda+1)T$, where λ is an integer. Then, we can assume $t = \varepsilon + \lambda T$ with $0 < \varepsilon = t - \lambda T < T$. In this case

$$\tilde{\varphi}_F(T,s,t) = \tilde{\varphi}_F(T,s,t-\lambda T) = \hat{\varphi}_F(T,s,t-\lambda T) = \hat{\varphi}_F(T,s,\varepsilon)\,. \tag{3.100}$$

From (3.100) and (3.79) it follows that

$$\tilde{\varphi}_F(T,s,t) = \sum_{i=1}^{\ell}\sum_{k=1}^{\nu_i}\frac{f_{ik}}{(k-1)!}\frac{\partial^{k-1}}{\partial s_i^{k-1}}\frac{\mathrm{e}^{(s_i-s)(t-\lambda T)}}{1-\mathrm{e}^{(s_i-s)T}}\,, \quad \lambda T < t < (\lambda+1)T\,. \tag{3.101}$$

The above expressions become greatly simplified if all poles of the image $F(s)$ are simple. In this case instead of (3.78) we have

$$F(s) = \sum_{i=1}^{p} \frac{f_i}{s - s_i} \tag{3.102}$$

and Eqs. (3.101) with (3.79) yields

$$\tilde{\varphi}_F(T,s,t) = \sum_{i=1}^{p} \frac{f_i \, \mathrm{e}^{(s_i - s)t}}{1 - \mathrm{e}^{(s_i - s)T}}, \quad 0 < t < T \tag{3.103}$$

$$\tilde{\varphi}_F(T,s,t) = \sum_{i=1}^{p} \frac{f_i \, \mathrm{e}^{(s_i - s)(t - \lambda T)}}{1 - \mathrm{e}^{(s_i - s)T}}, \quad \lambda T < t < (\lambda + 1)T. \tag{3.104}$$

We now formulate some general properties of the above DPFRs.

1. As follows from (3.101), the function $\tilde{\varphi}_F(T,s,t)$ is continuously differentiable by t for $t \neq \lambda T$. Therefore, the Fourier series (3.77) converges with respect to t on any subinterval inside the interval $\lambda T < t < (\lambda + 1)T$. Hence,

$$\varphi_F(T,s,t) = \tilde{\varphi}_F(T,s,t), \quad \lambda T < t < (\lambda + 1)T. \tag{3.105}$$

2. The behaviour of $\tilde{\varphi}_F(T,s,t)$ at the points $t = \lambda T$ depends on the properties of the image $F(s)$.

Theorem 3.8 *Let*

$$F(s) = \frac{m(s)}{d(s)} = \frac{m_1 s^{p-1} + m_2 s^{p-2} + \ldots + m_p}{s^p + d_1 s^{p-1} + \ldots + d_p} \tag{3.106}$$

and the partial fraction expansion (3.78) holds. If $m_1 \neq 0$, the function $\tilde{\varphi}_F(T,s,t)$ has discontinuities at the points $t = \lambda T$. If $m_1 = m_2 = \ldots = m_{\rho-1} = 0$ and $m_\rho \neq 0$, the function $\tilde{\varphi}_F(T,s,t)$ has derivatives up to the order $\rho - 1$ (inclusive). Moreover, the $(\rho-1)$-th derivative is piece-wise continuous, while the others are continuous.

Proof As follows from (3.80),

$$\begin{aligned} \tilde{\varphi}_F(T,s,+0) &= \sum_{i=1}^{\ell} \sum_{k=1}^{\nu_i} \frac{f_{ik}}{(k-1)!} \frac{\partial^{k-1}}{\partial s_i^{k-1}} \frac{1}{1 - \mathrm{e}^{(s_i - s)T}} \\ \tilde{\varphi}_F(T,s,-0) &= \tilde{\varphi}_F(T,s,T-0) \\ &= \sum_{i=1}^{\ell} \sum_{k=1}^{\nu_i} \frac{f_{ik}}{(k-1)!} \frac{\partial^{k-1}}{\partial s_i^{k-1}} \frac{\mathrm{e}^{(s_i - s)T}}{1 - \mathrm{e}^{(s_i - s)T}}. \end{aligned} \tag{3.107}$$

Taking into account that

$$\frac{e^{(s_i-s)T}}{1-e^{(s_i-s)T}} = -1 + \frac{1}{1-e^{(s_i-s)T}}$$

from (3.107) we find the value of discontinuity of the function $\tilde{\varphi}_F(T,s,t)$ at $t=0$

$$\Delta\tilde{\varphi}_F(0) \triangleq \tilde{\varphi}_F(T,s,+0) - \tilde{\varphi}_F(T,s,-0) = \sum_{i=1}^{\ell} f_{i1} = m_1 \,. \tag{3.108}$$

Thus, if $m_1 \neq 0$, the function $\tilde{\varphi}_F(T,s,t)$ has discontinuities at the points $t = \lambda T$. The value of this discontinuity is independent of s and equals m_1. If $m_1 = 0$, the function $\tilde{\varphi}_F(T,s,t)$ is continuous with respect to t. If $m_1 = 0$ and $m_2 \neq 0$, Eq. (3.101) shows that $\tilde{\varphi}_F(T,s,t)$ has a piece-wise continuous derivative. Next, let $m_1 = m_2 = 0$ and $m_3 \neq 0$. Then,

$$\frac{d\tilde{\varphi}_F(T,s,t)}{dt} = \frac{d\varphi_F(T,s,t)}{dt} = \frac{1}{T}\sum_{k=-\infty}^{\infty} kj\omega F(s+kj\omega)e^{kj\omega t} \,. \tag{3.109}$$

The series on the right-hand side of (3.109) converges uniformly, because its terms decrease as $|k|^{-2}$. Therefore, the derivatives appearing in (3.109) are continuous. Denote

$$F_1(s) = sF(s) \,.$$

Obviously,

$$\varphi_{F_1}(T,s,t) = \tilde{\varphi}_{F_1}(T,s,t) = \frac{1}{T}\sum_{k=-\infty}^{\infty} (s+kj\omega)F(s+kj\omega)e^{kj\omega t} \,. \tag{3.110}$$

As was proved above, the function $\tilde{\varphi}_{F_1}(T,s,t)$ has a piece-wise continuous derivative with respect to t, because the difference between the degrees of the numerator and denominator of $F_1(s)$ equals 2. From (3.109) and (3.110) we have

$$\frac{d\tilde{\varphi}_F(T,s,t)}{dt} = \tilde{\varphi}_{F_1}(T,s,t) - s\tilde{\varphi}_F(T,s,t) \,.$$

The right-hand side of this equation has a piece-wise continuous derivative with respect to t. Therefore, the left side also posseses this property. Thus, for $m_1 = m_2 = 0$ and $m_3 \neq 0$ the function $\tilde{\varphi}_F(T,s,t)$ has a continuous first derivative and a piece-wise continuous second derivative. Continuing the reasoning in this way, we arrive at the proposition of the theorem. ■

3. Consider the behaviour of the series (3.77) at the points $t = \lambda T$ for an integer λ. Since $\tilde{\varphi}_F(T,s,t)$ has bounded variation with respect to t, Eq. (3.31) yields

$$\varphi_F(T,s,\lambda T) = \varphi_F(T,s,0) = \frac{\tilde{\varphi}_F(T,s,+0) + \tilde{\varphi}_F(T,s,-0)}{2} \,. \tag{3.111}$$

Substituting (3.107), we find

$$\varphi_F(T,s,0) = \frac{1}{2}\sum_{i=1}^{\ell}\sum_{k=1}^{\nu_i}\frac{f_{ik}}{(k-1)!}\frac{\partial^{k-1}}{\partial s_i^{k-1}}\frac{1+\mathrm{e}^{(s_i-s)T}}{1-\mathrm{e}^{(s_i-s)T}}. \tag{3.112}$$

Taking into account that

$$\frac{1+\mathrm{e}^{(q-s)T}}{1-\mathrm{e}^{(q-s)T}} = \coth\frac{(s-q)T}{2}$$

equation (3.112) can be represented in the form

$$\varphi_F(T,s,0) = \frac{1}{2}\sum_{i=1}^{\ell}\sum_{k=1}^{\nu_i}\frac{f_{ik}}{(k-1)!}\frac{\partial^{k-1}}{\partial s_i^{k-1}}\coth\frac{(s-s_i)T}{2}. \tag{3.113}$$

We note, that Eqs. (3.112) and (3.113) hold independently of the value of m_1.

Example 3.4 The relations of section 3.5 should be applied to the bilateral Laplace transform

$$F(s) = \frac{1}{s-a} \tag{3.114}$$

with a constant a. The family of originals associated with this image consists of two functions:

$$f_+(t,a) = \begin{cases} \mathrm{e}^{at} & t>0 \\ 0 & t<0 \end{cases}, \qquad f_-(t,a) = \begin{cases} 0 & t>0 \\ -\mathrm{e}^{at} & t<0\,. \end{cases} \tag{3.115}$$

According to (3.101), the function $\tilde{\varphi}_F(T,s,t)$ has the form

$$\tilde{\varphi}_F(T,s,t) = \frac{\mathrm{e}^{(a-s)(t-\lambda T)}}{1-\mathrm{e}^{(a-s)T}}, \quad \lambda T < t < (\lambda+1)T\,. \tag{3.116}$$

For $\operatorname{Re} s > \operatorname{Re} a$ the function (3.116) coincides with the DPFR $\varphi_{f_+}(T,s,t)$, while for $\operatorname{Re} s < \operatorname{Re} a$ it coincides with the DPFR $\varphi_{f_-}(T,s,t)$. The periodic function (3.116) is discontinuous. Moreover,

$$\tilde{\varphi}_F(T,s,+0) = \frac{1}{1-\mathrm{e}^{(a-s)T}}, \quad \tilde{\varphi}_F(T,s,-0) = \frac{\mathrm{e}^{(a-s)T}}{1-\mathrm{e}^{(a-s)T}}.$$

Hence,

$$\Delta\tilde{\varphi}_F(0) = \tilde{\varphi}_F(T,s,+0) - \tilde{\varphi}_F(T,s,-0) = 1$$

which corresponds to (3.108). □

Example 3.5 The pecularities for developing the DPFR of a transform with multiple poles are considered for

$$F(s) = \frac{1}{(s-a)^2}. \tag{3.117}$$

The associated family of originals $\{f_m\}$ consists of the two functions

$$f_{1+}(t,a) = \begin{cases} te^{at} & t>0 \\ 0 & t<0 \end{cases}, \qquad f_{1-}(t,a) = \begin{cases} 0 & t>0 \\ -te^{at} & t<0 \end{cases}.$$

The functions $f_{1+}(t)$ and $f_{1-}(t)$ are continuous. Due to (3.80) and (3.115), the function $\hat{\varphi}_F(T,s,t)$ takes the form

$$\hat{\varphi}_F(T,s,t) = \frac{\partial}{\partial a}\frac{\mathrm{e}^{(a-s)t}}{1-\mathrm{e}^{(a-s)T}}, \quad 0 \le t \le T \tag{3.118}$$

or, in an expanded form,

$$\hat{\varphi}_F(T,s,t) = \mathrm{e}^{(a-s)t}\,\frac{t\left[1-\mathrm{e}^{(a-s)T}\right]+T\mathrm{e}^{(a-s)T}}{\left[1-\mathrm{e}^{(a-s)T}\right]^2}, \quad 0 \le t \le T. \tag{3.119}$$

It can be immediately verified that $\hat{\varphi}_F(T,s,+0) = \hat{\varphi}_F(T,s,-0)$ in accordance with the general theory. The corresponding function $\tilde{\varphi}_F(T,s,t)$ can be calculated by periodic continuation of (3.119) for the argument t. For $a=0$ we have $F(s) = s^{-2}$, and from (3.119) it follows that

$$\tilde{\varphi}_F(T,s,t) = \mathrm{e}^{-st}\,\frac{(1-\mathrm{e}^{-sT})t+T\mathrm{e}^{-sT}}{(1-\mathrm{e}^{-sT})^2}, \quad 0 \le t \le T. \tag{3.120}$$

For $\operatorname{Re} s > \operatorname{Re} a$ and $0 \le t \le T$ the function (3.119) coincides with $\varphi_{f_{1+}}(T,s,t)$, and for $\operatorname{Re} s < \operatorname{Re} a$ it coincides with $\varphi_{f_{1-}}(T,s,t)$. □

3.6 Variation of the sampling period

Let $F(s)$ be a Laplace image, which is associated, for a fixed T, with the corresponding DPFR

$$\varphi_F(T,s,t) = \frac{1}{T}\sum_{k=-\infty}^{\infty} F(s+kj\omega)\mathrm{e}^{kj\omega t}, \quad \omega = \frac{2\pi}{T}. \tag{3.121}$$

The value T can be considered as a parameter in Eq. (3.121). The aim of this section is to determine the connection between DPFRs given by (3.121) for different (multiple) T.
Let $N>1$ be an integer. Then, by definition,

$$\breve{\varphi}_F(NT,s,t) \triangleq \frac{1}{NT}\sum_{k=-\infty}^{\infty} F(s+\frac{k}{N}j\omega)\mathrm{e}^{\frac{k}{N}j\omega t}. \tag{3.122}$$

Theorem 3.9 *For all $-\infty < t < \infty$ the following equality holds*

$$\varphi_F(T,s,t) = \sum_{m=0}^{N-1} \breve{\varphi}_F(NT,s,t-mT). \tag{3.123}$$

Proof From (3.122) we have

$$\breve{\varphi}_F(NT,s,t-mT)=\frac{1}{NT}\sum_{k=-\infty}^{\infty}F\left(s+\frac{k}{N}\mathrm{j}\omega\right)\mathrm{e}^{\frac{k}{N}\mathrm{j}\omega t}\mathrm{e}^{-\frac{k}{N}m2\pi\mathrm{j}}\,. \tag{3.124}$$

Therefore,

$$\sum_{m=0}^{N-1}\breve{\varphi}_F(NT,s,t-mT)=\frac{1}{NT}\sum_{k=-\infty}^{\infty}F\left(s+\frac{k}{N}\mathrm{j}\omega\right)\mathrm{e}^{\frac{k}{N}\mathrm{j}\omega t}\sigma_{-kN} \tag{3.125}$$

where

$$\sigma_{\lambda N}\triangleq\sum_{m=0}^{N-1}\mathrm{e}^{\frac{\lambda}{N}2\pi m\mathrm{j}}\,. \tag{3.126}$$

Moreover,

$$\sigma_{\lambda N}=\begin{cases}N & \lambda=rN \quad (r-\text{integer})\\ 0 & \lambda\neq rN\end{cases}\,. \tag{3.127}$$

Indeed, if $\lambda = rN$ and r is an integer, all summands in (3.126) are equal to 1, so the first equation in (3.127) holds. If $\lambda \neq rN$, calculating the sum of the geometric progression (3.126), we have

$$\sigma_{\lambda N}=\frac{1-\mathrm{e}^{-2\pi\mathrm{j}\lambda}}{1-\mathrm{e}^{-2\pi\mathrm{j}\frac{\lambda}{N}}}=0\,. \tag{3.128}$$

Substituting (3.127) into (3.125), we obtain

$$\sum_{m=0}^{N-1}\breve{\varphi}_F(NT,s,t-mT)=\frac{1}{T}\sum_{r=-\infty}^{\infty}F(s+r\mathrm{j}\omega)\mathrm{e}^{r\mathrm{j}\omega t}=\varphi_F(T,s,t)\,.$$

■

Equation (3.123) defines the DPFR $\varphi_F(T,s,t)$ via the DPFRs corresponding to multiple sampling periods NT. The inverse relations are given by

Theorem 3.10 *The following formula holds*

$$\breve{\varphi}_F(NT,s,t)=\frac{1}{N}\sum_{m=0}^{N-1}\varphi_F\left(T,s+\frac{m\mathrm{j}\omega}{N},t\right)\mathrm{e}^{\frac{m\mathrm{j}\omega}{N}t}\,. \tag{3.129}$$

Proof Any integer k can be uniquely represented in the form

$$k=m+qN\,, \tag{3.130}$$

where m and q are integers and $0 \leq m < N$. Substituting this representation into (3.122), we obtain

$$\breve{\varphi}_F(NT,s,t) = \frac{1}{NT}\sum_{m=0}^{N-1}\sum_{q=-\infty}^{\infty} F\left(s+qj\omega+\frac{mj\omega}{N}\right) e^{qj\omega t} e^{\frac{mj\omega}{N}t}. \tag{3.131}$$

From (3.121) we have

$$\frac{1}{T}\sum_{q=-\infty}^{\infty} F\left(s+qj\omega+\frac{mj\omega}{N}\right) e^{mj\omega t} = \varphi_F(T, s+\frac{mj\omega}{N}, t). \tag{3.132}$$

Substituting (3.132) into (3.131), we obtain (3.129). ∎

Remark Similar formulae to (3.123) and (3.129) remain valid for the function $\varphi_f(T,s,t)$ (3.1), i.e.,

$$\varphi_f(T,s,t) = \sum_{m=0}^{N-1} \breve{\varphi}_f(NT,s,t-mT) \tag{3.133}$$

$$\breve{\varphi}_f(NT,s,t) = \frac{1}{N}\sum_{m=0}^{N-1} \varphi_f(T, s+\frac{mj\omega}{N}, t) e^{\frac{mj\omega}{N}t}. \tag{3.134}$$

Chapter 4

Discrete Laplace transforms for functions of a continuous argument

4.1 Basic relations

By the *discrete Laplace transform* ($\mathcal{D}-$transform) of a continuous function $f(t)$ we mean the sum of the series

$$\mathcal{D}_f(T,s,t) \triangleq \sum_{k=-\infty}^{\infty} f(t+kT)\mathrm{e}^{-ksT}, \quad -\infty < t < \infty. \tag{4.1}$$

Comparing (4.1) with (3.1), we find

$$\mathcal{D}_f(T,s,t) = \varphi_f(T,s,t)\mathrm{e}^{st}. \tag{4.2}$$

Using (4.2), all results of Chapter 3 can be extended to the series (4.1). Therefore, the following theorems relating to discrete Laplace transforms (4.1) are given without proofs.

Theorem 4.1 *Let $f(t) \in \Lambda(\alpha,\beta)$. Then, the series (4.1) converges absolutely and uniformly in any strip $\alpha < \alpha' \leq \operatorname{Re} s \leq \beta' < \beta$ and on any finite interval of t.* ∎

As immediately follows from (4.1),

$$\mathcal{D}_f(T,s,t) = \mathcal{D}_f(T,s+\mathrm{j}\omega,t), \quad \omega = 2\pi/T. \tag{4.3}$$

Moreover,

$$\mathcal{D}_f(T,s,t+T) = \mathcal{D}_f(T,s,t)\mathrm{e}^{sT}. \tag{4.4}$$

Equation (4.4) follows from the equality

$$\mathcal{D}_f(T,s,t+T) = \varphi_f(T,s,t+T)\mathrm{e}^{s(t+T)} = \mathcal{D}_f(T,s,t)\mathrm{e}^{sT}. \tag{4.5}$$

4.2 Connection with modified discrete Laplace transforms for sampled functions

If we consider in (4.1) the variation of the argument t only on the interval $0 < t < T$ and assume that $t = \varepsilon$, we obtain

$$\mathcal{D}_f(T, s, \varepsilon) = \sum_{k=-\infty}^{\infty} f(\varepsilon + kT)\mathrm{e}^{-ksT}, \quad 0 < \varepsilon < T. \tag{4.6}$$

Comparing (4.6) with (2.44), we find

$$\mathcal{D}_f(T, s, \varepsilon) = F^*(s, \varepsilon), \quad 0 < \varepsilon < T \tag{4.7}$$

i.e., for $0 < t < T$ the series (4.1) coincides with the modified discrete Laplace transform of the function $f(t)$.

A principal difference between the transforms $F^*(s,\varepsilon)$ and $\mathcal{D}_f(T,s,t)$ is that the function $\mathcal{D}_f(T,s,t)$ is defined over the whole axis $-\infty < t < \infty$, while the function $F^*(s,\varepsilon)$ is, by construction, given only in the interval $0 < \varepsilon < T$. Hence, the transformation $\mathcal{D}_f(T,s,t)$ can be considered as a continuation of $F^*(s,\varepsilon)$ over the time axis. This continuation is given by (4.4). The $\mathcal{D}$-transform $\mathcal{D}_f(T,s,t)$ can be constructed as follows. For a known $F^*(s,\varepsilon)$ we can find the DPFR

$$\varphi_f(T, s, \varepsilon) = F^*(s, \varepsilon)\mathrm{e}^{-s\varepsilon}, \quad 0 < \varepsilon < T. \tag{4.8}$$

A periodic continuation of (4.8) in time gives the DPFR $\varphi_f(T,s,t)$, and the discrete Laplace transform is given by (4.2). An explicit formula defining $\mathcal{D}_f(T,s,t)$ for any t has the form

$$\mathcal{D}_f(T, s, t) = F^*(s, t - \lambda T)\mathrm{e}^{\lambda sT}, \quad \lambda T < t < (\lambda + 1)T. \tag{4.9}$$

Example 4.1 Construct the discrete Laplace transform $\mathcal{D}_F(T,s,t)$ for $0 < t < \infty$ of the function

$$F(s) = \frac{1}{s - a}. \tag{4.10}$$

Using Eqs. (3.116) and (4.2), we find the discrete Laplace transform of the function (4.10) to be

$$\mathcal{D}_f(T, s, t) = \frac{\mathrm{e}^{at}\mathrm{e}^{-\lambda(a-s)T}}{1 - \mathrm{e}^{(a-s)T}}, \quad \lambda T < t < (\lambda + 1)T. \tag{4.11}$$

□

4.3 $\mathcal{D}$–transforms of discontinuous functions

Using the reasoning given in Section 3.2, we can obtain the corresponding results for $\mathcal{D}$-transforms of discontinuous functions. Let a function $f(t)$ have a finite number of singular points, and let $\epsilon > 0$ be a number such that the function $f(t)$ is continuous or has a single point of discontinuity on any interval with the length ϵ. Then, the following theorem can be formulated.

Theorem 4.2 *If the conditions of Theorem 4.1 hold, at any regular point of $f(t)$ the $\mathcal{D}$-transform $\mathcal{D}_f(T,s,t)$ is continuous with respect to t. Let t_0 be a singular point and $f_k^{\pm} = f(t_0 + kT \pm 0)$ be the corresponding limit values at the points $t_{0k} = t_0 + kT$. Then, the points t_{0k} are singular points of the function $\mathcal{D}_f(T,s,t)$. Moreover, there exist the following limits*

$$\begin{aligned} \mathcal{D}_f(T,s,t_0+0) &= \lim_{\tau\to+0} \mathcal{D}_f(T,s,t_0+\tau) &= \sum_{k=-\infty}^{\infty} f_k^{+} e^{-ksT} \\ \mathcal{D}_f(T,s,t_0-0) &= \lim_{\tau\to-0} \mathcal{D}_f(T,s,t_0+\tau) &= \sum_{k=-\infty}^{\infty} f_k^{-} e^{-ksT} \end{aligned} \tag{4.12}$$

where the right-hand sides represent the discrete Laplace transforms of the associated discrete sequences 'before' and 'after' the singular points. ■

Corollary Let $f(t) \in \Lambda_+(\alpha,\infty)$ and be continuous for $t \geq 0$, but $f(+0) \neq 0$. Consider the sampled function $\{f_k\}$:

$$f_0 = f(+0)\,;\; f_k = f(kT)\,, \quad (k > 0)\,. \tag{4.13}$$

Then, the discrete Laplace transform of the sampled function $\{f_k\}$ is given by

$$F^*(s) = \sum_{k=0}^{\infty} f_k e^{-ksT} = \mathcal{D}_f(T,s,+0)\,. \tag{4.14}$$

Example 4.2 Consider the properties of the discrete Laplace transform (4.11) at the singular points. With $\lambda = 0$ from (4.11) for $\operatorname{Re} s > \operatorname{Re} a$ we obtain

$$\mathcal{D}_f(T,s,t) = \frac{e^{at}}{1 - e^{(a-s)T}}\,, \quad 0 < t < T\,. \tag{4.15}$$

Hence,

$$\mathcal{D}_f(T,s,t+0) = \frac{1}{1 - e^{(a-s)T}}\,, \quad \operatorname{Re} s > \operatorname{Re} a\,. \tag{4.16}$$

The sampled function (4.10) associated with (4.13) has the values $f_k = e^{kaT}$, $(k \geq 0)$. Its discrete Laplace transform takes the form

$$F^*(s) = \sum_{k=0}^{\infty} e^{(a-s)kT} = \frac{1}{1 - e^{(a-s)T}}\,, \quad \operatorname{Re} s > \operatorname{Re} a\,. \tag{4.17}$$

□

Henceforth, to obtain the $\mathcal{D}$-transform of an arbitrary function $f(t) \in \Lambda(\alpha,\beta)$ we shall use the series (4.1). If the function $f(t)$ is discontinuous, the results of this section must be taken into account.

4.4 $\mathcal{D}$-transforms of finite functions

A function $f(t)$ will be called *finite*, if it vanishes outside a finite interval $a \le t \le b$. The least of such intervals is called the *support* of the function $f(t)$. Let, specifically, the support of a function $f(t)$ be the interval $0 \le t \le mT$ with an integer $m > 0$. Then, from (4.1) for $0 \le t < T$ we have the finite sum

$$\mathcal{D}_f(T,s,t) = \sum_{k=0}^{m-1} f(t+kT)\mathrm{e}^{-ksT}\,. \tag{4.18}$$

It can be verified that, conversely, if the $\mathcal{D}$-transform of the function $f(t)$ has the form (4.18) in the interval $0 \le t \le T$, then $f(t) = 0$ for $t < 0$ and $f(t) = 0$ for $t > mT$. Moreover, if

$$\mathcal{D}_f(T,s,\varepsilon) = \sum_{k=0}^{m-1} f_k(\varepsilon)\mathrm{e}^{-ksT} \tag{4.19}$$

for $0 < t = \varepsilon < T$, the associated original is defined by

$$f(t) = \begin{cases} 0 & t > mT \\ f_k(t-kT) & kT < t < (k+1)T, \quad (k = 0,1,\dots,m-1) \\ 0 & t < 0\,. \end{cases} \tag{4.20}$$

Example 4.3 Using (4.18), for the function (3.56) we have

$$\mathcal{D}_f(T,s,t) = 1 - \mathrm{e}^{-sT}\,, \quad 0 < t < T \tag{4.21}$$

which corresponds to (3.60). Continuation of this expression over the whole t-axis taking due account of (4.9) gives

$$\mathcal{D}_f(T,s,t) = \left(1 - \mathrm{e}^{-sT}\right)\mathrm{e}^{ksT}\,, \quad kT < t < (k+1)T\,. \tag{4.22}$$

□

Example 4.4 Build the discrete Laplace transform for the finite the function

$$f(t) = \begin{cases} 0 & t < 0 \\ 1 + t/T & 0 < t < T \\ 1 - t/T & T < t < 2T \\ 0 & t > 2T \end{cases} \tag{4.23}$$

shown in Figure 4.4. Calculating the $\mathcal{D}$-transform of the function (4.23) taking due account of (4.18), we find

$$\mathcal{D}_f(T,s,t) = 1 + \frac{t}{T} - \frac{t}{T}\mathrm{e}^{-sT}\,, \quad 0 < t < T\,. \tag{4.24}$$

Extending this relation over the whole t-axis by means of (4.9), we have

$$\mathcal{D}_f(T,s,t) = \left[1 + \left(1 - \mathrm{e}^{-sT}\right)\frac{t-kT}{T}\right]\mathrm{e}^{ksT}\,, \quad kT < t < (k+1)T\,. \tag{4.25}$$

□

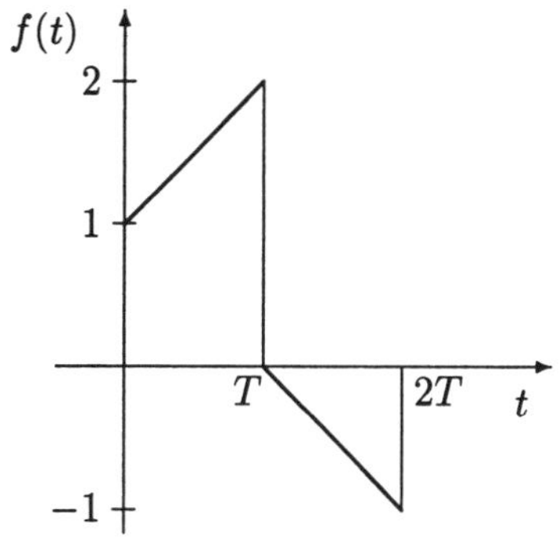

Figure 4.1: Original function for Example 4.4

4.5 Representation via Laplace transforms

Introduce the notation

$$\mathcal{D}_F(T,s,t) \stackrel{\triangle}{=} \varphi_F(T,s,t)\mathrm{e}^{st}\,, \quad -\infty < t < \infty \tag{4.26}$$

where $\varphi_F(T,s,t)$ is the DPFR (3.29). In this case, from (4.18) and (3.29) we have

$$\mathcal{D}_F(T,s,t) = \frac{1}{T}\sum_{k=-\infty}^{\infty} F(s+k\mathrm{j}\omega)\mathrm{e}^{(s+k\mathrm{j}\omega)t}\,, \quad \omega = 2\pi/T\,. \tag{4.27}$$

The series (4.27) will henceforth be called the *discrete Laplace transform ($\mathcal{D}$-transform) of the function $F(s)$.*
It can be immediately verified that if the series (4.27) converges, we have

$$\mathcal{D}_F(T,s,t) = \mathcal{D}_F(T,s+\mathrm{j}\omega,t) \tag{4.28}$$

and, moreover,

$$\mathcal{D}_F(T,s,t+T) = \mathcal{D}_F(T,s,t)\mathrm{e}^{sT}\,. \tag{4.29}$$

The relations between the transforms $\mathcal{D}_f(T,s,t)$ and $\mathcal{D}_F(T,s,t)$ are similar to those between $\varphi_f(T,s,t)$ and $\varphi_F(T,s,t)$. Under the same assumptions as in Section 3.3, at all regular points we have

$$\mathcal{D}_f(T,s,t) = \mathcal{D}_F(T,s,t)\,. \tag{4.30}$$

If t_0 is a singular point of $\mathcal{D}_f(T,s,t)$, then

$$\mathcal{D}_F(T,s,t_0) = \frac{1}{2}\left[\mathcal{D}_f(T,s,t_0+0) + \mathcal{D}_f(T,s,t_0-0)\right]\,. \tag{4.31}$$

Equation (4.30) certainly holds for all t under the conditions of Theorem 3.3, which can be reformulated as follows.

Theorem 4.3 *Let the function $F(s)$ be analytic and satisfy (1.33) in any stripe $\alpha < \alpha' \le \operatorname{Re} s \le \beta' < \beta$. Then, the following propositions hold.*

1. *The series (4.1) converges in the stripe $\alpha < \alpha' \le \operatorname{Re} s \le \beta' < \beta$ absolutely and uniformly with respect to s and t. Its sum $\mathcal{D}_f(T,s,t)$ is analytic with respect to s and continuous with respect to t.*
2. *For all $-\infty < t < \infty$*
$$\mathcal{D}_F(T,s,t) = \sum_{k=-\infty}^{\infty} f(t+kT)\mathrm{e}^{-ksT} = \mathcal{D}_f(T,s,t) \tag{4.32}$$
where $f(t)$ is the original for the image $F(s)$ in the stripe $\alpha < \operatorname{Re} s < \beta$.
3. *If the image $F(s)$ is analytic in $\operatorname{Re} s > \alpha$, then $f(t) = 0$ for $t < 0$ and we have*
$$\mathcal{D}_F(T,s,t) = \sum_{k=0}^{\infty} f(t+kT)\mathrm{e}^{-ksT} = \mathcal{D}_f(T,s,t)\,, \quad 0 \le t < T\,. \tag{4.33}$$

■

4.6 Inversion formulae

The inversion formulae for the transformations $\mathcal{D}_f(T,s,t)$ and $\mathcal{D}_F(T,s,t)$ and the conditions of their validity can be obtained by reformulating Theorems 3.4 and 3.5 taking due account of (4.2) and (4.18).

Theorem 4.4 *Let $f(t) \in \Lambda(\alpha,\beta)$. Then, for $\alpha < \operatorname{Re} s < \beta$ we have*

$$f(t) = \frac{T}{2\pi \mathrm{j}} \int_{c-\mathrm{j}\omega/2}^{c+\mathrm{j}\omega/2} \mathcal{D}_f(T,s,t)\,\mathrm{d}s\,, \quad \alpha < c < \beta \tag{4.34}$$

and the image $F(s)$ is given by

$$F(s) = \int_0^T \mathcal{D}_f(T,s,t)\mathrm{e}^{-st}\,\mathrm{d}s \tag{4.35}$$

■

Theorem 4.5 *If $f(t) \in \Lambda(\alpha,\beta)$, then*

$$F(s) = \int_0^T \mathcal{D}_F(T,s,t)\mathrm{e}^{-st}\,\mathrm{d}s\,. \tag{4.36}$$

If, moreover, the series (3.27) converges absolutely and uniformly with respect to s, the function $f(t)$ is continuous and

$$f(t) = \frac{T}{2\pi \mathrm{j}} \int_{c-\mathrm{j}\omega/2}^{c+\mathrm{j}\omega/2} \mathcal{D}_F(T,s,t)\,\mathrm{d}s\,, \quad \alpha < c < \beta\,. \tag{4.37}$$

■

It is noteworthy that Eq. (4.37) coincides with the discretized form of the inverse Laplace integral. Therefore, the Laplace transform and the discrete Laplace transform of a function of continuous argument are closely related. At the same time, Theorem 4.5 gives sufficient conditions of the validity of (1.32).

4.7 $\mathcal{D}$-transforms of real rational images

If the Laplace image $F(s)$ has the form (1.77) and (1.88), closed expressions can be obtained, and the properties of the functions $\mathcal{D}_f(T,s,t)$ and $\mathcal{D}_F(T,s,t)$ can be investigated in detail. The corresponding relations can be derived from equations given in Section 3.5. Under (3.77) we have

$$\mathcal{D}_F(T,s,t) = \frac{1}{T}\sum_{k=-\infty}^{\infty} \frac{m(s+kj\omega)}{d(s+kj\omega)}\,\mathrm{e}^{(s+kj\omega)t}\,. \tag{4.38}$$

In each regularity interval S_i: (μ_i, μ_{i+1}) the image $F(s)$ is associated with an original $f_i(t)$. Nevertheless, as for the DPFR, all $f_i(t)$ are associated with the same image independent of i

$$\tilde{\mathcal{D}}_F(T,s,t) \stackrel{\triangle}{=} =\tilde{\varphi}_F(T,s,t)\mathrm{e}^{st} \tag{4.39}$$

where $\tilde{\varphi}_F(T,s,t)$ is given by (3.79) and (3.80). In each interval S_i Eq. (4.38) defines the sum of the series

$$\mathcal{D}_{f_i}(T,s,t) = \sum_{k=-\infty}^{\infty} f_i(t+kT)\mathrm{e}^{-ksT} = \varphi_{f_i}(T,s,t)\mathrm{e}^{st}\,. \tag{4.40}$$

Closed expressions for the function $\tilde{\mathcal{D}}_F(T,s,t)$ for $0<t<T$ can be obtained by multiplication of (3.80) by e^{st}

$$\tilde{\mathcal{D}}_F(T,s,t) = \sum_{i=1}^{\ell}\sum_{k=1}^{\nu_i} \frac{f_{ik}}{(k-1)!}\frac{\partial^{k-1}}{\partial s_i^{k-1}}\frac{\mathrm{e}^{s_i t}}{1-\mathrm{e}^{(s_i-s)T}}\,, \quad 0<t<T\,. \tag{4.41}$$

In order to continue Eq. (4.41) over the whole t-axis we must multiply (3.101) by e^{st}, so that

$$\tilde{\mathcal{D}}_F(T,s,t) = \sum_{i=1}^{\ell}\sum_{k=1}^{\nu_i} \frac{f_{ik}}{(k-1)!}\frac{\partial^{k-1}}{\partial s_i^{k-1}}\frac{\mathrm{e}^{s_i t}\mathrm{e}^{\lambda(s-s_i)T}}{1-\mathrm{e}^{(s_i-s)T}}\,, \quad \lambda T<t<(\lambda+1)T\,. \tag{4.42}$$

In the special case, where $\lambda=-1$, i.e., $-T<t<0$, from (4.42) we have

$$\tilde{\mathcal{D}}_F(T,s,t) = \sum_{i=1}^{\ell}\sum_{k=1}^{\nu_i} \frac{f_{ik}}{(k-1)!}\frac{\partial^{k-1}}{\partial s_i^{k-1}}\frac{\mathrm{e}^{s_i t}}{\mathrm{e}^{(s-s_i)T}-1}\,. \tag{4.43}$$

The above formulae become greatly simplified if all poles of $F(s)$ are simple. In this case, using Eqs. (3.105) and (3.104), we obtain

$$\tilde{\mathcal{D}}_F(T,s,t) = \sum_{i=1}^{p} \frac{f_i\mathrm{e}^{s_i t}}{1-\mathrm{e}^{(s_i-s)T}}\,, \quad 0<t<T \tag{4.44}$$

$$\tilde{\mathcal{D}}_F(T,s,t) = \sum_{i=1}^{p} \frac{f_i\mathrm{e}^{s_i t}}{\mathrm{e}^{(s-s_i)T}-1}\,, \quad -T<t<0\,. \tag{4.45}$$

Example 4.5 The formulae of this section are applied to build the discrete Laplace transform of the image

$$F(s) = \frac{1}{s-a}. \tag{4.46}$$

Making use of (4.42) we have

$$\tilde{\mathcal{D}}_F(T,s,t) = \frac{e^{at}e^{\lambda(s-a)T}}{1-e^{(a-s)T}}, \quad \lambda T < t < (\lambda+1)T. \tag{4.47}$$

The function (4.47) is discontinuous in t. For $\lambda = 0$ and $\lambda = -1$ we have

$$\begin{aligned} \tilde{\mathcal{D}}_F(T,s,t) &= \frac{e^{at}}{1-e^{(a-s)T}}, \qquad 0 < t < T \\ \tilde{\mathcal{D}}_F(T,s,t) &= \frac{e^{at}e^{(a-s)T}}{1-e^{(a-s)T}}, \qquad -T < t < 0. \end{aligned} \tag{4.48}$$

Hence,

$$\tilde{\mathcal{D}}_F(T,s,+0) = \frac{1}{1-e^{(a-s)T}}, \quad \tilde{\mathcal{D}}_F(T,s,-0) = \frac{e^{(a-s)T}}{1-e^{(a-s)T}}. \tag{4.49}$$

Therefore, the value of discontinuity at the point $t = 0$ is equal to

$$\Delta\tilde{\mathcal{D}}_F(0) \stackrel{\Delta}{=} = \tilde{\mathcal{D}}_F(T,s,+0) - \tilde{\mathcal{D}}_F(T,s,-0) = 1. \tag{4.50}$$

□

Example 4.6 The pecularities for the application of (4.2) in the case of multiple poles of the image (3.117) are studied. Using (3.118) and (4.42), we have

$$\tilde{\mathcal{D}}_F(T,s,t) = e^{\lambda sT}\frac{\partial}{\partial a}\frac{e^{a(t-\lambda T)}}{1-e^{(a-s)T}}, \quad \lambda T \le t \le (\lambda+1)T. \tag{4.51}$$

Here, in contrast to (4.41) and (4.42), we use the symbol '$\le$', because the function (4.51) is continuous with respect to t . In the special case, when $a \to 0$, i.e., $F(s) = s^{-2}$, it follows from (4.51) that

$$\tilde{\mathcal{D}}_F(T,s,t) = e^{\lambda sT}\frac{\left(1-e^{-sT}\right)(t-\lambda T) + Te^{-sT}}{\left(1-e^{-sT}\right)^2}, \quad \lambda T \le t \le (\lambda+1)T. \tag{4.52}$$

With $\lambda = 0$ from (4.52) we get

$$\tilde{\mathcal{D}}_F(T,s,t) = \frac{\left(1-e^{-sT}\right)t + Te^{-sT}}{\left(1-e^{-sT}\right)^2}, \quad 0 \le t \le T. \tag{4.53}$$

Assuming that $t = \varepsilon$ in (4.53), we have

$$\tilde{\mathcal{D}}_F(T,s,\varepsilon) = F^*(s,\varepsilon) \tag{4.54}$$

where the function $F^*(s,\varepsilon)$ defines, for $\operatorname{Re} s > 0$, the modified discrete Laplace transform for the function

$$f_{1+}(t) = \begin{cases} t & t \geq 0 \\ 0 & t \leq 0 \end{cases} \tag{4.55}$$

while for $\operatorname{Re} s < 0$ it is associated with the function

$$f_{1-}(t) = \begin{cases} 0 & t \geq 0 \\ -t & t \leq 0 \,. \end{cases} \tag{4.56}$$

For $t = 0$ from (4.53) we obtain

$$\tilde{\mathcal{D}}_F(T,s,0) = \frac{T\mathrm{e}^{-sT}}{\left(1-\mathrm{e}^{-sT}\right)^2}\,. \tag{4.57}$$

According to the aforesaid, the function (4.57) for $\operatorname{Re} s > 0$ gives the discrete Laplace transform for the series $\{f_k^+\}$ with $f_k^+ = 0$, $(k \leq 0)$; $f_k^+ = k$, $(k \geq 0)$. For $\operatorname{Re} s < 0$ Eq. (4.57) is the discrete Laplace transform of the series $\{f_k^-\}$ with $f_k^- = -k$, $(k \leq 0)$; $f_k^- = 0$, $(k \geq 0)$. □

4.8 Limiting properties on the complex plane

In principle, the series (4.1) and (4.19) for $f(t) \in \Lambda(\alpha,\beta)$ define the corresponding $\mathcal{D}$-transforms in the stripe of convergency $\alpha < \operatorname{Re} s < \beta$. Nevertheless, in many cases the functions $\mathcal{D}_f(T,s,t)$ and $\mathcal{D}_F(T,s,t)$ allow analytical continuation outside this stripe. In particular, such a continuation is possible if the image $F(s)$ has the form (1.77) and (1.88). Indeed, in this case each original $f_m(t)$ associated with the image $F(s)$ in the regularity interval S_m has a $\mathcal{D}$-transform given by (4.42). This function is independent of m and allows analytical continuation over the whole complex plane, except for a finite number of poles.

In the following, the limits of $\tilde{\mathcal{D}}_F(T,s,t)$ as $\operatorname{Re} s \to \pm\infty$ play an important role. For the case considered in Section 4.7 these limits can be easily computed. Henceforth, for any function $G(s)$ in a complex variable we shall use the notation

$$\ell^+[G(s)] \triangleq \lim_{\operatorname{Re} s\to\infty} G(s)\,, \quad \ell^-[G(s)] \triangleq \lim_{\operatorname{Re} s\to-\infty} G(s)\,. \tag{4.58}$$

Then, calculate the limits $\ell^\pm[\tilde{\mathcal{D}}_F(T,s,t)]$ of the function (4.42) in various intervals of t. Taking the limit of (4.41) as $\operatorname{Re} s \to \infty$, we have

$$\ell^+\left[\tilde{\mathcal{D}}_f(T,s,t)\right] = \sum_{i=1}^{\ell}\sum_{k=1}^{\nu_i} \frac{f_{ik}}{(k-1)!}\,\frac{\partial^{k-1}}{\partial s_i^{k-1}}\,\mathrm{e}^{s_i t}\,, \quad 0 < t < T\,. \tag{4.59}$$

Comparing (4.59) with (1.99), we find

$$\ell^+\left[\tilde{\mathcal{D}}_f(T,s,t)\right] = f_+(t)\,, \quad +0 \le t < T \tag{4.60}$$

where $f_+(t)$ is the original associated with the image $F(s)$ in $\operatorname{Re} s > \mu_r$. Moreover, from (4.41) we have

$$\ell^-\left[\tilde{\mathcal{D}}_f(T,s,t)\right] = 0\,, \quad +0 \le t < T\,. \tag{4.61}$$

Similarly, taking the limit of (4.43) as $\operatorname{Re} s \to \infty$, for $-T < t < -0$ we find

$$\ell^+\left[\tilde{\mathcal{D}}_F(T,s,t)\right] = 0\,, \tag{4.62}$$

$$\ell^-\left[\tilde{\mathcal{D}}_F(T,s,t)\right] = f_-(t) \tag{4.63}$$

where $f_-(t)$ is given by (1.100). Note that for $t \le 0$ we have $t = -\nu T + \varepsilon$ with an integer $\nu > 0$ and $0 \le \varepsilon < T$. Hence, for $t < 0$ we have

$$\tilde{\mathcal{D}}_F(T,s,t) = \tilde{\mathcal{D}}_F(T,s,\varepsilon)\mathrm{e}^{-\nu sT}\,. \tag{4.64}$$

As follows from (4.62)–(4.64),

$$\ell^+\left[\tilde{\mathcal{D}}_F(T,s,t)\right] = 0\,, \quad t \le -0 \tag{4.65}$$

Similarly, for $t > 0$ we have $t = \mu T + \varepsilon$ with an integer $\mu \ge 0$, and

$$\tilde{\mathcal{D}}_F(T,s,t) = \tilde{\mathcal{D}}_F(T,s,\varepsilon)\mathrm{e}^{\mu sT}\,. \tag{4.66}$$

As follows from (4.66) and (4.61),

$$\ell^-\left[\tilde{\mathcal{D}}_F(T,s,t)\right] = 0\,, \quad t \ge +0\,. \tag{4.67}$$

Example 4.7 Find the limits (4.58) of the discrete Laplace transform (4.47) of Example 4.5. From (4.48) we have

$$\begin{aligned} &\ell^+\left[\tilde{\mathcal{D}}_F(T,s,t)\right] = \mathrm{e}^{at}\,, \quad \ell^-\left[\tilde{\mathcal{D}}_F(T,s,t)\right] = 0\,, \qquad +0 \le t < T \\ &\ell^+\left[\tilde{\mathcal{D}}_F(T,s,t)\right] = 0\,, \quad \ell^-\left[\tilde{\mathcal{D}}_F(T,s,t)\right] = -\mathrm{e}^{at}\,, \quad -T < t \le -0\,. \end{aligned} \tag{4.68}$$

The relations can easily be verified. □

Remark We note, without a proof, that in all equations of this section the tendency to limit is uniform with respect to $\operatorname{Im} s$ for $-\omega/2 \le \operatorname{Im} s \le \omega/2$.

4.9 Image of product of originals

Let us have two functions $f_i(t) \in \Lambda(\alpha, \beta)$ with $\alpha < 0$, $\beta > 0$. Then, it can be easily proved that $f_1(t)f_2(t) \in \Lambda(2\alpha, 2\beta)$. Hence, for $2\alpha < \operatorname{Re} s < 2\beta$ there exists the $\mathcal{D}$-transform

$$\mathcal{D}_{f_1 f_2}(T, s, t) = \sum_{k=-\infty}^{\infty} f_1(t + kT) f_2(t + kT) \mathrm{e}^{-ksT} . \tag{4.69}$$

From (4.34) we have

$$f_1(t) = \frac{T}{2\pi \mathrm{j}} \int_{c-\mathrm{j}\omega/2}^{c+\mathrm{j}\omega/2} \mathcal{D}_{f_1}(T, q, t)\, \mathrm{d}q , \quad \alpha < c = \operatorname{Re} q < \beta . \tag{4.70}$$

Substituting $t + kT$ for t and taking (4.4) into account, we obtain

$$f_1(t + kT) = \frac{T}{2\pi \mathrm{j}} \int_{c-\mathrm{j}\omega/2}^{c+\mathrm{j}\omega/2} \mathcal{D}_{f_1}(T, q, t)\, \mathrm{e}^{qkT}\, \mathrm{d}q , \quad \alpha < c = \operatorname{Re} q < \beta . \tag{4.71}$$

Next, Eqs. (4.71) and (4.69) yield

$$\mathcal{D}_{f_1 f_2}(T, s, t) = \frac{T}{2\pi \mathrm{j}} \sum_{k=-\infty}^{\infty} \int_{c-\mathrm{j}\omega/2}^{c+\mathrm{j}\omega/2} \mathcal{D}_{f_1}(T, q, t) f_2(t + kT) \mathrm{e}^{-k(s-q)T}\, \mathrm{d}q . \tag{4.72}$$

If $\alpha < \operatorname{Re} s - \operatorname{Re} q < \beta$, the series

$$\sum_{k=-\infty}^{\infty} f_2(t + kT) \mathrm{e}^{-k(s-q)T} = \mathcal{D}_{f_2}(T, s - q, t) \tag{4.73}$$

converges uniformly, and we can change the order of summation and integration in (4.72), so that

$$\mathcal{D}_{f_1 f_2}(T, s, t) = \frac{T}{2\pi \mathrm{j}} \int_{c-\mathrm{j}\omega/2}^{c+\mathrm{j}\omega/2} \mathcal{D}_{f_1}(T, q, t) \left[\sum_{k=-\infty}^{\infty} f_2(t + kT) \mathrm{e}^{-k(s-q)T} \right] \mathrm{d}q \tag{4.74}$$

or, equivalently,

$$\mathcal{D}_{f_1 f_2}(T, s, t) = \frac{T}{2\pi \mathrm{j}} \int_{c-\mathrm{j}\omega/2}^{c+\mathrm{j}\omega/2} \mathcal{D}_{f_1}(T, q, t) \mathcal{D}_{f_2}(T, s - q, t)\, \mathrm{d}q . \tag{4.75}$$

Due to symmetry, we also have

$$\mathcal{D}_{f_1 f_2}(T, s, t) = \frac{T}{2\pi \mathrm{j}} \int_{c-\mathrm{j}\omega/2}^{c+\mathrm{j}\omega/2} \mathcal{D}_{f_1}(T, s - q, t) \mathcal{D}_{f_2}(T, q, t)\, \mathrm{d}q . \tag{4.76}$$

4.10 Sum of squares of originals

Let $f(t) \in \Lambda(\alpha, \beta)$ with $\alpha < 0$, $\beta > 0$. Then, the convergence stripe of the $\mathcal{D}$-transform, $\alpha < \operatorname{Re} s < \beta$ contains the imaginary axis, and the function $\mathcal{D}_f(T, s, t)$ is defined for $s = 0$. Hence, from (4.1) we obtain

$$\sum_{k=-\infty}^{\infty} f(t + kT) = \mathcal{D}_f(T, 0, t) = \varphi_f(T, 0, t). \tag{4.77}$$

If t_0 is a singular point, instead of (4.77) we have, with reference to (4.12),

$$\sum_{k=-\infty}^{\infty} f(t_0 + kT \pm 0) = \mathcal{D}_f(T, 0, t_0 \pm 0) = \varphi_f(T, 0, t_0 \pm 0). \tag{4.78}$$

If $f(t) \in \Lambda_+(\alpha, \infty)$ in (4.78) with $\alpha < 0$ and $0 < t_0 < T$, then

$$\sum_{k=0}^{\infty} f(t_0 + kT \pm 0) = \mathcal{D}_f(T, 0, t_0 \pm 0) = \varphi_f(T, 0, t_0 \pm 0). \tag{4.79}$$

Example 4.8 By means of formula (4.79) find the sum of the series (4.77) for $t = +0$ in the case of

$$f(t) = \begin{cases} e^{at} & t > 0 \\ 0 & t < 0 \end{cases} \tag{4.80}$$

with $\operatorname{Re} a < 0$. From (4.11) we have

$$\mathcal{D}_f(T, s, t) = \frac{e^{at}}{1 - e^{(a-s)T}}, \quad 0 < t < T. \tag{4.81}$$

For $s = 0$ we obtain

$$\mathcal{D}_f(T, 0, t) = \frac{e^{at}}{1 - e^{aT}}, \quad 0 < t < T \tag{4.82}$$

and

$$\sum_{k=0}^{\infty} e^{a(t+kT)} = \frac{e^{at}}{1 - e^{aT}}, \quad 0 < t < T. \tag{4.83}$$

For $t \to +0$ from (4.83) it follows that

$$\sum_{k=0}^{\infty} e^{akT} = \frac{1}{1 - e^{aT}}. \tag{4.84}$$

The right-hand side of this formula is the sum of the values of the sampling $f(kT + 0)$. □

Consider Eq. (4.75). This image satisfies the conditions of applicability of Eq. (4.78). Hence, for $s = 0$ at regular points t from (4.75) we have

$$\sum_{k=-\infty}^{\infty} f_1(t+kT)f_2(t+kT) = \frac{T}{2\pi j}\int_{c-j\omega/2}^{c+j\omega/2} \mathcal{D}_{f_1}(T,q,t)\mathcal{D}_{f_2}(T,-q,t)\,dq \quad (4.85)$$

$$\alpha < c < \beta .$$

For singular values of t, instead of (4.85) we must use Eqs. (4.78) and (4.75). In the important special case, when $f_1(t) = f_2(t) = f(t)$, Eq. (4.85) takes the form

$$\sum_{k=-\infty}^{\infty} f^2(t+kT) = \frac{T}{2\pi j}\int_{c-j\omega/2}^{c+j\omega/2} \mathcal{D}_f(T,q,t)\mathcal{D}_f(T,-q,t)\,dq . \quad (4.86)$$

If the function has a singular point at $t = kT$, Eq. (4.86) yields

$$\sum_{k=-\infty}^{\infty} f^2(kT \pm 0) = \frac{T}{2\pi j}\int_{c-j\omega/2}^{c+j\omega/2} \mathcal{D}_f(T,q,\pm 0)\mathcal{D}_f(T,-q,\pm 0)\,dq \quad (4.87)$$

$$\alpha < c < \beta .$$

Because of the chosen stripe of convergence in Eqs. (4.85)–(4.87), we can take $c = \operatorname{Re} q = 0$. Then, taking account of

$$\mathcal{D}_f(T,s,t) = \varphi_f(T,s,t)e^{st}, \quad \mathcal{D}_f(T,-s,t) = \varphi_f(T,-s,t)e^{-st} \quad (4.88)$$

we obtain

$$\sum_{k=-\infty}^{\infty} f_1(t+kT\pm 0)f_2(t+kT\pm 0) = \frac{T}{2\pi j}\int_{-j\omega/2}^{j\omega/2} \varphi_{f_1}(T,q,t\pm 0)\varphi_{f_2}(T,-q,t\pm 0)\,dq \quad (4.89)$$

$$\sum_{k=-\infty}^{\infty} f^2(t+kT\pm 0) = \frac{T}{2\pi j}\int_{-j\omega/2}^{j\omega/2} \varphi_f(T,q,t\pm 0)\varphi_f(T,-q,t\pm 0)\,dq \quad (4.90)$$

$$\sum_{k=-\infty}^{\infty} f^2(kT\pm 0) = \frac{T}{2\pi j}\int_{-j\omega/2}^{j\omega/2} \varphi_f(T,q,\pm 0)\varphi_f(T,-q,\pm 0)\,dq . \quad (4.91)$$

4.11 Integrals from continuous functions and their squares

Let $f(t) \in \Lambda(\alpha,\beta)$ with $\alpha < 0$, $\beta > 0$. Then, the following integral converges absolutely

$$I_1 \triangleq \int_{-\infty}^{\infty} f(t)\,dt = F(0) . \quad (4.92)$$

Integral (4.92) can be expresssed via discrete Laplace transform and DPFR of the function $f(t)$. Indeed,

$$I_1 = \sum_{k=-\infty}^{\infty} \int_{kT}^{(k+1)T} f(t)\,\mathrm{d}t\,. \tag{4.93}$$

Since

$$\int_{kT}^{(k+1)T} f(t)\,\mathrm{d}t = \int_0^T f(u+kT)\,\mathrm{d}u \tag{4.94}$$

we obtain

$$I_1 = \sum_{k=-\infty}^{\infty} \int_0^T f(u+kT)\,\mathrm{d}u\,. \tag{4.95}$$

Due to Theorem 4.1 the series

$$\mathcal{D}_f(T,0,t) = \sum_{k=-\infty}^{\infty} f(t+kT) \tag{4.96}$$

converges uniformly in any finite interval of t. Therefore, the order of summation and integration in (4.95) can be changed, so that we obtain

$$I_1 = \int_{-\infty}^{\infty} f(t)\,\mathrm{d}t = \int_0^T \mathcal{D}_f(T,0,t)\,\mathrm{d}t\,. \tag{4.97}$$

As a special case, the above relations hold if $f(t) \in \Lambda_+(\alpha,\infty)$, $\alpha < 0$. In this case $f(t) = 0$ for $t < 0$, and we have

$$\int_0^{\infty} f(t)\,\mathrm{d}t = \int_0^T \mathcal{D}_f(T,0,t)\,\mathrm{d}t\,. \tag{4.98}$$

The result (4.98) can be used to calculate, under the above assumptions, the following integral

$$I_2 \triangleq \int_{-\infty}^{\infty} f^2(t)\,\mathrm{d}t\,. \tag{4.99}$$

In this case Eq. (4.97) yields

$$I_2 = \int_0^T \mathcal{D}_{f^2}(T,0,t)\,\mathrm{d}t\,. \tag{4.100}$$

But, proceeding from (4.75) for $0 < t < T$, with $c = 0$ (this holds due to the choice of α and β) we obtain

$$\mathcal{D}_{f^2}(T,0,t) = \frac{T}{2\pi\mathrm{j}} \int_{-\mathrm{j}\omega/2}^{\mathrm{j}\omega/2} \mathcal{D}_f(T,q,t)\mathcal{D}_f(T,-q,t)\,\mathrm{d}q\,. \tag{4.101}$$

Using (4.100) and (4.101), we find

$$I_2 = \frac{T}{2\pi\mathrm{j}} \int_{-\mathrm{j}\omega/2}^{+\mathrm{j}\omega/2} \left[\int_0^T \mathcal{D}_f(T,q,t)\mathcal{D}_f(T,-q,t)\,\mathrm{d}t\right] \mathrm{d}q\,. \tag{4.102}$$

Consider the internal integral in (4.102). Taking account of (4.88), it can be written in the form

$$\int_0^T \mathcal{D}_f(T,q,t)\mathcal{D}_f(T,-q,t)\,\mathrm{d}t = \int_0^T \varphi_f(T,q,t)\varphi_f(T,-q,t)\,\mathrm{d}t\,. \tag{4.103}$$

For $f(t) \in \Lambda(\alpha,\beta)$ with $\alpha < 0$ and $\beta > 0$, by Parseval's formula for Fourier series, at the right-hand side of (4.103) instead of the function $\varphi_f(T,s,t)$ we can use its Fourier series $\varphi_F(T,s,t)$ (3.29). Hence,

$$\int_0^T \mathcal{D}_f(T,q,t)\mathcal{D}_f(T,-q,t)\,\mathrm{d}t = \int_0^T \varphi_F(T,q,t)\varphi_F(T,-q,t)\,\mathrm{d}t\,. \tag{4.104}$$

As follows from (3.29),

$$\begin{aligned} \varphi_F(T,q,t) &= \frac{1}{T}\sum_{k=-\infty}^{\infty} F(q+k\mathrm{j}\omega)\mathrm{e}^{k\mathrm{j}\omega t} \\ \varphi_F(T,-q,t) &= \frac{1}{T}\sum_{k=-\infty}^{\infty} F(-q+k\mathrm{j}\omega)\mathrm{e}^{k\mathrm{j}\omega t}\,. \end{aligned} \tag{4.105}$$

Substituting (4.105) into (4.104) and integrating termwise (this is correct by Parseval's formula for Fourier series), we obtain

$$\int_0^T \mathcal{D}_f(T,q,t)\mathcal{D}_f(T,-q,t)\,\mathrm{d}t = \frac{1}{T}\sum_{k=-\infty}^{\infty} F(q+k\mathrm{j}\omega)F(-q-k\mathrm{j}\omega)\,. \tag{4.106}$$

Using the above notation and (4.106), we obtain

$$\int_0^T \mathcal{D}_f(T,q,t)\mathcal{D}_f(T,-q,t)\,\mathrm{d}t = \mathcal{D}_{F\underline{F}}(T,q,0) \tag{4.107}$$

where the underbar denotes the substitution of $-q$ for q. Using (4.107), from (4.102) we find

$$I_2 = \frac{T}{2\pi\mathrm{j}}\int_{-\mathrm{j}\omega/2}^{\mathrm{j}\omega/2} \mathcal{D}_{F\underline{F}}(T,q,0)\,\mathrm{d}q\,. \tag{4.108}$$

We note that Eq. (4.108) can also be obtained by discretization of the integral at the left-hand side of (1.67).

4.12 Variation of the sampling period

As in Section 3.6, we can find the connection between $\mathcal{D}$-transforms of the function $f(t)$ corresponding to multiple values of the parameter T. In the appropriate stripe of convergence we have

$$\mathcal{D}_F(T,s,t) = \frac{1}{T}\sum_{k=-\infty}^{\infty} F(s+kj\omega)e^{(s+kj\omega)t}, \quad \omega = 2\pi/T \tag{4.109}$$

$$\breve{\mathcal{D}}_F(NT,s,t) \triangleq \frac{1}{NT}\sum_{k=-\infty}^{\infty} F(s+\frac{kj\omega}{N})e^{(s+\frac{kj\omega}{N})t}. \tag{4.110}$$

Then, the following propositions hold.

Theorem 4.6 *The following relations are valid*

$$\mathcal{D}_F(T,s,t) = \sum_{m=0}^{N-1} \breve{\mathcal{D}}_F(NT,s,t-mT)e^{msT} \tag{4.111}$$

$$\breve{\mathcal{D}}_F(NT,s,t) = \frac{1}{N}\sum_{m=0}^{N-1} \mathcal{D}_F(T,s+\frac{mj\omega}{N},t). \tag{4.112}$$

Proof Taking into account that

$$\breve{\varphi}_F(NT,s,t) = \breve{\mathcal{D}}_F(NT,s,t)e^{-st} \tag{4.113}$$

we have

$$\breve{\varphi}_F(NT,s,t-mT) = \breve{\mathcal{D}}_F(NT,s,t-mT)e^{-s(t-mT)}. \tag{4.114}$$

Substituting (4.114) into (3.123), we find

$$\varphi_F(T,s,t) = \sum_{m=0}^{N-1} \breve{\mathcal{D}}_F(NT,s,t-mT)e^{-s(t-mT)} \tag{4.115}$$

whence, taking account of (4.26), follows (4.111). To prove (4.112) we note that

$$\varphi_F(T,s+\frac{mj\omega}{N},t) = \mathcal{D}_F(T,s+\frac{mj\omega}{N},t)e^{-(s+\frac{mj\omega}{N})t}. \tag{4.116}$$

Substituting (4.116) into (3.129), we have

$$\breve{\varphi}_F(NT,s,t) = \frac{1}{N}\sum_{m=0}^{N-1} \mathcal{D}_F(T,s+\frac{mj\omega}{N},t)e^{-st} \tag{4.117}$$

which is equivalent to (4.112) . ■

Remark 1 For practical computations Eq. (4.111) can be represented in another form. Substituting $t+NT$ for t in (4.111), we have

$$\mathcal{D}_F(T,s,t+NT)=\mathcal{D}_F(T,s,t)\mathrm{e}^{NsT}=\sum_{m=0}^{N-1}\breve{\mathcal{D}}_F(NT,s,t+NT-mT)\mathrm{e}^{msT}\,. \tag{4.118}$$

With the notation q for $N-m$, Eq. (4.118) can be represented in the form

$$\mathcal{D}_F(T,s,t)=\sum_{q=1}^{N}\breve{\mathcal{D}}_F(NT,s,t+qT)\mathrm{e}^{-qsT}\,. \tag{4.119}$$

The term associated with $q=N$ at the right-hand side of (4.119) can be represented in the form

$$\breve{\mathcal{D}}_F(NT,s,t+NT)\mathrm{e}^{-NsT}=\breve{\mathcal{D}}_F(NT,s,t)\,. \tag{4.120}$$

Then, we obtain the required formula as

$$\mathcal{D}_F(T,s,t)=\sum_{q=0}^{N-1}\breve{\mathcal{D}}_F(NT,s,t+qT)\mathrm{e}^{-qsT}\,. \tag{4.121}$$

Remark 2 All relations of this section hold also for the transform $\mathcal{D}_f(T,s,t)$.

Example 4.9 Remark 2 should be varified, for instance for the function (4.80). In $\operatorname{Re} s>a$ and $0<t<T$ we have, with reference to (4.81),

$$\breve{\mathcal{D}}_f(2T,s,t)=\frac{\mathrm{e}^{at}}{1-\mathrm{e}^{2(a-s)T}}\,,\quad 0<t<2T\,. \tag{4.122}$$

The analogue of Eq. (4.121) in this case has the form

$$\mathcal{D}_f(T,s,t)=\breve{\mathcal{D}}_f(2T,s,t)+\breve{\mathcal{D}}_f(2T,s,t+T)\mathrm{e}^{-sT}\,. \tag{4.123}$$

But for $0<t<T$ Eq. (4.122) yields

$$\breve{\mathcal{D}}_f(2T,s,t)=\frac{\mathrm{e}^{at}}{1-\mathrm{e}^{2(a-s)T}}\,,\quad \breve{\mathcal{D}}_f(2T,s,t+T)=\frac{\mathrm{e}^{a(t+T)}}{1-\mathrm{e}^{2(a-s)T}}\,. \tag{4.124}$$

The substitution of (4.124) into (4.123) for $0<t<T$ yields the known result (4.81).
□

Example 4.10 Let us verify the decomposition formula (4.112) for the function (4.80). For $0<t<T$ we have

$$\mathcal{D}_f(T,s,t)=\frac{\mathrm{e}^{at}}{1-\mathrm{e}^{(a-s)T}}\,,\quad \mathcal{D}_f(T,s+\frac{\mathrm{j}\omega}{2},t)=\frac{\mathrm{e}^{at}}{1+\mathrm{e}^{(a-s)T}}\,. \tag{4.125}$$

Adding these relations, we find

$$\frac{1}{2}\left[\mathcal{D}_f(T,s,t)+\mathcal{D}_f(T,s+\frac{\mathrm{j}\omega}{2},t)\right]=\frac{\mathrm{e}^{at}}{1-\mathrm{e}^{2(a-s)T}}=\breve{\mathcal{D}}_f(2T,s,t)\,. \tag{4.126}$$

To determine $\breve{\mathcal{D}}_f(2T,s,t)$ in the interval $T<t<2T$ we notice that, in accordance with (4.47),

$$\mathcal{D}_f(T,s,t)=\frac{\mathrm{e}^{at}\mathrm{e}^{(s-a)T}}{1-\mathrm{e}^{(a-s)T}}\,,\quad T<t<2T\,. \tag{4.127}$$

From (4.127) we also have

$$\mathcal{D}_f(T,s+\frac{\mathrm{j}\omega}{2},t)=\frac{-\mathrm{e}^{at}\mathrm{e}^{(s-a)T}}{1+\mathrm{e}^{(a-s)T}}\,,\quad T<t<2T\,. \tag{4.128}$$

Adding Eqs. (4.127) and (4.128) termwise, we find that Eq. (4.126) holds for $T<t<2T$ as well. □

Part II

Linear periodic operators and systems

Introduction Currently the operational approach is widely used for the investigation of linear control systems. A system under consideration with the input $x(t)$ and the output $y(t)$ is associated with a linear operator $y(t) = \mathsf{U}[x(t)]$, the properties of which define the properties of the considered system. For instance, the boundedness or the value of the norm of such an operator in various functional spaces can be used as one of these properties.

This part of the book investigates some general properties of linear systems with periodically varying parameters, linear periodic sampled-data systems being a special case of the latter. Some kinds of linear periodic operators generated by these systems are introduced. Moreover, two types of linear periodic operators are subjected to detailed investigation, namely, integral and sampled-data ones, both of which are most important for the theory of periodic sampled-data systems.

Chapter 5

Linear time-invariant operators and systems

5.1 Linear operators

As is commonly accepted in control theory, a control system is depicted as a connection of elements of directed action. By an element of directed action we mean a dynamic system with input $x(t)$ and output $y(t)$, such that the input $x(t)$ is independent of the output $y(t)$.

An element of directed action is defined mathematically by an equation, for example, differential, integral or difference equation. If there is a unique relation between the input and output, the element is defined by an *operator*, i.e., by a mathematical rule, that allows us to calculate $y(t)$ if the input $x(t)$ is known. Symbolically this can be written as

$$y(t) = \mathsf{U}[x(t)] \tag{5.1}$$

where U is the operator of the element, and t is an argument that will be henceforth called 'time'. By an operator U in the present book we mean a set of operations that are to be performed in order to calculate $y(t)$ for a known $x(t)$. From the functional analytical point of view, it is necessary to rigorously define the sets of functions for which (5.1) is considered.

An element defined by (5.1) is called *linear* if the operator U is linear, i.e., the sets of input and output signals are linear spaces, and for any admissible input signals and any complex constants α_1, α_2 we have

$$\mathsf{U}[\alpha_1 x_1(t) + \alpha_2 x_2(t)] = \alpha_1 \mathsf{U}[x_1(t)] + \alpha_2 \mathsf{U}[x_2(t)]\,. \tag{5.2}$$

In control theory linear operators are described by system functions that define the response of the operator to some standard input signals. The *Green function* is one of the most important characteristics of a linear operator. The Green function $h(t,\tau)$ of a linear operator is defined as its response to an input of the form $x(t) = \delta(t-\tau)$, where $\delta(t)$ is the Dirac delta-function and τ is a parameter, i.e.,

$$h(t,\tau) \triangleq \mathsf{U}[\delta(t-\tau)]\,. \tag{5.3}$$

The function

$$W(s,t) \triangleq \mathsf{U}\left[\mathrm{e}^{st}\right]\mathrm{e}^{-st} \tag{5.4}$$

with the complex variable s is another principal characteristic. In this case the operator is assumed to be such that (5.3) and (5.4) have sense. The function $W(s,t)$ will be called the *parametric transfer function* (PTF) of the operator U. In general, the PTF of an operator is a function in two arguments, namely, s and t. As follows from the definition (5.4), if a linear operator U has a PTF defined over a set $\mathcal{P}$ of the complex variable s, then, for $s \in \mathcal{P}$, we have

$$\mathsf{U}\left[\mathrm{e}^{st}\right] = W(s,t)\mathrm{e}^{st}\,. \tag{5.5}$$

Besides (5.1), to define a linear operator we shall in what follows use the notation

$$y(t) = W(s,t)x(t) \tag{5.6}$$

with or without specification of the set $\mathcal{P}$. Equation (5.6) will be called the *operator equation* of the element (5.1). At the same time, Eq. (5.6) shows that the response of the element to the input $x(t) = \mathrm{e}^{st}$, $s \in \mathcal{P}$, has the form (5.5). Equation (5.1) can be associated with the simplest block-diagram shown in Fig. 5.1.

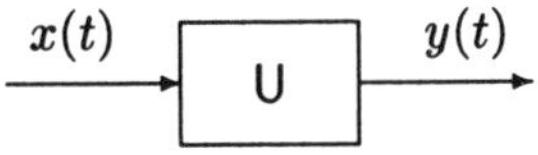

Figure 5.1: Block-diagram for (5.1)

It is possible to introduce summation and multiplication over the set of linear operators. If we have

$$y_1(t) = \mathsf{U}_1\left[x(t)\right]\,, \quad y_2(t) = \mathsf{U}_2\left[x(t)\right] \tag{5.7}$$

then, assuming that $y(t) = y_1(t) + y_2(t)$, we obtain the new operator

$$y = \mathsf{U}_\sigma\left[x(t)\right]\,, \tag{5.8}$$

which is called the sum of the operators U_1 and U_2 and is denoted by

$$\mathsf{U}_\sigma = \mathsf{U}_1 + \mathsf{U}_2\,. \tag{5.9}$$

Obviously, $\mathsf{U}_2 + \mathsf{U}_1 = \mathsf{U}_1 + \mathsf{U}_2$. The sum of operators is associated with the parallel connection of the corresponding elements (Fig. 5.2).
If we have

$$y(t) = \mathsf{U}_2\left[z(t)\right]\,, \quad z(t) = \mathsf{U}_1\left[x(t)\right] \tag{5.10}$$

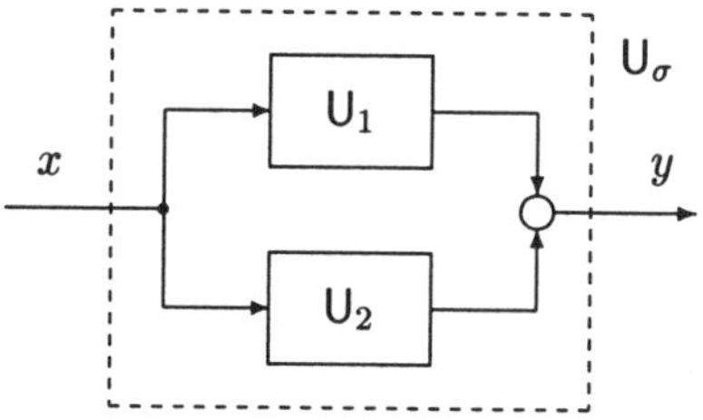

Figure 5.2: Sum of operators

the operator U, relating $y(t)$ to $x(t)$, is called the product of the operators U_1 and U_2 and it is denoted by

$$\mathsf{U}_\pi = \mathsf{U}_2\mathsf{U}_1 \,. \tag{5.11}$$

The structure of equation (5.10) is associated with the series connection of the elements shown in Fig. 5.3.

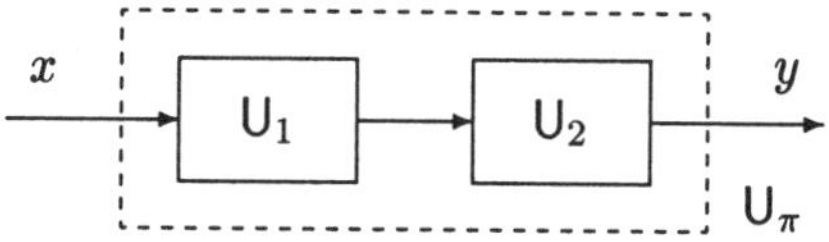

Figure 5.3: Product of operators

If the operators U_1 and U_2 can be connected in inverse order, so that

$$y(t) = \mathsf{U}_1\left[v(t)\right] \,, \quad v(t) = \mathsf{U}_2\left[x(t)\right] \tag{5.12}$$

this connection is associated with the product

$$\tilde{\mathsf{U}}_\pi = \mathsf{U}_1\mathsf{U}_2 \,. \tag{5.13}$$

The operators (5.11) and (5.13) are, in general, different. Moreover, if the operator U_π exists, the operator $\tilde{\mathsf{U}}_\pi$ must not have sense, and vice versa. If both of the operators U_π and $\tilde{\mathsf{U}}_\pi$ exist and coincide, i.e., $\mathsf{U}_\pi = \tilde{\mathsf{U}}_\pi$ is valid, the operators U_1 and U_2 are called *commutative*.

Example 5.1 An example of non-commutative operators is given. Let $\mathsf{U}_1 = \mathrm{d}/\,\mathrm{d}t$ be the operator of differentiation, and U_2 be the operator of multiplication by $\sin\omega t$, where ω is a constant. With $x(t) = 1$ we have

$$\mathsf{U}_2\mathsf{U}_1[x(t)] = 0 \,, \quad \mathsf{U}_1\mathsf{U}_2[x(t)] = \omega\cos\omega t \,,$$

i.e., the operators U_1 and U_2 are not commutative. □

5.2 Multivariable linear operators

The aforesaid can be generalized onto multivariable linear operators with a finite number of inputs and outputs. Henceforth, by vectors we shall mean column-vectors. Vectors and matrices are denoted by bold letters.
Let us consider the vector

$$\boldsymbol{x}(t) = \begin{bmatrix} x_1(t) \\ \vdots \\ x_m(t) \end{bmatrix} \tag{5.14}$$

where each of the functions $x_i(t)$ is defined for $-\infty < t < \infty$. The value m will be called the dimension of the vector $\boldsymbol{x}(t)$. If

$$\boldsymbol{x}_1(t) = \begin{bmatrix} x_{11}(t) \\ \vdots \\ x_{m1}(t) \end{bmatrix}, \quad \boldsymbol{x}_2(t) = \begin{bmatrix} x_{12}(t) \\ \vdots \\ x_{m2}(t) \end{bmatrix} \tag{5.15}$$

then, by definition,

$$\boldsymbol{x}_1(t) + \boldsymbol{x}_2(t) = \begin{bmatrix} x_{11}(t) + x_{12}(t) \\ \vdots \\ x_{m1}(t) + x_{m2}(t) \end{bmatrix} \tag{5.16}$$

and correspondingly, for a constant α we have

$$\alpha\boldsymbol{x}(t) = \begin{bmatrix} \alpha x_1(t) \\ \vdots \\ \alpha x_m(t) \end{bmatrix}. \tag{5.17}$$

If there is a rule, by which any vector $\boldsymbol{x}(t)$ of dimension m from the set $\mathcal{M}_{\boldsymbol{x}}$ is associated with a vector $\boldsymbol{y}(t)$ of the dimension n from the set $\mathcal{M}_{\boldsymbol{y}}$, we shall say that there is an operator

$$\boldsymbol{y}(t) = \mathsf{U}[\boldsymbol{x}(t)] \tag{5.18}$$

acting from $\mathcal{M}_{\boldsymbol{x}}$ to $\mathcal{M}_{\boldsymbol{y}}$. Henceforth, operators of the form (5.18) will be called multivariable. The operator (5.18) is called linear, if $\mathcal{M}_{\boldsymbol{x}}$ and $\mathcal{M}_{\boldsymbol{y}}$ are linear spaces and for any numbers α_1 and α_2 we have

$$\mathsf{U}\left[\alpha_1\boldsymbol{x}_1(t) + \alpha_2\boldsymbol{x}_2(t)\right] = \alpha_1\mathsf{U}\left[\boldsymbol{x}_1(t)\right] + \alpha_2\mathsf{U}\left[\boldsymbol{x}_2(t)\right]. \tag{5.19}$$

As in the SISO case, it is convenient to define multivariable operators by functions that determine the response of the operator to standard input signals. An important procedure in such definitions is the use of the parametric transfer matrix (PTM), which is introduced as follows. Consider the set of input vectors of the form

$$
\boldsymbol{x}_1(t) = \begin{bmatrix} e^{st} \\ 0 \\ \vdots \\ 0 \end{bmatrix}, \quad \boldsymbol{x}_2(t) = \begin{bmatrix} 0 \\ e^{st} \\ \vdots \\ 0 \end{bmatrix}, \ldots, \boldsymbol{x}_m(t) = \begin{bmatrix} 0 \\ 0 \\ \vdots \\ e^{st} \end{bmatrix} \tag{5.20}
$$

where s is a complex parameter belonging to a domain $\mathcal{P}$. Denote by $\boldsymbol{y}_i(s,t)$ the response of the operator (5.18) to the vector $\boldsymbol{x}_i(t)$. Let

$$
\boldsymbol{y}_i(s,t) \triangleq \mathsf{U}\left[\boldsymbol{x}_i(t)\right] = \begin{bmatrix} y_{1i}(s,t) \\ y_{2i}(s,t) \\ \vdots \\ y_{ni}(s,t) \end{bmatrix} . \tag{5.21}
$$

Next, using the vectors $\boldsymbol{y}_i(s,t)$, we construct the following rectangular $n \times m$ matrix

$$
\boldsymbol{H}(s,t) = \begin{bmatrix} y_{11}(s,t) & y_{12}(s,t) & \ldots & y_{1m}(s,t) \\ y_{21}(s,t) & y_{22}(s,t) & \ldots & y_{2m}(s,t) \\ \vdots & \vdots & \ddots & \vdots \\ y_{n1}(s,t) & y_{n2}(s,t) & \ldots & y_{nm}(s,t) \end{bmatrix} . \tag{5.22}
$$

Define

$$
w_{ik}(s,t) \triangleq y_{ik}(s,t)e^{-st} . \tag{5.23}
$$

The matrix

$$
\boldsymbol{W}(s,t) = \begin{bmatrix} w_{11}(s,t) & w_{12}(s,t) & \ldots & w_{1m}(s,t) \\ w_{21}(s,t) & w_{22}(s,t) & \ldots & w_{2m}(s,t) \\ \vdots & \vdots & \ddots & \vdots \\ w_{n1}(s,t) & w_{n2}(s,t) & \ldots & w_{nm}(s,t) \end{bmatrix} = \boldsymbol{H}(s,t)e^{-st} \tag{5.24}
$$

will be called the *parametric transfer matrix (PTM) of the operator* U. The above construction can be approached in another way. Let

$$
\boldsymbol{X}(t) = \begin{bmatrix} \boldsymbol{x}_1(t) & \boldsymbol{x}_2(t) & \ldots & \boldsymbol{x}_m(t) \end{bmatrix} \tag{5.25}
$$

be a matrix of dimensions $m \times m$ constructed of input vectors of dimensions $m \times 1$. Assuming the matrix (5.25) to be the input of the operator, we can consider the following matrix $\boldsymbol{Y}(t)$ of dimension $n \times m$ as the output

$$
\boldsymbol{Y}(t) = \begin{bmatrix} \mathsf{U}\left[\boldsymbol{x}_1(t)\right] & \mathsf{U}\left[\boldsymbol{x}_2(t)\right] & \ldots & \mathsf{U}\left[\boldsymbol{x}_m(t)\right] . \end{bmatrix} \tag{5.26}
$$

With such an approach, the PTM can be represented by

$$
\boldsymbol{W}(s,t) = \mathsf{U}\left[\boldsymbol{I}_m e^{st}\right] e^{-st} \tag{5.27}
$$

where $\boldsymbol{I}_m$ is the identity matrix of dimension $m \times m$. In the following, instead of (5.18) we shall use the equivalent form

$$y(t) = W(s,t)x(t) \tag{5.28}$$

or, in expanded form,

$$y_i(t) = \sum_{k=1}^{m} w_{ik}(s,t)x_k(t), \quad i = 1, \dots, n. \tag{5.29}$$

Equation (5.29) is associated with a multivariable system with m scalar inputs $x_k(t)$, $(k = 1, \dots, m)$ and n scalar outputs $y_i(t)$, $(i = 1, \dots, n)$. The connection between the input $x_k(t)$ and the output $y_i(t)$ is defined by the following linear operator

$$y_i(t) = \mathsf{U}_{ik}\left[x_k(t)\right]. \tag{5.30}$$

Moreover, by construction we have,

$$w_{ik}(s,t) = \mathsf{U}_{ik}\left[\mathrm{e}^{st}\right]\mathrm{e}^{-st} \tag{5.31}$$

i.e., the function $w_{ik}(s,t)$ is the PTF of the operator (5.30). It is easy to see that, conversely, Eq. (5.29) defines a multivariable linear operator from $\mathcal{M}_x$ to $\mathcal{M}_y$, and the matrix (5.24) constructed from the PTFs (5.31) is its parametric transfer matrix.

5.3 Linear time-invariant (LTI) operators

If it is not otherwise specified, the input and the output signals are here assumed to be defined on $-\infty < t < \infty$. For any such function we define the shift operator U_τ as

$$\mathsf{U}_\tau\left[x(t)\right] \triangleq x(t-\tau) \tag{5.32}$$

where τ is a constant.

Definition A linear operator U will be called *time-invariant*, if it is commutative with the shift operator U_τ for any τ, i.e., if for any τ we have

$$\mathsf{U}\mathsf{U}_\tau = \mathsf{U}_\tau\mathsf{U}. \tag{5.33}$$

In other words, a time-invariant operator is invariant with respect to a shift of the input in time.

It is easy to understand that (5.33) holds iff (if and if only) for any input $x(t)$ and for all τ we have

$$\mathsf{U}\left[x(t-\tau)\right] = y(t-\tau). \tag{5.34}$$

Then, we show that the PTF of a stationary operator is independent of t, i.e.,

$$W(s,t) = W(s). \tag{5.35}$$

Indeed, let us have (5.33) and let the operator U have the PTF $W(s,t)$ in $s \in \mathcal{P}$. Then, Eq. (5.4) does not change after substitution of $t-\tau$ for t, and

$$\mathsf{U}\left[e^{s(t-\tau)}\right] = e^{s(t-\tau)}W(s, t-\tau)\,.$$

On the other hand,

$$\mathsf{U}\left[e^{s(t-\tau)}\right] = e^{-s\tau}\mathsf{U}\left[e^{st}\right] = e^{s(t-\tau)}W(s,\tau)\,.$$

Comparison of the last two relation shows that $W(s,t) = W(s,t-\tau)$ for all τ and we have (5.35).
For an LTI operator instead of (5.5) we have

$$\mathsf{U}\left[e^{st}\right] = W(s)e^{st} \tag{5.36}$$

and the operator equation (5.6) reduces to

$$y(t) = W(s)x(t)\,. \tag{5.37}$$

Example 5.2 Find the transfer function of the differential operator

$$y(t) = \frac{dx(t)}{dt} \tag{5.38}$$

as an operator (5.1). Assuming that $x(t) = e^{st}$, we find $y(t) = se^{st}$ and Eq. (5.4) yields $W(s,t) = s$. Obviously, the operator is time-invariant, because for any $\tau = \text{const}$ we have

$$y(t-\tau) = \frac{dx(t-\tau)}{dt}$$

i.e., condition (5.34) is satisfied. The corresponding transfer function is defined for all s.
□

Example 5.3 Find the transfer function of the general differential operator

$$y(t) = a_0\frac{d^n x(t)}{dt^n} + a_1\frac{d^{n-1}x(t)}{dt^{n-1}} + \ldots + a_n x(t) \tag{5.39}$$

where a_i are constants. The operator is time-invariant and its transfer function

$$W(s) = a_0 s^n + a_1 s^{n-1} + \ldots + a_n \tag{5.40}$$

is defined for all s.
□

Example 5.4 Build the transfer function of the integral operator $\mathsf{U}[x(t)]$ given by

$$y(t) = \int_{-\infty}^{\infty} g(t-u)x(u)\,du = \mathsf{U}\left[x(t)\right] \tag{5.41}$$

where $g(t)$ is a known function. For $x(t) = \delta(t-\tau)$ we have $y(t) = g(t-\tau)$, i.e., the Green function of the operator (5.41). The operator (5.41) is time-invariant, because

$$\int_{-\infty}^{\infty} g(t-u)x(u-\tau)\,\mathrm{d}u = \int_{-\infty}^{\infty} g(t-\tau-\mu)x(\mu)\,\mathrm{d}\mu = y(t-\tau)\,. \qquad (5.42)$$

Next, we find the PTF of the operator (5.41), assuming that $g(t) \in \Lambda(\alpha,\beta)$. Then, for $x(t) = \mathrm{e}^{st}$, $\alpha < \operatorname{Re} s < \beta$ by (5.4) we have

$$W(s,t) = \int_{-\infty}^{\infty} g(t-u)\mathrm{e}^{-s(t-u)}\,\mathrm{d}u = \int_{-\infty}^{\infty} g(t)\mathrm{e}^{-st}\,\mathrm{d}t = G(s) \qquad (5.43)$$

where $G(s)$ is the Laplace transform of the function $g(t)$. Thus, in this example the PTF is defined by (5.43) in the stripe $\alpha < \operatorname{Re} s < \beta$. □

Henceforth, for brevity, we shall call PTFs of LTI operators simply transfer functions. Moreover, if it is not otherwise specified, we assume that operators U are such that the associated tranfer functions $W(s)$ take real values for real s.

5.4 Connections of LTI operators

Transfer functions of connections of LTI operators can be constructed on the basis of the statements of Section 5.1. Consider the series connection (5.10). Assuming that $x(t) = \mathrm{e}^{st}$ and using (5.35), we have $z(t) = W_1(s)\mathrm{e}^{st}$ in a domain $\mathcal{P}$. If the PTF of the operator U_2 is also defined in this domain, we have $y(t) = W_2(s)W_1(s)\mathrm{e}^{st}$, i.e.,

$$W(s,t) = W_2(s)W_1(s) = W(s)\,. \qquad (5.44)$$

Similar reasoning shows that for a parallel connection (5.7) the transfer function is

$$W(s) = W_1(s) + W_2(s) \qquad (5.45)$$

if there is a common domain of the functions $W_1(s)$ and $W_2(s)$.
For a feedback connection we find

$$y(t) = W_0(s)z(t)\,, \quad z(t) = x(t) - y(t)\,. \qquad (5.46)$$

Formal derivation of the transfer function from x to y gives

$$W(s) = \frac{W_0(s)}{1+W_0(s)}\,. \qquad (5.47)$$

5.5 Frequency response of LTI operators

Let $W(s)$ be the transfer function of a LTI operator. If $W(s)$ is defined on the imaginary axis, the function

$$\Phi(\mathrm{j}\nu) \triangleq W(s)\,|_{s=\mathrm{j}\nu} \qquad (5.48)$$

where ν is a real variable, will be called the *frequency response* of the operator. Using (5.37) and (5.48), we have

$$\mathsf{U}\left[e^{j\nu t}\right] = \Phi(j\nu)e^{j\nu t}\,. \tag{5.49}$$

Separating the real and imaginary part in (5.48), we obtain

$$\Phi(j\nu) = P(\nu) + jQ(\nu) \tag{5.50}$$

where the functions $P(\nu)$ and $Q(\nu)$ are called the real and the imaginary part of the frequency response, respectively. Then, assuming that $e^{j\nu t} = \cos\nu t + j\sin\nu t$ and separating the real and the imaginary part, we obtain

$$\begin{aligned} \mathsf{U}[\cos\nu t] &= P(\nu)\cos\nu t - Q(\nu)\sin\nu t \\ \mathsf{U}[\sin\nu t] &= P(\nu)\sin\nu t + Q(\nu)\cos\nu t\,. \end{aligned} \tag{5.51}$$

The functions

$$A(\nu) \triangleq \sqrt{P^2(\nu) + Q^2(\nu)}\,, \quad \psi(\nu) \triangleq \arctan\frac{Q(\nu)}{P(\nu)} \tag{5.52}$$

are called the *amplitude and phase response (characteristic)*, respectively. As follows from (5.50),

$$\Phi(j\nu) = A(\nu)e^{j\psi(\nu)}\,. \tag{5.53}$$

5.6 LTI systems

The notion of operators assumes a unique relation between input and output. For most real elements there is not such a relation. For the same inputs real elements can exhibit different motions dependent on some further conditions, for example, on the initial state of the elements. Therefore, many real elements are described by an operator equation (or system of equations) rather than by an operator (5.1).

Hereinafter, by an LTI system we shall mean any material system, in which the relation between the input $x(t)$ and the output $y(t)$ is defined by an operator equation of the form

$$\mathsf{U}_1\left[y(t)\right] = \mathsf{U}_2\left[x(t)\right] \tag{5.54}$$

where U_1 and U_2 are linear stationary operators, possibly multivariable. If the operators U_i $(i = 1, 2)$ have transfer functions $W_i(s)$ $(i = 1, 2)$ in a common domain $\mathcal{P}$, with reference to (5.37) we can write (5.54) in the form

$$W_1(s)y(t) = W_2(s)x(t)\,. \tag{5.55}$$

Example 5.5 Consider the set of operators defined by the differential equation

$$\frac{dy}{dt} = x \tag{5.56}$$

which is, in the form of (5.55), equivalent to

$$sy = x\,.$$

The general solution of Eq. (5.56) can be represented in the form

$$y(t) = \int_0^t x(\tau)\,\mathrm{d}\tau + c[x(t)] \tag{5.57}$$

where c is a constant that depends, in the general case, on $x(t)$. The relationship between $y(t)$ and $x(t)$ ia not unique and depends on the choice of the functional $c[x(t)]$. If the functional is fixed, Eq. (5.57) defines an operator. Moreover, if the functional $c[x(t)]$ is nonlinear, the operator will be nonlinear as well. □

As follows from Example 5.5, a linear functional (operator) equation (5.54) defines a set Ξ of operators (nonlinear in the general case). Nevertheless, the set Ξ may include LTI operators. All of them have the same transfer function, which is easy to define.
We shall say that an operator $y(t) = \mathsf{U}\,[x(t)]$ is generated by equation (system) (5.54) if for any $x(t)$ the relation $\mathsf{U}\,[x(t)]$ defines a particular solution of Eq. (5.54), i.e., for all admissible $x(t)$ the following equality holds

$$\mathsf{U}_1\mathsf{U}\,[x(t)] = \mathsf{U}_2\,[x(t)]\,. \tag{5.58}$$

Suppose that there is a LTI operator U satisfying the identity (5.58) with tranfer function $W(s)$ defined in a domain $\mathcal{P}$. Assume that the transfer functions $W_1(s)$ and $W_2(s)$ of the operators U_1 and U_2, respectively, are also defined in $\mathcal{P}$. Then, for $x(t) = \mathrm{e}^{st}$, $s \in \mathcal{P}$ we have

$$\begin{aligned} \mathsf{U}_2\,[x(t)] &= \mathrm{e}^{st}W_2(s)\,, & \mathsf{U}\,[x(t)] &= \mathrm{e}^{st}W(s) \\ \mathsf{U}_1\mathsf{U}_2\,[x(t)] &= \mathrm{e}^{st}W_1(s)W_2(s)\,. \end{aligned} \tag{5.59}$$

Substituting $x(t) = \mathrm{e}^{st}$, $s \in \mathcal{P}$ in (5.58) and taking (5.59) into account, we find

$$W(s) = \frac{W_2(s)}{W_1(s)}\,. \tag{5.60}$$

Thus, the transfer function $W(s)$ is defined uniquely. Henceforth we shall call this function the *transfer function of the system* (5.54).
We note, that the system transfer function $W(s)$ can be determined independently of the properties of the operators defined by the system (5.54). Indeed, assuming in (5.54)

$$x = \mathrm{e}^{st}\,, \quad y = W(s)\mathrm{e}^{st} \tag{5.61}$$

where $W(s)$ is an unknown function independent of t, we can pass immediately to (5.60). In the following, therefore, the transfer function of a LTI system will be used to mean the function $W(s)$ derived from (5.60).

5.7 Application to the finite-dimensional case

Let us apply the above reasoning to the investigation of a system described by the following differential equation with constant coefficients

$$\frac{d^p y}{dt^p} + d_1 \frac{d^{p-1} y}{dt^{p-1}} + \ldots + d_p y = m_1 \frac{d^{p-1} x}{dt^{p-1}} + m_2 \frac{d^{p-2} x}{dt^{p-2}} + \ldots + m_p x . \quad (5.62)$$

Systems described by such an equation (or systems of equations) we shall call *finite-dimensional.* Equation (5.62) can be written in the form (5.55)

$$d(s)y = m(s)x \quad (5.63)$$

where

$$\begin{aligned} m(s) &= m_1 s^{p-1} + m_2 s^{p-2} + \ldots + m_p \\ d(s) &= s^p + d_1 s^{p-1} + \ldots + d_p . \end{aligned} \quad (5.64)$$

According to (5.60), the transfer function of the equation (system) (5.62) is a strictly proper real rational function

$$W(s) = \frac{m(s)}{d(s)} . \quad (5.65)$$

Let s_i, $(i = 1, 2, \ldots \ell)$ be different roots of the equation

$$d(s) = s^p + d_1 s^{p-1} + \ldots + d_p = 0 \quad (5.66)$$

with multiplicities ν_i, $(i = 1, 2, \ldots \ell)$, and let the following partial fraction expansion (3.78) hold

$$W(s) = \sum_{i=1}^{\ell} \sum_{k=1}^{\nu_i} \frac{w_{ik}}{(s - s_i)^k} . \quad (5.67)$$

The set of LTI operators generated by Eq. (5.62) is characterized by

Theorem 5.1 *Let $S_m : (\mu_m, \mu_{m+1})$ be the set of regularity intervals for the function (5.65) and let $g_m(t)$ be the associated originals*

$$g_m(t) = \frac{1}{2\pi \mathrm{j}} \int_{c-\mathrm{j}\infty}^{c+\mathrm{j}\infty} W(s) \mathrm{e}^{st} \, \mathrm{d}s , \quad \mu_m < c < \mu_{m+1} . \quad (5.68)$$

In that case the set of LTI operators U_m generated by equation (system) (5.62) is defined by

$$\mathsf{U}_m [x(t)] = \int_{-\infty}^{\infty} g_m(t - \tau) x(\tau) \, \mathrm{d}\tau \quad (5.69)$$

for all $x(t)$, for which the corresponding integrals in (5.69) converge.

The proof of Theorem 5.1 was given in Rosenwasser (1977). ∎

The function $g_m(t)$ (5.68) will be hereinafter called the *impulse response* of the system, while the function $h_m(t, \tau) \triangleq g_m(t - \tau)$ is its *Green function.* Theorem 5.1 is of prime importance for the further exposition. Note that this theorem yields the following conclusions.

Remark 1 Let μ_m, $(m = 1, \ldots, r)$ be the different real parts of the poles of the transfer function (5.67) and $S_m : (\mu_m, \mu_{m+1})$ be the corresponding regularity intervals. The numbers μ_m will be henceforth called the *characteristic indices* of the system (5.62). The set $\{\mathsf{U}_m\}$ of operators (5.69) will be called the family of linear time-invariant operators generated by the equation (system) (5.62). The whole set $\{\mathsf{U}_m\}$ can be determined by a single analytic function $W(s)$ (5.65) with the help of Eqs. (5.68) and (5.69).

Remark 2 Using the above property of the transfer function (5.65) we can derive a general method for defining the set of LTI operators generated by a LTI system, which consists of networks with real rational transfer functions. With this aim in view, by the well-known methods we should find the transfer function $W_0(s)$ from an arbitrary input $x(t)$ to an arbitrary output $y(t)$. If the function $W_0(s)$ is strictly proper, on this basis we can find the set of characteristic indices $\{\mu_i\}$ and the set of the corresponding operators (5.69).

Remark 3 As was proved above, $g_m(t) \in \Lambda(\mu_m, \mu_{m+1})$. Using this fact, we can prove that in a special case when

$$x(t) = \mathrm{e}^{\lambda t} x_1(t)\,, \quad |x_1(t)| < M = \text{const.}\,, \quad -\infty < t < \infty \tag{5.70}$$

and $\operatorname{Re}\lambda \in S_m$, the integral (5.69) converges. Moreover, the associated particular solution $y(t)$ of Eq. (5.62) can be found in the form

$$y(t) = \mathrm{e}^{\lambda t} y_1(t)\,, \quad |y_1(t)| < M_1 = \text{const.}\,, \quad -\infty < t < \infty\,. \tag{5.71}$$

This solution is unique.

Remark 4 If zero is not in the set of characteristic indices μ_m, $(m = 1, \ldots, r)$, Eq.(5.62) will be called *exponentially dichotomic (e-dichotomic).* For an e-dichotomic system (5.62) there exists an interval $S_q : (\mu_q, \mu_{q+1})$ such that $\mu_q < 0 < \mu_{q+1}$. The impulse response $g_q(t)$ corresponding to the interval S_q will be called *primary impulse response.* The primary impulse response satisfies the estimate

$$|g_q(t)| < N\mathrm{e}^{-\chi|t|} \tag{5.72}$$

with constants $\chi > 0$ and $N > 0$. In this case the function $g_q(t-\tau)$ will be called the *primary Green function.* If Eq. (5.62) is e-dichotomic and $|x(t)| < M$ =const., $-\infty < t < \infty$, the formula

$$y(t) = \int_{-\infty}^{\infty} g_q(t-\tau)x(\tau)\,\mathrm{d}\tau \tag{5.73}$$

defines uniquely a bounded solution of (5.62) such that

$$|y(t)| < M_1 = \text{const.}\,, \quad -\infty < t < \infty\,. \tag{5.74}$$

As a special case, for $x(t) = \mathrm{e}^{\mathrm{j}\nu t}$ with a real ν from (5.73) we find

$$y(t) = \int_{-\infty}^{\infty} g_q(t-\tau)\mathrm{e}^{\mathrm{j}\nu\tau}\,\mathrm{d}\tau = \Phi(\mathrm{j}\nu)\mathrm{e}^{\mathrm{j}\nu t} \tag{5.75}$$

where $\Phi(\mathrm{j}\nu)$ is the frequency response (5.48).

Corollary 5 The impulse response $g_r(t)$ associated with a regularity interval $S_r = S_+ : (\mu_r, \infty)$ will be denoted by $g_+(t)$. According to (1.99), we have

$$g_+(t) = \begin{cases} \displaystyle\sum_{i=1}^{\ell}\sum_{k=1}^{\nu_i} \frac{w_{ik}}{(k-1)!} t^{k-1}\mathrm{e}^{s_i t} & t > 0 \\ 0 & t < 0\,. \end{cases} \tag{5.76}$$

Due to (5.76), the operator U_r obtained from (5.69) for $m = r$ has the form

$$y(t) = \int_{-\infty}^{t} g_+(t-\tau)x(\tau)\,\mathrm{d}\tau \tag{5.77}$$

and is causal, i.e, physically realizable. The function $g_+(t-\tau)$ will be called the *weight function* of the system. The other operators (5.69) are not causal. We note that the weight function $g_+(t-\tau)$ is the primary Green function iff $\mu_r < 0$, i.e., all poles of the transfer function (5.65) are in the left half-plane.

Example 5.6 Build the set of LTI operators generated by the differential equation

$$\frac{\mathrm{d}y}{\mathrm{d}t} + 5y = x \tag{5.78}$$

or, in operator notation,

$$(s+5)y = x\,. \tag{5.79}$$

In this case $W_1(s) = d(s) = s+5$ and $W_2(s) = m(s) = 1$, so that Eq. (5.65) yields the transfer function of the system

$$W(s) = \frac{1}{s+5}\,. \tag{5.80}$$

Next, we find LTI operators defined by Eq. (5.78) (or, by the transfer function (5.80)). In this case we have two regularity intervals $S_- = S_0 : (-\infty, -5)$ and $S_+ = S_1 : (-5, \infty)$. By virtue of (1.99) and (1.100), the corresponding impulse responses are given by

$$g_1(t) = \frac{1}{2\pi\mathrm{j}}\int_{c-\mathrm{j}\infty}^{c+\mathrm{j}\infty} W(s)\mathrm{e}^{st}\,\mathrm{d}s = \begin{cases} \mathrm{e}^{-5t} & t > 0 \\ 0 & t < 0 \end{cases} \qquad c > -5 \tag{5.81}$$

$$g_0(t) = \frac{1}{2\pi\mathrm{j}}\int_{c-\mathrm{j}\infty}^{c+\mathrm{j}\infty} W(s)\mathrm{e}^{st}\,\mathrm{d}s = \begin{cases} 0 & t > 0 \\ -\mathrm{e}^{-5t} & t < 0 \end{cases} \qquad c < -5\,. \tag{5.82}$$

The function $g_1(t-\tau)$ is the weight function of the system (5.78) and, at the same time, its primary Green function. The associated operators (5.69) have the form

$$y(t) = \mathsf{U}_+ [x(t)] = \int_{-\infty}^{t} \mathrm{e}^{-5(t-\tau)} x(\tau) \, \mathrm{d}\tau \tag{5.83}$$

$$y(t) = \mathsf{U}_- [x(t)] = -\int_{t}^{\infty} \mathrm{e}^{-5(t-\tau)} x(\tau) \, \mathrm{d}\tau \tag{5.84}$$

for all $x(t)$, for which the integrals in (5.83) and (5.84) converge. These integrals determine particular solutions of Eq. (5.78). Indeed, Eqs. (5.83) and (5.84) can be represented in the form

$$\begin{aligned} y(t) &= \mathsf{U}_+ [x(t)] = \int_{0}^{t} \mathrm{e}^{-5(t-\tau)} x(\tau) \, \mathrm{d}\tau + \mathrm{e}^{-5t} c_+[x(t)] \\ y(t) &= \mathsf{U}_- [x(t)] = \int_{0}^{t} \mathrm{e}^{-5(t-\tau)} x(\tau) \, \mathrm{d}\tau + \mathrm{e}^{-5t} c_-[x(t)] \end{aligned} \tag{5.85}$$

where $c_+[x(t)]$ and $c_-[x(t)]$ are constants dependent on the input signal

$$c_+[x(t)] = \int_{-\infty}^{0} \mathrm{e}^{5\tau} x(\tau) \, \mathrm{d}\tau \,, \quad c_-[x(t)] = -\int_{0}^{\infty} \mathrm{e}^{5\tau} x(\tau) \, \mathrm{d}\tau \,. \tag{5.86}$$

The operator (5.83) is causal in accordance with the above theory. □

5.8 Generalization to the multivariable case

All the aforesaid can be, without major modifications, extended to systems of the form

$$\boldsymbol{D}(s)\boldsymbol{y} = \boldsymbol{M}(s)\boldsymbol{x} \tag{5.87}$$

where $\boldsymbol{y}$ and $\boldsymbol{x}$ are vectors of dimensions n and m, respectively. Here $\boldsymbol{D}(s)$ and $\boldsymbol{M}(s)$ are polynomial matrices of dimensions $n \times n$ and $n \times m$, respectively,

$$\begin{aligned} \boldsymbol{D}(s) &= \boldsymbol{I}_n s^p + \boldsymbol{D}_1 s^{p-1} + \boldsymbol{D}_2 s^{p-2} + \ldots + \boldsymbol{D}_p \\ \boldsymbol{M}(s) &= \boldsymbol{M}_1 s^{p-1} + \boldsymbol{M}_2 s^{p-2} + \ldots + \boldsymbol{M}_p \end{aligned} \tag{5.88}$$

where $\boldsymbol{D}_i$ and $\boldsymbol{M}_i$ are constant matrices of dimensions $n \times n$ and $n \times m$. Let

$$d(s) \triangleq \det \boldsymbol{D}(s) \tag{5.89}$$

and

$$\boldsymbol{W}(s) = \boldsymbol{D}^{-1}(s)\boldsymbol{M}(s) = \frac{\boldsymbol{N}(s)}{d(s)} \tag{5.90}$$

be valid. If the roots of the equation

$$d(s) = 0 \tag{5.91}$$

are $s_1, s_2, \ldots, s_\ell$, then the ratio (5.90) is called *irreducible* if $\boldsymbol{N}(s_i) \neq \boldsymbol{0}$ is valid for all s_i, where $\boldsymbol{0}$ is the zero matrix of appropriate type. Under the above assumptions the transfer matrix (5.90) is irreducible and strictly proper. Let the roots s_i have the multiplicities ν_i. Then, as in the SISO case, the following partial fraction expansion is possible

$$\boldsymbol{W}(s) = \sum_{i=1}^{\ell} \sum_{k=1}^{\nu_i} \frac{\boldsymbol{W}_{ik}}{(s - s_i)^k} \tag{5.92}$$

where $\boldsymbol{W}_{ik}$ are constant matrices. Moreover, in analogy with the SISO case, we can introduce the set of characteristic indices μ_m, $(m = 1, \ldots, r)$ and the set of the corresponding regularity intervals S_m, $(m = 0, 1, \ldots, r)$. In this case the integrals

$$\boldsymbol{G}_m(t) = \frac{1}{2\pi \mathrm{j}} \int_{c-\mathrm{j}\infty}^{c+\mathrm{j}\infty} \boldsymbol{W}(s) \mathrm{e}^{st} \, \mathrm{d}s, \quad \mu_m < c < \mu_{m+1} \tag{5.93}$$

define the impulse responses of the system (5.87), while the integrals

$$\boldsymbol{y}_m(t) = \mathsf{U}_m [\boldsymbol{x}(t)] \triangleq \int_{-\infty}^{\infty} \boldsymbol{G}_m(t - \tau) \boldsymbol{x}(\tau) \, \mathrm{d}\tau \tag{5.94}$$

define the set of LTI operators generated by the system (5.87).

Chapter 6

Linear periodic operators and systems

6.1 Linear periodic operators

Linear operators, that are not time-invariant and do not satisfy (5.34) at least for a single τ or for a single $x(t)$ will be called *time-variable.* For a time-variable operator the property of invariance does not take place and the response to a shifted signal will not, in the general case, coincide with the shifted response. PTFs of time-variable operators depend explicitely on two arguments, s and t.

Example 6.1 The operator

$$y(t) = \int_0^t x(\tau)\,\mathrm{d}\tau \triangleq \mathsf{U}[x(t)] \tag{6.1}$$

is shown to be time-variable. Using (5.4), we find

$$W(s,t) = \mathrm{e}^{-st} \int_0^t \mathrm{e}^{s\tau}\,\mathrm{d}\tau = \frac{1-\mathrm{e}^{-st}}{s}\,. \tag{6.2}$$

Since $W(s,t)$ depends explicitely on s and t, the operator (6.1) is time-variable. It can be easily verified that the PTF (6.2) is defined for all s and t. □

Example 6.2 The differential operator

$$y(t) = a_0(t)\frac{\mathrm{d}^p x(t)}{\mathrm{d}t^p} + a_1(t)\frac{\mathrm{d}^{p-1} x(t)}{\mathrm{d}t^{p-1}} + \ldots + a_p(t)x(t) \tag{6.3}$$

where $a_i(t)$ are known functions of time, is shown to be time-variable. The operator has the PTF

$$W(s,t) = a_0(t)s^p + a_1(t)s^{p-1} + \ldots + a_p(t) \tag{6.4}$$

which is defined over the whole complex s-plane. □

A special class of linear time-variable operators, called periodic, plays the leading part in the present book.

Definition A linear time-variable operator U (5.1) is called *periodic ($T-$periodic)*, if there exists a real positive number T such that

$$\mathsf{U}_T\mathsf{U} = \mathsf{U}\mathsf{U}_T \tag{6.5}$$

where U_T is the shift operator by T.
Equation (6.5) is equivalent to the proposition that for any $x(t)$ from

$$y(t) = \mathsf{U}\left[x(t)\right] \tag{6.6}$$

it follows that

$$y(t-T) = \mathsf{U}\left[x(t-T)\right] \tag{6.7}$$

i.e., a shift of the input by T causes the same shift of the output. In other words, periodic operators behave like time-invariant ones with respect to a shift by T. The value T appearing in (6.5) will be called a *period* of the operator U. Obviously, the numbers $T_k = kT$ for any integer k are also periods. Henceforth, if it is not specified otherwise, by the period of the operator we mean the least positive T for which (6.5) holds.

Example 6.3 We show that the operator

$$y(t) = \mathsf{U}\left[x(t)\right] = \sum_{i=1}^{\ell} a_i(t)\mathsf{U}_i\left[x(t)\right] \tag{6.8}$$

where $a_i(t) = a_i(t+T)$ are known functions and U_i are LTI operators, is $T-$periodic. Indeed, substituting $t-T$ for T on the right-hand side of (6.8), we have

$$\mathsf{U}\left[x(t-T)\right] = \sum_{i=1}^{\ell} a_i(t-T)\mathsf{U}_i\left[x(t-T)\right] = y(t-T)\,. \tag{6.9}$$

□

Under conditions that are normally valid, the periodicity of the operator can be established on the basis of the properties of its PTF.

Theorem 6.1 *If a $T-$periodic operator U has the PTF $W(s,t)$ in a domain $\mathcal{P}$, then*

$$W(s,t) = W(s,t+T)\,, \quad s \in \mathcal{P}\,. \tag{6.10}$$

Proof Let $x(t) = e^{st}$, $s \in \mathcal{P}$ in (6.6). Then by virtue of (5.5),

$$y(t) = \mathsf{U}\left[e^{st}\right] = e^{st}W(s,t)\,. \tag{6.11}$$

Substituting $t - T$ for t and taking (6.7) into account, we have

$$\mathsf{U}\left[e^{s(t-T)}\right] = e^{s(t-T)}W(s,t-T)\,. \tag{6.12}$$

On the other hand,

$$\mathsf{U}\left[e^{s(t-T)}\right] = e^{-sT}\mathsf{U}\left[e^{st}\right] = e^{s(t-T)}W(s,t)\,. \tag{6.13}$$

Comparing the last two equations, we obtain (6.10). ■

As will be clear from the further exposition, the condition (6.10) is, generally speaking, sufficient for an operator to be periodic.

6.2 Linear periodic systems

A system described by the operator equation

$$\mathsf{U}_1\left[y(t)\right] = \mathsf{U}_2\left[x(t)\right] \tag{6.14}$$

with known $T-$periodic operators U_1, U_2 will be called a linear *periodic ($T-$periodic) system.* If the operators U_1 and U_2 have PTFs $W_i(s,t)$, $(i = 1,2)$ in a common domain $\mathcal{P}$, we can use the following expression for Eq. (6.14)

$$W_1(s,t)y = W_2(s,t)x\,. \tag{6.15}$$

As in the time-invariant case, Eq. (6.15) can generate the set of operators $y(t) = \mathsf{U}\left[x(t)\right]$, including nonlinear ones. Nevertheless, in most practical cases this equation can generate a *family of linear $T-$periodic operators.* The PTF of these operators must satisfy an operator equation derived below.
Let a $T-$periodic operator U be generated by Eq. (6.14) Then, the following identity holds for all admissible $x(t)$

$$\mathsf{U}_1\mathsf{U}\left[x(t)\right] = \mathsf{U}_2\left[x(t)\right]\,. \tag{6.16}$$

Let $x(t) = e^{\lambda t}$, $\lambda \in \mathcal{P}$, and suppose that the operator U also has a PTF for $\lambda \in \mathcal{P}$. Then,

$$\mathsf{U}_2\left[x(t)\right] = e^{\lambda t}W_2(\lambda,t)\,, \quad \mathsf{U}\left[x(t)\right] = e^{\lambda t}W(\lambda,t)\,, \tag{6.17}$$

where $W_2(\lambda,t) = W_2(\lambda,t+T)$ is the PTF of the operator U_2 and $W(\lambda,t)$ is the required PTF of the periodic operator U. With due account for (6.17), Eq.(6.16) yield

$$\mathsf{U}_1\left[e^{\lambda t}W(\lambda,t)\right] = e^{\lambda t}W_2(\lambda,t)\,. \tag{6.18}$$

Equation (6.18) will be henceforth called the Zadeh equation. Thus, the problem of determination of the PTF of the operator became reduced to the determination of solutions of the Zadeh equation having period T with respect to t. In what follows we shall consider the Zadeh equation independently of the properties of the operators generated by the system. In this case a periodic solution $W(\lambda, t)$ of the Zadeh equation will be called the parametric transfer function (PTF) of the system (6.14). Moreover, any periodic operator generated by the system (6.14) will have, in the corresponding domain, a PTF which coincides with the PTF of the system $W(\lambda, t)$.
From the aforesaid follows a general method of determination of the PTF for linear periodic systems described by an equation of the form (6.14), or by a system of such equations, or by a block-diagram. For a chosen input $x(t)$ and output $y(t)$ we assume that

$$x = \mathrm{e}^{\lambda t}, \quad y = \mathrm{e}^{\lambda t} W(\lambda, t), \quad W(\lambda, t) = W(\lambda, t+T) \tag{6.19}$$

where $W(\lambda, t)$ is an unknown function periodic with respect to t, and λ is a complex parameter. If such a solution exists and is unique, the function $W(\lambda, t)$ with $\lambda = s$ is the PTF of the corresponding system from the input $x(t)$ to the output $y(t)$. It should be emphasized that the methods of determination of the PTF of a linear periodic system based on Eq. (6.19) are independent of the properties of the operators generated by the system, and are determined only by Eq. (6.14).

Example 6.4 We construct the Zadeh equation for the PTF of a linear periodic system described by the linear differential equation with $T-$periodic coefficients

$$y^{(p)} + d_1(t) y^{(p-1)} + \ldots + d_p(t) y = m_1(t) x^{(p-1)} + \ldots + d_p(t) x \tag{6.20}$$

where $d_i(t) = d_i(t+T)$ and $m_i(t) = m_i(t+T)$ are known functions. Using (6.3) and (6.4), we can represent Eq. (6.20) in the operator form (6.15), where

$$\begin{aligned} W_1(s,t) &\triangleq d(s,t) = s^n + d_1(t) s^{p-1} + \ldots + d_p(t) \\ W_2(s,t) &\triangleq m(s,t) = m_1(t) s^{p-1} + m_2(t) s^{p-2} + \ldots + m_p(t) \end{aligned} \tag{6.21}$$

are polynomials in s. Direct calculation gives

$$d(s,t)\left[\mathrm{e}^{\lambda t} W(\lambda, t)\right] = \mathrm{e}^{\lambda t} d(s+\lambda, t)\left[W(\lambda, t)\right]. \tag{6.22}$$

From (6.18) and (6.22) we can derive the Zadeh equation for the PTF

$$d(s+\lambda, t)[W(\lambda, t)] = m(\lambda, t). \tag{6.23}$$

In this case the Zadeh equation (6.23) is a linear differential equation with periodic coefficients dependent on the parameter λ. The problem of determining the PTF of the system (6.20) reduces to the determination of the $T-$periodic solution of (6.23). In some exceptional cases it is possible to find the solution of (6.23) in a closed form. □

Example 6.5 By means of the Zadeh equation find the exact formula for the linear periodic system

$$\frac{\mathrm{d}y}{\mathrm{d}t} - \frac{\sin t}{2+\cos t} y = x . \tag{6.24}$$

In this case

$$d(s,t) = s - \frac{\sin t}{2+\cos t}, \quad m(s,t) = 1 . \tag{6.25}$$

Since

$$d(s+\lambda,t) = s + \lambda - \frac{\sin t}{2+\cos t} \tag{6.26}$$

the Zadeh equation (6.23) has the form

$$\frac{\mathrm{d}W}{\mathrm{d}t} + \lambda W - \frac{\sin t}{2+\cos t} W = 1 . \tag{6.27}$$

Except for singular values of λ, a solution of this equation exists, is unique and has the form

$$W(\lambda,t) = \frac{1}{2+\cos t}\left(\frac{2}{\lambda} + \frac{\lambda\cos t + \sin t}{1+\lambda^2}\right) . \tag{6.28}$$

After substituting s for λ, relation (6.28) gives the PTF $W(s,t)$ of the system (6.24), which is defined over the whole complex plane except for the points $s=0$ and $s=\pm\mathrm{j}$. □

6.3 PTF of connections of linear periodic operators

To obtain PTFs of complex systems constructed by connections of initial periodic operators, we consider some formal properties of the simplest connections (5.7) and (5.10), and the feedback connection.

1. For the parallel connection (5.7) and (5.8), the PTF appears as

$$W(s,t) = W_1(s,t) + W_2(s,t) \tag{6.29}$$

where it is assumed that the PTFs are defined in a common domain $\mathcal{P}$. Indeed, taking $x = \mathrm{e}^{st}$, $s \in \mathcal{P}$ in (5.7), we obtain

$$\mathsf{U}\left[\mathrm{e}^{st}\right] = \mathsf{U}_1\left[\mathrm{e}^{st}\right] + \mathsf{U}_2\left[\mathrm{e}^{st}\right] . \tag{6.30}$$

Using (5.5), we obtain after cancellation (6.29).
We note that formula (6.29) does not make sense if there is no common domain $\mathcal{P}$. This formula is also meaningful if one of the operators U_i, for example U_1, is time-invariant. In this case instead of (6.29) we have

$$W(s,t) = W_1(s) + W_2(s,t) \tag{6.31}$$

in the common domain.

2. Consider the series connection (5.10) assuming that both operators U_1 and U_2 are T−periodic. Denote the PTF of the connection by $W_\pi(s,t) = W_\pi(s,t+T)$. According to the general scheme (6.19), we take

$$x = \mathrm{e}^{st}\,, \quad y = W_\pi(s,t)\mathrm{e}^{st}\,, \quad W_\pi(s,t) = W_\pi(s,t+T)\,. \tag{6.32}$$

Assume that all the PTFs $W_1(s,t)$, $W_2(s,t)$ and $W_\pi(s,t)$ are defined for $s \in \mathcal{P}$. As a result, we have

$$z(t) = W_1(s,t)\mathrm{e}^{st}\,. \tag{6.33}$$

Taking into account that $W_1(s,t) = W_1(s,t+T)$, we assume that the following Fourier expansion holds

$$W_1(s,t) = \sum_{k=-\infty}^{\infty} w_{1k}(s)\mathrm{e}^{kj\omega t}\,, \quad \omega = \frac{2\pi}{T} \tag{6.34}$$

where

$$w_{1k}(s) = \frac{1}{T}\int_0^T W_1(s,t)\mathrm{e}^{-kj\omega t}\,\mathrm{d}t\,. \tag{6.35}$$

As follows from (6.34) and (6.33),

$$z(t) = \sum_{k=-\infty}^{\infty} w_{1k}(s)\mathrm{e}^{(s+kj\omega)t}\,. \tag{6.36}$$

Assume that from the condition $s \in \mathcal{P}$ follows that $s + kj\omega \in \mathcal{P}$ for any integer k. Then, Eq. (6.11) gives

$$\mathsf{U}_2\left[\mathrm{e}^{(s+kj\omega)t}\right] = \mathrm{e}^{(s+kj\omega)t}\,W_2(s+kj\omega,t) \tag{6.37}$$

and the response of the operator U_2 to the input (6.36) has the form

$$y(t) = \sum_{k=-\infty}^{\infty} W_2(s+kj\omega,t)w_{1k}(s)\mathrm{e}^{(s+kj\omega)t}\,. \tag{6.38}$$

Comparing the expressions (6.32) and (6.38) for $y(t)$, we find the required PTF

$$W_\pi(s,t) = \sum_{k=-\infty}^{\infty} W_2(s+kj\omega,t)w_{1k}(s)\mathrm{e}^{kj\omega t}\,. \tag{6.39}$$

From (6.39) we can derive an important special case. Let the operator U_1 in the connection (5.10) be time-invariant with the PTF $W_1(s)$. Then, in (6.38) we have

$$w_{10}(s) = W_1(s)\,, \quad w_{1k}(s) = 0\,,\ (k \neq 0) \tag{6.40}$$

and from (6.39) it follows that

$$W_\pi(s,t) = W_2(s,t)W_1(s) \tag{6.41}$$

which is similar to the case with LTI operators.
In many cases Eq. (6.39) can be represented in an integral form. To derive this equation, we substitute the expression (6.35) for the Fourier coefficients into (6.39)

$$W_\pi(s,t) = \frac{1}{T}\sum_{k=-\infty}^{\infty} W_2(s+kj\omega,t)\int_0^T W_1(s,\tau)e^{-kj\omega\tau}\,d\tau\, e^{kj\omega t}\,. \tag{6.42}$$

Assuming that the order of summation and integration can be changed, we obtain

$$W_\pi(s,t) = \int_0^T \mathcal{D}_2(T,s,t,\tau)W_1(s,\tau)\,d\tau \tag{6.43}$$

where

$$\mathcal{D}_2(T,s,t,\tau) \triangleq \frac{1}{T}\sum_{k=-\infty}^{\infty} W_2(s+kj\omega,t)e^{kj\omega(t-\tau)} \tag{6.44}$$

and the series is assumed to converge. It should be noted that, in the general case, the series connection of linear periodic operators is not commutative. This is substantiated by Example 5.1.

Example 6.6 Under the conditions of Example 5.1 the PTF of the operator $\mathsf{U}_{21} = \mathsf{U}_2\mathsf{U}_1$ appears as $W_{21}(s,t) = s\sin\omega t$, while the PTF of the operator $\mathsf{U}_{12} = \mathsf{U}_1\mathsf{U}_2$ has the form $W_{12} = s\sin\omega t + \omega\cos\omega t$. □

Next, we demonstrate the special case when the PTFs of the connections (5.10) and (5.12) coincide.

Lemma 6.1 *Let the operator* U_2 *be* T*−periodic and the operator* U_1 *be time-invariant with a PTF* $W_1(s)$ *such that*

$$W_1(s) = W_1(s+j\omega)\,, \quad \omega = \frac{2\pi}{T}\,. \tag{6.45}$$

Then, the PTFs of the operators (6.10) and (6.12) coincide.

Proof The PTF $W_{21}(s,t)$ of the system (6.10) is $W_{21}(s,t) = W_2(s,t)W_1(s)$. To find the PTF $W_{12}(s,t)$ of the system (5.12) we use Eq. (6.39), which gives

$$W_{12}(s,t) = \sum_{k=-\infty}^{\infty} W_1(s+kj\omega)w_{2k}(s)e^{kj\omega t} \tag{6.46}$$

where

$$w_{2k}(s) = \frac{1}{T}\int_0^T W_2(s,\tau)e^{-kj\omega\tau}\,d\tau\,. \tag{6.47}$$

With due account for (6.45), Eq. (6.46) yields

$$W_{12}(s,t) = W_1(s)\sum_{k=-\infty}^{\infty} w_{2k}(s)e^{kj\omega t} = W_1(s)W_2(s,t)\,. \tag{6.48}$$

■

3. The operator $y(t) = \mathsf{U}_0\,[x(t)]$, corresponding to the relationships

$$y(t) = \mathsf{U}_\rho\,[v(t)]\,, \quad v(t) = x(t) - \mu y(t) \tag{6.49}$$

where μ is a real constant (Fig. 6.1), will be called the *feedback operator.*

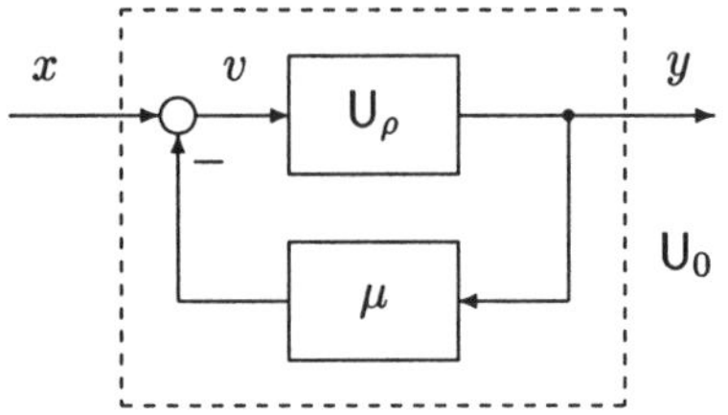

Figure 6.1: Feedback operator

The T−periodic operator U_ρ appearing in (6.49) will be called the *open-loop operator.* To determine the PTF of the operator (6.49) we use the general scheme (6.19) with

$$x = \mathrm{e}^{st}\,, \quad y = W_0(s,t)\mathrm{e}^{st} \quad W_0(s,t) = W_0(s,t+T) \tag{6.50}$$

where $W_0(s,t)$ is the required PTF of the feedback operator. Let

$$W_0(s,t) = \sum_{k=-\infty}^{\infty} w_{0k}(s)\mathrm{e}^{kj\omega t} \tag{6.51}$$

where

$$w_{0k}(s) = \frac{1}{T}\int_0^T W_0(s,\tau)\mathrm{e}^{-kj\omega\tau}\,\mathrm{d}\tau\,. \tag{6.52}$$

From (6.49), (6.50) and (6.51) we have

$$v(t) = \mathrm{e}^{st} - \mu \sum_{k=-\infty}^{\infty} w_{0k}(s)\mathrm{e}^{(s+kj\omega)t}\,. \tag{6.53}$$

Hence, with reference to Fig. 6.1, we obtain

$$y(t) = W_\rho(s,t)\mathrm{e}^{st} - \mu \sum_{k=-\infty}^{\infty} W_\rho(s+kj\omega,t)w_{0k}(s)\mathrm{e}^{(s+kj\omega)t} \tag{6.54}$$

where $W_\rho(s,t)$ is the PTF of the operator U_ρ. Comparing the expressions (6.50) and (6.54) for $y(t)$, we derive the functional equation

$$W_0(s,t) = W_\rho(s,t) - \mu \sum_{k=-\infty}^{\infty} W_\rho(s+kj\omega,t)w_{0k}(s)\mathrm{e}^{kj\omega t}\,. \tag{6.55}$$

Using (6.43) and (6.44), this equation can be written in the form of an integral equation with respect to the PTF $W_0(s,t)$

$$W_0(s,t) + \mu \int_0^T \mathcal{D}_\rho(T,s,t,\tau) W_0(s,\tau)\,\mathrm{d}\tau = W_\rho(s,t) \tag{6.56}$$

where the function $\mathcal{D}_\rho(T,s,t,\tau)$ can be calculated by (6.44).
A detailed investigation of integral equations of the form (6.56) for linear continuous systems with periodically varying parameters is given in Rosenwasser (1970, 1973).

6.4 Parametric frequency response

If the PTF $W(s,t)$ of the operator U is defined on the imaginary axis, the function

$$\Phi(\mathrm{j}\nu,t) \triangleq W(s,t)\,|_{s=\mathrm{j}\nu} \tag{6.57}$$

with the real argument ν will be called, by analogy with the time-invariant case, the *parametric frequency response* of the operator U. Separating the real and imaginary parts in (6.57), we have

$$\Phi(\mathrm{j}\nu,t) = P(\nu,t) + \mathrm{j}Q(\nu,t)\,. \tag{6.58}$$

The functions $P(\nu,t)$ and $Q(\nu,t)$ will be called parametric real and imaginary frequency responses, respectively. Moreover, the functions

$$A(\nu,t) \triangleq \sqrt{P^2(\nu,t)+Q^2(\nu,t)}\,, \quad \psi(\nu,t) \triangleq \arctan\frac{Q(\nu,t)}{P(\nu,t)} \tag{6.59}$$

can naturally be called the parametric amplitude and phase frequency responses, respectively. As follows from (6.58) and (6.59),

$$\Phi(\mathrm{j}\nu,t) = A(\nu,t)\mathrm{e}^{\mathrm{j}\psi(\nu,t)}\,. \tag{6.60}$$

If the operator U is periodic, all of these frequency responses are periodic with respect to t. The use of the parametric frequency responses makes it possible to calculate the response of a linear periodic operator to a harmonic input. Let the PTF from the input $x(t)$ to the output $y(t)$ be equal to $W(s,t)$ and be defined over the imaginary axis. Then,

$$\mathsf{U}\left[\mathrm{e}^{\mathrm{j}\nu t}\right] = \Phi(\mathrm{j}\nu,t)\mathrm{e}^{\mathrm{j}\nu t} \tag{6.61}$$

where U is the corresponding periodic operator. Taking into account that $\mathrm{e}^{\mathrm{j}\nu t} = \cos\nu t + \mathrm{j}\sin\nu t$ and using (6.58), we can separate the real and imaginary parts

$$\begin{aligned} \mathsf{U}[\cos\nu t] &= P(\nu,t)\cos\nu t - Q(\nu,t)\sin\nu t \\ \mathsf{U}[\sin\nu t] &= P(\nu,t)\sin\nu t + Q(\nu,t)\cos\nu t\,. \end{aligned} \tag{6.62}$$

If the input signal has the form $x(t) = x_0\mathrm{e}^{\mathrm{j}\nu t}$ with a constant x_0, by (6.61) the output is

$$y(t) = \Phi(\mathrm{j}\nu,t)\mathrm{e}^{\mathrm{j}\nu t}x_0\,. \tag{6.63}$$

In this case, the value

$$\overline{E} \triangleq \lim_{a\to\infty} \frac{1}{2a} \int_{-a}^{a} |y(t)|^2 \, dt \tag{6.64}$$

defines the average power of the output, Perry et al. (1991). Substituting (6.63) into (6.64), we find

$$\overline{E} = |x_0|^2 B(\nu) \tag{6.65}$$

where

$$B(\nu) \triangleq \lim_{a\to\infty} \frac{1}{2a} \int_{-a}^{a} |\Phi(\mathrm{j}\nu, t)|^2 \, dt\,. \tag{6.66}$$

Since $\Phi(\mathrm{j}\nu, t) = \Phi(\mathrm{j}\nu, t+T)$, Eq. (6.66) can be written in a simpler form

$$B(\nu) = \frac{1}{T} \int_{0}^{T} |\Phi(\mathrm{j}\nu, t)|^2 \, dt\,. \tag{6.67}$$

Note that for an LTI operator $\Phi(\mathrm{j}\nu, t) = \Phi(\mathrm{j}\nu)$ is valid. In this case Eq. (6.67) yields

$$B(\nu) = |\Phi(\mathrm{j}\nu)|^2 = A^2(\nu) \tag{6.68}$$

where $A(\nu)$ is the amplitude frequency response.
The notion of the parametric frequency response can be generalized onto the system (6.14). If the PTF of this system $W(s,t)$ satisfies (6.19), the parametric frequency response is also defined as the function $\Phi(\mathrm{j}\nu, t)$ given by (6.57). In this case the function $\Phi(\mathrm{j}\nu, t)$ may have singular points for some ν.

6.5 Periodic integral operators

An important class of linear operators with wide practical applications is presented by integral operators of the form

$$y(t) = \mathsf{U}[x(t)] = \int_{-\infty}^{\infty} h(t,\tau) x(\tau)\, d\tau \tag{6.69}$$

where $h(t,\tau)$ is a known function, which will be called the *Green function of the integral operators.* If

$$h(t,\tau) = 0 \quad \text{for} \quad t < \tau \tag{6.70}$$

the operator (6.69) has the form

$$y(t) = \int_{-\infty}^{t} h(t,\tau) x(\tau)\, d\tau \tag{6.71}$$

and is, therefore, *causal.* With the change of variable $u = t - \tau$ we can represent the operator in the form

$$y(t) = \int_{-\infty}^{\infty} g(t,u)x(t-u)\,\mathrm{d}u \tag{6.72}$$

where

$$g(t,u) \triangleq h(t,\tau)\,|_{\tau=t-u}\ . \tag{6.73}$$

The inverse relation is also valid

$$h(t,\tau) = g(t,u)\,|_{u=t-\tau}\ . \tag{6.74}$$

The function $g(t,u)$ will henceforth be called the *impulse response* of the operator (6.69). The causality condition (6.70) can be expressed in terms of the impulse response $g(t,u)$ as follows

$$g(t,u) = 0 \quad \text{for} \quad u < 0\,. \tag{6.75}$$

If (6.75) holds, the operator (6.72) takes the form

$$y(t) = \int_{0}^{\infty} g(t,u)x(t-u)\,\mathrm{d}u\,. \tag{6.76}$$

Formally, from (6.69) it follows that

$$\mathsf{U}[\mathrm{e}^{st}] = \int_{-\infty}^{\infty} h(t,\tau)\mathrm{e}^{s\tau}\,\mathrm{d}\tau\,. \tag{6.77}$$

Therefore, using (5.4), we obtain an expression for the PTF of the integral operator

$$W(s,t) = \int_{-\infty}^{\infty} h(t,\tau)\mathrm{e}^{-s(t-\tau)}\,\mathrm{d}\tau\,. \tag{6.78}$$

Assuming $u = t - \tau$ in (6.78), we find

$$W(s,t) = \int_{-\infty}^{\infty} g(t,u)\mathrm{e}^{-su}\,\mathrm{d}u\,. \tag{6.79}$$

Hence, it follows that, in the general case, the PTF $W(s,t)$ is the Laplace transform of the impulse response $g(t,u)$ with respect to u. If the causality condition holds, from (6.79) we obtain the unilateral transform

$$W(s,t) = \int_{0}^{\infty} g(t,u)\mathrm{e}^{-su}\,\mathrm{d}u\,. \tag{6.80}$$

Next, we formulate the conditions under which Eqs. (6.78)–(6.80) are correct from the mathematical point of view. Before that we introduce some definitions.

We shall say that a function of two variables $f(t,u)$ is *complete continuous* inside the square $a \le t,\ u \le b$ with constants $a < b$, if for a given $\epsilon > 0$ there exists a $\delta > 0$ such that from $|t_1 - t_2| < \delta$ and $a \le t_1,\ t_2 \le b$ it follows that

$$\int_a^b |f(t_1,u) - f(t_2,u)|\,\mathrm{d}u < \epsilon\,. \tag{6.81}$$

As was shown in Michlin (1953), for a function $f(t,u)$ to be completely continuous it is sufficient that the following conditions hold.

A) We have

$$\int_a^b |f(t,u)|^2\,\mathrm{d}u < K = \text{const.}\,, \quad a \le t \le b\,. \tag{6.82}$$

B) The function $f(t,u)$ has in the square $a \le t,\ u \le b$ discontinuities only along a finite number of curves which can be described by equations of the form $u = g_k(t)$ with continuous functions $g_k(t)$.

Theorem 6.2 *Let the following conditions hold.*

i) *The function $g(t,u)$ is complete continuous in any square $a \le t,\ u \le b$, where a and b are finite numbers.*
ii) *For any fixed $\tilde{t}$ $(a \le \tilde{t} \le b)$ the function $g(\tilde{t},u)$ has finite variation on the interval $a \le u \le b$.*
iii) *The function $g(t,u)$ satisfies for any s with $\alpha < \operatorname{Re} s < \beta$ the estimate*

$$\left|g(t,u)\mathrm{e}^{-su}\right| < M\mathrm{e}^{-\gamma|u|} \tag{6.83}$$

where M and γ are positive constants.

Then, the following propositions hold:

1. *The integral (6.79) converges absolutely and uniformly with respect to s and t.*
2. *For $\alpha < \operatorname{Re} s < \beta$ the function $W(s,t)$ is analytic with respect to s and continuous with respect to t.*
3. *In any stripe $\alpha < \alpha' \le \operatorname{Re} s \le \beta' < \beta$ for $|s| \to \infty$ and any fixed t the following estimate holds*

$$|W(s,t)| < L|s|^{-1} \tag{6.84}$$

where $L > 0$ is a constant.

4. *For any t the following inversion formula holds*

$$g(t,u) = \frac{1}{2\pi \mathrm{j}} \int_{c-\mathrm{j}\infty}^{c+\mathrm{j}\infty} W(s,t)\mathrm{e}^{su}\,\mathrm{d}s \quad \alpha < \operatorname{Re} s < \beta \tag{6.85}$$

where the integral converges at least in the sense of the principal value.

5. *The Green function $h(t,\tau)$ is defined by the integral*

$$h(t,\tau) = \frac{1}{2\pi \mathrm{j}} \int_{c-\mathrm{j}\infty}^{c+\mathrm{j}\infty} W(s,t)\mathrm{e}^{s(t-\tau)}\,\mathrm{d}s\,, \quad \alpha < \operatorname{Re} s < \beta\,. \tag{6.86}$$

Proof 1. The proposition on the convergence follows from the estimate (6.83) and Section 1.2.

2. The analyticity with respect to s follows from Section 1.6. Let us prove that $W(s,t)$ is continuous with respect to t for $\alpha < \alpha' \leq \operatorname{Re} s \leq \beta' < \beta$. Using (6.80), we find

$$|W(s,t_1) - W(s,t_2)| \leq \int_{-\infty}^{\infty} |g(t_1,u) - g(t_2,u)| \, \mathrm{e}^{-\operatorname{Re} su} \, \mathrm{d}u \,. \tag{6.87}$$

Due to the estimate (6.83) there exists an $a > 0$ such that for any $\epsilon > 0$ we have

$$\int_{a}^{\infty} |g(t,u)| \, \mathrm{e}^{-\operatorname{Re} su} \, \mathrm{d}u < \frac{\epsilon}{6}\,, \qquad \int_{-\infty}^{-a} |g(t,u)| \, \mathrm{e}^{-\operatorname{Re} su} \, \mathrm{d}u < \frac{\epsilon}{6} \tag{6.88}$$

Hence, from (6.87) it follows that

$$|W(s,t_1) - W(s,t_2)| \leq \int_{-a}^{a} |g(t_1,u) - g(t_2,u)| \, \mathrm{e}^{-\operatorname{Re} su} \, \mathrm{d}u + \frac{4\epsilon}{6} \,. \tag{6.89}$$

The integral at the right-hand side of (6.89) is a continuous function in t_1 and t_2, as follows from Assumption i). In other words, for any given $\epsilon/3$ we can find a $\delta > 0$ such that for $|t_1 - t_2| < \delta$ we obtain

$$\int_{-a}^{a} |g(t_1,u) - g(t_2,u)| \mathrm{e}^{-\operatorname{Re} su} \, \mathrm{d}u < \frac{\epsilon}{3} \,. \tag{6.90}$$

Therefore, for any $\epsilon > 0$ there is an $\delta > 0$ such that for $|t_1 - t_2| < \delta$ we have $|W(s,t_1) - W(s,t_2)| < \epsilon$, i.e., the PTF $W(s,t)$ is continuous with respect to t.

3. The estimate (6.84) follows immediately from the estimate (1.45).

4. For a fixed t Eq. (1.24) follows from Eq. (6.85).

5. Equation (6.86) is obtained from (6.85) and (6.74) after the substitution $t - \tau$ for u. ∎

Moreover, the following propositions, which are, in a certain sense, the inverse of Theorem 6.2, also hold.

Theorem 6.3 *Let us have a function $W(s,t)$ continuous with respect to t and analytic with respect to s for $\alpha < \operatorname{Re} s < \beta$. Assume that $W(s,t)$ satisfies the conditions of Theorem 1.4 or 1.5 and the function $g(t,u)$ is defined by Eq. (6.85). Moreover, let $g(t,u)$ satisfy the conditions* i) – iii) *of Theorem 6.2. Then, $W(s,t)$ is the PTF of the integral operator (6.69), the Green function $h(t,\tau)$ of which is given by (6.74). If, in this case, $\beta = \infty$, the condition (6.75) holds and the corresponding operator has the form (6.71).*

Henceforth, the set of linear periodic integral operator satisfying the conditions of Theorem 6.1, will be called *regular* in the interval (α,β) and denoted by $\mathcal{U}(\alpha,\beta)$. Hence, for a time-invariant integral operator (5.41) we have

$\mathsf{U} \in \mathcal{U}(\alpha, \beta)$, if $g(t) \in \Lambda(\alpha, \beta)$. An integral operator $\mathsf{U} \in \mathcal{U}(-\lambda, \lambda)$ with a positive λ will be called *exponentially-dichotomic (e-dichotomic)*. As follows from (6.83), the impulse response of an e-dichotomic operator satisfies the estimate

$$|g(t, u)| < M \, \mathrm{e}^{-\gamma|u|} \tag{6.91}$$

where M and γ are positive constants.
We note that a generalization of the results obtained in this section is given in Rosenwasser (1977).
Conditions for the periodicity of the integral operator (6.69) are given by

Theorem 6.4 *For the operator (6.69) to be T-periodic it is sufficient that*

$$h(t, \tau) = h(t + T, \tau + T) \tag{6.92}$$

or, equivalently,

$$g(t + T, u) = g(t, u) \, . \tag{6.93}$$

Proof Let (6.69) be valid. Then,

$$y(t - T) = \int_{-\infty}^{\infty} h(t - T, \tau) x(\tau) \, \mathrm{d}\tau \, . \tag{6.94}$$

On the other hand,

$$\mathsf{U}\left[x(t - T)\right] = \int_{-\infty}^{\infty} h(t, \tau) x(\tau - T) \, \mathrm{d}\tau = \int_{-\infty}^{\infty} h(t, \tau + T) x(\tau) \, \mathrm{d}\tau \, . \tag{6.95}$$

For the operator (6.69) to be $T-$periodic, it is sufficient that the right-hand sides of (6.94) and (6.95) coincide for any input $x(t)$. This leads to $h(t-T, \tau) = h(t, \tau + T)$, which is equivalent to (6.92). Relation (6.93) follows from (6.92) by virtue of (6.73). ■

6.6 Linear periodic pulse operators (LPPO)

In this section we consider a general class of linear periodic operators which cannot be reduced to integral ones. These operators are important for the description of processes in sampled-data (SD) systems in continuous time.

1. Let $x(t)$ be a function of finite variation defined for $-\infty < t < \infty$. Let $T > 0$ be an arbitrary real number. In this case the values $x_k = x(kT + 0)$ for any k define the numerical sequence x_k. We shall say that a *linear periodic pulse operator* (LPPO)

$$y(t) = \mathsf{U}_d \left[x(t)\right] \tag{6.96}$$

is defined, if the following formula holds

$$y(t) = \sum_{k=-\infty}^{\infty} q(t - kT)x_k \tag{6.97}$$

where $q(t)$ is a known function called the *pulse response* of the LPPO. In the following, we assume that the function $q(t)$ is real and has bounded variation. The form (6.97) shows the character of processes in actual SD control systems under exogenous disturbances acting upon sampling units. The output of such systems is affected only by the input values at the sampling points $t_k = kT$, though the output is changing in continuous time.

2. The LPPO (6.97) is called *causal (or physically realizable)* if

$$q(t) = 0 \quad \text{for } t < 0 . \tag{6.98}$$

For a causal LPPO Eq. (6.97) takes the form

$$y(t) = \sum_{k=-\infty}^{m} q(t - kT)x_k , \quad mT < t < (m+1)T . \tag{6.99}$$

3. The LPPO (6.97) will be called *regular* in the interval (α, β), if $q(t) \in \Lambda(\alpha, \beta)$. The set of LPPOs that are regular in the interval (α, β) will be henceforth denoted by $\mathcal{U}_d(\alpha, \beta)$. If $\mathsf{U}_d \in \mathcal{U}_d(\alpha, \beta)$ there exists the transform

$$Q(s) = \int_{-\infty}^{\infty} q(t)\mathrm{e}^{-st}\,\mathrm{d}t , \quad \alpha < \operatorname{Re} s < \beta \tag{6.100}$$

which will be called the *equivalent transfer function* (ETF) of the LPPO (6.97). Under the above assumptions Eq. (6.100) gives

$$q(t) = \frac{1}{2\pi \mathrm{j}} \int_{c-\mathrm{j}\infty}^{c+\mathrm{j}\infty} Q(s)\mathrm{e}^{st}\,\mathrm{d}s , \quad \alpha < c < \beta . \tag{6.101}$$

If $q(t) \in \Lambda_+(\alpha, \infty)$, we have

$$Q(s) = \int_{0}^{\infty} q(t)\mathrm{e}^{-st}\,\mathrm{d}t , \quad \operatorname{Re} s > \alpha . \tag{6.102}$$

4. Next, we calculate the PTF $W_d(s,t)$ of the operator $\mathsf{U}_d \in \mathcal{U}_d(\alpha, \beta)$. Using the general definition of the PTF (5.4), we find

$$W_d(s,t) = \mathsf{U}_d\left[\mathrm{e}^{st}\right]\mathrm{e}^{-st} . \tag{6.103}$$

Taking into account the explicit expression for the operator U_d, from (6.97) and (6.103) we obtain

$$W_d(s,t) = \sum_{k=-\infty}^{\infty} q(t + kT)\mathrm{e}^{-s(t+kT)} . \tag{6.104}$$

Using the notation (3.1), we obtain

$$W_d(s,t) = \varphi_q(T,s,t) \tag{6.105}$$

i.e., the PTF $W_d(s,t)$ is equal to the DPFR of the impulse response $q(t)$. Henceforth, it is assumed that the right-hand side of (6.105) is a function of bounded variation with respect to t. Representing the right-hand side of (6.104) in the form of a Fourier series, at regular points we have

$$W_d(s,t) = \varphi_Q(T,s,t) \tag{6.106}$$

where

$$\varphi_Q(T,s,t) = \frac{1}{T}\sum_{k=-\infty}^{\infty} Q(s+kj\omega)\mathrm{e}^{kj\omega t}\,. \tag{6.107}$$

We note that under some assumptions that are commonly valid the DPFR of the form (6.107) can be associated with a set of regular LPPOs (6.97). Indeed, for an image $Q(s)$ let be the original (6.101) in the stripe $\alpha < \operatorname{Re} s < \beta$, such that $q(t) \in \Lambda(\alpha,\beta)$. Then the series (6.107) can be uniquely associated with the LPPO (6.97). Usually the transform $Q(s)$ allows analytical continuation outside the stripe $\alpha < \operatorname{Re} s < \beta$ and is, in some stripes $\alpha_i < \operatorname{Re} s < \beta_i$, the image of functions $q_i(t) \in \Lambda(\alpha_i,\beta_i)$. Each of these functions generates an LPPO U_{d_i}. These LPPOs $\{\mathsf{U}_{d_i}\}$ taken in the aggregate are called the *family of LPPO* generated by the function $Q(s)$.

5. Besides the PTF $W_d(T,s,t)$, to describe an LPPO (6.97) we can use the function

$$D_d(s,t) \triangleq W_d(s,t)\mathrm{e}^{st} \tag{6.108}$$

which will be called the *discrete transfer function* (DTF) of the operator U_d. As follows from (6.108) and (6.103),

$$D_d(s,t) = \mathsf{U}_d\left[\mathrm{e}^{st}\right]\,. \tag{6.109}$$

Using (6.108) and (6.104), we have

$$D_d(s,t) = \sum_{k=-\infty}^{\infty} q(t+kT)\mathrm{e}^{-skT} = \mathcal{D}_q(T,s,t) \tag{6.110}$$

i.e., the discrete transfer function (DTF) $D_d(s,t)$ is equal to the discrete Laplace transform of the impulse response $q(t)$. The DTF $D_d(s,t)$ can be expressed via the Laplace transform $Q(s)$ by the series

$$D_d(s,t) = \frac{1}{T}\sum_{k=-\infty}^{\infty} Q(s+kj\omega)\mathrm{e}^{(s+kj\omega)t} = \mathcal{D}_Q(T,s,t)\,. \tag{6.111}$$

All the relations derived in this section can, without modification, be extended to the case of multivariable LPPO, where $x(t)$ and $y(t)$ in (6.97) are vectors of dimensions $m \times 1$ and $n \times 1$, respectively, and $q(t)$ is an $n \times m$ matrix, Rosenwasser (1996a).

6.7 LPPO with phase shift

We shall say that the relation

$$y(t) = \sum_{k=-\infty}^{\infty} q(t - kT - \phi)x(kT + \phi + 0) \triangleq \mathsf{U}_d^{\phi}[x(t)] \tag{6.112}$$

where ϕ is a known constant, defines an *LPPO with phase shift* by ϕ with respect to the LPPO (6.97). For $\phi = 0$ the LPPO with phase shift reduces to (6.97). We show that, without loss of generality, we can assume that $0 \le \phi < T$ in (6.112). Indeed, any ϕ can be represented in the form $\phi = \tilde{\phi} + mT$ with integer m and $0 \le \tilde{\phi} < T$. Then, from (6.112) we obtain

$$y(t) = \sum_{k=-\infty}^{\infty} q(t - kT - mT - \tilde{\phi})x(kT + mT + \tilde{\phi} + 0)\,.$$

Taking $k + m = n$, we find

$$y(t) = \sum_{n=-\infty}^{\infty} q(t - nT - \tilde{\phi})x(nT + \tilde{\phi} + 0)\,.$$

Because of the aforesaid, we henceforth assume $0 \le \phi < T$ in (6.112).
Next, we find the PTF $W_d^{\phi}(s,t)$ of the operator (6.112), by using the general formula (5.4). Obviously,

$$\mathsf{U}_d^{\phi}\left[\mathrm{e}^{st}\right] = \sum_{k=-\infty}^{\infty} q(t - kT - \phi)\mathrm{e}^{s(kT+\phi)}\,.$$

Hence,

$$W_d^{\phi}(s,t) = \mathsf{U}_d^{\phi}\left[\mathrm{e}^{st}\right]\mathrm{e}^{-st} = \sum_{k=-\infty}^{\infty} q(t - kT - \phi)\mathrm{e}^{-s(t-kT-\phi)}\,. \tag{6.113}$$

Using (3.1), from (6.113) we obtain

$$W_d^{\phi}(s,t) = \varphi_q(T, s, t - \phi)\,. \tag{6.114}$$

In accordance with (6.107), the PTF (6.114) can be associated with the series

$$W_d^{\phi}(s,t) = \frac{1}{T}\sum_{k=-\infty}^{\infty} Q(s + kj\omega)\mathrm{e}^{kj\omega(t-\phi)}\,. \tag{6.115}$$

6.8 LPPO connected with LTI operators

While investigating SD systems it is also necessary to analyse connections of LPPOs (6.97) with time-invariant integral operators of the form (5.41). In this section we consider the most important kinds of such connections.

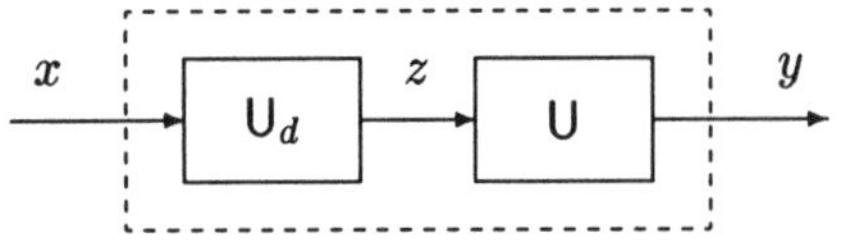

Figure 6.2: Series connection of operators

1. Consider the series connection of an LPPO and a LTI operator U shown in Fig. 6.2. The equations of the obtained operator have the form

$$z(t) = \mathsf{U}_d[x(t)]\,, \quad y(t) = \mathsf{U}[z(t)] = \int_{-\infty}^{\infty} g(t-\tau)z(\tau)\,\mathrm{d}\tau\,. \tag{6.116}$$

Let $g(t) \in \Lambda(\alpha,\beta)$ and $\mathsf{U}_d \in \mathcal{U}_d(\alpha,\beta)$. The equations (6.116) taken in the aggregate define a linear periodic operator $y(t) = \mathsf{U}_1[x(t)]$. Its PTF can be found with the help of Eq. (5.4), assuming that $x = \mathrm{e}^{st}$ in $\alpha < \mathrm{Re}\, s < \beta$. Then, the representation of $z(t)$ in terms of the ETF (6.110) has the form

$$z(t) = \frac{1}{T}\sum_{k=-\infty}^{\infty} Q(s+k\mathrm{j}\omega)\mathrm{e}^{(s+k\mathrm{j}\omega)t}\,. \tag{6.117}$$

Obviously,

$$\mathsf{U}\left[\mathrm{e}^{(s+k\mathrm{j}\omega)t}\right] = G(s+k\mathrm{j}\omega)\mathrm{e}^{(s+k\mathrm{j}\omega)t} \tag{6.118}$$

where

$$G(s) = \int_{-\infty}^{\infty} g(t)\mathrm{e}^{-st}\,\mathrm{d}t\,, \quad \alpha < \mathrm{Re}\, s < \beta \tag{6.119}$$

is the transfer function of the operator U. Therefore, determining the response of the continuous element to each term of the series (6.117), we find

$$y(t) = \mathrm{e}^{st}\frac{1}{T}\sum_{k=-\infty}^{\infty} Q_1(s+k\mathrm{j}\omega)\mathrm{e}^{k\mathrm{j}\omega t} \tag{6.120}$$

with

$$Q_1(s) = Q(s)G(s)\,. \tag{6.121}$$

Multiplying (6.120) by e^{-st}, we obtain the required PTF

$$W_1(s,t) = \frac{1}{T}\sum_{k=-\infty}^{\infty} Q_1(s+k\mathrm{j}\omega)\mathrm{e}^{k\mathrm{j}\omega t} \tag{6.122}$$

whence follows that the operator U_1 is a LPPO with the equivalent transfer function (6.121). As follows from (6.121), the impulse response $q_1(t)$ of the LPPO

$$\mathsf{U}_1 \triangleq \mathsf{U}_{d1}$$

is given by

$$q_1(t) = \int_{-\infty}^{\infty} g(t-\tau)q(\tau)\,\mathrm{d}\tau\,. \tag{6.123}$$

From the properties of the convolution defined above it follows that, under the given assumptions, $q_1(t) \in \Lambda(\alpha,\beta)$, and therefore $\mathsf{U}_{d1} \in \mathcal{U}_d(\alpha,\beta)$.

2. Consider the series connection of an integral operator and an LPPO shown in Fig. 6.3. Suppose that the assumptions made at the beginning of this section

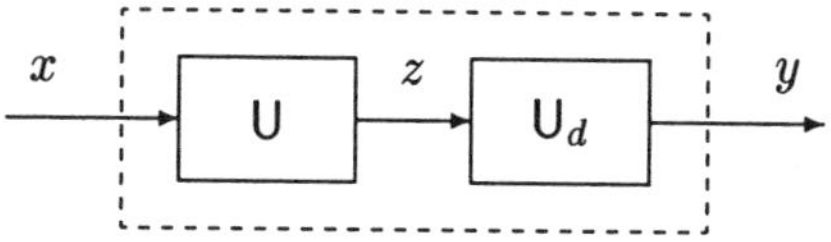

Figure 6.3: Series connection of LTI operator and LPPO

hold. The operator $y(t) = \mathsf{U}_2[x(t)]$ corresponding to Fig. 6.3 will be henceforth called *continuous-discrete.* Using (6.41) and (6.105), we find the PTF $W_2(s,t)$ of the operator U_2 as follows

$$W_2(s,t) = G(s)\varphi_q(T,s,t)\,. \tag{6.124}$$

Let us show that, under fairly general assumptions, the operator under consideration reduces to an integral operator. Indeed, in this case we have

$$z(t) = \int_{-\infty}^{\infty} g(t-\tau)x(\tau)\,\mathrm{d}\tau\,. \tag{6.125}$$

For $x(t) \in \Lambda(\alpha,\beta)$ we have $z(t) \in \Lambda(\alpha,\beta)$ and the function $z(t)$ is continuous due to the properties of the convolution. Therefore, for any integer k we have

$$z_k \stackrel{\triangle}{=} z(kT) = \int_{-\infty}^{\infty} g(kT-\tau)x(\tau)\,\mathrm{d}\tau\,. \tag{6.126}$$

Using (6.126) and (6.97), we find

$$y(t) = \sum_{k=-\infty}^{\infty} q(t-kT)\int_{-\infty}^{\infty} g(kT-\tau)x(\tau)\,\mathrm{d}\tau\,. \tag{6.127}$$

Assuming that the order of summation and integration can be changed in (6.127), we obtain the integral operator

$$y(t) = \int_{-\infty}^{\infty} h_2(t,\tau)x(\tau)\,\mathrm{d}\tau \tag{6.128}$$

with the Green function

$$h_2(t,\tau) = \sum_{k=-\infty}^{\infty} q(t-kT)g(kT-\tau)\,. \tag{6.129}$$

Assuming that $\tau = t-u$, we find the corresponding impulse response

$$g_2(t,u) = \sum_{k=-\infty}^{\infty} q(t-kT)g(kT-t+u)\,. \tag{6.130}$$

It can be immediately verified that

$$g_2(t+T,u) = g_2(t,u) \tag{6.131}$$

i.e., the operator (6.128) is periodic.

Theorem 6.5 *If $q(t) \in \Lambda(\alpha,\beta)$ and $g(t) \in \Lambda(\alpha,\beta)$, for any s in the stripe $\alpha < \operatorname{Re} s < \beta$ the following estimate holds*

$$\left|g_2(t,u)\mathrm{e}^{-su}\right| < M \tag{6.132}$$

where M is a constant independent of t.

Proof From (6.130) we find

$$g_2(t,u)\mathrm{e}^{-su} = \sum_{k=-\infty}^{\infty} \left[q(t+kT)\mathrm{e}^{-s(t-kT)}\right]\left[g(kT-t+u)\mathrm{e}^{-s(kT-t+u)}\right]. \tag{6.133}$$

From Theorem 1.2 it follows that

$$\begin{aligned} \left|q(t+kT)\mathrm{e}^{-s(t-kT)}\right| &< M_1\mathrm{e}^{-\delta|t-kT|} \\ \left|g(kT-t+u)\mathrm{e}^{-s(kT-t+u)}\right| &< N_1\mathrm{e}^{-\gamma|kT-t+u|} \le N_1 \end{aligned} \tag{6.134}$$

where M_1, N_1, δ and γ are positive constants. Using (6.133) and (6.134), we find

$$\left|g_2(t,u)\mathrm{e}^{-su}\right| < M_1N_1 \sum_{k=-\infty}^{\infty} \mathrm{e}^{-\delta|t-kT|}. \tag{6.135}$$

Consider the function

$$\chi(t) \triangleq \sum_{k=-\infty}^{\infty} \mathrm{e}^{-\delta|t-kT|} \tag{6.136}$$

which is continuous and T-periodic, i.e., $\chi(t) = \chi(t+T)$. Therefore, it suffices to investigate the properties of the function $\chi(t)$ in the interval $0 \le t < T$. For $0 \le t < T$ we have

$$\begin{aligned} |t-kT| &= -t+kT \quad \text{for} \quad k>0 \\ |t-kT| &= t-kT \quad \text{for} \quad k \le 0\,. \end{aligned}$$

Hence,

$$\chi(t) = \sum_{k=-\infty}^{0} \mathrm{e}^{-\delta(t-kT)} + \sum_{k=1}^{\infty} \mathrm{e}^{-\delta(-t+kT)}$$

with the estimate

$$|\chi(t)| < \frac{2}{1-\mathrm{e}^{-\delta T}}\,. \tag{6.137}$$

From (6.135) and (6.137) we obtain the estimate (6.132). ∎

As follows from Theorem 6.5, the estimate (6.83) holds for the impulse response $g(t, u)$. In applications considered in the present book it is taken to be true that the impulse response (6.130) satisfies all assumptions of Theorem 6.2. Therefore, the operator (6.128) possesses all the properties specified in this theorem. It can be proved that for $x(t) \in \Lambda(\alpha, \beta)$ the change of the order of summation and integration in (6.127) is legal. The validity of such transformation can be substantiated for all applications investigated below.

Chapter 7

Analysis of linear periodic operators and systems

7.1 Analysis by Laplace transformed inputs

Let us have a linear operator

$$y(t) = \mathsf{U}[x(t)] \tag{7.1}$$

with the PTF

$$W(s,t) = \mathsf{U}\left[\mathrm{e}^{st}\right]\mathrm{e}^{-st} \tag{7.2}$$

defined in the stripe $\alpha < \operatorname{Re} s < \beta$ or in the half-plane $\operatorname{Re} s > \alpha$ (in this case we take $\beta = \infty$). Assume that $x(t) \in \Lambda(\alpha, \beta)$ and

$$X(s) = \int_{-\infty}^{\infty} x(t)\mathrm{e}^{-st}\,\mathrm{d}t\,, \quad \alpha < \operatorname{Re} s < \beta\,. \tag{7.3}$$

In the given case the inversion formula (1.24) holds

$$x(t) = \frac{1}{2\pi\mathrm{j}}\int_{c-\mathrm{j}\infty}^{c+\mathrm{j}\infty} X(s)\mathrm{e}^{st}\,\mathrm{d}s\,, \quad \alpha < c < \beta\,. \tag{7.4}$$

Substituting (7.4) into (7.1), we obtain

$$y(t) = \mathsf{U}\left[\frac{1}{2\pi\mathrm{j}}\int_{c-\mathrm{j}\infty}^{c+\mathrm{j}\infty} X(s)\mathrm{e}^{st}\,\mathrm{d}s\right]. \tag{7.5}$$

Assume that the operator U and the integration in (7.5) may be interchanged. Then, Eq. (7.5) yields

$$y(t) = \frac{1}{2\pi\mathrm{j}}\int_{c-\mathrm{j}\infty}^{c+\mathrm{j}\infty} \mathsf{U}\left[\mathrm{e}^{st}\right]X(s)\,\mathrm{d}s\,. \tag{7.6}$$

Using (7.2) in the last equation, we find

$$y(t) = \frac{1}{2\pi \mathrm{j}} \int_{c-\mathrm{j}\infty}^{c+\mathrm{j}\infty} W(s,t)X(s)\mathrm{e}^{st}\,\mathrm{d}s\,, \quad \alpha < c < \beta\,. \tag{7.7}$$

The integral (7.7) is similar to the inversion integral (1.24) and differs from the latter by a more complicated dependence of the integrand on the parameter t. If the operator U is time-invariant, we have $W(s,t) = W(s)$ and Eq. (7.7) takes the well-known form

$$y(t) = \frac{1}{2\pi \mathrm{j}} \int_{c-\mathrm{j}\infty}^{c+\mathrm{j}\infty} W(s)X(s)\mathrm{e}^{st}\,\mathrm{d}s\,. \tag{7.8}$$

Remark We show that if Eq. (7.7) is applicable, the independence of $W(s,t)$ of t ensures that the operator U is time-invariant, while from the condition $W(s,t) = W(s,t+T)$ follows the periodicity of the operator U.

1. Let (7.8) hold. Then, the response $y_\tau(t)$ of the operator U to an input signal $x_\tau(t) = x(t-\tau)$ is equal, with respect to the shift rule (1.53), to

$$y_\tau(t) = \frac{1}{2\pi \mathrm{j}} \int_{c-\mathrm{j}\infty}^{c+\mathrm{j}\infty} W(s)X(s)\mathrm{e}^{s(t-\tau)}\,\mathrm{d}s = y(t-\tau) \tag{7.9}$$

i.e., formula (7.8) defines an LTI operator.

2. Let $W(s,t) = W(s,t+T)$. Then, the response $y_\tau(t)$ of the operator U to an input signal $x_T(t) = x(t-T)$ is equal, with account for the shift rule, to

$$\begin{aligned} y_T(t) &= \frac{1}{2\pi \mathrm{j}} \int_{c-\mathrm{j}\infty}^{c+\mathrm{j}\infty} W(s,t)X(s)\mathrm{e}^{s(t-T)}\,\mathrm{d}s = \\ &= \frac{1}{2\pi \mathrm{j}} \int_{c-\mathrm{j}\infty}^{c+\mathrm{j}\infty} W(s,t-T)X(s)\mathrm{e}^{s(t-T)}\,\mathrm{d}s = y(t-T) \end{aligned} \tag{7.10}$$

whence follows that the operator defined by Eq. (7.7) is periodic. ■

In most applications it suffices to calculate the output of a linear operator under exogenous disturbance having the transform

$$X(s) = \frac{1}{(s-b)^\ell} = \frac{1}{(\ell-1)!}\frac{\partial^{\ell-1}}{\partial b^{\ell-1}}\frac{1}{s-b}\,, \quad \mathrm{Re}\, s > b \tag{7.11}$$

where b is a constant and ℓ is an integer. As follows from (1.99), the image (7.11) is associated with the original

$$x(t) = \begin{cases} \frac{1}{(\ell-1)!}\, t^{\ell-1}\,\mathrm{e}^{bt} & t > 0 \\ 0 & t < 0\,. \end{cases} \tag{7.12}$$

In the case of (7.11), from (7.7) we find

$$y(t) = \frac{1}{2\pi\mathrm{j}} \int_{c-\mathrm{j}\infty}^{c+\mathrm{j}\infty} W(s,t) \left[\frac{1}{(\ell-1)!} \frac{\partial^{\ell-1}}{\partial b^{\ell-1}} \frac{1}{s-b} \right] \mathrm{e}^{st}\, \mathrm{d}s\,. \tag{7.13}$$

Equation (7.13) can normally be written in the form

$$y(t) = \frac{1}{(\ell-1)!} \frac{\partial^{\ell-1}}{\partial b^{\ell-1}} y_1(t,b) \tag{7.14}$$

where

$$y_1(t,b) \triangleq \frac{1}{2\pi\mathrm{j}} \int_{c-\mathrm{j}\infty}^{c+\mathrm{j}\infty} \frac{W(s,t)}{s-b} \mathrm{e}^{st}\, \mathrm{d}s\,. \tag{7.15}$$

In the problems considered below the PTF $W(s,t)$ is a meromorphic function in the argument s, i.e., its singular points s_i are poles. Moreover, the set of the poles s_i appears to be independent of t. In many cases the integral (7.15) can be calculated by the residue theorem. If the number b is not a pole of $W(s,t)$, the integrand in (7.15) has poles at the points s_i and b. Therefore,

$$\operatorname*{Res}_{b} \left[\frac{W(s,t)}{s-b} \mathrm{e}^{st} \right] = W(b,t)\mathrm{e}^{bt}\,. \tag{7.16}$$

In the following, 'Res' denotes a residue. If the residue theorem is applicable, Eq. (7.15) yields

$$y_1(t,b) = \sum_i \operatorname*{Res}_{s_i} \left[\frac{W(s,t)}{s-b} \mathrm{e}^{st} \right] + W(b,t)\mathrm{e}^{bt} \tag{7.17}$$

where the sum is taken over all poles of the PTF. The first term in (7.17) presents the transient $y_a(t)$, while the second one is the steady-state process $y_\infty(t)$. Using (7.17) and (7.14), under the given assumptions, we find that the steady-state process in the general case (7.11) is given by

$$y_\infty(t) = \frac{1}{(\ell-1)!} \frac{\partial^{\ell-1}}{\partial b^{\ell-1}} \left[W(b,t)\, \mathrm{e}^{bt} \right]. \tag{7.18}$$

A more strict investigation shows that the operator $\frac{\partial}{\partial b}$ in (7.18) should be understood as

$$\frac{\partial}{\partial b} = \lim_{s\to b} \frac{\partial}{\partial s}\,. \tag{7.19}$$

In the special case, when $b = 0$ and

$$x(t) = \begin{cases} \frac{1}{(\ell-1)!} t^{\ell-1} & t > 0 \\ 0 & t < 0 \end{cases} \tag{7.20}$$

the steady-state process is given by

$$y_\infty(t) = \frac{1}{(\ell-1)!} \lim_{s\to 0} \frac{\partial^{\ell-1}}{\partial s^{\ell-1}} \left[W(s,t)\, e^{st} \right] . \tag{7.21}$$

Equations (7.7), (7.17) and (7.18) show that the PTF has approximately the same properties as the transfer function of an LTI operator used in classical control theory. In particular, if the PTF $W(s,t)$ is known, it is, in principle, possible to investigate transient and steady-state processes at the output of linear time-variable operators under deterministic disturbances.
It should be emphasized that the formulae of this section require strict mathematical substantiation in any special case. This will be done below while considering various types of operators and systems.

7.2 Analysis by $\mathcal{D}$-transformed inputs

If the operator (7.1) is T−periodic, i.e.,

$$y(t-T) = \mathsf{U}[x(t-T)] \tag{7.22}$$

the following method, based on the use of the $\mathcal{D}$-transform of the input, is a convenient way to calculate the output. Let the operator (7.1) be T−periodic. Then, for any integer k we have

$$y(t+kT) = \mathsf{U}[x(t+kT)] . \tag{7.23}$$

Multiplying (7.23) by e^{-skT} and adding the corresponding expressions for all integers k, we obtain

$$\sum_{k=-\infty}^{\infty} y(t+kT)e^{-skT} = \sum_{k=-\infty}^{\infty} \mathsf{U}\left[x(t+kT)e^{-skT}\right] . \tag{7.24}$$

If the operators $\sum$ and U are commutative, we have

$$\sum_{k=-\infty}^{\infty} y(t+kT)e^{-skT} = \mathsf{U}\left[\sum_{k=-\infty}^{\infty} x(t+kT)e^{-skT} \right] . \tag{7.25}$$

With the notation (4.1), we can write

$$\mathcal{D}_y(T,s,t) = \mathsf{U}\left[\mathcal{D}_x(T,s,t)\right] . \tag{7.26}$$

The function (7.26), which is henceforth called the *discrete model* of the linear periodic operator, plays a major part in the further exposition.

If Eq. (7.26) is applicable, it gives the following new method for the construction of operational transforms of the output of a linear periodic operator (system): It is assumed that, instead of a real signal $x(t)$ the input is affected by a fictitious signal

$$\tilde{x}(t) = \mathcal{D}_x(T,s,t) = \mathrm{e}^{st}\varphi_x(T,s,t) \tag{7.27}$$

where

$$\varphi_x(T,s,t) = \sum_{k=-\infty}^{\infty} x(t+kT)\mathrm{e}^{-s(t+kT)} \tag{7.28}$$

is the DPFR of the input. By construction,

$$\varphi_x(T,s,t) = \varphi_x(T,s,t+T)\,. \tag{7.29}$$

It is assumed that the output of the system has the form

$$\tilde{y}(t) = \mathrm{e}^{st}y_0(s,t)\,, \quad y_0(s,t) = y_0(s,t+T)\,. \tag{7.30}$$

For a wide class of systems such a solution exists and is unique. Therefore, a comparison of Eqs. (7.26), (7.27) and (7.30) shows that

$$\tilde{y}(t) = \mathcal{D}_y(T,s,t)\,, \quad y_0(s,t) = \varphi_y(T,s,t)\,. \tag{7.31}$$

Using (4.34), we obtain

$$y(t) = \frac{T}{2\pi\mathrm{j}}\int_{c-\mathrm{j}\omega/2}^{c+\mathrm{j}\omega/2} \mathsf{U}\left[\mathcal{D}_x(T,s,t)\right]\,\mathrm{d}s \tag{7.32}$$

with an appropriately chosen constant c. Equation (7.26) can be expressed in terms of the PTF of the operator U and the Laplace transform of the input $X(s)$. With this aim in view, we assume that, together with (7.26), the following relation holds

$$\mathcal{D}_Y(T,s,t) = \mathsf{U}\left[\mathcal{D}_X(T,s,t)\right] \tag{7.33}$$

or, in an expanded form,

$$\mathcal{D}_Y(T,s,t) = \mathsf{U}\left[\frac{1}{T}\sum_{k=-\infty}^{\infty} X(s+k\mathrm{j}\omega)\mathrm{e}^{(s+k\mathrm{j}\omega)t}\right]\,. \tag{7.34}$$

Assuming that the order of the operations U and $\sum$ can be changed, we obtain

$$\mathcal{D}_Y(T,s,t) = \frac{1}{T}\sum_{k=-\infty}^{\infty} X(s+k\mathrm{j}\omega)\,\mathsf{U}\left[\mathrm{e}^{(s+k\mathrm{j}\omega)t}\right]\,. \tag{7.35}$$

From (7.2) it follows that

$$\mathsf{U}\left[\mathrm{e}^{(s+k\mathrm{j}\omega)t}\right] = W(s+k\mathrm{j}\omega,t)\mathrm{e}^{(s+k\mathrm{j}\omega)t}\,. \tag{7.36}$$

Hence, Eq. (7.33) takes the form

$$\mathcal{D}_Y(T,s,t) = \frac{1}{T}\sum_{k=-\infty}^{\infty} W(s+k\mathrm{j}\omega,t)X(s+k\mathrm{j}\omega)\,\mathrm{e}^{(s+k\mathrm{j}\omega)t}\,. \tag{7.37}$$

Using the inversion formula (4.37), from (7.37) we obtain the output in the form

$$y(t) = \frac{1}{2\pi \mathrm{j}} \int_{c-\mathrm{j}\omega/2}^{c+\mathrm{j}\omega/2} \left[\frac{1}{T} \sum_{k=-\infty}^{\infty} W(s + k\mathrm{j}\omega, t) X(s + k\mathrm{j}\omega)\, \mathrm{e}^{(s+k\mathrm{j}\omega)t} \right] \mathrm{d}s\,. \quad (7.38)$$

The derivation presented in this section is not mathematically strict, i.e., for any special case these formulae must be substantiated.

7.3 Analysis of linear integral operators

In this section we consider linear integral operators of the form (6.69).

Theorem 7.1 *Let an operator* U *satisfy the condition of Theorem 6.2, i.e.,* $\mathsf{U} \in \mathcal{U}(\alpha, \beta)$. *Let also* $x(t) \in \Lambda(\alpha, \beta)$. *Therefore, we assume that for* $|s| \to \infty$ *and* $\alpha < \operatorname{Re} s < \beta$ *the estimate (1.45) holds:*

$$|x(s)| < A|s|^{-1}\,. \quad (7.39)$$

Then, the following statements are valid:

1. *The formula (7.7) holds,*

$$y(t) = \frac{1}{2\pi \mathrm{j}} \int_{c-\mathrm{j}\infty}^{c+\mathrm{j}\infty} W(s, t) X(s) \mathrm{e}^{st}\, \mathrm{d}s\,, \quad \alpha < c < \beta\,. \quad (7.40)$$

2. *Equations (7.37) and (7.38) hold.*

Proof 1. We consider an integral of the form

$$\tilde{y}(t, \mu) \triangleq \int_{-\infty}^{\infty} g(\mu, u) x(t - u)\, \mathrm{d}u \quad (7.41)$$

where μ and t are independent parameters. For a fixed μ, the right-hand side of (7.41) is a convolution, which, under the given assumptions, exists due to Section 1.10. Using Eq. (1.63), we find

$$\tilde{y}(t, \mu) = \frac{1}{2\pi \mathrm{j}} \int_{c-\mathrm{j}\infty}^{c+\mathrm{j}\infty} W(s, \mu) X(s) \mathrm{e}^{st}\, \mathrm{d}s\,, \quad \alpha < c < \beta\,. \quad (7.42)$$

As follows from the estimates (7.39) and (6.81), for $|s| \to \infty$, $\alpha < \operatorname{Re} s < \beta$ in any finite interval of t we have

$$|W(s, \mu) X(s)| < K_1 |s|^{-2}\,. \quad (7.43)$$

Since the integral

$$\int_{c-\mathrm{j}\infty}^{c+\mathrm{j}\infty} |s|^{-2}\, \mathrm{d}s \quad (7.44)$$

converges for $c \neq 0$, the integral (7.42) converges uniformly with respect to μ and t. In this case, since the integrand on the right-hand side of (7.42) is continuous with respect to μ and t, the left-hand side also possesses this property. Therefore, we can take $\mu = t$ in (7.42), which yields

$$\tilde{y}(t,\mu)\,|_{\mu=t} = \frac{1}{2\pi \mathrm{j}} \int_{c-\mathrm{j}\infty}^{c+\mathrm{j}\infty} W(s,t)X(s)\mathrm{e}^{st}\,\mathrm{d}s\,, \quad \alpha < c < \beta\,. \tag{7.45}$$

Moreover, for $t = \mu$ Eq. (7.41) gives (6.69). Therefore, Eq. (7.45) is equivalent to (7.40).

2. By the estimate (7.43), the integral (7.40) converges absolutely and can be discretized as

$$y(t) = \frac{T}{2\pi \mathrm{j}} \int_{c-\mathrm{j}\omega/2}^{c+\mathrm{j}\omega/2} \left[\frac{1}{T} \sum_{k=-\infty}^{\infty} W(s+k\mathrm{j}\omega,t)X(s+k\mathrm{j}\omega)\,\mathrm{e}^{(s+k\mathrm{j}\omega)t} \right] \mathrm{d}s \tag{7.46}$$

i.e., Eq. (7.38) is proved. Then, we have to prove Eq. (7.37). For a fixed μ, the product $W(s,\mu)X(s)$ is the image of the function $\tilde{y}(t,\mu)$ (7.41) with respect to t. As follows from the estimate (7.43) and Theorem 4.3,

$$\sum_{k=-\infty}^{\infty} \tilde{y}(t+kT,\mu)\mathrm{e}^{-skT} = \frac{1}{T} \sum_{k=-\infty}^{\infty} W(s+k\mathrm{j}\omega,\mu)X(s+k\mathrm{j}\omega)\,\mathrm{e}^{(s+k\mathrm{j}\omega)t}\,. \tag{7.47}$$

Taking into account that $\tilde{y}(t,\mu) = \tilde{y}(t,\mu+T)$, we can write the last relation in the form

$$\sum_{k=-\infty}^{\infty} \tilde{y}(t+kT,\mu+kT)\mathrm{e}^{-skT} = \frac{1}{T} \sum_{k=-\infty}^{\infty} W(s+k\mathrm{j}\omega,\mu)X(s+k\mathrm{j}\omega)\,\mathrm{e}^{(s+k\mathrm{j}\omega)t}\,. \tag{7.48}$$

The left and right-hand sides of (7.48) are continuous with respect to t and μ, therefore, we can assume $\mu = t$ on both sides. Thus,

$$\sum_{k=-\infty}^{\infty} \tilde{y}(t+kT,t+kT)\mathrm{e}^{-skT} = \sum_{k=-\infty}^{\infty} y(t+kT)\mathrm{e}^{-skT} = \tag{7.49}$$

$$= \mathcal{D}_y(T,s,t) = \mathcal{D}_Y(T,s,t)$$

which, with due account for (7.48), yields (7.37). ∎

Theorem 7.1 presents a substantiation of the operational method for investigations of the operator (6.69) under exogenous disturbances having its Laplace transform in the corresponding stripe of convergence. Nevertheless, in reality these operators are defined for a much wider class of functions. We demonstrate this fact for an e-dichotomic operator (6.69), the pulse response of which satisfies the estimate (6.91). Introduce a new class of functions. Denote by $\mathcal{R}_\beta$ the set of functions $f(t)$ integrable on a certain finite interval and satisfying the estimate

$$|f(t)| < K\mathrm{e}^{\beta|t|}, \quad -\infty < t < \infty, \quad K = \text{const.} \tag{7.50}$$

For $\beta > 0$ the set $\mathcal{R}_\beta$ includes functions that increase exponentially for $|s| \to \infty$.

Theorem 7.2 *Let the pulse response $g(t,u)$ of the operator (6.69)*

$$y(t) = \int_{-\infty}^{\infty} h(t,\tau)x(\tau)\,\mathrm{d}\tau \tag{7.51}$$

satisfy the estimate

$$|g(t,u)| < M\mathrm{e}^{-\gamma|u|} \tag{7.52}$$

with $\gamma > 0$. Then, the integral in (7.51) converges for any $x(t) \in \mathcal{R}_\beta$ with $0 \le \beta < \gamma$. Moreover, $y(t) \in \mathcal{R}_\beta$.

Proof As follows from the estimate (7.50),

$$|x(t)| < K\left(\mathrm{e}^{\beta t} + \mathrm{e}^{-\beta t}\right), \quad -\infty < t < \infty. \tag{7.53}$$

Moreover, Eqs. (7.52) and (6.74) give

$$|h(t,\tau)| < M\mathrm{e}^{-\gamma|t-\tau|}. \tag{7.54}$$

From (7.51), taking due account of (7.53), we have

$$|y(t)| \le \int_{-\infty}^{\infty} |h(t,\tau)| \cdot |x(\tau)|\,\mathrm{d}\tau < KM\psi(t) \tag{7.55}$$

with the notation

$$\psi(t) \triangleq \int_{-\infty}^{\infty} \mathrm{e}^{-\gamma|t-\tau|}\left(\mathrm{e}^{\beta\tau} + \mathrm{e}^{-\beta\tau}\right)\,\mathrm{d}\tau. \tag{7.56}$$

We will show that the integral in (7.56) converges and $\psi(t) \in \mathcal{R}_\beta$. With this aim in view, let

$$\psi(t) = \psi_+(t) + \psi_-(t)$$

where

$$\psi_-(t) \triangleq \int_{-\infty}^{t} \mathrm{e}^{-\gamma|t-\tau|}\left(\mathrm{e}^{\beta\tau} + \mathrm{e}^{-\beta\tau}\right)\,\mathrm{d}\tau$$

$$\psi_+(t) \triangleq \int_{t}^{\infty} \mathrm{e}^{-\gamma|t-\tau|}\left(\mathrm{e}^{\beta\tau} + \mathrm{e}^{-\beta\tau}\right)\,\mathrm{d}\tau.$$

Consider the integral $\psi_-(t)$. Since $\tau \le t$, we can write

$$\psi_-(t) = \int_{-\infty}^{t} \mathrm{e}^{-\gamma(t-\tau)}\left(\mathrm{e}^{\beta\tau} + \mathrm{e}^{-\beta\tau}\right)\,\mathrm{d}\tau =$$

$$= e^{-\gamma t} \int_{-\infty}^{t} e^{(\gamma-\beta)\tau} \, d\tau + e^{-\gamma t} \int_{-\infty}^{t} e^{(\gamma+\beta)\tau} \, d\tau .$$

Since $|\beta| < \gamma$, both integrals at the right-hand side converge and

$$|\psi_-(t)| \le \frac{1}{\gamma - \beta} \left(e^{\beta t} + e^{-\beta t} \right) < \frac{2}{\gamma - \beta} e^{\beta |t|}$$

i.e., $\psi_-(t) \in \mathcal{R}_\beta$. Similarly, it can be shown that $\psi_+(t) \in \mathcal{R}_\beta$. Then, we have $\psi(t) \in \mathcal{R}_\beta$, and $y(t) \in \mathcal{R}_\beta$ due to (7.55). ∎

7.4 Analysis of linear periodic pulse operators

In this section we consider the applicability of the relations obtained in Sections 7.1 and 7.2 for the analysis of the LPPO

$$y(t) = \mathsf{U}_d[x(t)] = \sum_{k=-\infty}^{\infty} q(t - kT) x_k . \tag{7.57}$$

Theorem 7.3 *Let $\mathsf{U} \in \mathcal{U}_d(\alpha, \beta)$ and the input sequence $x_k \in \Lambda_d(\alpha, \beta)$. Then, the following statements hold.*

1.
$$y(t) \in \mathrm{H}(\alpha, \beta) , \tag{7.58}$$
where the set of functions $\mathrm{H}(\alpha, \beta)$ is defined in Section 1.1.

2. *The discrete Laplace transform of the output*
$$\mathcal{D}_y(T, s, t) = \sum_{k=-\infty}^{\infty} y(t + kT) e^{-skT} \tag{7.59}$$
is given by
$$\mathcal{D}_y(T, s, t) = D_d(s, t) X^*(s) , \quad \alpha < \operatorname{Re} s < \beta \tag{7.60}$$
where
$$D_d(s, t) = \sum_{k=-\infty}^{\infty} q(t + kT) e^{-skT} = \mathcal{D}_q(T, s, t) \tag{7.61}$$
is the discrete transfer function (6.110) and
$$X^*(s) = \sum_{k=-\infty}^{\infty} x_k e^{-ksT} \tag{7.62}$$
is the discrete Laplace transform of the input sequence $\{x_k\}$.

3. *The Laplace transform of the output*
$$Y(s) = \int_{-\infty}^{\infty} y(t) e^{-st} \, dt , \quad \alpha < \operatorname{Re} s < \beta$$
is given by
$$Y(s) = Q(s) X^*(s) , \quad \alpha < \operatorname{Re} s < \beta . \tag{7.63}$$

Proof 1. To prove (7.58), we note that

$$y(t)\mathrm{e}^{-st} = \sum_{k=-\infty}^{\infty} q(t-kT)x_k\,\mathrm{e}^{-st} \tag{7.64}$$

which can be represented in the form

$$y(t)\mathrm{e}^{-st} = \sum_{k=-\infty}^{\infty} \left[q(t-kT)\mathrm{e}^{-s(t-kT)}\right]\left(x_k\,\mathrm{e}^{-ksT}\right). \tag{7.65}$$

Due to the given assumptions, the following estimates hold for $\alpha < \alpha' \leq \operatorname{Re} s \leq \beta' < \beta$

$$\begin{aligned} \left|q(t-kT)\mathrm{e}^{-s(t-kT)}\right| &< M\mathrm{e}^{-\gamma|t-kT|} \\ \left|x_k\mathrm{e}^{-ksT}\right| &< N\mathrm{e}^{-\delta|k|T} \leq N \end{aligned} \tag{7.66}$$

where M, N, γ, and δ are positive constants. From (7.65) and (7.66) we have

$$\left|y(t)\mathrm{e}^{-st}\right| < NM\mathrm{e}^{-\gamma|t-kT|}. \tag{7.67}$$

Using (6.136) and (6.137), we obtain

$$\left|y(t)\mathrm{e}^{-st}\right| < N_1 = \text{const.}, \quad \alpha < \alpha' \leq \operatorname{Re} s \leq \beta' < \beta. \tag{7.68}$$

Hence, by Theorem 1.1, we obtain (7.58).

2. From (7.58) it follows that the following Laplace transform exists:

$$\mathcal{D}_y(T,s,t) = \sum_{m=-\infty}^{\infty} y(t+mT)\mathrm{e}^{-msT}, \quad \alpha < \operatorname{Re} s < \beta. \tag{7.69}$$

To calculate $\mathcal{D}_y(T,s,t)$, we note that Eq. (7.57) yields

$$y(t+mT) = \sum_{k=-\infty}^{\infty} q(t+mT-kT)x_k. \tag{7.70}$$

Substituting this formula into (7.69), we find

$$\mathcal{D}_y(T,s,t) = \sum_{m=-\infty}^{\infty}\sum_{k=-\infty}^{\infty} q(t+mT-kT)x_k\mathrm{e}^{-msT}. \tag{7.71}$$

Assuming that $m-k=r$, we obtain

$$\mathcal{D}_y(T,s,t) = \left[\sum_{r=-\infty}^{\infty} q(t+rT)\mathrm{e}^{-rsT}\right]\left[\sum_{k=-\infty}^{\infty} x_k\mathrm{e}^{-ksT}\right] \tag{7.72}$$

which is equivalent to (7.60). The passage from (7.71) to (7.72) is justified, because the series at the right-hand side of (7.72) converge absolutely.

3. Integrating (7.64) by t, we obtain

$$\int_{-\infty}^{\infty} y(t)\mathrm{e}^{-st}\,\mathrm{d}t = \int_{-\infty}^{\infty}\left[\sum_{k=-\infty}^{\infty} q(t-kT)x_k\right]\mathrm{e}^{-st}\,\mathrm{d}t\,. \tag{7.73}$$

Changing the order of summation and integration, we formally obtain

$$Y(s) = \sum_{k=-\infty}^{\infty} x_k \int_{-\infty}^{\infty} q(t-kT)\mathrm{e}^{-st}\,\mathrm{d}t = Q(s)\sum_{k=-\infty}^{\infty} x_k \mathrm{e}^{-ksT} \tag{7.74}$$

which is equivalent to (7.63). To substantiate the passage from (7.73) to (7.74), we denote

$$Y_a(s) \triangleq \int_{-\infty}^{\infty} |y(t)|\mathrm{e}^{-st}\,\mathrm{d}t\,, \quad \alpha < \operatorname{Re} s < \beta\,. \tag{7.75}$$

The convergence of the integral (7.75) follows from (7.58). Consider the series of absolute values

$$I_a(t) \triangleq \sum_{k=-\infty}^{\infty} \left|h(t-kT)\mathrm{e}^{-st}\right| |x_k|\,. \tag{7.76}$$

Integrating (7.76) by t and assuming that $\operatorname{Re} s = \lambda$, we have

$$\int_{-\infty}^{\infty} I_a(t)\,\mathrm{d}t = Y_a(\lambda)\sum_{k=-\infty}^{\infty} |x_k|\mathrm{e}^{-k\lambda T} < N = \text{const.} \tag{7.77}$$

Since the series (7.77) converges, the passage from (7.73) to (7.74) is legitimate and Eq. (7.63) is valid. ■

Corollary 1 From (7.60) we have

$$D_d(s,t) = \frac{\mathcal{D}_y(T,s,t)}{X^*(s)} \tag{7.78}$$

i.e., the discrete transfer function $D_d(s,t)$ of the LPPO is equal to the ratio of the discrete Laplace transform of the output and that of the input sequence.

Corollary 2 From (7.63) we find

$$Q(s) = \frac{Y(s)}{X^*(s)} \tag{7.79}$$

i.e., the equivalent transfer function $Q(s)$ is equal to the ratio of the Laplace transform of the output in continuous time and the discrete Laplace transform of the input sequence.

Next, we assume that the input of the LPPO is affected by a signal $x(t) \in \Lambda(\alpha, \beta)$. Henceforth it is assumed that the input sequence $\{x_k\}$ is formed as $x_k = x(kT)$ if $t_k = kT$ is a point of continuity, and as $x_k = x(kT+0)$ for points of discontinuity. Then, according to relations given in Section 4.3, we have

$$X^*(s) = \mathcal{D}_x(T, s, +0) \tag{7.80}$$

where $\mathcal{D}_x(T, s, t)$ is the discrete Laplace transform of the input signal $x(t)$. In this case, Eqs. (7.60) and (7.63) take the form

$$\mathcal{D}_y(T, s, t) = D_d(s, t)\mathcal{D}_x(T, s, +0) = \mathcal{D}_q(T, s, t)\mathcal{D}_x(T, s, +0) \tag{7.81}$$

and, correspondingly,

$$Y(s) = Q(s)\mathcal{D}_x(T, s, +0)\,. \tag{7.82}$$

For the further exposition it is necessary to state an important property of the LPPO (7.57). Let the input of the LPPO (7.75) be acted upon by an input signal of the form

$$x(t) = f(t)g(t)\,, \quad g(t) = g(t+T)\,. \tag{7.83}$$

Then, the input signal $\{x_k\} = \{x(kT+0)\}$ does not change if we consider, instead of (7.83), the input

$$\hat{x}(t) = f(t)g(+0)\,. \tag{7.84}$$

Therefore, the change of the input (7.83) by (7.84) causes no modifications in the output $y(t)$ of the LPPO. This property will be called the ***stroboscopic effect*** of the LPPO. Using (7.83) and (7.84), we show that Eq. (7.81) can be considered as a substantiation of the general formula (7.26) for the LPPO. Indeed, let $x(t) \in \Lambda(\alpha, \beta)$ and the input of the LPPO be acted upon by the fictitious signal

$$\tilde{x}(t) \triangleq \mathcal{D}_x(T, s, t) = \mathrm{e}^{st}\varphi_x(T, s, t)\,. \tag{7.85}$$

The input (7.85) has the form (7.83), where $f(t) = \mathrm{e}^{st}$ and $g(t) = \varphi_x(T, s, t)$. Using the stroboscopic effect and taking account of (7.84), we can substitute the following equivalent signal for (7.84)

$$\hat{x}(t) = \mathrm{e}^{st}\varphi_x(T, s, +0) = \mathrm{e}^{st}\mathcal{D}_x(T, s, +0)\,. \tag{7.86}$$

Taking due account of (6.109) and (6.110), it is evident that

$$\mathsf{U}_d[\tilde{x}(t)] = \mathsf{U}_d[\hat{x}(t)] = \mathsf{U}_d\left[\mathrm{e}^{st}\right]\mathcal{D}_x(T, s, +0) = \mathcal{D}_q(T, s, t)\mathcal{D}_x(T, s, +0)\,.$$

Thus, under the given assumptions Eq. (7.26) holds for the LPPO (7.57).

Using the inversion formulae for operational transformations given in Chapters 1 and 4, we can derive general formulae for the determination of the image of the output by known transforms (7.81) and (7.82).

1. Using the inversion formula for Laplace transform (4.34), from (7.81) we have

$$y(t) = \frac{T}{2\pi \mathrm{j}} \int_{c-\mathrm{j}\omega/2}^{c+\mathrm{j}\omega/2} \mathcal{D}_q(T,s,t)\mathcal{D}_x(T,s,+0)\,\mathrm{d}s\,, \quad \alpha < c < \beta\,. \tag{7.87}$$

2. Under the above assumptions let the output of the LPPO $y(t)$ have bounded variation on any finite interval. Then, from (7.82) we have

$$y(t) = \frac{1}{2\pi \mathrm{j}} \int_{c-\infty}^{c+\infty} Q(s)\mathcal{D}_x(T,s,+0)\mathrm{e}^{st}\,\mathrm{d}s\,, \quad \alpha < c < \beta\,. \tag{7.88}$$

Next, we formally discretize the integral (7.88) by means of (1.32) with the step $\mathrm{j}\omega$. Taking into account that

$$\mathcal{D}_x(T,s,+0) = \mathcal{D}_x(T,s+\mathrm{j}\omega,+0) \tag{7.89}$$

and using (1.32), we find

$$y(t) = \frac{T}{2\pi \mathrm{j}} \int_{c-\mathrm{j}\omega/2}^{c+\mathrm{j}\omega/2} \mathcal{D}_Q(T,s,t)\mathcal{D}_x(T,s,+0)\,\mathrm{d}s\,, \quad \alpha < c < \beta\,. \tag{7.90}$$

For all t, for which $\mathcal{D}_q(T,s,t) = \mathcal{D}_Q(T,s,t)$, Eqs. (7.87) and (7.90) yield the same result.

3. We present sufficient conditions for Eqs. (7.87), (7.88) and (7.90) to be equivalent for all t.

Theorem 7.4 *Let the following estimate hold for $\alpha < \alpha' \le \operatorname{Re} s \le \beta' < \beta$ and $|s| \to \infty$*

$$|Q(s)| < K|s|^{-1-\lambda} \tag{7.91}$$

where K and λ are positive constants. Then,

1. *Formulas (7.87), (7.88) and (7.90) yield the same result.*
2. *The corresponding output is continuous, and $y(t) \in \Lambda(\alpha,\beta)$.*

Proof 1. Since $x(t) \in \Lambda(\alpha,\beta)$, from Theorem 3.1 we have $|\mathcal{D}_x(T,s,+0)| = |\varphi_x(T,s,+0)| < L =$const. for $\alpha' \le \operatorname{Re} s \le \beta'$. Therefore, in this stripe the function $Q(s)\mathcal{D}_x(T,s,+0)$ satisfies an estimate similar to (1.33). Then, by Theorem 1.4, the output is continuous, and $y(t) \in \Lambda(\alpha,\beta)$.

2. Under the given assumptions the integral (7.88) can be discretized to the form (7.90). Moreover, $\mathcal{D}_q(T,s,t) = \mathcal{D}_Q(T,s,t)$ for all t, therefore, the integrals (7.87) and (7.90) coincide. Thus, all the statements of the theorem are proved. ■

For the LPPO (7.57) we can also obtain the inversion integral of the form (7.7), which involves the PTF $W_d(s,t)$. With this aim in mind, we assume that the input $x(t) \in \Lambda(\alpha, \beta)$, and that it is continuous, and we rewrite the equation of LPPO in the form

$$y(t) = \sum_{k=-\infty}^{\infty} q(t-kT)x(kT)\,. \tag{7.92}$$

Under the above assumptions, the following inversion formula holds.

$$x(t) = \frac{1}{2\pi \mathrm{j}} \int_{c-\mathrm{j}\infty}^{c+\mathrm{j}\infty} X(s)\mathrm{e}^{st}\,\mathrm{d}s\,, \quad \alpha < c < \beta \tag{7.93}$$

so that for all k

$$x(kT) = \frac{1}{2\pi \mathrm{j}} \int_{c-\mathrm{j}\infty}^{c+\mathrm{j}\infty} X(s)\mathrm{e}^{ksT}\,\mathrm{d}s\,, \quad \alpha < c < \beta\,. \tag{7.94}$$

Substituting (7.94) into (7.92), we have

$$y(t) = \frac{1}{2\pi \mathrm{j}} \sum_{k=-\infty}^{\infty} q(t-kT) \int_{c-\mathrm{j}\infty}^{c+\mathrm{j}\infty} X(s)\mathrm{e}^{ksT}\,\mathrm{d}s\,, \quad \alpha < c < \beta\,. \tag{7.95}$$

After changing the order of summation and integration, we formally obtain

$$y(t) = \frac{1}{2\pi \mathrm{j}} \int_{c-\mathrm{j}\infty}^{c+\mathrm{j}\infty} X(s) \left[\sum_{k=-\infty}^{\infty} q(t-kT)\mathrm{e}^{ksT} \right] \mathrm{d}s\,, \quad \alpha < c < \beta\,. \tag{7.96}$$

Substituting $-k$ for k on the right-hand side of (7.96), taking due account of Eqs. (6.104) and (6.110) we obtain the integral

$$y(t) = \frac{1}{2\pi \mathrm{j}} \int_{c-\mathrm{j}\infty}^{c+\mathrm{j}\infty} W_d(s,t)X(s)\mathrm{e}^{st}\,\mathrm{d}s = \frac{1}{2\pi \mathrm{j}} \int_{c-\mathrm{j}\infty}^{c+\mathrm{j}\infty} D_d(s,t)X(s)\,\mathrm{d}s\,. \tag{7.97}$$

To justify the above derivation we can employ the following theorem.

Theorem 7.5 *Let* $q(t) \in \Lambda(\alpha, \beta)$ *and for* $\alpha < \alpha' \le \operatorname{Re} s \le \beta' < \beta$, $|s| \to \infty$ *the following estimate may hold*

$$|X(s)| < L|s|^{-1-\lambda} \tag{7.98}$$

where L *and* λ *are positive constants. Then, Eq. (7.97) defines the same function* $y(t)$ *as Eq. (7.87).*

Proof In the given case, by virtue of Corollary 1 of Theorem 3.1, we have

$$|W_d(s,t)| = |\varphi_q(T,s,t)| < M\,, \quad \alpha < \alpha' \le \operatorname{Re} s \le \beta' < \beta\,. \tag{7.99}$$

Therefore, the function $X(s)W_d(s,t)$ satisfies an estimate similar to (1.33). Hence, the integral (7.97) can be discretized with the step $\mathrm{j}\omega$. Then, taking account of

$$D_d(s,t) = D_d(s+\mathrm{j}\omega, t) \tag{7.100}$$

we find

$$y(t) = \frac{T}{2\pi\mathrm{j}} \int_{c-\mathrm{j}\omega/2}^{c+\mathrm{j}\omega/2} D_d(s,t)\mathcal{D}_X(T,s,0)\,\mathrm{d}s \tag{7.101}$$

where

$$\mathcal{D}_X(T,s,0) = \frac{1}{T}\sum_{k=-\infty}^{\infty} X(s+k\mathrm{j}\omega)\,. \tag{7.102}$$

If (7.98) holds, the function $x(t)$ is continuous and $x(t) \in \Lambda(\alpha,\beta)$, and we have

$$\mathcal{D}_x(T,s,+0) = \mathcal{D}_x(T,s,0) = \mathcal{D}_X(T,s,0)\,. \tag{7.103}$$

Therefore, under the given assumptions the integral (7.101) coincides with (7.87). ∎

Remark The condition (7.98) is too burdensome for many applications, because it restricts the input signals to the class of continuous functions. For instance, the condition (7.98) is false for a unit step (1.42) with $X(s) = s^{-1}$. Nevertheless, all difficulties are removed if, instead of the function (7.102), we use

$$\mathcal{D}_x(T,s,+0) = \lim_{t\to+0} \mathcal{D}_x(T,s,t) \tag{7.104}$$

in the integral (7.101) obtained from (7.97) by discretization.

This technique will be employed further for the investigation of a general class of integrals of the form

$$y(t) = \frac{1}{2\pi\mathrm{j}} \int_{c-\mathrm{j}\infty}^{c+\mathrm{j}\infty} \mathcal{D}_Q(T,s,t)X(s)\,\mathrm{d}s \tag{7.105}$$

where

$$\mathcal{D}_Q(T,s,t) = \frac{1}{T}\sum_{k=-\infty}^{\infty} Q(s+k\mathrm{j}\omega)\mathrm{e}^{(s+k\mathrm{j}\omega)t} \tag{7.106}$$

and $Q(s)$ and $X(s)$ are the images of the functions $q(t) \in \Lambda(\alpha,\beta)$ and $x(t) \in \Lambda(\alpha,\beta)$, respectively, in the stripe $\alpha < \operatorname{Re} s < \beta$. Considering the series (7.106) as a discrete transfer function of an LPPO U_d, we can regard the integral (7.105) as a result of the passage of the input through the operator U_d . In this case, assuming that for a discontinuous input the operator fixes the values $x_k = x(kT+0)$, from the integral (7.105) we can pass to

$$y(t) = \frac{T}{2\pi\mathrm{j}} \int_{c-\mathrm{j}\omega/2}^{c+\mathrm{j}\omega/2} \mathcal{D}_q(T,s,t)\mathcal{D}_x(T,s,+0)\,\mathrm{d}s\,. \tag{7.107}$$

We note that the relations of this section can, without major modification, be extended to LPPO with phase shift U_d^{ϕ} defined by (6.112). If the input of the operator U_d^{ϕ} has the form (7.83), with respect to (6.112) it can be substituted by the equivalent signal

$$\hat{x}_{\phi}(t) = f(t)g(\phi + 0)\,. \tag{7.108}$$

Equation (7.108) defines the stroboscopic property of an LPPO with phase shift. Using (7.108), we can immediately prove that for an LPPO with phase shift formula (7.81) should be changed to

$$\mathcal{D}_y(T,s,t) = \mathcal{D}_q(T,s,t-\phi)\cdot\mathcal{D}_x(T,s,\phi+0)\,. \tag{7.109}$$

Taking due account of (4.35), we find the Laplace transform of the output

$$Y(s) = \int_0^T \mathcal{D}_q(T,s,t-\phi)\mathrm{e}^{-st}\,\mathrm{d}t\cdot\mathcal{D}_x(T,s,\phi+0) \tag{7.110}$$

where

$$\mathcal{D}_q(T,s,t-\phi)\mathrm{e}^{-st} = \varphi_q(T,s,t-\phi)\mathrm{e}^{-s\phi} \tag{7.111}$$

i.e.,

$$Y(s) = \int_0^T \varphi_q(T,s,t-\phi)\,\mathrm{d}t\cdot\mathrm{e}^{-s\phi}\,\mathcal{D}_x(T,s,\phi+0)\,. \tag{7.112}$$

Using (6.115), we obtain

$$\int_0^T \varphi_q(T,s,t-\phi)\,\mathrm{d}t = \int_0^T \varphi_Q(T,s,t-\phi)\,\mathrm{d}t = Q(s)\,. \tag{7.113}$$

Hence, taking into account that

$$\mathrm{e}^{-s\phi}\mathcal{D}_x(T,s,\phi+0) = \varphi_x(T,s,\phi+0) \tag{7.114}$$

from (7.112) we find

$$Y(s) = Q(s)\varphi_x(T,s,\phi+0)\,. \tag{7.115}$$

Chapter 8

Stochastic analysis and $\mathcal{H}_2$–norm of linear periodic operators

8.1 Stochastic functions of continuous or discrete arguments

In this chapter we investigate the response of a linear periodic operator to an input signal $x(t)$, that is a centered, and in a loose sense, stationary stochastic process. The last property means that, Åström (1970)

$$\mathsf{E}[x(t)] = 0 \tag{8.1}$$

$$\mathsf{E}\left[x(t_1)x(t_2)\right] = K_x(t_2 - t_1) \tag{8.2}$$

where E denotes the operator of mathematical expectation and $K_x(t)$ is the autocorrelation function of the signal $x(t)$. If we take $t_1 = t$ and $t_2 = t + \tau$, then

$$\mathsf{E}\left[x(t)x(t+\tau)\right] = K_x(\tau) \tag{8.3}$$

and, as is well known,

$$K_x(t) = K_x(-t)\,. \tag{8.4}$$

If the following estimate holds

$$|K_x(t)| < N\mathrm{e}^{-\lambda|t|} \tag{8.5}$$

with positive constants N and λ, for $|\operatorname{Re} s| < \lambda$ there exists a bilateral Laplace transform

$$R_x(s) \triangleq \int_{-\infty}^{\infty} K_x(t)\mathrm{e}^{-st}\,\mathrm{d}t \tag{8.6}$$

which will be called the *spectral density* of the signal $x(t)$. In what follows it is always assumed that the function $K_x(t)$ is a function of bounded variation. Then, Eq. (8.6) yields

$$K_x(t) = \frac{1}{2\pi \mathrm{j}} \int_{c-\mathrm{j}\infty}^{c+\mathrm{j}\infty} R_x(s) \mathrm{e}^{st} \, \mathrm{d}s \,, \quad -\lambda < c < \lambda \,. \tag{8.7}$$

Because of symmetry of (8.4), the correlation function $K_x(t)$ is continuous for $t = 0$, so that the value

$$d_x \triangleq K_x(0) \tag{8.8}$$

makes sense and is called the *variance* of the signal $x(t)$. Since $K_x(t)$ is continuous at $t = 0$, we can take $t = 0$ in (8.7), which gives, taking due account of (8.8),

$$d_x = \frac{1}{2\pi \mathrm{j}} \int_{c-\mathrm{j}\infty}^{c+\mathrm{j}\infty} R_x(s) \, \mathrm{d}s \,. \tag{8.9}$$

We note that the symmetric property of (8.4) and (8.6) yields

$$R_x(s) = R_x(-s) \tag{8.10}$$

i.e., the function $R_x(s)$ is symmetric. Hence, the function $R_x(\mathrm{j}\nu)$ is real at the imaginary axis $s = \mathrm{j}\nu$. In technical applications we are often dealing with cases in which the spectral density is a strictly proper real rational function

$$R_x(s) = \frac{p(s)}{q(s)} \tag{8.11}$$

with polynomials $p(s)$ and $q(s)$. Henceforth, we always assume that all poles of $R_x(s)$ are simple. Then, using the fact that $R_x(s)$ is strictly proper and symmetric, the following partial fraction expansion holds

$$R_x(s) = \sum_{i=1}^{m} \beta_i \left(\frac{1}{s - s_i} - \frac{1}{s + s_i} \right) \tag{8.12}$$

where β_i are constants and s_i are the poles of $R_x(s)$. As follows from (8.12), the polynomials $p(s)$ and $q(s)$ in (8.11) have the form

$$\begin{aligned} p(s) &= p_1 s^{2m-2} + p_2 s^{2m-4} + \ldots + p_m \\ q(s) &= s^{2m} + q_1 s^{2m-2} + \ldots + q_m \end{aligned} \tag{8.13}$$

where p_i and q_i are constants. From (8.13) it follows that for $|s| \to \infty$ the following estimate holds

$$|R_x(s)| < L|s|^{-2} \tag{8.14}$$

where L is a constant.

Similarly, we can investigate the properties of stochastic functions of discrete arguments, i.e., stochastic sequences. Let us consider a centered stochastic sequence $\{x_k\}$ that is stationary in a loose sense. This means that

$$\mathsf{E}\,[x_k] = 0 \tag{8.15}$$

$$\mathsf{E}\,[x_{k_1} x_{k_2}] = K_x^*(k_2 - k_1) \tag{8.16}$$

where $K_x^*(k)$, $(k = 0, \pm 1, \pm 2, \dots)$ is the autocorrelation function of the sequence $\{x_k\}$. With $k_1 = k$, $k_2 = k + n$ Eq. (8.16) gives

$$\mathsf{E}\,[x_k x_{k+n}] = K_x^*(n) \tag{8.17}$$

i.e., the correlation function of a stationary sequence is in fact defined by the single sequence $\{K_x^*(n)\}$. This function is symmetric, i.e.,

$$K_x^*(n) = K_x^*(-n)\,. \tag{8.18}$$

Assume that the estimate (2.25) holds

$$\left|K_x^*(n)\mathrm{e}^{-nsT}\right| < M_d\,, \quad (n = 0, \pm 1, \pm 2, \dots)\,, \quad -\lambda < \operatorname{Re} s < \lambda \tag{8.19}$$

where M_d, T and λ are positive constants. Then, for $|\operatorname{Re} s| < \lambda$ the following bilateral discrete Laplace transform converges

$$R_x^*(s) = \sum_{n=-\infty}^{\infty} K_x^*(n)\mathrm{e}^{-nsT}\,. \tag{8.20}$$

The function (8.20) is called the *discrete spectral density* of the sequence $\{x_k\}$. Assume that the stochastic sequence $\{x_k\}$ is formed as $x_k = x(kT)$, where $x(t)$ is a centered stationary stochastic process, all realizations of which are continuous. Let the correlation function of this process $K_x(t)$ and its spectral density $R_x(s)$ also be continuous. In this case there is a simple relation between the autocorrelation functions $K_x(t)$ and $K_x^*(n)$ on the one hand, and spectral densities $R_x(s)$ and $R_x^*(s)$ on the other hand. Indeed, for any integer k we have

$$K_x^*(n) = K_x(nT)\,. \tag{8.21}$$

Hence, Eq. (8.20) gives

$$R_x^*(s) = \sum_{n=-\infty}^{\infty} K_x(nT)\mathrm{e}^{-nsT} = \mathcal{D}_{K_x}(T, s, 0)\,. \tag{8.22}$$

Assume that the function $R_x(s)$ decreases as $|s|^{-2}$ for $|s| \to \infty$ and $|\operatorname{Re} s| < \lambda$. Then, for $|\operatorname{Re} s| < \lambda$ the conditions of Theorem 4.3 are satisfied, and, therefore,

$$\mathcal{D}_{R_x}(T, s, t) = \mathcal{D}_{K_x}(T, s, t) \tag{8.23}$$

where

$$\mathcal{D}_{R_x}(T,s,t) = \frac{1}{T}\sum_{n=-\infty}^{\infty} R_x(s+nj\omega)\mathrm{e}^{(s+nj\omega)t} \tag{8.24}$$

$$\mathcal{D}_{K_x}(T,s,t) = \sum_{n=-\infty}^{\infty} K_x(t+nT)\mathrm{e}^{-nsT}\,. \tag{8.25}$$

The sum of the series (8.24) and (8.25) are continuous with respect to t. For $t = 0$, taking due account of (8.22), from (8.23)–(8.25) we find

$$R_x^*(s) = \frac{1}{T}\sum_{n=-\infty}^{\infty} R_x(s+nj\omega) = \mathcal{D}_{R_x}(T,s,0)\,. \tag{8.26}$$

Formula (8.26) defines the connection between the discrete spectral density $R_x^*(s)$ of the sampled signal and the spectral density $R_x(s)$ of the initial process $x(t)$.
In case of (8.12) we can derive a closed expression for the discrete spectral density (8.26). Indeed, using (4.41) and (8.12), for $0 \le t < T$ we have

$$\mathcal{D}_{R_x}(T,s,t) = \sum_{i=1}^{m}\beta_i\left(\frac{\mathrm{e}^{s_i t}}{1-\mathrm{e}^{(s_i-s)T}} - \frac{\mathrm{e}^{-s_i t}}{1-\mathrm{e}^{-(s_i+s)T}}\right).$$

Hence, taking due account of (8.26), for $t = 0$ we have

$$R_x^*(s) = \sum_{i=1}^{m}\beta_i\,\frac{2\mathrm{e}^{-sT}\sinh(s_iT)}{1-2\mathrm{e}^{-sT}\cosh(s_iT)-\mathrm{e}^{-2sT}}\,. \tag{8.27}$$

8.2 Correlation function and output variance of integral operators

Let us have an e-dichotomic integral operator

$$y(t) = \int_{-\infty}^{\infty} h(t,\tau)x(\tau)\,\mathrm{d}\tau = \mathsf{U}[x(t)] \tag{8.28}$$

with the Green function satisfying the estimate

$$|h(t,\tau)| < L\mathrm{e}^{-\lambda|t-\tau|} \tag{8.29}$$

where L and λ are positive constants. It is also assumed that the Green function satisfies the conditions of Theorem 6.2.

With regard to the input $x(t)$, we assume that it is a centered and, in a loose sense, stationary stochastic process, such that the integral (8.28) converges for each of its realizations. Then, the stochastic process $y(t)$ is centered, i.e.,

$$\mathsf{E}\left[y(t)\right] = \int_{-\infty}^{\infty} h(t,\tau)\mathsf{E}\left[x(\tau)\right]\,\mathrm{d}\tau = 0\,. \tag{8.30}$$

For arbitrary $t = t_1$ and $t = t_2$ we have

$$y(t_1) = \int_{-\infty}^{\infty} h(t_1, u)x(u)\,\mathrm{d}u \tag{8.31}$$

$$y(t_2) = \int_{-\infty}^{\infty} h(t_1, v)x(v)\,\mathrm{d}v$$

substituting these equalities, we find

$$y(t_1)y(t_2) = \int_{-\infty}^{\infty}\int_{-\infty}^{\infty} h(t_1, u)h(t_2, v)x(u)x(v)\,\mathrm{d}u\,\mathrm{d}v \tag{8.32}$$

where the right-hand side is assumed to be independent of the order of integration by u and v. Using the operation of mathematical expectation at both sides of (8.32), we find

$$\mathsf{E}\left[y(t_1)y(t_2)\right] = \mathsf{E}\left[\int_{-\infty}^{\infty}\int_{-\infty}^{\infty} h(t_1, u)h(t_2, v)x(u)x(v)\,\mathrm{d}u\,\mathrm{d}v\right]. \tag{8.33}$$

Assume that the operation of integration and mathematical expectation in (8.33) can be interchanged. Then,

$$\mathsf{E}\left[y(t_1)y(t_2)\right] = \int_{-\infty}^{\infty}\int_{-\infty}^{\infty} h(t_1, u)h(t_2, v)\mathsf{E}\left[x(u)x(v)\right]\,\mathrm{d}u\,\mathrm{d}v\,. \tag{8.34}$$

Denote

$$\mathsf{E}\left[y(t_1)y(t_2)\right] \triangleq K_y(t_1, t_2) \tag{8.35}$$

where $K_y(t_1, t_2)$ is the corresponding output correlation function. Moreover, from (8.2) we have

$$\mathsf{E}\left[x(u)x(v)\right] = K_x(v - u) = K_x(u - v)\,. \tag{8.36}$$

Taking due account of (8.7), for $c = 0$ we can represent the last equation in the form

$$\mathsf{E}\left[x(u)x(v)\right] = \frac{1}{2\pi \mathrm{j}}\int_{c-\mathrm{j}\infty}^{c+\mathrm{j}\infty} R_x(s)\mathrm{e}^{s(v-u)}\,\mathrm{d}s\,. \tag{8.37}$$

Substituting (8.37) into (8.34) and taking into account (8.35), we have

$$K_y(t_1, t_2) = \frac{1}{2\pi \mathrm{j}}\int_{-\infty}^{\infty}\int_{-\infty}^{\infty}\int_{-\mathrm{j}\infty}^{\mathrm{j}\infty} h(t_1, u)h(t_2, v)R_x(s)\mathrm{e}^{s(v-u)}\,\mathrm{d}s\,\mathrm{d}v\,\mathrm{d}u\,. \tag{8.38}$$

Assuming that the order of integration by various variables can be interchanged, we have

$$K_y(t_1, t_2) = \frac{1}{2\pi \mathrm{j}}\int_{-\mathrm{j}\infty}^{\mathrm{j}\infty} R_x(s)\left[\int_{-\infty}^{\infty} h(t_1, u)\mathrm{e}^{-su}\,\mathrm{d}u\right]\left[\int_{-\infty}^{\infty} h(t_2, v)\mathrm{e}^{sv}\,\mathrm{d}v\right]\mathrm{d}s\,. \tag{8.39}$$

As follows from (6.78),

$$\int_{-\infty}^{\infty} h(t_1,u)\mathrm{e}^{-su}\,\mathrm{d}u = \mathrm{e}^{-st_1}W(-s,t_1)\,, \qquad \int_{-\infty}^{\infty} h(t_2,v)\mathrm{e}^{sv}\,\mathrm{d}v = \mathrm{e}^{st_2}W(s,t_2)\,. \tag{8.40}$$

Using (8.40) we can represent (8.39) in the form

$$K_y(t_1,t_2) = \frac{1}{2\pi\mathrm{j}}\int_{-\mathrm{j}\infty}^{\mathrm{j}\infty} R_x(s)W(-s,t_1)W(s,t_2)\mathrm{e}^{s(t_2-t_1)}\,\mathrm{d}s\,. \tag{8.41}$$

As follows from (8.41), the correlation function depends, in general, on the variables t_1 and t_2 separately rather than on the difference between them. Therefore, the response of a linear integral operator to a stationary stochastic process is not, in general, a stationary stochastic process. Under the conditions of Theorem 6.1, the PTF $W(s,t)$ is continuous with respect to t and decreases as $|s|^{-1}$ for $|s| \to \infty$. Therefore, if the spectral density $R_x(s)$ is bounded on the imaginary axis, the integrand in (8.41) decreases as $|s|^{-2}$ for $|s| \to \infty$. Hence, in this case the function $K_y(t_1,t_2)$ is continuous with respect to t_1 and t_2. Therefore, we can take $t_1 = t_2 = t$ in (8.41). As it is known, in this case the value

$$d_y(t) \triangleq K_y(t_1,t_2)\,|_{t_1=t_2=t} = \mathsf{E}\left[y^2(t)\right] \tag{8.42}$$

is the output variance which depends on time. Therefore, assuming $t_1 = t_2 = t$ in (8.41) and using (8.42), we find

$$d_y(t) = \frac{1}{2\pi\mathrm{j}}\int_{-\mathrm{j}\infty}^{\mathrm{j}\infty} R_x(s)W(-s,t)W(s,t)\,\mathrm{d}s\,. \tag{8.43}$$

Equation (8.43) defines the instantaneous variance for any t. In many practical cases it is useful to evaluate the *mean output variance* $\overline{d}_y$ defined as

$$\overline{d}_y \triangleq \lim_{t\to\infty}\frac{1}{2t}\int_{-t}^{t} d_y(\tau)\,\mathrm{d}\tau\,. \tag{8.44}$$

Using (8.43) and (8.44), we find

$$\overline{d}_y = \frac{1}{2\pi\mathrm{j}}\int_{-\mathrm{j}\infty}^{\mathrm{j}\infty} R_x(s)B_0(s)\,\mathrm{d}s \tag{8.45}$$

where

$$B_0(s) \triangleq \lim_{t\to\infty}\frac{1}{2t}\int_{-t}^{t} W(s,\tau)W(-s,\tau)\,\mathrm{d}\tau\,. \tag{8.46}$$

The above formulae are valid, under the given assumptions, for an arbitrary linear integral operator U. If the operator is periodic, i.e., the condition (6.92) holds, we can add to these formulae some important relations. Indeed, if the operator U is T−periodic, then

$$W(s,t) = W(s,t+T) \tag{8.47}$$

and we immediately have

$$K_y(t_1,t_2) = K_y(t_1+T, t_2+T). \tag{8.48}$$

Indeed, if (8.47) holds, from (8.41) we obtain (8.48). Moreover, from (8.43) taking due account of (8.47) we have

$$d_y(t) = d_y(t+T) \tag{8.49}$$

i.e., output variance is periodic with respect to t with a period equal to the period T of the operator.

Stochastic signals satisfying (8.48) will be called *periodically non-stationary.* Thus, Eq. (8.49) means that the response of a periodic operator (8.28) to a stationary stochastic signal is periodically non-stationary.

We note that in the case of a periodic operator instead of (8.44) we can express the mean variance in the form

$$\bar{d}_y = \frac{1}{T}\int_0^T d_y(t)\,\mathrm{d}t. \tag{8.50}$$

Formula (8.45) remains valid in this case, but, instead of (8.46), we can obtain a simpler relation for $B_0(s)$

$$B_0(s) = \frac{1}{T}\int_0^T W(s,t)W(-s,t)\,\mathrm{d}t. \tag{8.51}$$

Consider the important special case of Eq. (8.43), when $R_x(s) = 1$. As is known, this corresponds to the case when the input signal is a unit white noise. In this situation, to denote the output variance we use the following special notation

$$r_y(t) \triangleq d_y(t)\,\big|_{R_x(s)=1}\,. \tag{8.52}$$

Hence, Eq. (8.43) takes the form

$$r_y(t) = \frac{1}{2\pi \mathrm{j}}\int_{-\mathrm{j}\infty}^{\mathrm{j}\infty} W(-s,t)W(s,t)\,\mathrm{d}s. \tag{8.53}$$

The function $r_y(t)$ is independent of the characteristics of the input signals and is determinded only by the properties of the operator (8.28), i.e., by the properties of the Green function $h(t,\tau)$. Since for a fixed t we have

$$W(s,t) = \int_{-\infty}^{\infty} g(t,u)\mathrm{e}^{-su}\,\mathrm{d}u$$

where $g(t, u)$ is the pulse response of the operator (8.28). Applying Parseval's formula (1.67) gives

$$r_y(t) = \int_{-\infty}^{\infty} [g(t,u)]^2 \, du \,. \tag{8.54}$$

Using (6.73), we obtain

$$r_y(t) = \int_{-\infty}^{\infty} [h(t, t-v)]^2 \, dv \,. \tag{8.55}$$

Performing the change of variable $t - v = \tau$, we express $r_y(t)$ in terms of the Green function

$$r_y(t) = \int_{-\infty}^{\infty} [h(t,\tau)]^2 \, d\tau \,. \tag{8.56}$$

If the operator is causal, i.e. (6.70) holds, we obtain from (8.56)

$$r_y(t) = \int_{-\infty}^{t} [h(t,\tau)]^2 \, d\tau \,. \tag{8.57}$$

We denote the average value of the function $r_y(t)$ by $\overline{r}_y$, i.e.,

$$\overline{r}_y \triangleq \frac{1}{T} \int_0^T r_y(t) \, dt \,. \tag{8.58}$$

Then, from (8.56) it follows that

$$\overline{r}_y = \frac{1}{T} \int_0^T \int_{-\infty}^{\infty} [h(t,\tau)]^2 \, d\tau \, dt \,. \tag{8.59}$$

For a causal operator Eq. (8.59) takes the form

$$\overline{r}_y = \frac{1}{T} \int_0^T \int_{-\infty}^{t} [h(t,\tau)]^2 \, d\tau \, dt \tag{8.60}$$

The number

$$\|\mathsf{U}\|_2 \triangleq +\sqrt{\overline{r}_y} \tag{8.61}$$

will be henceforth called the *$\mathcal{H}_2$-norm* of the operator under consideration. The $\mathcal{H}_2$-norm of the operator can be expressed in terms of the function $B_0(s)$ (8.51). Indeed, using (8.53), we have

$$\overline{r}_y = \frac{1}{T} \int_0^T \left[\frac{1}{2\pi \mathrm{j}} \int_{-\mathrm{j}\infty}^{\mathrm{j}\infty} W(s,t) W(-s,t) \, ds \right] dt \,.$$

Interchanging the order of integration by s and t and using (8.51), we find

$$\overline{r}_y = \frac{1}{2\pi \mathrm{j}} \int_{-\mathrm{j}\infty}^{\mathrm{j}\infty} B_0(s) \, ds \,. \tag{8.62}$$

Correspondingly, the $\mathcal{H}_2$-norm appears as

$$\|\mathsf{U}\|_2^2 = \frac{1}{2\pi \mathrm{j}} \int_{-\mathrm{j}\infty}^{\mathrm{j}\infty} B_0(s)\, \mathrm{d}s\,. \tag{8.63}$$

Since $\mathsf{U} \in \mathcal{U}(-\lambda, \lambda)$ by assumption, the integrand in (8.62) and (8.63) decreases as $|s|^{-2}$ for $|s| \to \infty$. Therefore, the integrals can be discretized with period T (i.e., with step $\mathrm{j}\omega$). As a special case, Eq. (8.63) takes the form

$$\|\mathsf{U}\|_2^2 = \frac{T}{2\pi \mathrm{j}} \int_{-\mathrm{j}\omega/2}^{\mathrm{j}\omega/2} \mathcal{D}_{B_0}(T, s, 0)\, \mathrm{d}s \tag{8.64}$$

where

$$\mathcal{D}_{B_0}(T, s, 0) = \frac{1}{T} \sum_{k=-\infty}^{\infty} B_0(s + k\mathrm{j}\omega)\,, \quad \omega = 2\pi/T\,. \tag{8.65}$$

8.3 Stochastic analysis and $\mathcal{H}_2$–norm of multivariable operators

The relations of Sections 8.1 and 8.2 can be, with minor modifications, extended to the multivariable case, Rosenwasser (1977). Let us take the multivariable operator

$$\boldsymbol{y}(t) = \int_{-\infty}^{\infty} \boldsymbol{h}(t,\tau)\boldsymbol{x}(\tau)\, \mathrm{d}\tau = \mathsf{U}[\boldsymbol{x}(t)] \tag{8.66}$$

where $\boldsymbol{y}$ and $\boldsymbol{x}$ are vectors of dimensions $n \times 1$ and $m \times 1$, respectively, and $\boldsymbol{h}(t,\tau)$ is an $n \times m-$ matrix, that will be called the *Green matrix* of the operator (8.66). Henceforth, it is assumed that each component of the operator U belongs to the space $\mathcal{U}(-\lambda, \lambda)$ with a positive λ. In this case the operator (8.66) will be called *e-dichotomic.*

Let the input $\boldsymbol{x}(t)$ be a centered stationary stochastic vector, i.e., the components (5.14) of the vector $\boldsymbol{x}(t)$ are centered stationary and stationary connected scalar stochastic processes. The mathematical expectation of the vector $\boldsymbol{x}(t)$ equals

$$\mathsf{E}\left[\boldsymbol{x}(t)\right] = \begin{bmatrix} \mathsf{E}\left[x_1(t)\right] \\ \mathsf{E}\left[x_2(t)\right] \\ \vdots \\ \mathsf{E}\left[x_m(t)\right] \end{bmatrix}. \tag{8.67}$$

Since the process is centered,

$$\mathsf{E}\left[\boldsymbol{x}(t)\right] = \mathbf{0}_{m1}\,, \tag{8.68}$$

where $\mathbf{0}_{nm}$ denotes the $n \times m$ zero matrix. The matrix

$$\boldsymbol{K}_{\boldsymbol{x}\boldsymbol{x}}(t_1,t_2) \triangleq \mathsf{E}\left[\boldsymbol{x}(t_1)\boldsymbol{x}'(t_2)\right] \tag{8.69}$$

is called the *autocorrelation matrix* of the vector $\boldsymbol{x}(t)$. In (8.69) and henceforth the prime denotes the transpose operator. If there is a centered stochastic vector $\boldsymbol{z}(t)$ of the same dimension, the matrix

$$\boldsymbol{K}_{\boldsymbol{x}\boldsymbol{z}}(t_1,t_2) \triangleq \mathsf{E}\left[\boldsymbol{x}(t_1)\boldsymbol{z}'(t_2)\right] \tag{8.70}$$

is the *cross-correlation matrix* of the vectors $\boldsymbol{x}(t)$ and $\boldsymbol{z}(t)$. From (8.69) and (8.70) it immediately follows that

$$\boldsymbol{K}_{\boldsymbol{x}\boldsymbol{x}}(t_1,t_2) = \boldsymbol{K}_{\boldsymbol{x}\boldsymbol{x}}{}'(t_2,t_1)\,, \quad \boldsymbol{K}_{\boldsymbol{x}\boldsymbol{z}}(t_1,t_2) = \boldsymbol{K}_{\boldsymbol{z}\boldsymbol{x}}{}'(t_2,t_1)\,. \tag{8.71}$$

If the vector $\boldsymbol{x}(t)$ is stationary, then

$$\boldsymbol{K}_{\boldsymbol{x}\boldsymbol{x}}(t_1,t_2) = \boldsymbol{K}_{\boldsymbol{x}\boldsymbol{x}}(t_2-t_1)\,. \tag{8.72}$$

Two stationary vector processes are called stationary connected if

$$\boldsymbol{K}_{\boldsymbol{x}\boldsymbol{z}}(t_1,t_2) = \boldsymbol{K}_{\boldsymbol{x}\boldsymbol{z}}(t_2-t_1)\,. \tag{8.73}$$

Taking $t_1 = t$ and $t_2 = t+\tau$ in (8.72) and (8.73), with due account for (8.69) and (8.70) we find

$$\begin{aligned} \boldsymbol{K}_{\boldsymbol{x}\boldsymbol{x}}(t,t+\tau) &= \mathsf{E}\left[\boldsymbol{x}(t)\boldsymbol{x}'(t+\tau)\right] = \boldsymbol{K}_{\boldsymbol{x}\boldsymbol{x}}(\tau) \\ \boldsymbol{K}_{\boldsymbol{x}\boldsymbol{z}}(t,t+\tau) &= \mathsf{E}\left[\boldsymbol{x}(t)\boldsymbol{z}'(t+\tau)\right] = \boldsymbol{K}_{\boldsymbol{x}\boldsymbol{z}}(\tau)\,. \end{aligned} \tag{8.74}$$

Then, as follows from (8.71),

$$\boldsymbol{K}_{\boldsymbol{x}\boldsymbol{x}}(\tau) = \boldsymbol{K}_{\boldsymbol{x}\boldsymbol{x}}{}'(-\tau)\,, \quad \boldsymbol{K}_{\boldsymbol{x}\boldsymbol{z}}(\tau) = \boldsymbol{K}_{\boldsymbol{z}\boldsymbol{x}}{}'(-\tau)\,. \tag{8.75}$$

The spectral characteristics of vector stochastic processes are of prime importance together with correlation functions (matrices). Let the following estimate hold

$$\|\boldsymbol{K}_{\boldsymbol{x}\boldsymbol{x}}(\tau)\| < N\mathrm{e}^{-\lambda|\tau|} \tag{8.76}$$

where N and λ are positive constants and $\|\cdot\|$ denotes a numerical norm of a finite-dimensional vector or matrix. In this case, there exists the Laplace transform

$$\boldsymbol{R}_{\boldsymbol{x}\boldsymbol{x}}(s) = \int_{-\infty}^{\infty} \boldsymbol{K}_{\boldsymbol{x}\boldsymbol{x}}(\tau)\mathrm{e}^{-s\tau}\,\mathrm{d}\tau \tag{8.77}$$

which is called the *spectral density matrix* of the vector $\boldsymbol{x}(t)$. The matrix

$$\boldsymbol{R}_{\boldsymbol{x}\boldsymbol{z}}(s) = \int_{-\infty}^{\infty} \boldsymbol{K}_{\boldsymbol{x}\boldsymbol{z}}(\tau)\mathrm{e}^{-s\tau}\,\mathrm{d}\tau \tag{8.78}$$

is called, under the same assumptions about convergence, the *cross spectral matrix* of the vectors $\boldsymbol{x}(t)$ and $\boldsymbol{z}(t)$. Assuming that $\boldsymbol{K}_{\boldsymbol{x}\boldsymbol{x}} \in \Lambda(-\lambda,\lambda)$ from (8.77) we obtain, using the inversion formula (1.24),

$$\boldsymbol{K_{xx}}(\tau) = \frac{1}{2\pi\mathrm{j}} \int_{-\mathrm{j}\infty}^{\mathrm{j}\infty} \boldsymbol{R_{xx}}(s)\mathrm{e}^{s\tau}\,\mathrm{d}s\,. \tag{8.79}$$

The inversion of Eq.(8.78) under similar assumptions yields

$$\boldsymbol{K_{xz}}(\tau) = \frac{1}{2\pi\mathrm{j}} \int_{-\mathrm{j}\infty}^{\mathrm{j}\infty} \boldsymbol{R_{xz}}(s)\mathrm{e}^{s\tau}\,\mathrm{d}s\,. \tag{8.80}$$

From (8.77) and (8.78) taking due account of (8.75) it follows that

$$\boldsymbol{R_{xx}}(s) = \boldsymbol{R_{xx}}'(-s)\,, \quad \boldsymbol{R_{xz}}(s) = \boldsymbol{R_{xz}}'(-s)\,. \tag{8.81}$$

Using the above relations, we can determine statistical characteristics of the output of the operator (8.66), assuming that it is e-dichotomic. The input vector $\boldsymbol{x}(t)$ is assumed to be stationary and centered. In this case the output $\boldsymbol{y}(t)$ is a centered stochastic vector. Assuming that the integral (8.66) converges for any realization of the input, we have

$$\boldsymbol{y}(t_1) = \int_{-\infty}^{\infty} \boldsymbol{h}(t_1,u)\boldsymbol{x}(u)\,\mathrm{d}u\,, \qquad \boldsymbol{y}(t_2) = \int_{-\infty}^{\infty} \boldsymbol{h}(t_2,v)\boldsymbol{x}(v)\,\mathrm{d}v\,. \tag{8.82}$$

Hence,

$$\boldsymbol{y}(t_1)\boldsymbol{y}'(t_2) = \int_{-\infty}^{\infty}\int_{-\infty}^{\infty} \boldsymbol{h}(t_1,u)\boldsymbol{x}(u)\boldsymbol{x}'(v)\boldsymbol{h}'(t_2,v)\,\mathrm{d}u\,\mathrm{d}v\,. \tag{8.83}$$

Using the mathematical expectation operator on both sides of (8.83) and assuming that it is commutative with the integration operators by u and v, we obtain

$$\mathsf{E}\left[\boldsymbol{y}(t_1)\boldsymbol{y}'(t_2)\right] = \int_{-\infty}^{\infty}\int_{-\infty}^{\infty} \boldsymbol{h}(t_1,u)\,\mathsf{E}\left[\boldsymbol{x}(u)\boldsymbol{x}'(v)\right]\boldsymbol{h}'(t_2,v)\,\mathrm{d}u\,\mathrm{d}v\,. \tag{8.84}$$

Using (8.69) and (8.72), we find

$$\boldsymbol{K_{yy}}(t_1,t_2) = \int_{-\infty}^{\infty}\int_{-\infty}^{\infty} \boldsymbol{h}(t_1,u)\,\boldsymbol{K_{xx}}(v-u)\boldsymbol{h}'(t_2,v)\,\mathrm{d}u\,\mathrm{d}v\,. \tag{8.85}$$

From (8.79) it follows that

$$\boldsymbol{K_{xx}}(v-u) = \frac{1}{2\pi\mathrm{j}} \int_{-\mathrm{j}\infty}^{\mathrm{j}\infty} \boldsymbol{R_{xx}}(s)\mathrm{e}^{s(v-u)}\,\mathrm{d}s\,. \tag{8.86}$$

Substituting (8.86) into (8.85) and changing the order of integration by different variables, we find

$$\boldsymbol{K_{yy}}(t_1,t_2) = \frac{1}{2\pi\mathrm{j}} \int_{-\mathrm{j}\infty}^{\mathrm{j}\infty} \left[\int_{-\infty}^{\infty} \boldsymbol{h}(t_1,u)\mathrm{e}^{-su}\,\mathrm{d}u\right] \boldsymbol{R_{xx}}(s) \left[\int_{-\infty}^{\infty} \boldsymbol{h}'(t_2,v)\mathrm{e}^{sv}\,\mathrm{d}v\right] \mathrm{d}s\,. \tag{8.87}$$

But, by definition of the PTM (5.27), we have

$$\begin{aligned}\int_{-\infty}^{\infty} \boldsymbol{h}(t_1,u)\mathrm{e}^{-su}\,\mathrm{d}u &= \boldsymbol{W}(-s,t_1)\mathrm{e}^{-st_1}\\ \int_{-\infty}^{\infty} \boldsymbol{h}'(t_2,v)\mathrm{e}^{sv}\,\mathrm{d}v &= \boldsymbol{W}'(s,t_2)\mathrm{e}^{st_2}\end{aligned} \tag{8.88}$$

where $\boldsymbol{W}(s,t)$ is the PTM of the operator (8.66). Substituting (8.88) into (8.87), we obtain the output correlation matrix

$$\boldsymbol{K_{yy}}(t_1,t_2) = \frac{1}{2\pi\mathrm{j}}\int_{-\mathrm{j}\infty}^{\mathrm{j}\infty} \boldsymbol{W}(-s,t_1)\boldsymbol{R_{xx}}(s)\boldsymbol{W}'(s,t_2)\mathrm{e}^{s(t_2-t_1)}\,\mathrm{d}s\,. \tag{8.89}$$

If the operator (8.66) is T–periodic, i.e.,

$$\boldsymbol{h}(t+T,\tau+T) = \boldsymbol{h}(t,\tau)$$

the corresponding PTM is given by

$$\boldsymbol{W}(s,t+T) = \boldsymbol{W}(s,t) \tag{8.90}$$

and Eq. (6.69) yields

$$\boldsymbol{K_{yy}}(t_1+T,t_2+T) = \boldsymbol{K_{yy}}(t_1,t_2) \tag{8.91}$$

i.e., the output $\boldsymbol{y}(t)$ is a periodically non-stationary process. By the above assumptions, for $s=\mathrm{j}\nu$ and $|\nu|\to\infty$ the PTM $\boldsymbol{W}(s,t)$ satisfies the following estimate, similar to (6.84)

$$\|\boldsymbol{W}(\mathrm{j}\nu,t)\| < L|\nu|^{-1} \tag{8.92}$$

where $L>0$ is a constant. Hence, if the elements of the matrix $\boldsymbol{R_{xx}}(s)$ are uniformly bounded on the imaginary axis, then the elements of the integrand in (8.89) decrease as $|\nu|^{-2}$ for $|\nu|\to\infty$. Therefore, the correlation matrix $\boldsymbol{K_{yy}}(t_1,t_2)$ is continuous with respect to both arguments, t_1 and t_2, and we can take $t_1=t_2=t$ in (8.89). Thus, we obtain the matrix

$$\boldsymbol{K_y}(t) \triangleq \boldsymbol{K_{yy}}(t_1,t_2)\,|_{t_1=t_2=t} = \frac{1}{2\pi\mathrm{j}}\int_{-\mathrm{j}\infty}^{\mathrm{j}\infty} \boldsymbol{W}(-s,t)\boldsymbol{R_{xx}}(s)\boldsymbol{W}'(s,t)\,\mathrm{d}s \tag{8.93}$$

which is continuous with respect to t. In this case Eq. (8.90) yields

$$\boldsymbol{K_y}(t) = \boldsymbol{K_y}(t+T)\,. \tag{8.94}$$

The trace of the matrix (8.93)

$$\operatorname{tr}\boldsymbol{K_y}(t) \triangleq \boldsymbol{d_y}(t) \tag{8.95}$$

will be called the variance of the output, and

$$\overline{d}_{\boldsymbol{y}} \triangleq \frac{1}{T}\int_0^T \boldsymbol{d}_{\boldsymbol{y}}(t)\,\mathrm{d}t \tag{8.96}$$

will be called the mean variance of the vector output $\boldsymbol{y}(t)$. Using (8.93), we find

$$\boldsymbol{d}_{\boldsymbol{y}}(t) = \frac{1}{2\pi \mathrm{j}}\int_{-\mathrm{j}\infty}^{\mathrm{j}\infty} \mathrm{tr}\left[\boldsymbol{W}(-s,t)\boldsymbol{R}_{\boldsymbol{x}\boldsymbol{x}}(s)\boldsymbol{W}'(s,t)\right]\,\mathrm{d}s\,. \tag{8.97}$$

Since

$$\mathrm{tr}\,\boldsymbol{A}\boldsymbol{B} = \mathrm{tr}\,\boldsymbol{B}\boldsymbol{A} \tag{8.98}$$

Eq. (8.97) can be written in the form

$$\begin{aligned}\boldsymbol{d}_{\boldsymbol{y}}(t) &= \frac{1}{2\pi \mathrm{j}}\int_{-\mathrm{j}\infty}^{\mathrm{j}\infty} \mathrm{tr}\left[\boldsymbol{W}'(s,t)\boldsymbol{W}(-s,t)\boldsymbol{R}_{\boldsymbol{x}\boldsymbol{x}}(s)\right]\,\mathrm{d}s \\ &= \frac{1}{2\pi \mathrm{j}}\int_{-\mathrm{j}\infty}^{\mathrm{j}\infty} \mathrm{tr}\left[\boldsymbol{R}_{\boldsymbol{x}\boldsymbol{x}}(s)\boldsymbol{W}'(s,t)\boldsymbol{W}(-s,t)\right]\,\mathrm{d}s\,.\end{aligned} \tag{8.99}$$

Substituting (8.99) into (8.96) and changing the order of integration by s and t, we find

$$\overline{d}_{\boldsymbol{y}} = \frac{1}{2\pi \mathrm{j}}\int_{-\mathrm{j}\infty}^{\mathrm{j}\infty} \mathrm{tr}\left[\boldsymbol{B}_0(s)\boldsymbol{R}_{\boldsymbol{x}\boldsymbol{x}}(s)\right]\,\mathrm{d}s = \frac{1}{2\pi \mathrm{j}}\int_{-\mathrm{j}\infty}^{\mathrm{j}\infty} \mathrm{tr}\left[\boldsymbol{R}_{\boldsymbol{x}\boldsymbol{x}}(s)\boldsymbol{B}_0(s)\right]\,\mathrm{d}s \tag{8.100}$$

where

$$\boldsymbol{B}_0(s) = \frac{1}{T}\int_0^T \boldsymbol{W}'(s,t)\boldsymbol{W}(-s,t)\,\mathrm{d}t \tag{8.101}$$

is an $m \times m$ matrix.
Consider the important special case of Eq.(8.97), when

$$\boldsymbol{R}_{\boldsymbol{x}\boldsymbol{x}}(s) = \boldsymbol{I}_m\,. \tag{8.102}$$

In this case, using the notation (8.52), from (8.97) we obtain

$$\boldsymbol{r}_{\boldsymbol{y}}(t) = \frac{1}{2\pi \mathrm{j}}\int_{-\mathrm{j}\infty}^{\mathrm{j}\infty} \mathrm{tr}\left[\boldsymbol{W}(-s,t)\boldsymbol{W}'(s,t)\right]\,\mathrm{d}s = \frac{1}{2\pi \mathrm{j}}\int_{-\mathrm{j}\infty}^{\mathrm{j}\infty} \mathrm{tr}\left[\boldsymbol{W}'(s,t)\boldsymbol{W}(-s,t)\right]\,\mathrm{d}s\,. \tag{8.103}$$

In this case, the value $\overline{\boldsymbol{r}}_{\boldsymbol{y}}$ (8.58) is given by

$$\overline{\boldsymbol{r}}_{\boldsymbol{y}} = \frac{1}{2\pi \mathrm{j}}\int_{-\mathrm{j}\infty}^{\mathrm{j}\infty} \mathrm{tr}\,\boldsymbol{B}_0(s)\,\mathrm{d}s\,. \tag{8.104}$$

The value

$$\|\mathsf{U}\|_2 \triangleq \sqrt{\overline{\boldsymbol{r}}_{\boldsymbol{y}}} \tag{8.105}$$

will be called the $\mathcal{H}_2$-norm of the integral operator (8.66). As follows from the above relations,

$$\|\mathsf{U}\|_2^2 = \frac{1}{2\pi \mathrm{j}} \int_{-\mathrm{j}\infty}^{\mathrm{j}\infty} \operatorname{tr} \boldsymbol{B}_0(s)\,\mathrm{d}s\,. \tag{8.106}$$

Under the above assumptions each component of the PTM $\boldsymbol{W}(s,t)$ decreases, in the stripe of convergence, as $|s|^{-2}$ for $|s| \to \infty$. Therefore, the integral (8.106) can be discretized with the period T as

$$\|\mathsf{U}\|_2^2 = \frac{T}{2\pi \mathrm{j}} \int_{-\mathrm{j}\omega/2}^{\mathrm{j}\omega/2} \operatorname{tr} \mathcal{D}_{\boldsymbol{B}_0}(T,s,0)\,\mathrm{d}s \tag{8.107}$$

where

$$\mathcal{D}_{\boldsymbol{B}_0}(T,s,0) = \frac{1}{T} \sum_{k=-\infty}^{\infty} \boldsymbol{B}_0(s+k\mathrm{j}\omega)\,, \quad \omega = 2\pi/T\,. \tag{8.108}$$

Next, we show that the $\mathcal{H}_2$-norm of the operator (8.66) can be expressed in terms of the Green function. With this aim in view, we consider the integral

$$\tilde{\boldsymbol{K}}_{\boldsymbol{y}}(t) = \frac{1}{2\pi \mathrm{j}} \int_{-\mathrm{j}\infty}^{\mathrm{j}\infty} \boldsymbol{W}(-s,t)\boldsymbol{W}'(s,t)\,\mathrm{d}s \tag{8.109}$$

obtained from (8.93) for $\boldsymbol{R}_{\boldsymbol{xx}}(s) = \boldsymbol{I}_m$. Similarly to the SISO case, we have

$$\boldsymbol{W}(s,t) = \int_{-\infty}^{\infty} \boldsymbol{g}(t,u)\mathrm{e}^{-su}\,\mathrm{d}u \tag{8.110}$$

where

$$\boldsymbol{g}(t,u) \stackrel{\triangle}{=} \boldsymbol{h}(t,\tau)\,|_{\tau=t-u}\ . \tag{8.111}$$

Therefore, using Parseval's formula in the form (1.66), we can find the following equation similar to (8.54)

$$\tilde{\boldsymbol{K}}_{\boldsymbol{y}}(t) = \int_{-\infty}^{\infty} \boldsymbol{g}(t,u)\boldsymbol{g}'(t,u)\,\mathrm{d}u\,. \tag{8.112}$$

Taking $u = t - \tau$, we obtain

$$\tilde{\boldsymbol{K}}_{\boldsymbol{y}}(t) = \int_{-\infty}^{\infty} \boldsymbol{h}(t,\tau)\boldsymbol{h}'(t,\tau)\,\mathrm{d}\tau\,. \tag{8.113}$$

Therefore, using the above equations, we obtain the formula for the $\mathcal{H}_2$-norm of the operator U

$$\|\mathsf{U}\|_2^2 = \frac{1}{T} \int_0^T \int_{-\infty}^{\infty} \operatorname{tr}\left[\boldsymbol{h}(t,\tau)\boldsymbol{h}'(t,\tau)\right]\,\mathrm{d}\tau\,\mathrm{d}t\,. \tag{8.114}$$

If, as a special case,

$$\boldsymbol{h}(t,\tau) = \boldsymbol{0}_{nm} \qquad \text{for} \quad t < \tau \tag{8.115}$$

where $\boldsymbol{0}_{nm}$ denotes the $n \times m$ zero matrix, i.e. the operator (8.66) is causal, it follows from (8.113) that

$$\tilde{\boldsymbol{K}}_{\boldsymbol{y}}(t) = \int_{-\infty}^{t} \boldsymbol{h}(t,\tau)\boldsymbol{h}'(t,\tau)\,\mathrm{d}\tau \tag{8.116}$$

and Eq. (8.114) takes the form

$$\|\,\mathsf{U}\|_2^2 = \frac{1}{T}\int_0^T \int_{-\infty}^{t} \operatorname{tr}\boldsymbol{h}(t,\tau)\boldsymbol{h}'(t,\tau)\,\mathrm{d}\tau\,\mathrm{d}t\,. \tag{8.117}$$

8.4 Stochastic analysis of continuous-discrete operators and LPPO

In this section we consider the transition process of a stationary stochastic signal through a continuous-discrete operator U_2 shown in Fig. 6.3. This operator can be defined by the equations

$$y(t) = \mathsf{U}_d[z(t)]\,, \quad z(t) = \mathsf{U}[x(t)] \tag{8.118}$$

under the assumptions of Section 6.3. In this case Eq. (8.118) reduces to the integral operator (6.128)

$$y(t) = \mathsf{U}_2[x(t)] = \int_{-\infty}^{\infty} h_2(t,\tau)x(\tau)\,\mathrm{d}\tau \tag{8.119}$$

where the Green function $h_2(t,\tau)$ is given by (6.129). Under these assumptions, all the relations obtained in Section 8.2 are valid for the integral operator (8.119). Nevertheless, since an LPPO (6.97) is a part of the operator U_d (8.118), we can obtain some more important relations. Let $\mathsf{U} \in \mathcal{U}(-\lambda,\lambda)$ and $\mathsf{U}_d \in \mathcal{U}_d(-\lambda,\lambda)$, where λ is a positive number. Then, as follows from (6.132), the continuous-discrete operator U_2 is e-dichotomic, i.e., $\mathsf{U}_2 \in \mathcal{U}(-\lambda,\lambda)$. In this case the transfer function $G(s)$ of the operator U and the equivalent transfer function $Q(s)$ of the operator U_d are analytic in the stripe $|\operatorname{Re} s| < \lambda$ and decrease as $|s|^{-1}$ for $|s| \to \infty$.
Therefore, using (8.41) and the formula for the PTF (6.124), we find the output correlation function of the operator U_2

$$K_y(t_1,t_2) = \frac{1}{2\pi\mathrm{j}}\int_{-\mathrm{j}\infty}^{+\mathrm{j}\infty} \varphi_q(T,-s,t_1)\varphi_q(T,s,t_2)G(s)G(-s)R_x(s)\mathrm{e}^{s(t_2-t_1)}\,\mathrm{d}s\,. \tag{8.120}$$

Taking due account of (4.2), we can represent (8.120) in the form

$$K_y(t_1,t_2) = \frac{1}{2\pi\mathrm{j}}\int_{-\mathrm{j}\infty}^{\mathrm{j}\infty} \mathcal{D}_q(T,-s,t_1)\mathcal{D}_q(T,s,t_2)R_z(s)\,\mathrm{d}s \tag{8.121}$$

where

$$\mathcal{D}_q(T,s,t) = \varphi_q(T,s,t)\mathrm{e}^{st}\,, \quad R_z(s) \triangleq G(s)G(-s)R_x(s)\,. \tag{8.122}$$

If the spectral density $R_x(s)$ is bounded on the imaginary axis, due to the estimate (1.45) the integrands in (8.120) and (8.121) decrease as $|s|^{-2}$ for $|s| \to \infty$. Therefore, the integral (8.121) can be discretized with the period $T = 2\pi/\omega$. Then, taking into account that

$$\mathcal{D}_q(T,s,t) = \mathcal{D}_q(T,s+\mathrm{j}\omega,t)$$

from (8.121) we find

$$K_y(t_1,t_2) = \frac{T}{2\pi\mathrm{j}}\int_{-\mathrm{j}\omega/2}^{\mathrm{j}\omega/2} \mathcal{D}_q(T,-s,t_1)\mathcal{D}_q(T,s,t_2)\mathcal{D}_{R_z}(T,s,0)\,\mathrm{d}s \tag{8.123}$$

where

$$\mathcal{D}_{R_z}(T,s,0) = \frac{1}{T}\sum_{k=-\infty}^{\infty} G(s+k\mathrm{j}\omega)G(-s-k\mathrm{j}\omega)R_x(s+k\mathrm{j}\omega)\,. \tag{8.124}$$

Under the given assumptions, the function $\mathcal{D}_q(T,s,t)$ is continuous with respect to t. Therefore, for $t_1 = t_2 = t$ Eq. (8.123) yield the expression for the output variance

$$d_y(t) = \frac{T}{2\pi\mathrm{j}}\int_{-\mathrm{j}\omega/2}^{\mathrm{j}\omega/2} \mathcal{D}_q(T,-s,t)\mathcal{D}_q(T,s,t)\mathcal{D}_{R_z}(T,s,0)\,\mathrm{d}s\,. \tag{8.125}$$

The above results can be interpreted in the following way. Really, the function $R_z(s)$ in (8.122) is the spectral density of the signal $z(t)$ acting on the input of the LPPO U_d. Therefore, under the appropriate restrictions on the spectral density of the input signal, Eqs. (8.121) and (8.125) are valid for the statistical analysis of linear periodic pulse operators (LPPO).

Let

$$K_z(t) \triangleq \frac{1}{2\pi\mathrm{j}}\int_{-\mathrm{j}\infty}^{\mathrm{j}\infty} R_z(s)\mathrm{e}^{st}\,\mathrm{d}s \tag{8.126}$$

be the correlation function of the input signal of a LPPO. In this case, if the function $R_z(s)$ decreases sufficiently fast for $|s| \to \infty$, then the function $K_z(t)$ is continuous and $K_z(t) \in \Lambda(-\lambda,\lambda)$. Then, using (8.22) and (8.24), we have

$$\mathcal{D}_{R_z}(T,s,0) = \sum_{i=-\infty}^{\infty} K_z(iT)\mathrm{e}^{-isT}\,. \tag{8.127}$$

Comparing (8.127) with (8.22), we find

$$\mathcal{D}_{R_z}(T,s,0) = R_z^*(s)$$

where $R_z^*(s)$ is the discrete spectral density of the stochastic sequence $\{z_m\} = \{z(mT)\}$. Then, Eqs. (8.123) and (8.125) can be represented in the form

$$K_y(t_1, t_2) = \frac{T}{2\pi \mathrm{j}} \int_{-\mathrm{j}\omega/2}^{\mathrm{j}\omega/2} \mathcal{D}_q(T, -s, t_1) \mathcal{D}_q(T, s, t_2) R_z^*(s) \, \mathrm{d}s \tag{8.128}$$

and, correspondingly,

$$d_y(t) = \frac{T}{2\pi \mathrm{j}} \int_{-\mathrm{j}\omega/2}^{\mathrm{j}\omega/2} \mathcal{D}_q(T, -s, t) \mathcal{D}_q(T, s, t) R_z^*(s) \, \mathrm{d}s \,. \tag{8.129}$$

Equations (8.128) and (8.129) have a simple physical meaning, because the output of the LPPO depends only on the stochastic sequence $\{z_m\}$. Considering the integrals (8.128) and (8.129) as initial, we can go further rejecting the assumptions that the input sequence $\{z_m\}$ is a sampling of a continuous stochastic process $z(t)$. Instead, we can assume that there is a stochastic sequence $\{z_m\}$ with correlation function $K_z^*(iT)$ and spectral density

$$R_z^*(s) = \sum_{i=-\infty}^{\infty} K_z(iT) \mathrm{e}^{-isT} \,. \tag{8.130}$$

With such an interpretation, the integrals (8.128) and (8.129) determine properties of the response of the LPPO to an arbitrary stationary discrete input sequence having discrete spectral density (8.130). As a special case, these formulae hold when

$$K_z^*(iT) = 0 \qquad \text{for} \quad |i| > N \tag{8.131}$$

where $N \geq 0$ is an integer. Moreover, the discrete spectral density is a finite sum

$$R_z^*(s) = \sum_{i=-N}^{N} K_z(iT) \mathrm{e}^{-isT} \,. \tag{8.132}$$

In the special case, when there is not a correlation between the consecutive values of $z(mT)$, i.e., the input signal is a discrete white noise, we have

$$K_z^*(0) = d_z^* \,, \quad K_z^*(iT) = 0 \quad (i \neq 0) \tag{8.133}$$

where d_z^* is the variance of the input signal. If (8.133) holds, from (8.132) we obtain

$$R_z^*(s) = d_z^* \tag{8.134}$$

and Eq. (8.129) yields

$$d_y(t) = \frac{T d_z^*}{2\pi \mathrm{j}} \int_{-\mathrm{j}\omega/2}^{\mathrm{j}\omega/2} \mathcal{D}_q(T, -s, t) \mathcal{D}_q(T, s, t) \, \mathrm{d}s \,. \tag{8.135}$$

Using (8.129) and (8.135), we can derive an expression for the mean output variance $\bar{d}_y$ defined, as before, by (8.50). From (8.129) and (8.50) we have

$$\overline{d}_y = \frac{T}{2\pi \mathrm{j}} \int_{-\mathrm{j}\omega/2}^{\mathrm{j}\omega/2} B_0^*(s) R_z^*(s)\,\mathrm{d}s \tag{8.136}$$

where

$$B_0^*(s) = \frac{1}{T} \int_0^T \mathcal{D}_q(T,-s,t)\mathcal{D}_q(T,s,t)\,\mathrm{d}t =$$
$$= \frac{1}{T} \int_0^T \varphi_q(T,-s,t)\varphi_q(T,s,t)\,\mathrm{d}t\,. \tag{8.137}$$

Moreover, by Parseval's formula, we have for the Fourier series

$$B_0^*(s) = \frac{1}{T} \int_0^T \varphi_Q(T,-s,t)\varphi_Q(T,s,t)\,\mathrm{d}t \tag{8.138}$$

where

$$\varphi_Q(T,s,t) = \frac{1}{T} \sum_{k=-\infty}^{\infty} Q(s+k\mathrm{j}\omega)\mathrm{e}^{k\mathrm{j}\omega t} \tag{8.139}$$

and $Q(s)$ is the equivalent transfer function of the LPPO defined by (6.100). If $R_z^*(s) = 1$, we shall use the notation (8.52) and (8.58) for the variance and the mean variance, respectively. Moreover, the value

$$\|\mathsf{U}_d\|_2 \stackrel{\triangle}{=} +\sqrt{\overline{r}_y} \tag{8.140}$$

will be called the $\mathcal{H}_2$−norm of the LPPO (6.97). From (8.136) it follows that

$$\overline{r}_y = \|\mathsf{U}_d\|_2^2 = \frac{T}{2\pi \mathrm{j}} \int_{-\mathrm{j}\omega/2}^{\mathrm{j}\omega/2} B_0^*(s)\,\mathrm{d}s\,. \tag{8.141}$$

Then, we show that the $\mathcal{H}_2$−norm of the LPPO U_d can be expressed in terms of the pulse response $q(t)$. Indeed, from (8.141) and (8.137) it follows that

$$\|\mathsf{U}_d\|_2^2 = \frac{T}{2\pi \mathrm{j}} \int_{-\mathrm{j}\omega/2}^{\mathrm{j}\omega/2} \left[\frac{1}{T} \int_0^T \mathcal{D}_q(T,-s,t)\mathcal{D}_q(T,s,t)\,\mathrm{d}t\right] \mathrm{d}s\,. \tag{8.142}$$

Using Eq. (4.102), we can represent the last equation in the form

$$\|\mathsf{U}_d\|_2^2 = \frac{1}{T} \int_{-\infty}^{\infty} q^2(t)\,\mathrm{d}t\,. \tag{8.143}$$

In the special case, when $q(t) = 0$ for $t < 0$, i.e., the LPPO U_d is causal, we obtain

$$\|\mathsf{U}_d\|_2^2 = \frac{1}{T} \int_0^{\infty} q^2(t)\,\mathrm{d}t\,. \tag{8.144}$$

Part III

Mathematical description of sampled-data systems in continuous time

Introduction In this part we present, on the basis of the general theory developed above, a general frequency domain description of sampled-data systems as linear non-stationary systems with periodically varying parameters that operate in continuous time. To investigate the processes in the infinite time interval $-\infty < t < \infty$, we use the parametric transfer function (PTF) technique. To determine the PTF $W(s,t)$ of a system of any structure, we use the following formal method. For a system with chosen input $x(t)$ and output $y(t)$ we take $x(t) = e^{st}$, where s is a complex parameter. Then, we find the output in the form

$$y(t) = W(s,t)e^{st}\,, \quad W(s,t) = W(s,t+T)$$

where T is the period of the system. For the problems considered in the present book such a solution is always unique and can be obtained in a closed form. In this case the function $W(s,t)$ is the required PTF.

With a known PTF it appears possible to construct the set of linear periodic operators generated by the system under consideration, and to find the corresponding operational transform of the output. To investigate the process under given initial conditions, we use the bilateral Laplace transformation in continuous time.

Chapter 9

Open-loop SD systems

9.1 Sample and zero-order hold

By a *sampling unit of zero-order* we shall mean a system where the input $x(t)$ and output $y(t)$ processes are related by

$$y(t) = \mu(t - kT)x(kT + 0)\,, \quad kT < t < (k+1)T \tag{9.1}$$

where $T > 0$ is a constant called the sampling period, and $\mu(t)$ is a function that defines the shape of the output pulses. Henceforth, it is assumed that the function $\mu(t)$ is defined on $0 \leq t \leq T - 0$, and is piece-wise smooth in this interval. For $t < 0$ and $t > T$ we have $\mu(t) = 0$. Figure 9.1 presents the scheme of signal transformation in accordance to Eq. (9.1).
Equation (9.1) can be represented in the form

$$y(t) = m(t)x(kT + 0)\,, \quad kT < t < (k+1)T \tag{9.2}$$

where $m(t) = m(t + T)$ is a periodic function defined by

$$m(t) \stackrel{\triangle}{=} \sum_{k=-\infty}^{\infty} \mu(t - kT) \tag{9.3}$$

The function $m(t)$ will be called *modulating function.*
If we take $\mu(t) = 1$ in (9.1), the sampling unit (or amplitude pulse element) is called a *sample with zero-order hold.* The equation of a sample with zero-order hold has the form

$$y(t) = x(kT + 0)\,, \quad kT < t < (k+1)T\,. \tag{9.4}$$

A sample and hold element of general form (9.1) can be represented as a series connection of a sample with zero-order hold and a multiplier by the modulating function $m(t)$. The corresponding block-diagram is given in Fig. 9.2, where H_0 denotes the sample and zero-order hold, and $\otimes$ is the multiplication operator.

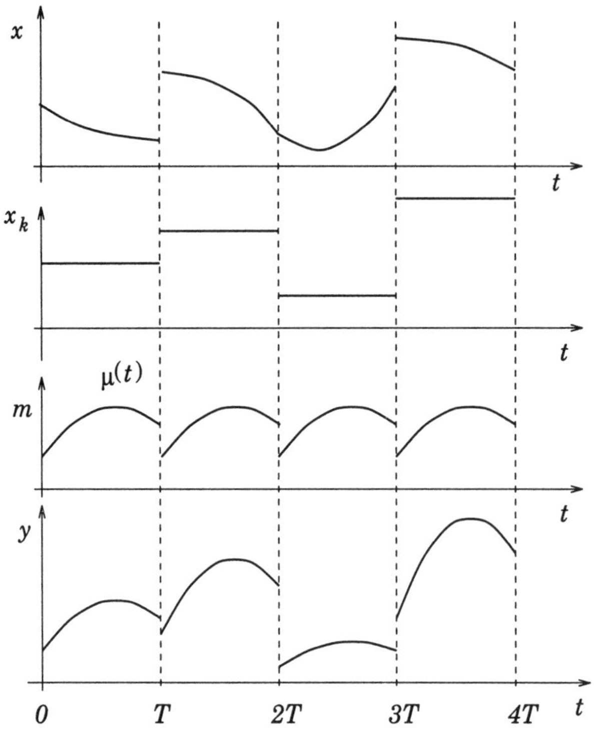

Figure 9.1: Effect of a sampling unit of zero order

Equation (9.1) assumes that the input is real, though it can be extended onto complex inputs as well. In this case, if $x(t) = a(t) + \mathrm{j}b(t)$, where $a(t)$ and $b(t)$ are real functions, we have

$$y(t) = \mu(t - kT)[a(kT + 0) + \mathrm{j}b(kT + 0)], \quad kT < t < (k+1)T. \tag{9.5}$$

With such an approach, Eq.(9.1) defines an operator

$$y(t) = \mathsf{U}_a [x(t)] \tag{9.6}$$

in the space of complex-valued functions of a real argument. Comparing (9.5) with (6.97), we arrive at the conclusion that the operator (9.6) is, from the mathematical point of view, a linear periodic pulse operator (LPPO), the

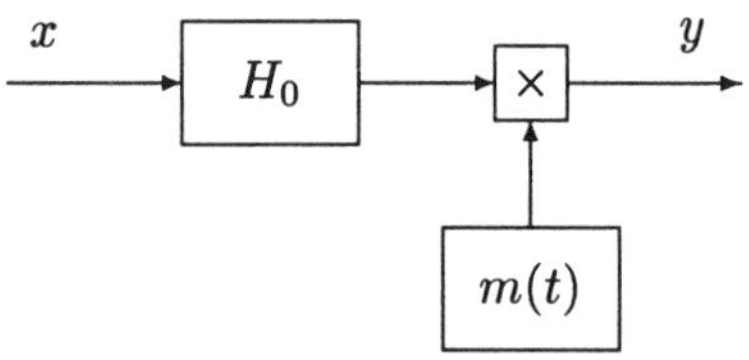

Figure 9.2: Structure of general sampling unit of zero order

impulse response $q(t)$ of which is defined by

$$q(t) = \mu(t)\,. \tag{9.7}$$

Therefore, the PTF $W_a(s,t)$ of the operator (9.6), defined by (6.104) takes the form

$$W_a(s,t) = \sum_{k=-\infty}^{\infty} \mu(t+kT)\mathrm{e}^{-s(t+kT)}\,. \tag{9.8}$$

Taking due account of (9.7), we can represent $W_a(s,t)$ in the form

$$W_a(s,t) = \mu(t)\mathrm{e}^{-st}\,,\ 0 < t < T\,;\quad W_a(s,t) = W_a(s,t+T)\,. \tag{9.9}$$

Taking account of (9.7), the equivalent transfer function (6.100) of the operator U_a takes the form

$$Q(s) = \int_0^T \mu(t)\mathrm{e}^{-st}\,\mathrm{d}t \triangleq M(s)\,. \tag{9.10}$$

The function $M(s)$ will henceforth be called the *transfer function of the forming element.* As follows from (9.10), $M(s)$ is an entire function in the complex variable s. Moreover, since $\mu(t)$ is a function of bounded variation that vanish for $t < 0$, in any half-plane $\operatorname{Re} s > \alpha =$const for $|s| \to \infty$ the following estimate holds

$$|M(s)| < K|s|^{-1} \tag{9.11}$$

where K is a positive constant.

For a zero-order hold, substituting $\mu(t) = 1$ in (9.10), we have

$$M(s) = \int_0^T \mathrm{e}^{-st}\,\mathrm{d}t = \frac{1-\mathrm{e}^{-sT}}{s} \triangleq M_0(s)\,. \tag{9.12}$$

Taking due account of (9.10), the PTF (9.9) can be represented as a Fourier series (6.107) of the form

$$W_a(s,t) = \frac{1}{T}\sum_{k=-\infty}^{\infty} M(s+kj\omega)\mathrm{e}^{kj\omega t} = \varphi_M(T,s,t)\,. \tag{9.13}$$

Multiplying (9.13) by e^{st} and using (6.111), we find the discrete transfer function $D_a(s,t)$ of the operator U_a as

$$D_a(s,t) = \frac{1}{T}\sum_{k=-\infty}^{\infty} M(s+kj\omega)\mathrm{e}^{(s+kj\omega)t} = \mathcal{D}_M(T,s,t)\,. \tag{9.14}$$

Using the series (6.104), with reference to (9.8) we have

$$D_a(s,t) = \mathcal{D}_\mu(T,s,t)\,. \tag{9.15}$$

Equation (9.15) can be represented in a closed form as

$$D_a(s,t) = \mu(t-kT)\mathrm{e}^{ksT}\,, \quad kT < t < (k+1)T\,. \tag{9.16}$$

If the input signal of the sample and hold is such that $x(t) \in \Lambda(\alpha,\beta)$ and $X(s)$ is its Laplace transform, the Laplace image of the output can be found with the use of (7.82) as

$$Y(s) = M(s)\mathcal{D}_x(T,s,+0) \tag{9.17}$$

where

$$\mathcal{D}_x(T,s,+0) = \sum_{k=-\infty}^{\infty} x(kT+0)\mathrm{e}^{-ksT} = \varphi_x(T,s,+0)\,. \tag{9.18}$$

In this case, in accordance with (7.60) and (7.80), the discrete Laplace transform of the output $\mathcal{D}_y(T,s,t)$ is given by

$$\mathcal{D}_y(T,s,t) = \mathcal{D}_\mu(T,s,t)\mathcal{D}_x(T,s,+0)\,. \tag{9.19}$$

Since $\mu(t)$ is piece-wise smooth, from (7.58) it follows that for $x(t) \in \Lambda(\alpha,\beta)$ we have also $y(t) \in \Lambda(\alpha,\beta)$. Therefore, at the points of continuity of $y(t)$ the inversion formula (1.24) holds. Hence,

$$y(t) = \frac{1}{2\pi\mathrm{j}}\int_{c-\mathrm{j}\infty}^{c+\mathrm{j}\infty} M(s)\mathcal{D}_x(T,s,+0)\mathrm{e}^{st}\,\mathrm{d}s\,, \quad \alpha < c < \beta\,. \tag{9.20}$$

Moreover, the inversion formulae for the discrete Laplace transformation (4.34) and (9.19) yield

$$y(t) = \frac{T}{2\pi\mathrm{j}}\int_{c-\mathrm{j}\omega/2}^{c+\mathrm{j}\omega/2} \mathcal{D}_\mu(T,s,t)\mathcal{D}_x(T,s,+0)\mathrm{e}^{st}\,\mathrm{d}s\,, \quad \alpha < c < \beta\,. \tag{9.21}$$

If, in addition, the image of the input signal $X(s)$ satisfies the estimate (7.98), we can use the inversion integral appearing in (7.97), which in the given case has the form

$$y(t) = \frac{1}{2\pi\mathrm{j}}\int_{c-\mathrm{j}\infty}^{c+\mathrm{j}\infty} W_a(s,t)X(s)\mathrm{e}^{st}\,\mathrm{d}s = \frac{1}{2\pi\mathrm{j}}\int_{c-\mathrm{j}\infty}^{c+\mathrm{j}\infty} D_a(s,t)X(s)\,\mathrm{d}s\,, \quad \alpha < c < \beta\,. \tag{9.22}$$

9.2 Sampling units with phase shift

The general equation of a sampling unit of zero order with phase shift has the form

$$y(t) = x(kT+\phi+0)\mu(t-kT-\phi)\,, \quad kT+\phi < t < (k+1)T+\phi\,, \tag{9.23}$$

where ϕ is the phase shift, and it is assumed, without loss of generality, that $0 \le \phi < T$. As regards the function $\mu(t)$, the same assumptions as in Section 9.1

are taken to be true. Denote by U_a^ϕ the operator defined by (9.23). Then, comparing (9.23) with (6.112), we find that the operator U_d^ϕ defines an LPPO shifted in phase by ϕ with respect to the operator U_a of the sampling unit. Therefore, using (6.114) and (9.9), for the PTF $W_a^\phi(s,t)$ of the operator U_d^ϕ we obtain

$$W_a^\phi(s,t) = W_a(s,t-\phi) = \varphi_\mu(T,s,t-\phi) \tag{9.24}$$

which is, in closed form, equivalent to

$$\begin{aligned} W_a^\phi(s,t) &= \mu(t-\phi)\mathrm{e}^{-s(t-\phi)}\,, \quad \phi < t < \phi + T \\ W_a^\phi(s,t) &= W_a^\phi(s,t+T)\,. \end{aligned} \tag{9.25}$$

Using the representation of $W_a(s,t)$ in the form of a Fourier series (9.13), we have

$$W_a^\phi(s,t) = \frac{1}{T}\sum_{k=-\infty}^{\infty} M(s+k\mathrm{j}\omega)\mathrm{e}^{k\mathrm{j}\omega(t-\phi)} = \varphi_M(T,s,t-\phi)\,. \tag{9.26}$$

We note that, in the given case, the notion of discrete transfer function is not applicable directly.
If $x(t) \in \Lambda(\alpha,\beta)$, with the use of (7.115) we obtain the output Laplace transform

$$Y(s) = M(s)\varphi_x(T,s,\phi+0)\,, \quad \alpha < \operatorname{Re} s < \beta\,. \tag{9.27}$$

Taking due account of (7.109), the discrete Laplace transform of the output is given by

$$\mathcal{D}_y(T,s,t) = \mathcal{D}_\mu(T,s,t-\phi)\mathcal{D}_x(T,s,\phi+0)\,, \quad \alpha < \operatorname{Re} s < \beta\,. \tag{9.28}$$

9.3 Linear periodic holds

By a *periodic hold* we mean a unit that transforms discrete values of the input $x(t)$ at the points $t_k = kT$ into a function of continuous argument t. A hold is called linear, if its output depends linearly on the measured values of the input signal. As the simplest linear periodic hold, we will consider the zero-order hold (9.2). A wide class of linear periodic holds can be defined analytically as follows. It is assumed that there are periodic functions $m_\nu(t) = m_\nu(t+T)\,,\ (\nu = 1,\ldots,r)$, and

$$\mu_\nu(t) = m_\nu(t)\,, \quad 0 < t < T \tag{9.29}$$

where $\mu_\nu(t) = 0$ for $t < 0$ and $t > T$. Also let the values $a_{n\nu}$ for any integer n be defined by

$$a_{n\nu} = \sum_{\lambda=0}^{\ell_\nu} \beta_{\nu\lambda} x(nT - \lambda T + 0) \tag{9.30}$$

where $\beta_{\nu\lambda}$ are known constants and ℓ_ν are natural numbers. Then, a unit realizing the relation

$$y(t) = \sum_{\nu=1}^{r} a_{n\nu}\mu_\nu(t-nT)\,, \quad nT < t < (n+1)T \tag{9.31}$$

is called a *periodic (T-periodic) feedforward hold* or *periodic FIR hold*, Zypkin (1964); Petersohn et al. (1994). The function (9.29) defines the type of extrapolation. If, for example,

$$\mu_\nu(t) = t^\nu\,, \quad 0 < t < T \tag{9.32}$$

the hold (9.31) realizes polynomial extrapolation. Equations (9.29)–(9.31) define an operator

$$y(t) = \mathsf{U}_h[x(t)]\,. \tag{9.33}$$

It can be easily verified that the operator (9.33) is linear and periodic. Moreover, this operator is a sampling unit, because the output $y(t)$ depends only on the input values at $t_k = kT$. To determine the PTF $W_h(s,t)$ of the operator U_h, we use the general formula (5.4). For $x(t) = \mathrm{e}^{st}$ Eq.(9.30) yields

$$a_{n\nu} = \sum_{\lambda=0}^{\ell_\nu} \beta_{\nu\lambda}\mathrm{e}^{(n-\lambda)sT} = \mathrm{e}^{nsT}\sum_{\lambda=0}^{\ell_\nu} \beta_{\nu\lambda}\mathrm{e}^{-\lambda sT}\,. \tag{9.34}$$

From (9.31) we obtain

$$\mathsf{U}_h\left[\mathrm{e}^{st}\right] = \mathrm{e}^{nsT}\sum_{\nu=1}^{r}\sum_{\lambda=0}^{\ell_\nu} \beta_{\nu\lambda}\mathrm{e}^{-\lambda sT}\mu_\nu(t-nT)\,, \quad nT < t < (n+1)T\,. \tag{9.35}$$

As a special case, for $n = 0$ Eq. (9.35) yields

$$\mathsf{U}_h\left[\mathrm{e}^{st}\right] = \sum_{\nu=1}^{r}\sum_{\lambda=0}^{\ell_\nu} \beta_{\nu\lambda}\mathrm{e}^{-\lambda sT}\mu_\nu(t)\,, \quad 0 < t < T\,. \tag{9.36}$$

Multiplying (9.36) by e^{-st}, we find the PTF $W_h(s,t)$ in the interval $0 < t < T$

$$W_h(s,t) = \sum_{\nu=1}^{r} w_{h\nu}(s)\mu(t)\mathrm{e}^{-st} \tag{9.37}$$

where

$$w_{h\nu}(s) \triangleq \sum_{\lambda=1}^{\ell_\nu} \beta_{\nu\lambda}\mathrm{e}^{-\lambda sT} \tag{9.38}$$

are transfer functions of superpositioned pure delay elements. Extending Eq. (9.37) over the whole $t-$axis, we obtain

$$W_h(s,t) = \sum_{\nu=1}^{r} w_{h\nu}(s) W_{a\nu}(s,t) \tag{9.39}$$

where $W_{a\nu}(s,t)$ are the PTFs of sampling units (9.1) with pulse forms $\mu_\nu(t)$. Similarly to (9.10), we denote

$$M_\nu(s) \triangleq \int_0^T \mu_\nu(t) \mathrm{e}^{-st} \, \mathrm{d}t = \int_0^T W_{a\nu}(s,t) \, \mathrm{d}t \, . \tag{9.40}$$

The function

$$Q_h(s) \triangleq \int_0^T W_h(s,t) \, \mathrm{d}t \tag{9.41}$$

will be called the *equivalent transfer function of the sample and hold.* Substituting (9.39) into (9.41) and using (9.40), we find

$$Q_h(s) = \sum_{\nu=1}^{r} w_{h\nu}(s) M_\nu(s) \, . \tag{9.42}$$

Moreover, using the Fourier series expansion

$$W_{a\nu}(s,t) = \frac{1}{T} \sum_{k=-\infty}^{\infty} M_\nu(s + k\mathrm{j}\omega) \mathrm{e}^{k\mathrm{j}\omega t} \tag{9.43}$$

and the relation

$$w_{h\nu}(s) = w_{h\nu}(s + \mathrm{j}\omega) \, , \quad \omega = 2\pi/T \tag{9.44}$$

which follows from (9.38), we obtain the Fourier series for $W_h(s,t)$ in the form

$$W_h(s,t) = \frac{1}{T} \sum_{k=-\infty}^{\infty} Q_h(s + k\mathrm{j}\omega) \mathrm{e}^{k\mathrm{j}\omega t} \, . \tag{9.45}$$

Comparing (9.45) with (6.107), we find that, from the mathematical point of view, Eq. (9.29)–(9.31) define an LPPO with the equivalent transfer function (9.42). Hence, if $x(t) \in \Lambda(\alpha, \beta)$, in analogy with (9.17) we obtain the Laplace transform of the output

$$Y(s) = Q_h(s) \mathcal{D}_x(T, s, +0) \, . \tag{9.46}$$

We note also, that the discrete Laplace transform of the output is given by

$$\mathcal{D}_y(T,s,t) = \sum_{\nu=1}^{r} w_{h\nu}(s) \mathcal{D}_{\mu_\nu}(T,s,t) \, \mathcal{D}_x(T,s,+0) \, . \tag{9.47}$$

Example 9.1 As an example for the use of the above relations we consider a first-order hold described in Fig. 9.3. According to Fig. 9.3, we have

$$y(t) = x_n + \frac{x_n - x_{n-1}}{T} (t - nT) \, , \quad nT < t < (n+1)T \tag{9.48}$$

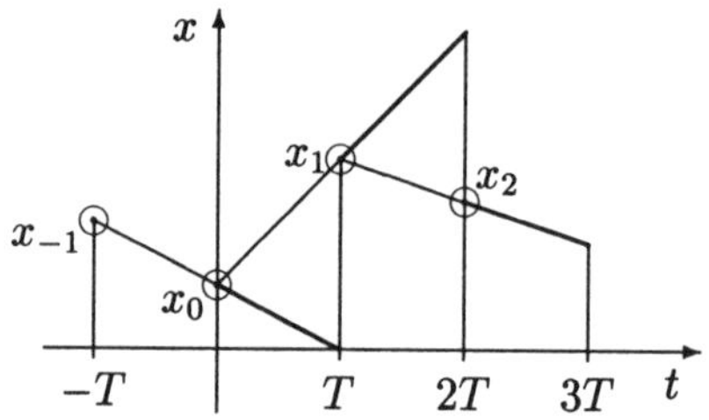

Figure 9.3: First-order hold

where $x_n = x(nT+0)$. Comparing (9.48) and (9.31), we find

$$r = 1\,, \quad a_{n0} = x_n\,, \quad a_{n1} = \frac{1}{T}(x_n - x_{n-1}) \tag{9.49}$$

and, taking account of (9.30),

$$\ell_0 = 0\,, \quad \ell_1 = 1\,, \quad \beta_{00} = 1\,, \quad \beta_{10} = \frac{1}{T}\,, \quad \beta_{11} = -\frac{1}{T}\,. \tag{9.50}$$

Moreover, from (9.48) and (9.31) it follows that

$$\mu_0(t) = 1\,, \quad \mu_1(t) = t \tag{9.51}$$

and Eqs. (9.50) and (9.38) yield

$$w_{h0}(s) = 1\,, \quad w_{h1}(s) = \beta_{10} - \beta_{11}\mathrm{e}^{-sT} = \frac{1}{T}\left(1 - \mathrm{e}^{-sT}\right)\,. \tag{9.52}$$

The transfer functions of the forming elements associated with the functions (9.51) are given by

$$M_0(s) = \frac{1 - \mathrm{e}^{-sT}}{s}\,, \quad M_1(s) = \frac{1 - \mathrm{e}^{-sT} - sT\mathrm{e}^{-sT}}{s^2}\,. \tag{9.53}$$

Substituting (9.52) and (9.53) into (9.42), we find the equivalent transfer function

$$Q_1(s) = \frac{\left(1 - \mathrm{e}^{-sT}\right)^2 (Ts + 1)}{Ts^2}\,. \tag{9.54}$$

Next, we find the impulse response $q_1(t)$ of a first-order hold. As follows from (6.101), it suffices to build the inverse Laplace transform for the image (9.54). We note that from (6.97) it follows that the impulse response $q(t)$ is the response of the LPPO to the input sequence $\{x_k\}$ with $x_0 = 1$ and $x_k = 0$ for $k \neq 0$. Then, for a first-order hold we have $q(t) = f(t)$, where the function $f(t)$ is given by (4.23) and shown in Fig. 4.4. It can be immediately verified, that the Laplace image of the function (4.23) has the form (9.54). Moreover, from (9.54) and (9.46) we find the Laplace transform of the output

$$Y(s) = \frac{\left(1 - \mathrm{e}^{-sT}\right)^2 (Ts + 1)}{Ts^2}\mathcal{D}_x(T, s, +0)\,. \tag{9.55}$$

To derive the discrete Laplace transform of the output, we use the general formula (7.81). According to (4.25), the discrete Laplace transform of the function (4.23) equals

$$\mathcal{D}_q(T,s,t) = \left[1 + \left(1 - \mathrm{e}^{-sT}\right) \frac{t-kT}{T}\right] \mathrm{e}^{ksT}, \quad kT < t < (k+1)T. \tag{9.56}$$

Therefore,

$$\mathcal{D}_y(T,s,t) = \left[1 + \left(1 - \mathrm{e}^{-sT}\right) \frac{t-kT}{T}\right] \mathrm{e}^{ksT} \mathcal{D}_x(T,s,+0) \tag{9.57}$$

$$kT < t < (k+1)T. \quad \square$$

9.4 Elementary open-loop SD systems

In this section we study the system shown in Fig. 9.4. The equations linking

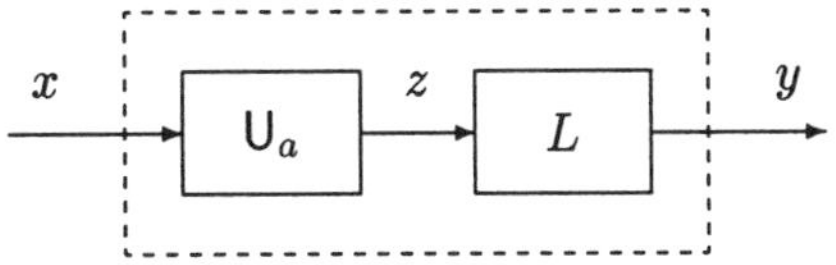

Figure 9.4: Elementary open-loop SD system

the input $x(t)$ and $z(t)$ have the form

$$z(t) = \mathsf{U}_a[x(t)] \tag{9.58}$$

where U_a is the operator of the sampling unit (9.6). The output $y(t)$ and $z(t)$ are related by a linear differential equation with constant coefficients

$$y^{(n)} + d_1 y^{(n-1)} + \ldots + d_n y = m_1 z^{(n-1)} + \ldots + m_n z. \tag{9.59}$$

Using the operator polynomials

$$\begin{aligned} d(s) &= s^n + d_1 s^{n-1} + \ldots + d_n \\ m(s) &= m_1 s^{n-1} + \ldots + m_n \end{aligned} \tag{9.60}$$

we can write (9.59) in the form

$$d(s)y = m(s)z. \tag{9.61}$$

It should be noted that, from a formal viewpoint, the system of equations (9.58) and (9.59) is incorrect, because the function $z(t)$ is, as a rule, discontinuous, and its derivatives at the right-hand side of (9.59), are meaningless in the classical theory. To overcome this hindrance, we pass from the differential equation

(9.59) to an equivalent integral one. First, we consider the processes for $t > 0$. We assume that the functions $y(t)$ and $z(t)$ in Eq. (9.59) can be differentiated as many times as necessary, and have, together with the corresponding derivatives, absolutely convergent Laplace transforms in the half-plane $\operatorname{Re} s > \alpha$ with a sufficiently large α. Using (1.73), we find the associated unilateral Laplace transforms $Y(s)$ and $Z(s)$

$$\begin{aligned} y^{(i)}(t) &\stackrel{\cdot}{\longrightarrow} s^i Y(s) - s^{i-1} y(+0) - \ldots - y^{(i-1)}(+0) \\ z^{(i)}(t) &\stackrel{\cdot}{\longrightarrow} s^i Z(s) - s^{i-1} z(+0) - \ldots - z^{(i-1)}(+0)\,. \end{aligned} \tag{9.62}$$

Performing the Laplace transform in (9.59) and using (9.62), we find the equation in images

$$d(s)Y(s) + d_0(s) = m(s)Z(s) + \tilde{m}_0(s) \tag{9.63}$$

where the polynomials $d_0(s)$ and $\tilde{m}_0(s)$ depend on initial conditions, and

$$\deg d_0(s) \le n-1\,, \quad \deg \tilde{m}_0(s) \le n-2\,. \tag{9.64}$$

Since the coefficients of $d_0(s)$ depend on n arbitrary constants $y(+0), \ldots, y^{(n-1)}(+0)$, Eq. (9.63) can be represented in the form

$$Y(s) = F(s)Z(s) + \frac{m_0(s)}{d(s)} \tag{9.65}$$

where $m_0(s)$ is an arbitrary polynomial such that $\deg m_0(s) \le n-1$, and

$$F(s) = \frac{m(s)}{d(s)} \tag{9.66}$$

is the transfer function of Eq. (9.59). Let

$$d(s) = (s-s_1)^{\nu_1} \cdot \ldots \cdot (s-s_\ell)^{\nu_\ell}\,, \quad \nu_1 + \ldots + \nu_\ell = n\,. \tag{9.67}$$

Then, the partial fraction expansion (1.88) holds

$$F(s) = \sum_{i=1}^{\ell} \sum_{k=1}^{\nu_i} \frac{f_{ik}}{(s-s_i)^k}\,. \tag{9.68}$$

In a similar way we can obtain the expansion

$$Y_0(s) \triangleq \frac{m_0(s)}{d(s)} = \sum_{i=1}^{\ell} \sum_{k=1}^{\nu_i} \frac{m_{ik}}{(s-s_i)^k} \tag{9.69}$$

where the coefficients m_{ik} can be considered as arbitrary constants. The function (1.99)

$$f(t) = \begin{cases} \displaystyle\sum_{i=1}^{\ell} \sum_{k=1}^{\nu_i} \frac{f_{ik}}{(k-1)!} t^{k-1} e^{s_i t}\,, & t > 0 \\ 0 & t < 0\,. \end{cases} \tag{9.70}$$

is the original of the image (9.68) in the half-plane $\operatorname{Re} s > \alpha$. Similarly, the original for $Y_0(s)$ appears as

$$y_0(t) \triangleq \begin{cases} \sum_{i=1}^{\ell} \sum_{k=1}^{\nu_i} c_{ik} \, t^{k-1} \, e^{s_i t} \, , & t > 0 \\ 0 & t < 0 \end{cases} \tag{9.71}$$

where c_{ik} are certain constants. Therefore, passing from (9.65) to the originals in $\operatorname{Re} s > \alpha$ and using the convolution theorem, for $t > 0$ we find

$$y(t) = \int_0^t f(t-\tau) z(\tau) \, d\tau + y_0(t) \, . \tag{9.72}$$

Nevertheless, it can be immediately shown that, with the correspondingly chosen c_{ik}, Eq. (9.72) gives the general solution to Eq. (9.59) for all t. Therefore, we will henceforth use (9.72) instead of the differential equation (9.59). In contrast to (9.59), Eq. (9.72) does not contain derivatives of the input $z(t)$. Therefore, we shall use Eq. (9.59) without assumptions on the continuity of the input signal and its derivatives. This approach can be rigorously substantiated using the theory of generalized functions, Schwartz (1961). We also note that if we have $c_{ik} = 0$ in Eq. (9.71) for all i and k, then

$$y(t) = \int_0^t f(t-\tau) z(\tau) \, d\tau \, . \tag{9.73}$$

This solution will henceforth be called the solution with zero initial energy.

9.5 Transfer functions of open-loop systems

In contrast to the problems considered in Sections 9.1–9.3, the relation between the input $x(t)$ and the output $z(t)$ in (9.58) and (9.59) is not unique and depends on the choice of constants c_{ik}. Hence, we have a system in the sense of Section 6.2 rather than an operator. This system is linear and periodic. To find the PTF $W_p(s,t)$ of this system we take $x = e^{st}$ in Eqs. (9.58) and (9.59), considering s as a parameter. If we find a solution of Eqs. (9.58) and (9.59) having the form

$$y(s,t) = W_p(s,t) e^{st} \, , \quad W_p(s,t) = W_p(s, t+T) \tag{9.74}$$

the function $W_p(s,t)$ will be the required PTF. In the given case this solution can be easily found with the use of Fourier series. Let $x(t) = e^{st}$ in (9.58). Then, for any s we have

$$z(t) = W_a(s,t) e^{st} = \frac{1}{T} \sum_{k=-\infty}^{\infty} M(s + kj\omega) e^{(s+kj\omega)t} \, . \tag{9.75}$$

Then, we have to find a solution of the differential equation (9.59) having the form (9.74) with $z(t)$ given by (9.75). We note, that for $z(t) = e^{\lambda t}$, $\lambda = \text{const.}$, Eq. (9.59) has the solution

$$\tilde{y}(\lambda,t) = \frac{m(\lambda)}{d(\lambda)}\,\mathrm{e}^{\lambda t} = W(\lambda)\mathrm{e}^{\lambda t} \tag{9.76}$$

which is meaningful for all λ except for $\lambda = s_i$, where s_i are the roots of the polynomial (9.67). Using (9.76), we find the solution of (9.75) as the sum of particular solutions of the form (9.76) associated with different terms in the series (9.75). As follows from (9.76), if

$$z(t) = z_k(t) \triangleq \frac{1}{T} M(s+k\mathrm{j}\omega)\mathrm{e}^{(s+k\mathrm{j}\omega)t} \tag{9.77}$$

the corresponding solution of the type (9.76) is given by

$$\tilde{y}_k(s,t) = \frac{1}{T} F(s+k\mathrm{j}\omega)M(s+k\mathrm{j}\omega)\mathrm{e}^{(s+k\mathrm{j}\omega)t}\,. \tag{9.78}$$

Summing up the solutions (9.78) for all k, we find the solution in the form (9.74)

$$y(s,t) \triangleq \frac{1}{T}\sum_{k=-\infty}^{\infty} F(s+k\mathrm{j}\omega)M(s+k\mathrm{j}\omega)\mathrm{e}^{(s+k\mathrm{j}\omega)t}\,. \tag{9.79}$$

Next, we show that this solution is unique. Since (9.79) is a particular solution of (9.59), the corresponding general solution can be written as

$$y(t) = y(s,t) + \sum_{i=1}^{\ell}\sum_{k=1}^{\nu_i} c_{ik}\, t^{k-1}\, \mathrm{e}^{s_i t}\,.$$

It can be immediately verified that the function $y(t)$ has the form (9.74) only for $c_{ik} = 0$. Multiplying (9.79) by e^{-st}, we find the required PTF

$$W_p(s,t) = \frac{1}{T}\sum_{k=-\infty}^{\infty} F(s+k\mathrm{j}\omega)M(s+k\mathrm{j}\omega)\mathrm{e}^{k\mathrm{j}\omega t} \triangleq \varphi_{FM}(T,s,t)\,. \tag{9.80}$$

Since the solution (9.79) is unique, the PTF $W_p(s,t)$ is unique as well. The function

$$D_p(s,t) \triangleq W_p(s,t)\mathrm{e}^{st} = y(s,t) \tag{9.81}$$

will be called the *discrete transfer function* (DTF) of the system (9.58)–(9.59). As follows from (9.80) and (9.81),

$$D_p(s,t) = \frac{1}{T}\sum_{k=-\infty}^{\infty} F(s+k\mathrm{j}\omega)M(s+k\mathrm{j}\omega)\mathrm{e}^{(s+k\mathrm{j}\omega)t} \triangleq \mathcal{D}_{FM}(T,s,t)\,. \tag{9.82}$$

Taking $t = \varepsilon$ and $0 \le \varepsilon < T$, we obtain the function

$$\tilde{D}_p(s,\varepsilon) \triangleq \mathcal{D}_{FM}(T,s,\varepsilon) \tag{9.83}$$

which coincides with the transfer function of the system in the sense of the modified Laplace transformation, Zypkin (1964). As will be shown, the DTF $D_p(s,t)$ is continuous with respect to t. Moreover, the function

$$D_p(s,0) = \frac{1}{T} \sum_{k=-\infty}^{\infty} F(s+kj\omega)M(s+kj\omega) \tag{9.84}$$

coincides. with the transfer function of the system under investigation in the sense of the discrete Laplace transformation. Hence, the function

$$\tilde{D}_p(z) \triangleq D_p(s,0)\,|_{e^{sT}=z} \tag{9.85}$$

is the transfer function of the system in the sense of the z−transformation.

9.6 Closed expressions for DTF and PTF of systems

If (9.68) holds, closed expressions can be obtained for the sum of the series (9.80) and (9.82). We note that, as follows from (9.10),

$$M(s+kj\omega) = \int_0^T \mu(\tau)e^{-(s+kj\omega)\tau}\, d\tau\,. \tag{9.86}$$

Substituting (9.86) into (9.82), we have

$$D_p(s,t) = \frac{1}{T} \sum_{k=-\infty}^{\infty} F(s+kj\omega) \int_0^T \mu(\tau)e^{-(s+kj\omega)\tau}\, d\tau\; e^{(s+kj\omega)t}\,.$$

Changing the order of summation and integration, we find

$$D_p(s,t) = \int_0^T \mathcal{D}_F(T,s,t-\tau)\mu(\tau)\, d\tau \tag{9.87}$$

where

$$\mathcal{D}_F(T,s,t) = \frac{1}{T} \sum_{k=-\infty}^{\infty} F(s+kj\omega)e^{(s+kj\omega)t} \tag{9.88}$$

is the discrete Laplace transform of the transfer function $F(s)$. Let us represent Eq. (9.87) in the form

$$D_p(s,t) = \int_0^t \mathcal{D}_F(T,s,t-\tau)\mu(\tau)\, d\tau + \int_t^T \mathcal{D}_F(T,s,t-\tau)\mu(\tau)\, d\tau\,. \tag{9.89}$$

Taking $0 < t < T$, we obtain $0 < t-\tau < T$ in the first integral and $-T < t-\tau < 0$ in the second one. Then, using (4.41) and (4.43), we have

$$\mathcal{D}_F(T,s,t-\tau) = \begin{cases} \displaystyle\sum_{i=1}^{\ell}\sum_{k=1}^{\nu_i} \frac{f_{ik}}{(k-1)!}\frac{\partial^{k-1}}{\partial s_i^{k-1}}\frac{e^{s_i(t-\tau)}}{1-e^{(s_i-s)T}}\,, & 0<t-\tau<T \\ \displaystyle\sum_{i=1}^{\ell}\sum_{k=1}^{\nu_i} \frac{f_{ik}}{(k-1)!}\frac{\partial^{k-1}}{\partial s_i^{k-1}}\frac{e^{s_i(t-\tau)}}{e^{(s-s_i)T}-1}\,, & -T<t-\tau<0\,. \end{cases} \tag{9.90}$$

Since

$$\frac{1}{1-\mathrm{e}^{(s_i-s)T}} = 1 + \frac{1}{\mathrm{e}^{(s-s_i)T}-1} \tag{9.91}$$

from (9.90) we obtain

$$\mathcal{D}_F(T,s,t-\tau) = \begin{cases} \displaystyle\sum_{i=1}^{\ell}\sum_{k=1}^{\nu_i} \frac{f_{ik}}{(k-1)!}\frac{\partial^{k-1}}{\partial s_i^{k-1}}\frac{\mathrm{e}^{s_i(t-\tau)}}{\mathrm{e}^{(s-s_i)T}-1} + \\ \displaystyle\qquad + \sum_{i=1}^{\ell}\sum_{k=1}^{\nu_i} \frac{f_{ik}}{(k-1)!}\frac{\partial^{k-1}}{\partial s_i^{k-1}}\mathrm{e}^{s_i(t-\tau)}, & 0 < t-\tau < T \\ \displaystyle\sum_{i=1}^{\ell}\sum_{k=1}^{\nu_i} \frac{f_{ik}}{(k-1)!}\frac{\partial^{k-1}}{\partial s_i^{k-1}}\frac{\mathrm{e}^{s_i(t-\tau)}}{\mathrm{e}^{(s-s_i)T}-1}, & -T < t-\tau < 0. \end{cases} \tag{9.92}$$

Instead of (9.92) we can also present the equivalent formulae

$$\mathcal{D}_F(T,s,t-\tau) = \begin{cases} \displaystyle\sum_{i=1}^{\ell}\sum_{k=1}^{\nu_i} \frac{f_{ik}}{(k-1)!}\frac{\partial^{k-1}}{\partial s_i^{k-1}}\frac{\mathrm{e}^{s_i(t-\tau)}}{1-\mathrm{e}^{(s_i-s)T}}, & 0 < t-\tau < T \\ \displaystyle\sum_{i=1}^{\ell}\sum_{k=1}^{\nu_i} \frac{f_{ik}}{(k-1)!}\frac{\partial^{k-1}}{\partial s_i^{k-1}}\frac{\mathrm{e}^{s_i(t-\tau)}}{1-\mathrm{e}^{(s_i-s)T}} - \\ \displaystyle\qquad - \sum_{i=1}^{\ell}\sum_{k=1}^{\nu_i} \frac{f_{ik}}{(k-1)!}\frac{\partial^{k-1}}{\partial s_i^{k-1}}\mathrm{e}^{s_i(t-\tau)}, & -T < t-\tau < 0. \end{cases} \tag{9.93}$$

Substituting (9.92) into (9.89), we obtain the formula for $0 \le t \le T$

$$D_p(s,t) = \sum_{i=1}^{\ell}\sum_{k=1}^{\nu_i} \frac{f_{ik}}{(k-1)!}\frac{\partial^{k-1}}{\partial s_i^{k-1}}\frac{\mathrm{e}^{s_i t}}{\mathrm{e}^{(s-s_i)T}-1}\int\limits_0^T \mu(\tau)\mathrm{e}^{-s_i\tau}\,\mathrm{d}\tau + \int\limits_0^t h(t-\tau)\mu(\tau)\,\mathrm{d}\tau \tag{9.94}$$

where

$$h(t) = \sum_{i=1}^{\ell}\sum_{k=1}^{\nu_i} \frac{f_{ik}}{(k-1)!}t^{k-1}\,\mathrm{e}^{s_i t}. \tag{9.95}$$

Similarly, substituting (9.93) into (9.89), we obtain the equivalent formulae for $0 \le t \le T$

$$D_p(s,t) = \sum_{i=1}^{\ell}\sum_{k=1}^{\nu_i} \frac{f_{ik}}{(k-1)!}\frac{\partial^{k-1}}{\partial s_i^{k-1}}\frac{\mathrm{e}^{s_i t}}{1-\mathrm{e}^{(s_i-s)T}}\int\limits_0^T \mu(\tau)\mathrm{e}^{-s_i\tau}\,\mathrm{d}\tau - \int\limits_t^T h(t-\tau)\mu(\tau)\,\mathrm{d}\tau\,. \tag{9.96}$$

Introduce the notation

$$h_p(t) \triangleq \int_0^t h(t-\tau)\mu(\tau)\,\mathrm{d}\tau \tag{9.97}$$

$$h_p^*(t) \triangleq -\int_t^T h(t-\tau)\mu(\tau)\,\mathrm{d}\tau\,. \tag{9.98}$$

Taking due account of (9.10), from (9.94) and (9.96) we obtain the following equivalent formulae

$$D_p(s,t) = \sum_{i=1}^{\ell}\sum_{k=1}^{\nu_i} \frac{f_{ik}}{(k-1)!}\frac{\partial^{k-1}}{\partial s_i^{k-1}}\frac{\mathrm{e}^{s_i t}M(s_i)}{\mathrm{e}^{(s-s_i)T}-1} + h_p(t)\,, \quad 0 \le t \le T \tag{9.99}$$

or

$$D_p(s,t) = \sum_{i=1}^{\ell}\sum_{k=1}^{\nu_i} \frac{f_{ik}}{(k-1)!}\frac{\partial^{k-1}}{\partial s_i^{k-1}}\frac{\mathrm{e}^{s_i t}M(s_i)}{1-\mathrm{e}^{(s_i-s)T}} + h_p^*(t)\,, \quad 0 \le t \le T\,. \tag{9.100}$$

As will be shown below, the function $D_p(s,t)$ is continuous with respect to t. In what follows, we therefore, consider the closed interval $0 \le t \le T$. In the important case, when all poles of the polynomial (9.67) are simple, the above formulae become greatly simplified. Indeed, in this case the transfer function $F(s)$ can be expanded into partial fractions as

$$F(s) = \sum_{i=1}^{n} \frac{f_i}{s-s_i}\,. \tag{9.101}$$

Hence, from (9.99) and (9.100) we obtain the following less cumbersome formulae

$$D_p(s,t) = \sum_{i=1}^{n} \frac{f_i M(s_i)\mathrm{e}^{s_i t}}{\mathrm{e}^{(s-s_i)T}-1} + h_p(t)\,, \quad 0 \le t \le T \tag{9.102}$$

$$D_p(s,t) = \sum_{i=1}^{n} \frac{f_i M(s_i)\mathrm{e}^{s_i t}}{1-\mathrm{e}^{(s_i-s)T}} + h_p^*(t)\,, \quad 0 \le t \le T \tag{9.103}$$

where the functions $h_p(t)$ and $h_p^*(t)$ are given by (9.97) and (9.98) with

$$h(t) = \sum_{i=1}^{n} f_i \mathrm{e}^{s_i t}\,. \tag{9.104}$$

Equations (9.99), (9.100), (9.102) and (9.103) define the required PTF in the interval $0 \le t \le T$.

To calculate $D_p(s,t)$ on the whole $t-$axis, we use the formula

$$D_p(s,t) = D_p(s,t-mT)\mathrm{e}^{msT}\,, \quad mT \le t \le (m+1)T\,. \tag{9.105}$$

Using (9.105), from (9.99) and (9.100) we find the formulae

$$D_p(s,t) = \left[\sum_{i=1}^{\ell}\sum_{k=1}^{\nu_i} \frac{f_{ik}}{(k-1)!}\frac{\partial^{k-1}}{\partial s_i^{k-1}}\frac{\mathrm{e}^{s_i(t-mT)}M(s_i)}{\mathrm{e}^{(s-s_i)T}-1} + h_p(t-mT)\right]\mathrm{e}^{msT} \tag{9.106}$$

$$D_p(s,t) = \left[\sum_{i=1}^{\ell}\sum_{k=1}^{\nu_i} \frac{f_{ik}}{(k-1)!}\frac{\partial^{k-1}}{\partial s_i^{k-1}}\frac{e^{s_i(t-mT)}M(s_i)}{1-e^{(s_i-s)T}} + h_p^*(t-mT)\right] e^{msT} \tag{9.107}$$

which hold for $mT \le t \le (m+1)T$. For the case of simple roots we obtain

$$D_p(s,t) = \left[\sum_{i=1}^{n} \frac{f_i M(s_i)e^{s_i(t-mT)}}{e^{(s-s_i)T}-1} + h_p(t-mT)\right] e^{msT} \tag{9.108}$$

$$D_p(s,t) = \left[\sum_{i=1}^{n} \frac{f_i M(s_i)e^{s_i(t-mT)}}{1-e^{(s_i-s)T}} + h_p^*(t-mT)\right] e^{msT}. \tag{9.109}$$

Using the above expressions in (9.81), we obtain closed formulae for the PTF $W_p(s,t)$. For example, for $0 \le t \le T$ from (9.108) and (9.109) we have

$$W_p(s,t) = \sum_{i=1}^{n} \frac{f_i M(s_i)e^{(s_i-s)t}}{e^{(s-s_i)T}-1} + e^{-st}h_p(t) \tag{9.110}$$

$$W_p(s,t) = \sum_{i=1}^{n} \frac{f_i M(s_i)e^{(s_i-s)t}}{1-e^{(s_i-s)T}} + e^{-st}h_p^*(t). \tag{9.111}$$

Example 9.2 Let us consider the application of (10.8), (10.9) to the system of Figure 9.4 for

$$F(s) = \frac{1}{s-a} \tag{9.112}$$

i.e., the transfer function has a single pole $s_1 = a$. It is assumed that the sampling unit is a zero-order hold with $\mu(t) = 1$ and the transfer function of the forming element (9.12). In this case, for $n = 1$ Eqs. (9.102) and (9.103) yield

$$\begin{aligned} D_p(s,t) &= \frac{M(a)e^{at}}{e^{(s-a)T}-1} + h_p(t) \\ D_p(s,t) &= \frac{M(a)e^{at}}{1-e^{(a-s)T}} + h_p^*(t). \end{aligned} \tag{9.113}$$

To calculate the functions $h_p(t)$ and $h_p^*(t)$, we note that in the given case it follows from (9.109) that

$$h(t) = e^{at}.$$

Therefore, using (9.97) and (9.98), we find

$$h_p(t) = \frac{e^{at}-1}{a}, \quad h_p^*(t) = \frac{e^{a(t-T)}-1}{a}. \tag{9.114}$$

Taking due account of (9.113), we obtain the following closed expressions for $0 \le t \le T$

$$
\begin{aligned}
D_p(s,t) &= \frac{1-\mathrm{e}^{-aT}}{a}\,\frac{\mathrm{e}^{at}}{\mathrm{e}^{(s-a)T}-1} + \frac{\mathrm{e}^{at}-1}{a} \triangleq D_{p1}(s,a,t) \\
D_p(s,t) &= \frac{1-\mathrm{e}^{-aT}}{a}\,\frac{\mathrm{e}^{at}}{1-\mathrm{e}^{(a-s)T}} + \frac{\mathrm{e}^{a(t-T)}-1}{a} = D_{p1}(s,a,t)\,.
\end{aligned}
\tag{9.115}
$$

It can be immediately verified that these equations are equivalent. If $a=0$ in (9.112), i.e., $F(s) = s^{-1}$, we have to pass to the limit as $a \to 0$ in (9.115). Since

$$
\lim_{a\to 0}\frac{1-\mathrm{e}^{-aT}}{a} = T\,, \quad \lim_{a\to 0}\frac{\mathrm{e}^{at}-1}{a} = t\,, \quad \lim_{a\to 0}\frac{\mathrm{e}^{a(t-T)}-1}{a} = t-T \tag{9.116}
$$

from (9.115) we find the following equivalent formulae for $D_p(s,t)$

$$
\begin{aligned}
D_p(s,t) &= D_{p1}(s,0,t) = \frac{T}{\mathrm{e}^{sT}-1} + t \\
D_p(s,t) &= \frac{T}{1-\mathrm{e}^{-sT}} + t - T\,.
\end{aligned}
\tag{9.117}
$$

To calculate the PTF of the system we multiply Eq. (9.115) by e^{-st}, thus obtaining for $0 \le t \le T$

$$
\begin{aligned}
W_p(s,t) &= \frac{1-\mathrm{e}^{-aT}}{a}\,\frac{\mathrm{e}^{(a-s)t}}{\mathrm{e}^{(s-a)T}-1} + \mathrm{e}^{-st}\,\frac{\mathrm{e}^{at}-1}{a} \\
W_p(s,t) &= \frac{1-\mathrm{e}^{-aT}}{a}\,\frac{\mathrm{e}^{(a-s)t}}{1-\mathrm{e}^{(a-s)T}} + \mathrm{e}^{-st}\,\frac{\mathrm{e}^{a(t-T)}-1}{a}
\end{aligned}
\tag{9.118}
$$

□

Example 9.3 Let (9.112) hold and let us have the sampling unit with

$$
\mu(t) = \begin{cases} 1 & 0 < t < T/2 \\ 0 & T/2 < t < T\,. \end{cases} \tag{9.119}
$$

As follows from (9.10), in this case

$$
M(s) = \frac{1-\mathrm{e}^{-sT/2}}{s}\,. \tag{9.120}
$$

Hence,

$$
M(a) = \frac{1-\mathrm{e}^{-aT/2}}{a}\,. \tag{9.121}
$$

Next, we find $h_p(t)$ and $h_p^*(t)$ for the case under consideration. Using (9.95) and (9.112), from (9.97) and (9.98) we find

$$
h_p(t) = \int_0^t \mathrm{e}^{a(t-\tau)}\mu(\tau)\,\mathrm{d}\tau = \begin{cases} \frac{1}{a}(\mathrm{e}^{at}-1) & 0 \le t \le T/2 \\ \frac{1}{a}\mathrm{e}^{at}(1-\mathrm{e}^{-aT/2}) & T/2 \le t \le T\,, \end{cases} \tag{9.122}
$$

$$h_p^*(t) = \int_t^T e^{a(t-\tau)}\mu(\tau)\,d\tau = \begin{cases} \frac{1}{a}(e^{a(t-T/2)} - 1) & 0 \le t \le T/2 \\ 0 & T/2 \le t \le T\,. \end{cases} \tag{9.123}$$

Substituting (9.121) and (9.123) into (9.113), we find closed expressions for the DTF on the interval $0 \le t \le T$. These relations can be extended over the whole t–axis by means of Eq. (9.105). □

Example 9.4 The example should show that for $t \to 0$ the formulae (9.106), (9.107) develop into relations that can be found by the common classical z–transform methods. Let, Ackermann (1988)

$$F(s) = \frac{1}{s(s+1)} = \frac{1}{s} - \frac{1}{s+1}\,, \quad M(s) = M_0(s) = \frac{1-e^{-sT}}{s}\,. \tag{9.124}$$

In this case $f_1 = 1$, $f_2 = -1$, $s_1 = 0$, $s_2 = -1$. From (9.115) and (9.124), for $0 \le t \le T$ we have

$$D_p(s,t) = D_{p1}(s,0,t) - D_{p1}(s,-1,t) \tag{9.125}$$

which, taking account of (9.115) and (9.117), is equivalent to

$$D_p(s,t) = \mathcal{D}_{FM_0}(T,s,t) = \frac{T}{e^{sT}-1} + t - \left[\frac{(e^T - 1)\,e^{-t}}{e^{sT}e^T - 1} + 1 - e^{-t}\right], \quad 0 \le t \le T. \tag{9.126}$$

For $t = 0$ from (9.126) we have

$$\mathcal{D}_{FM_0}(T,s,0) = \frac{T}{e^{sT}-1} - \frac{e^T - 1}{e^{sT}e^T - 1}\,. \tag{9.127}$$

Assuming $e^{sT} = z$ in (9.127), we have

$$D_p^*(z) \triangleq \mathcal{D}_{FM_0}(T,s,0)\,|_{e^{sT}=z} = \frac{z(T-1+e^{-T}) + 1 - Te^{-T} - e^{-T}}{(z-1)(z-e^{-T})} \tag{9.128}$$

which coincides with the result obtained in Ackermann (1988). □

Example 9.5 Let us construct closed expressions for the DTF under (9.112) for an arbitrary form of the controlling pulse. Using (9.102) and (9.103), for $0 \le t \le T$ we obtain the following equivalent expressions

$$\begin{aligned} D_p(s,t) &\triangleq D_{p1}(s,a,t) = \frac{M(a)e^{at}}{e^{(s-a)T}-1} + \int_0^t e^{a(t-\tau)}\mu(\tau)\,d\tau \\ D_p(s,t) &= D_{p1}(s,a,t) = \frac{M(a)e^{at}}{1-e^{(a-s)T}} - \int_t^T e^{a(t-\tau)}\mu(\tau)\,d\tau\,. \end{aligned} \tag{9.129}$$

□

Using (9.129), we can obtain yet another general closed form for $D_p(s,t)$. Substituting (9.95), (9.97) and (9.98) into (9.99) and (9.100), we obtain for $0 \le t \le T$

$$D_p(s,t) = \sum_{i=1}^{\ell}\sum_{k=1}^{\nu_i} \frac{f_{ik}}{(k-1)!}\frac{\partial^{k-1}}{\partial s_i^{k-1}}\left[\frac{M(s_i)e^{s_i t}}{e^{(s-s_i)T}-1} + \int_0^t e^{s_i(t-\tau)}\mu(\tau)\,d\tau\right] \tag{9.130}$$

and

$$D_p(s,t) = \sum_{i=1}^{\ell}\sum_{k=1}^{\nu_i} \frac{f_{ik}}{(k-1)!}\frac{\partial^{k-1}}{\partial s_i^{k-1}}\left[\frac{M(s_i)e^{s_i t}}{1-e^{(s_i-s)T}} - \int_t^T e^{s_i(t-\tau)}\mu(\tau)\,d\tau\right]. \tag{9.131}$$

Taking account of (9.129), we can write these two formulae as a single equation

$$D_p(s,t) = \sum_{i=1}^{\ell}\sum_{k=1}^{\nu_i} \frac{f_{ik}}{(k-1)!}\frac{\partial^{k-1}}{\partial s_i^{k-1}} D_{p1}(s,s_i,t). \tag{9.132}$$

Equation (9.132) is derived for $0 \le t \le T$. Nevertheless, it is valid for all t. Indeed, let $mT \le t \le (m+1)T$ where m is an integer. Then, $0 \le t - mT \le T$, and from (9.132) we find

$$D_p(s,t-mT) = \sum_{i=1}^{\ell}\sum_{k=1}^{\nu_i} \frac{f_{ik}}{(k-1)!}\frac{\partial^{k-1}}{\partial s_i^{k-1}} D_{p1}(s,s_i,t-mT). \tag{9.133}$$

Multiplying (9.132) by e^{msT} and taking into account that

$$D_p(s,t-mT)e^{msT} = D_p(s,t), \quad D_{p1}(s,s_i,t-mT)e^{msT} = D_{p1}(s,s_i,t)$$

we find that (9.132) holds for all t. This equation is convenient to use if the function $F(s)$ has multiple poles.

Example 9.6 We use formula (9.132) to construct the discrete transfer function under the conditions

$$F(s) = \frac{1}{(s-a)^2}, \quad M(s) = M_0(s) = \frac{1-e^{-T}}{s}. \tag{9.134}$$

As follows from (9.132), in this case

$$D_p(s,t) = \frac{\partial}{\partial a} D_{p1}(T,a,t) \tag{9.135}$$

where the function $D_{p1}(T,a,t)$ is given in the interval $0 \le t \le T$ by (9.115). As a result, we obtain

$$D_p(s,t) = \frac{\partial}{\partial a}\left[\frac{1-\mathrm{e}^{-aT}}{a}\frac{\mathrm{e}^{at}}{\mathrm{e}^{(s-a)T}-1} + \frac{\mathrm{e}^{at}-1}{a}\right]. \tag{9.136}$$

Performing differentiation in (9.136) and taking the limit as $a \to 0$, for $F(s) = s^{-2}$ and $0 \le t \le T$ we find that

$$D_p(s,t) = \frac{T^2}{2}\frac{\mathrm{e}^{sT}+1}{(\mathrm{e}^{sT}-1)^2} + \frac{Tt}{\mathrm{e}^{sT}-1} + \frac{t^2}{2}. \tag{9.137}$$

□

9.7 Limit relations

Using the formulae from the previous section, we can obtain limit relations, which define the behaviour of the DTF $D_p(s,t)$ for $\operatorname{Re} s \to \pm\infty$. These relations will be important for the analysis of processes in SD systems. Henceforth, as in Section 4.8, for any function $F(s)$ we shall denote

$$\ell^+[F(s)] \triangleq \lim_{\operatorname{Re} s \to \infty} F(s), \qquad \ell^-[F(s)] \triangleq \lim_{\operatorname{Re} s \to -\infty} F(s). \tag{9.138}$$

Taking the limit as $\operatorname{Re} s \to \infty$ in (9.99), we immediately obtain

$$\ell^+[D_p(s,t)] = h_p(t), \quad 0 \le t \le T. \tag{9.139}$$

Similarly, from (9.100) for $\operatorname{Re} s \to -\infty$ we have

$$\ell^-[D_p(s,t)] = h_p^*(t), \quad 0 \le t \le T. \tag{9.140}$$

Since there exist finite limits (9.139) and (9.140), for $0 \le t \le T$ the function $D_p(s,T)$ is a *limited rational periodic function* (see Appendix A).

Let $mT \le t \le (m+1)T$, where m is an integer. Then,

$$D_p(s,t) = D_p(s,t-mT)\mathrm{e}^{msT} \tag{9.141}$$

where $0 \le t - mT \le T$. Taking account of (9.139) and (9.140), from (9.141) we find

$$\ell^-[D_p(s,t)] = 0, \quad t \ge T \tag{9.142}$$

$$\ell^+[D_p(s,t)] = 0, \quad t \le 0. \tag{9.143}$$

From (9.100) it follows that

$$\ell^-[D_p(s,0)] = h_p^*(0) = -\int_0^T h(-\tau)\mu(\tau)\,\mathrm{d}\tau. \tag{9.144}$$

Using (9.95) and (9.100), we find

$$\ell^+[D_p(s,0)] = -\sum_{i=1}^{\ell}\sum_{k=1}^{\nu_i}\frac{f_{ik}}{(k-1)!}\frac{\partial^{k-1}}{\partial s_i^{k-1}}M(s_i). \tag{9.145}$$

9.8 Non-pathological conditions

As follows from the formulae obtained in Section 9.6, the DTF $D_p(s,t)$ is, for any t, a real rational function in $z = \mathrm{e}^{sT}$ or $\zeta = \mathrm{e}^{-sT}$. For further exposition we have to analyse some properties of the function obtained from $D_p(s,t)$ as a result of the change of variable to ζ. Henceforth, for any function $F(s)$ that is, in fact, dependent on e^{-sT} we use the notation

$$F^o(\zeta) \triangleq F(s)\,\big|_{\mathrm{e}^{-sT}=\zeta}\ . \tag{9.146}$$

Using the change of variable $\zeta = \mathrm{e}^{-sT}$ in (9.99) and (9.100), with the notation (1.46) for $0 \le t \le T$ we obtain the following equivalent formulae

$$D_p^o(\zeta,t) = \zeta \sum_{i=1}^{\ell} \sum_{k=1}^{\nu_i} \frac{f_{ik}}{(k-1)!} \frac{\partial^{k-1}}{\partial s_i^{k-1}} \frac{\mathrm{e}^{s_i(t+T)} M(s_i)}{1 - \mathrm{e}^{s_i T}\zeta} + h_p(t) \tag{9.147}$$

$$D_p^o(\zeta,t) = \sum_{i=1}^{\ell} \sum_{k=1}^{\nu_i} \frac{f_{ik}}{(k-1)!} \frac{\partial^{k-1}}{\partial s_i^{k-1}} \frac{\mathrm{e}^{s_i t} M(s_i)}{1 - \mathrm{e}^{s_i T}\zeta} + h_p^*(t)\,. \tag{9.148}$$

Performing differentiation, after transformations we obtain

$$D_p^o(\zeta,t) = \frac{\beta^o(\zeta,t)}{\alpha^o(\zeta)}\,, \quad 0 \le t \le T \tag{9.149}$$

where

$$\beta^o(\zeta,t) = \beta_0(t) + \beta_1(t)\zeta + \ldots + \beta_n(t)\zeta^n \tag{9.150}$$

$$\alpha^o(\zeta) = \prod_{i=1}^{\ell} \left(1 - \mathrm{e}^{s_i T}\zeta\right)^{\nu_i}\ . \tag{9.151}$$

The coefficients $\beta_0(t)$ and $\beta_n(t)$ of the polynomial $\beta^o(t)$ can be found immediately. To calculate $\beta_0(t)$ we note that, due to (9.149)–(9.151),

$$\beta_0(t) = D_p^o(\zeta,t)\,|_{\zeta=0}\ . \tag{9.152}$$

Obviously,

$$D_p^o(\zeta,t)\,|_{\zeta=0} = \ell^+[D_p(s,t)]\,. \tag{9.153}$$

Therefore, using (9.139), we obtain

$$\beta_0(t) = h_p(t)\,. \tag{9.154}$$

To calculate the coefficient $\beta_n(t)$ we note, that the coefficient by ζ^n in the numerator of (9.149) coincides with the coefficient by ζ^n in the product

$$\left(1 - \mathrm{e}^{s_1 T}\zeta\right)^{\nu_1} \cdot \ldots \cdot \left(1 - \mathrm{e}^{s_\ell T}\zeta\right)^{\nu_\ell} h_p^*(t)\,. \tag{9.155}$$

Since $\nu_1 + \ldots + \nu_\ell = n$, we have

$$\beta_n(t) = (-1)^n \mathrm{e}^{T(s_1\nu_1+\ldots+s_\ell\nu_\ell)} h_p^*(t)\,. \tag{9.156}$$

But, as follows from (9.67),

$$s_1\nu_1 + \ldots + s_\ell \nu \ell = -d_1$$

where d_1 is the coefficient by s^{n-1} in the polynomial $d(s)$ in (9.60). Therefore, Eq. (9.156) takes the form

$$\beta_n(t) = (-1)^n \mathrm{e}^{-Td_1} h_p^*(t)\,. \tag{9.157}$$

Example 9.7 Build the function $D_p^o(\zeta, t)$ under the conditions of Example 9.2. Then, we have

$$d(s) = s - a\,, \quad d_1 = -a\,, \quad n = 1\,. \tag{9.158}$$

The functions $h_p(t)$ and $h_p^*(t)$ are defined by (9.114). Then, as a result of (9.154) and (9.157), we have

$$\beta_0(t) = \frac{\mathrm{e}^{at} - 1}{a}\,, \quad \beta_1(t) = -\mathrm{e}^{aT}\,\frac{\mathrm{e}^{a(t-T)} - 1}{a}\,.$$

Hence,

$$\beta^o(\zeta, t) = \frac{\mathrm{e}^{at} - 1}{a} - \frac{\mathrm{e}^{at} - \mathrm{e}^{aT}}{a}\zeta$$

$$\alpha^o(\zeta) = 1 - \mathrm{e}^{aT}\zeta$$

and, using (9.149), we obtain

$$W_p^o(\zeta, t) = \frac{1}{a(1 - \mathrm{e}^{aT}\zeta)}\left[\mathrm{e}^{at} - 1 + (\mathrm{e}^{aT} - \mathrm{e}^{at})\zeta\right]\,. \tag{9.159}$$

It can be immediately verified that (9.159) transforms itself into (9.115) for $\zeta = \mathrm{e}^{-sT}$. □

Since $h_p(0) = 0$, from (9.149) and (9.154) it follows that

$$D_p^o(\zeta, 0) = \frac{\zeta\chi^o(\zeta)}{\alpha^o(\zeta)} \tag{9.160}$$

where $\chi^o(\zeta)$ is a polynomial such that $\deg \chi^o \le n - 1$.
Using the above relations, we can introduce a notion that will be important for further discussion.

Definition The open-loop SD system (9.58), (9.59) will be called *non-pathological*, if the ratio (9.160) is irreducible, i.e., the polynomials $\chi^o(\zeta)$ and $\alpha^o(\zeta)$ have no common roots.

Theorem 9.1 *If Eq. (9.67) holds, the system under consideration is non-pathological iff*

$$e^{s_i T} \neq e^{s_k T}, \quad \text{for}\, i \neq k;\ (i, k = 1, \ldots, \ell) \tag{9.161}$$

and

$$M(s_i) \neq 0, \quad (i = 1, 2, \ldots, \ell). \tag{9.162}$$

Proof *Sufficiency:* Performing differentiation in (9.147), we can find $D_p^o(\zeta, t)$ as the sum of terms that have different powers of the polynomial $q_i(\zeta) = 1 - e^{s_i T}\zeta$ as their denominators. The largest power of $q_i(\zeta)$ can be found in the term

$$U_{\nu_i} = e^{s_i t} M(s_i) \frac{\partial^{\nu_i - 1}}{\partial s_i^{\nu_i - 1}} \frac{1}{1 - e^{s_i T}\zeta}. \tag{9.163}$$

If $M(s_i) \neq 0$, the function U_{ν_i} has a pole of multiplicity ν_i at $\zeta = \zeta_i = e^{-s_i T}$. If (9.161) holds, all poles ζ_i are different. Therefore, if (9.161) and (9.162) are valid, the ratio (9.149) has poles at the point $\zeta = \zeta_i$ with multiplicities ν_i. The total number of poles (taking account of their multiplicities) is equal to $\nu_1 + \ldots + \nu_\ell = n$, therefore, the ratio is irreducible.
Necessity: If at least one of the conditions (9.61) and (9.62) is not valid, then, either some of the poles ζ_i coincide, or, due to (9.163), the multiplicities of these poles decreases so that the total number of poles of the function (9.149) becomes less than n, i.e., the ratio (9.160) is reducible. ∎

Corollary 1 As follows from the proof, if the non-pathological conditions are satisfied, the ratio $D_p^o(\zeta, t)$ in (9.149) is irreducible for all t.

Corollary 2 If the non-pathological conditions are satisfied, the DTF $D_p(s, t)$ can be represented in the form

$$D_p(s, t) = \frac{1}{T} F(s) M(s) e^{st} + C(s, t) \tag{9.164}$$

where the function $C(s, t)$ is analytic at $s = s_i$, where s_i are the poles of the transfer function $F(s)$, and the product $F(s)M(s)$ has at the points $s = s_i$ poles of the same multiplicities as $F(s)$.

Proof As follows from (9.164) and (9.82),

$$C(s,t) = \frac{1}{T} \sum_{\substack{k=-\infty \\ k\neq 0}}^{\infty} F(s+kj\omega)M(s+kj\omega)e^{(s+kj\omega)t}\,. \tag{9.165}$$

As follows from (9.165), the function $C(s,t)$ can have poles only at $s_i + mj\omega \triangleq s_{im}$ for $m \neq 0$. But, due to (9.161) none of the numbers s_{im} is a pole of $F(s)$. Moreover, from (9.162) it follows that the functions $F(s)$ and $F(s)M(s)$ have, taking due account of multiplicities, the same number of poles. ■

Remark If the sampling unit (9.58) is a sample with zero-order hold, the non-pathological conditions (9.162) is excessive. Indeed, using (9.12), we obtain

$$M_0(s) = \frac{1 - e^{-sT}}{s} = 0$$

for $s = s_k = kj\omega$ with any $k \neq 0$. But for these $s = s_k$ the conditions (9.161) are not valid. Therefore, for a system with a zero-order hold the non-pathological conditions reduce to (9.161).

Example 9.8 An example is given, where the non-pathological condition (9.162) is violated. Let

$$F(s) = \frac{1}{s^2 + \frac{64\pi^2}{9T^2}}$$

and

$$\mu(t) = \begin{cases} 1 & 0 < t < \frac{3}{4}T \\ 0 & \frac{3}{4}T < t < T \end{cases}$$

so that

$$M(s) = \frac{1 - e^{-\frac{3}{4}sT}}{s}\,.$$

The poles of $F(s)$ are $s_1 = 8\pi j/3T$ and $s_2 = -8\pi j/3T$. Obviously, $e^{s_1 T} \neq e^{s_2 T}$, i.e., the conditions (9.161) hold. At the same time, $M(s_1) = M(s_2) = 0$, and the system is pathological. □

9.9 General properties of transfer functions and operator description of systems

Starting from the above relations, we consider some general properties of the PTF $W_p(s,t)$ and the DTF $D_p(s,t)$ in detail. The non-pathological conditions (9.161) and (9.162) will assumed to be valid.

1. The function

$$Q_p(s) \triangleq F(s)M(s) \tag{9.166}$$

will be called the *equivalent transfer function* (ETF) of the system (9.58)–(9.59). Due to (9.164), the ETF $Q_p(s)$ has the same poles as $F(s)$. Denote by $\mu_0 = -\infty$, $\mu_1, \ldots, \mu_r$, $\mu_{r+1} = \infty$ the characteristic indices of the transfer function $F(s)$. Since $F(s)$ is strictly proper and the estimate (9.11) holds, in any stripe $\mu_i < \operatorname{Re} s < \mu_{i+1}$ for $|s| \to \infty$ the following estimate holds

$$|Q_p(s)| < L|s|^{-2} \tag{9.167}$$

where $L > 0$ is a constant. From the estimate (9.167) and Theorem 1.4 it follows that there exist the originals

$$q_{pi}(t) = \frac{1}{2\pi \mathrm{j}} \int_{c-\mathrm{j}\infty}^{c+\mathrm{j}\infty} Q_p(s) \mathrm{e}^{st} \, \mathrm{d}s\,, \quad \mu_i < c < \mu_{i+1} \tag{9.168}$$

where the function $q_{pi}(t)$ is continuous and $q_{pi}(t) \in \Lambda(\mu_i, \mu_{i+1})$. Due to (9.167), the integral (9.168) can be discretized, taking due account of (9.166), as

$$q_{pi}(t) = \frac{T}{2\pi \mathrm{j}} \int\limits_{c-\mathrm{j}\omega/2}^{c+\mathrm{j}\omega/2} \mathcal{D}_{FM}(T, s, t) \, \mathrm{d}s = \frac{T}{2\pi \mathrm{j}} \int\limits_{c-\mathrm{j}\omega/2}^{c+\mathrm{j}\omega/2} D_p(s, t) \, \mathrm{d}s\,, \quad \mu_i < c < \mu_{i+1}\,. \tag{9.169}$$

With the notation

$$f_i(t) = \frac{1}{2\pi \mathrm{j}} \int_{c-\mathrm{j}\infty}^{c+\mathrm{j}\infty} F(s) \mathrm{e}^{st} \, \mathrm{d}s\,, \quad \mu_i < c < \mu_{i+1} \tag{9.170}$$

from (9.166) and the properties of the image of convolution we have

$$q_{pi}(t) = \int_{-\infty}^{\infty} f_i(t - \tau) \mu(\tau) \, \mathrm{d}\tau = \int_0^T f_i(t - \tau) \mu(\tau) \, \mathrm{d}\tau\,. \tag{9.171}$$

Closed formulae for $f_i(t)$ are given in Section 1.14.

2. The PTF (9.80) can be represented in the form

$$W_p(s, t) = \frac{1}{T} \sum_{k=-\infty}^{\infty} Q_p(s + k\mathrm{j}\omega) \mathrm{e}^{k\mathrm{j}\omega t} = \varphi_{Q_p}(T, s, t)\,. \tag{9.172}$$

From the estimate (9.167) and Theorem 3.3 it can be obtained that the PTF $W_p(s, t)$ is continuous with respect to t for all s except for its poles.

3. Consider the DTF $D_p(s, t)$ (9.82), which can be written in the form

$$D_p(s, t) = \frac{1}{T} \sum_{k=-\infty}^{\infty} Q_p(s + k\mathrm{j}\omega) \mathrm{e}^{(s + k\mathrm{j}\omega)t}\,. \tag{9.173}$$

The function $D_p(s,t)$ is continuous with respect to t together with $W_p(s,t)$. Moreover, as follows from Eqs. (9.139) and (9.140), that the DTF $D_p(s,t)$ is a limited rational periodic function for $0 \le t \le T$. If the non-pathological conditions are valid, the set of characteristic indices of $D_p(s,t)$ for any t coincides with the set of characteristic indices $\{\mu_i\}$, $(i = 1, \ldots, r)$ of the function $F(s)$. Then, in any stripe $\mu_i < \operatorname{Re} s < \mu_{i+1}$ we have

$$D_p(s,t) = \sum_{k=-\infty}^{\infty} q_{pi}(t+kT)e^{-ksT}. \tag{9.174}$$

As follows from (9.174), in any interval $S_i\,(\mu_i, \mu_{i+1})$ the DTF $D_p(s,t)$ of the system can be considered as the DTF of an LPPO U_{pi} with equivalent transfer function (9.166) and impulse response (9.171).

4. The set of LPPO U_{pi} constructed above will be called the *family of LPPO* associated with the system (9.58), (9.59). Next, we show that each LPPO

$$y(t) = \mathsf{U}_{pi}[x(t)] \tag{9.175}$$

for $x(t) \in \Lambda(\mu_i, \mu_{i+1})$ defines a particular solution of the Eqs. (6.58), (6.59). Indeed, due to Section 5.7, the operator

$$y(t) = \int_{-\infty}^{\infty} g_i(t-\tau)z(\tau)\,d\tau \tag{9.176}$$

defines a particular solution of Eq. (9.59) over the corresponding set of input signals. Then, assuming that $z(t)$ and $x(t)$ are related by Eq. (9.58), we obtain an operator of the form (6.116), the PTF of which is given by (6.122). Thus, the set of LPPO U_{pi} constructed above coincides with the set of LPPO associated with the operator equation

$$y(t) = \int_{-\infty}^{\infty} g_i(t-\tau)z(\tau)\,d\tau\,, \quad z(t) = \mathsf{U}_a[x(t)]\,. \tag{9.177}$$

It is noteworthy that in this section the family of LPPO U_{pi} was constructed directly from the PTF $W_p(s,t)$. Such a situation will also take place in more general cases.

9.10 Equations for images

Using the relations of the previous sections, we can obtain relations for different solutions of the system of equations (9.58), (9.59).

First, consider the solution associated with the operator equations (9.177). These equations define, in the stripe $\mu_i < \operatorname{Re} s < \mu_{i+1}$, an LPPO U_{pi} with the ETF (9.166). Therefore, if $x(t) \in \Lambda(\mu_i, \mu_{i+1})$, using (7.82), we can obtain the Laplace transform of the corresponding output

$$Y(s) = F(s)M(s)\mathcal{D}_x(T, s, +0), \quad \mu_i < \operatorname{Re} s < \mu_{i+1}. \tag{9.178}$$

Moreover, from (9.81) we find the discrete Laplace transform of the output

$$\mathcal{D}_y(T, s, t) = D_p(s, t)\mathcal{D}_x(T, s, +0), \quad \mu_i < \operatorname{Re} s < \mu_{i+1} \tag{9.179}$$

where the DTF $D_p(s,t)$ can be calculated using the closed formulae given in Section 9.6.

The originals for the images (9.178) and (9.179) are some steady-state modes of the system (9.58), (9.59) defined, in the general case, for $-\infty < t < \infty$.

For the analysis of transients in the system under consideration for $t > 0$ it is convenient to use the Laplace transform in the half-plane $\operatorname{Re} s > \mu_r$, where μ_r is the largest characteristic index of the function $F(s)$. In this case, assuming that $x(t) \in \Lambda_+(\mu_r, \infty)$, taking due account of (9.17) we have

$$Z(s) = M(s)\mathcal{D}_x(T, s, +0). \tag{9.180}$$

Substituting this equation into (9.65), we find the Laplace transform of the output

$$Y(s) = F(s)M(s)\mathcal{D}_x(T, s, +0) + \frac{m_0(s)}{d(s)}, \quad \mu_i < \operatorname{Re} s < \mu_{i+1}. \tag{9.181}$$

9.11 Systems with phase-shifted samplers

If the sampling unit in the system shown in Fig. 9.4 operates with a phase shift, instead of (9.59) we have

$$z(t) = \mathsf{U}_a^\phi[x(t)] \tag{9.182}$$

where the Operator U_a^ϕ is given by (9.23). For the determination of the PTF $W_p^\phi(s,t)$ of the system (9.59) and (9.182) we use the same technique as in Section 9.5. Assume that $x(t) = \mathrm{e}^{st}$ and find the solution of Eqs. (9.59) and (9.182) having the form (9.74). Then, using (9.26), we have

$$z(s,t) = \mathrm{e}^{st}\varphi_M(T, s, t-\phi) = \frac{1}{T}\sum_{k=-\infty}^{\infty} M(s + k\mathrm{j}\omega)\mathrm{e}^{k\mathrm{j}\omega(t-\phi)}\mathrm{e}^{st}. \tag{9.183}$$

Similarly to (9.79), the corresponding solution $y(s,t)$ of Eq. (6.59) has the form

$$y(s,t) = \frac{1}{T}\sum_{k=-\infty}^{\infty} F(s + k\mathrm{j}\omega)M(s + k\mathrm{j}\omega)\mathrm{e}^{(s+k\mathrm{j}\omega)t}\mathrm{e}^{-k\mathrm{j}\omega\phi}. \tag{9.184}$$

Multiplying (9.184) by e^{-st}, we find the PTF

$$W_p^\phi(s,t) = \frac{1}{T}\sum_{k=-\infty}^{\infty} F(s + k\mathrm{j}\omega)M(s + k\mathrm{j}\omega)\mathrm{e}^{k\mathrm{j}\omega(t-\phi)} = W_p(s, t-\phi). \tag{9.185}$$

It can be immediately verified that the PTF (9.185) defines a family of LPPO with phase shift ϕ with respect to the LPPO U_{pi} given by (9.177). Then, using the relations of Section 7.4, we can directly obtain the associated Laplace transforms. For $x(t) \in \Lambda(\mu_i, \mu_{i+1})$ from (7.115) we have

$$Y(s) = F(s)M(s)\varphi_x(T,s,\phi+0)\,, \quad \mu_i < \operatorname{Re} s < \mu_{i+1}\,. \tag{9.186}$$

Due to (7.115), the corresponding discrete Laplace transform has the form

$$\mathcal{D}_y(T,s,t) = D_p(s,t-\phi)\mathcal{D}_x(T,s,\phi+0)\,, \quad \mu_i < \operatorname{Re} s < \mu_{i+1}\,. \tag{9.187}$$

Taking into account that, in the given case,

$$Z(s) = M(s)\varphi_x(T,s,\phi+0)$$

from (9.65) we find the Laplace transform for the output in the half-plane $\operatorname{Re} s > \mu_r$

$$Y(s) = F(s)M(s)\varphi_x(T,s,\phi+0) + \frac{m_0(s)}{d(s)}\,. \tag{9.188}$$

Using (9.185) and the relations of Section 9.6, we find closed formulae for the PTF $W_p^\phi(s,t)$. We note that for $0 \le t \le T$ and $0 \le \phi \le T$ we have $|t-\phi| < T$. Therefore, for $t \ge \phi$ we obtain $0 \le t-\phi \le T$ and and for $t \le \phi$ we have $-T \le t-\phi \le 0$. Taking these relations into account, we shall use the formula

$$W_p^\phi(s,t) = \begin{cases} W_p(s,t-\phi)\,, & 0 \le t-\phi \le T \\ W_p(s,t-\phi+T)\,, & -T \le t-\phi \le 0\,. \end{cases}$$

Consider the special case, when all poles of $F(s)$ are simple. Then, using (9.111), we obtain

$$W_p^\phi(s,t) = \begin{cases} \mathrm{e}^{-s(t-\phi)} \left[\sum_{i=1}^{n} \dfrac{f_i M(s_i)\mathrm{e}^{s_i(t-\phi)}}{1-\mathrm{e}^{(s_i-s)T}} + h_p^*(t-\phi) \right], & 0 \le t-\phi \le T \\ \mathrm{e}^{-s(t-\phi+T)} \left[\sum_{i=1}^{n} \dfrac{f_i M(s_i)\mathrm{e}^{s_i(t-\phi+T)}}{1-\mathrm{e}^{(s_i-s)T}} + h_p^*(t-\phi+T) \right], & \\ & \phi - T \le t \le \phi. \end{cases} \tag{9.189}$$

9.12 Pure delay in the continuous part

If the continuous part of the system shown in Fig. 9.4 contains a pure delay τ, instead of (9.66) we shall have the transfer function

$$F_\tau(s) \triangleq F(s)\mathrm{e}^{-s\tau} = \frac{m(s)}{d(s)}\mathrm{e}^{-s\tau}\,. \tag{9.190}$$

The PTF $W_p^\tau(s,t)$ of the system

$$y = F_\tau(s)z\,, \quad z = \mathsf{U}_a[x] \tag{9.191}$$

can be obtained in a similar way. As a result, in analogy with (9.80) we find

$$W_p^\tau(s,t) = \frac{1}{T}\sum_{k=-\infty}^{\infty} F_\tau(s+kj\omega)M(s+kj\omega)e^{kj\omega t}\,.$$

Using (9.190), we obtain

$$W_p^\tau(s,t) = \frac{1}{T}\sum_{k=-\infty}^{\infty} F(s+kj\omega)M(s+kj\omega)e^{kj\omega(t-\tau)}e^{-s\tau} = \varphi_{FM}(T,s,t-\tau)e^{-s\tau}. \tag{9.192}$$

Multiplying the last equation by e^{st}, we find the discrete transfer function $D_p^\tau(s,t)$ for the system with pure delay

$$D_p^\tau(s,t) = \frac{1}{T}\sum_{k=-\infty}^{\infty} F(s+kj\omega)M(s+kj\omega)e^{(s+kj\omega)(t-\tau)} = \mathcal{D}_{FM}(T,s,t-\tau)\,. \tag{9.193}$$

For $t=0$ Eq. (9.193) yields the transfer function of the time-delayed system in the sense of the discrete Laplace transformation

$$D_p^\tau(s,0) = \mathcal{D}_{FM}(T,s,-\tau) \triangleq D_{p\tau}^*(s)\,. \tag{9.194}$$

Let m be an integer such that

$$0 \le mT - \tau < T \tag{9.195}$$

which is equivalent to

$$\tau = mT - \delta T\,, \quad 0 \le \delta < 1\,. \tag{9.196}$$

Obviously, for a given τ the values of m and δ are defined uniquely. Substituting (9.196) into (9.194) and using (9.141), we obtain

$$D_{p\tau}^*(s) = \mathcal{D}_{FM}(T,s,\delta T - mT) = \mathcal{D}_{FM}(T,s,\delta T)\,e^{-msT}\,. \tag{9.197}$$

Example 9.9 Find the transfer function in the delayed discrete Laplace transform and in the z−transform if conditions (9.124) are valid and the arbitrary delay time satisfies (9.196). Then, if the conditions of Example 9.4 hold, we have

$$D_{p\tau}^*(s) = \left[\frac{T}{e^{sT}-1} + \delta T - \frac{(e^T-1)e^{-\delta T}}{e^{sT}e^T-1} - 1 + e^{-\delta T}\right]e^{-msT}\,. \tag{9.198}$$

Assuming that $e^{sT} = z$, we find the transfer function of the time-delayed system in the sense of the z−transformation

$$D_{p\tau}(z) = \left[\frac{T}{z-1} + \delta T - \frac{(e^T-1)e^{-\delta T}}{ze^T-1} - 1 + e^{-\delta T}\right]z^{-m}\,. \tag{9.199}$$

□

As follows from the formulae of Section 9.6, for any τ the function $D_p^\tau(s,t)$ is rational periodic with respect to s. Moreover, using Corollary 1 of Theorem 9.1, it can be shown that the set of regularity intervals for the rational periodic function $D_p^\tau(s,t)$ for any τ is the same as for $\tau = 0$. Therefore, the functions $W_p^\tau(s,t)$ and $D_p^\tau(s,t)$ define a family of LPPO with a single ETF

$$Q_p^\tau(s) \triangleq \int_0^T W_p^\tau(s,t)\,\mathrm{d}t = F(s)M(s)\mathrm{e}^{-s\tau}. \tag{9.200}$$

For any regularity interval $\mu_i < \operatorname{Re} s < \mu_{i+1}$ we have the Laplace transform of the output

$$Y_\tau(s) = F(s)M(s)\mathrm{e}^{-s\tau}\varphi_x(T,s,+0), \quad \mu_i < \operatorname{Re} s < \mu_{i+1} \tag{9.201}$$

and the corresponding discrete Laplace transform

$$\mathcal{D}_y(T,s,t) = \mathcal{D}_{FM}(T,s,t-\tau)\mathcal{D}_x(T,s,+0), \quad \mu_i < \operatorname{Re} s < \mu_{i+1}. \tag{9.202}$$

It is easy to consider the case when the sampling unit in the time-delayed system has a phase shift. In this case, instead of (9.201) we have

$$Y_{\tau\phi}(s) = F(s)M(s)\mathrm{e}^{-s\tau}\varphi_x(T,s,\phi+0), \quad \mu_i < \operatorname{Re} s < \mu_{i+1}. \tag{9.203}$$

We note that, using (9.193) and the formulae from Section 9.6, we can obtain closed formulae for the DTF and PTF of the time-delayed system. For instance, in the case of (9.108), for $mT < t-\tau < (m+1)T$ Eq. (9.141) yields

$$D_p^\tau(s,t)\mathrm{e}^{-msT} = \sum_{i=1}^{n} \frac{f_i M(s_i)\mathrm{e}^{s_i(t-\tau-mT)}}{\mathrm{e}^{(s-s_i)T}-1} + h_p(t-\tau-mT). \tag{9.204}$$

The corresponding PTF is given by

$$W_p^\tau(s,t) = D_p^\tau(s,t)\mathrm{e}^{-st} \tag{9.205}$$

9.13 Open-loop systems with sample and hold

If, instead of a sampling unit (9.1), we have

$$z(t) = \mathsf{U}_h[x(t)] \tag{9.206}$$

in the system shown in Fig. 9.4, then, as follows from (9.42), this is equivalent to the presence of an LPPO with ETF

$$Q_h(s) = \sum_{\nu=1}^{r} w_{h\nu}M_\nu(s) \tag{9.207}$$

at the input of the continuous-time part of the system. Then, in a similar way, we obtain that the PTF $W_h(s,t)$ of the open-loop system with a hold is given by

$$W_{hp}(s) = \sum_{\nu=1}^{r} w_{h\nu}\varphi_{FM_\nu}(T,s,t)\,. \tag{9.208}$$

For $x(t) \in \Lambda(\mu_i, \mu_{i+1})$ the Laplace transform of the output of the associated LPPO has the form

$$Y(s) = F(s)Q_h(s)\mathcal{D}_x(T,s,+0)\,, \quad \mu_i < \operatorname{Re} s < \mu_{i+1}\,. \tag{9.209}$$

Chapter 10

Open-loop systems with a computer

10.1 Continuous-time description of computers

A simplified block-diagram of a computer with analogue interface is shown in Fig. 10.1, where

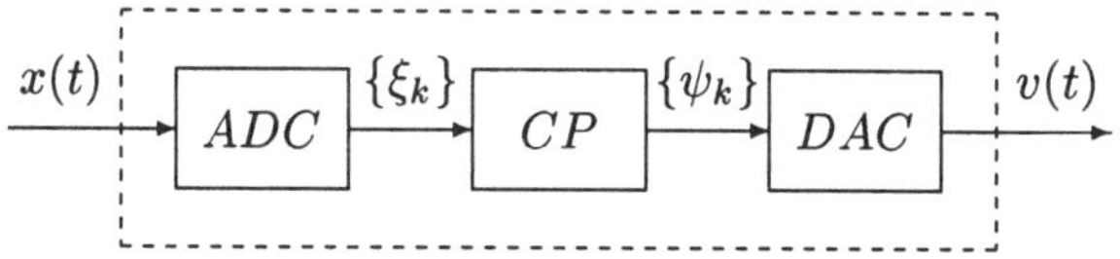

Figure 10.1: Computer with analog interface

ADC is an analog-to-digital converter that transforms the input signal $x(t)$ into a numerical sequence $\{\xi_k\}$

CP is a computer program that transforms the input sequence $\{\xi_k\}$ into the control sequence $\{\psi_k\}$

DAC is a digital-to-analog converter that reconstructs a continuous-time function $v(t)$ on the basis of the sequence $\{\psi_k\}$.

Computers in this book will always have analogue interfaces and therefore, will be simply called computers. Henceforth it is assumed that the ADC generates an equidistant sampling from the input signal

$$\xi_k = Rx(kT + 0) \tag{10.1}$$

where R is the constant ADC gain.

The computer program accepts a sequence (10.1) and transforms it into the sequence $\{\psi_k\}$ as a solution of a linear difference equation

$$a_0\psi_n + a_1\psi_{n-1} + \ldots + a_p\psi_{n-p} = b_0\xi_n + b_1\xi_{n-1} + \ldots + b_p\xi_{n-p} \tag{10.2}$$

with constant coefficients a_i and b_i ($i = 1, \ldots p$), where $a_0 \neq 0$ and at least one of the numbers a_p or b_p is non-zero. Equation (10.2) will henceforth be called the *control program* and the number p is the *order* of the program. Henceforth, without loss of generality, we assume that $R = 1$ in (10.1). If $R \neq 1$, the corresponding multiplier can be included in the coefficients b_i.
A digital-to-analog converter that transforms a discrete sequence $\{\psi_k\}$ into a continuous-time function $v(t)$, is, from the mathematical point of view, a hold device. In the present book it is assumed that the output signal $v(t)$ is defined by

$$v(t) = \mu(t - kT)\psi_k\,, \quad kT < t < (k+1)T \tag{10.3}$$

where the function $\mu(t)$ defines the form of the controlling pulses. Henceforth it is assumed that the function $\mu(t)$ is piece-wise smooth in the interval $0 \leq t \leq T$, and it is zero outside this interval. As follows from (10.3), the value of the output $v(t)$ in the interval $kT < t < (k+1)T$ is uniquely defined by the number ψ_k. This kind of extrapolation is the simplest. We note that, if the control program reduces to the identity

$$\xi_k = \psi_k \tag{10.4}$$

for $R = 1$ Eqs. (10.1)–(10.3) yield

$$v(t) = \mu(t - kT)x(kT + 0)\,, \quad kT < t < (k+1)T$$

which coincides with the equation of the sampling unit (9.1). Equations (10.1)–(10.3) taken in the aggregate describe a linearized mathematical model of the continuous-time processes in a computer. This model is approximative, at least because it ignores the amplitude quantization. Moreover, a more detailed investigation Phillips and Nagle (1995); Åström and Wittenmark (1984); Stearns (1988) shows that processes in actual ADC and DAC are of a considerably more complex nature than that described by Eqs. (10.1)–(10.3). Nevertheless, such a model is commonly accepted and is applicable for many theoretical and technical problems.

10.2 Operational analysis of control programs

To determine the relations between the numerical sequences $\{\xi_k\}$ and $\{\psi_k\}$ appearing at the input and output of the computer, we use the discrete Laplace transformation (see Sec. 2.4). Let us take a unilateral numerical sequence

$$\{\xi\} \triangleq \begin{cases} \xi_k & k \geq 0 \\ 0 & k < 0\,. \end{cases} \tag{10.5}$$

Denote by $\{\xi\}_m$ the sequence

$$\{\xi\}_m \triangleq \begin{cases} \xi_{k+m} & k \geq 0 \\ 0 & k < 0 \end{cases} \quad m \geq 0 \,. \tag{10.6}$$

Then, by definition, $\{\xi\} = \{\xi\}_0$. Next, we find the relation between the discrete Laplace transforms (DLT) of the sequences (10.5) and (10.6). Let

$$\xi_m^*(s) \triangleq \sum_{k=0}^{\infty} \xi_{k+m} \mathrm{e}^{-ksT} \tag{10.7}$$

be the DLT of the sequence $\{\xi\}_m$. As a special case, the DLT of the sequence $\{\xi\}_1$ has the form

$$\xi_1^*(s) \triangleq \sum_{k=0}^{\infty} \xi_{k+1} \mathrm{e}^{-ksT} \,. \tag{10.8}$$

Denoting $\xi^*(s) \triangleq \xi_0^*(s)$, from (10.8) and (10.7) for $m = 0$ we obtain

$$\mathrm{e}^{-sT} \xi_1^*(s) = \xi^*(s) - \xi_0$$

i.e.,

$$\xi_1^*(s) = \mathrm{e}^{sT} \left[\xi^*(s) - \xi_0\right] \,. \tag{10.9}$$

Using (10.9) to the sequence $\{\xi\}_2$, we find

$$\xi_2^*(s) = \mathrm{e}^{sT} \left[\xi_1^*(s) - \xi_1\right]$$

which yields

$$\xi_2^*(s) = \mathrm{e}^{2sT} \left[\xi^*(s) - \xi_0\right] - \mathrm{e}^{sT} \xi_1$$

Performing similar calculations, for any $m > 0$ we obtain

$$\xi_m^*(s) = \mathrm{e}^{msT} \xi^*(s) - \mathrm{e}^{msT} \xi_0 - \mathrm{e}^{(m-1)sT} \xi_1 - \ldots - \mathrm{e}^{sT} \xi_{m-1} \,. \tag{10.10}$$

Equation (10.10) can be used for the determination of the DLT of the unilateral sequence that is a solution of Eq. (10.2). We represent Eq. (10.2) in the form

$$a_0 \psi_{n+p} + a_1 \psi_{n+p-1} + \ldots + a_p \psi_n = b_0 \xi_{n+p} + b_1 \xi_{n+p-1} + \ldots + b_p \xi_n \,. \tag{10.11}$$

Assuming that $n = 0, 1, 2, \ldots$, sequentially, we have

$$\begin{aligned} a_0 \psi_p + a_1 \psi_{p-1} + \ldots + a_p \psi_0 &= b_0 \xi_p + b_1 \xi_{p-1} + \ldots + b_p \xi_0 \\ a_0 \psi_{1+p} + a_1 \psi_p + \ldots + a_p \psi_1 &= b_0 \xi_{1+p} + b_1 \xi_p + \ldots + b_p \xi_1 \\ \ldots &= \ldots \\ a_0 \psi_{\ell+p} + a_1 \psi_{\ell+p-1} + \ldots + a_p \psi_\ell &= b_0 \xi_{\ell+p} + b_1 \xi_{\ell+p-1} + \ldots + b_p \xi_\ell \,. \end{aligned} \tag{10.12}$$

If we multiply the k-th equation from (10.12) by e^{-ksT} and take the sum with respect to (10.7), we obtain

$$a_0 \psi_p^*(s) + a_1 \psi_{p-1}^*(s) + \ldots + a_p \psi^*(s) = b_0 \xi_p^*(s) + b_1 \xi_{p-1}^*(s) + \ldots + b_p \xi^*(s) \,. \tag{10.13}$$

Taking account of (10.10), from (10.13) we find

$$\left[a_0 e^{psT} + a_1 e^{(p-1)sT} + \ldots + a_p\right] \psi^*(s) + \mu_1(s)\psi_{p-1} + \ldots + \mu_p(s)\psi_0 =$$

$$= \left[b_0 e^{psT} + b_1 e^{(p-1)sT} + \ldots + b_p\right] \xi^*(s) + \nu_1(s)\xi_{p-1} + \ldots + \nu_p(s)\xi_0 \quad (10.14)$$

where $\mu_i(s)$, $\nu_i(s)$ are polynomials in $z = e^{sT}$, with not higher degree than p. With the notation

$$\begin{aligned} a_0 e^{psT} + a_1 e^{(p-1)sT} + \ldots + a_p &\stackrel{\triangle}{=} \tilde{a}(s) \\ b_0 e^{psT} + b_1 e^{(p-1)sT} + \ldots + b_p &\stackrel{\triangle}{=} \tilde{b}(s) \end{aligned} \quad (10.15)$$

from (10.14) we obtain

$$\psi^*(s) = \frac{\tilde{b}(s)}{\tilde{a}(s)} \xi^*(s) + \frac{\tilde{b}_0(s)}{\tilde{a}(s)} e^{sT} \quad (10.16)$$

where $\tilde{b}_0(s)$ is a polynomial in $z = e^{sT}$ having a degree not higher than $p-1$. Relation (10.16) can be represented in the form

$$\psi^*(s) = W_d(s)\xi^*(s) + \frac{b_0(s)}{a(s)} \quad (10.17)$$

where

$$W_d(s) = \frac{b(s)}{a(s)} \quad (10.18)$$

with quasipolynomials $a(s)$ and $b(s)$

$$\begin{aligned} a(s) &= a_0 + a_1 e^{-sT} + \ldots + a_p e^{-psT} \\ b(s) &= b_0 + b_1 e^{-sT} + \ldots + b_p e^{-psT} . \end{aligned} \quad (10.19)$$

Here $b_0(s)$ is a polynomial in $\zeta = e^{-sT}$ having a degree not higher than $p-1$. Since $\psi_0, \ldots, \psi_{p-1}$ are arbitrary numbers, the coefficients of the polynimial

$$b_1^o(\zeta) = b_0(s)\left|_{e^{-sT}=\zeta}\right. \quad (10.20)$$

can be considered as arbitrary constants. The function $W_d(s)$ will be called the *transfer function of the control program* (10.2). If all coefficients of the polynomial $b_0^o(\zeta)$ are zero, the corresponding solution of the difference equation (10.2) will be called the *solution with zero initial energy*. For this solution we have

$$\psi_0^*(s) = W_d(s)\xi^*(s) . \quad (10.21)$$

Hence, the transfer function of the control program can be defined as a ratio of discrete Laplace transforms of the output and input sequences with zero initial energy.

Example 10.1 The above formulae are used to construct the transfer function and the transformed output, where the input and output are connected by the control program

$$\psi_n + 3\psi_{n-1} + 2\psi_{n-2} = \xi_{n-1} + 5\xi_{n-2}\,.$$

In this case, using (10.19), we have

$$a(s) = 1 + 3e^{-sT} + 2e^{-2sT}, \qquad b(s) = e^{-sT} + 5e^{-2sT}.$$

The transfer function $W_d(s)$ of the control program appears as

$$W_d(s) = \frac{e^{-sT} + 5e^{-2sT}}{1 + 3e^{-sT} + 2e^{-2sT}}\,.$$

The equation (10.17) of the images has the form

$$\psi^*(s) = \frac{e^{-sT} + 5e^{-2sT}}{1 + 3e^{-sT} + 2e^{-2sT}}\,\xi^*(s) + \frac{\tilde{b}_0 + \tilde{b}_1 e^{-sT}}{1 + 3e^{-sT} + 2e^{-2sT}}$$

where $\tilde{b}_0$ and $\tilde{b}_1$ are arbitrary constants. □

Taking account of (10.1), we can express the transform $\psi^*(s)$ in terms of the input signal $x(t)$. Assuming that $x(t) \in \Lambda_+(\alpha, \infty)$, where α is a sufficiently large number, we have

$$\xi^*(s) = \mathcal{D}_x(T, s, +0)\,, \tag{10.22}$$

where $\mathcal{D}_x(T, s, +0)$ is the DLT of the input signal. From (10.22) and (10.17), we find

$$\psi^*(s) = W_d(s)\mathcal{D}_x(T, s, +0) + \frac{b_0(s)}{a(s)}\,. \tag{10.23}$$

Using the inversion formulae for the DLT in (10.23), for a given input $x(t)$ we can find the control sequence $\{\psi_k\}$. By (10.3), this sequence defines the continuous-time output $v(t)$ of the computer. Hence, the relation between the input $x(t)$ and the output $v(t)$ is, in general, not unique. This means that, using the terminology accepted in the book, the system of equations (10.1)–(10.3) defines a system. Moreover, since Eqs. (10.1)–(10.3) are linear and the operators appearing in these equations are periodic, this system is linear and T-periodic.

10.3 Transfer functions of computers

To find the PTF of the computer as a linear periodic system, we use the general approach described above. Assume that $x(t) = e^{st}$ where s is a complex parameter, and find a solution of Eqs. (10.1)–(10.3) having the form

$$v(s,t) = W_d(s,t)e^{st}, \qquad W_d(s,t) = W_d(s, t+T)\,. \tag{10.24}$$

The function $W_d(s,t)$ obtained in such a way will be called the *PTF of the computer*. For a practical calculation, we notice that for $x(t) = e^{st}$ we have

$$\xi_k = e^{ksT}. \tag{10.25}$$

Assuming that $\mu(t) = 1$, we obtain

$$v(t) = \psi_k, \qquad kT < t < (k+1)T. \tag{10.26}$$

Comparing (10.24) and (10.26), we find

$$v(s,t) = W_d(s, kT+0)e^{ksT}, \qquad kT < t < (k+1)T. \tag{10.27}$$

Denote

$$G_d(s) \triangleq W_d(s, kT+0) = W_d(s, +0). \tag{10.28}$$

Then, the sequence $\psi_k(s)$ associated with the solution of (10.24) has the form

$$\psi_k(s) = G_d(s)e^{ksT} \tag{10.29}$$

where $G_d(s)$ is an unknown function to be found. To determine this function we use Eqs. (10.25) and (10.29) together with the program (10.2). Obviously,

$$\xi_{n-q} = e^{(n-q)sT}, \qquad \psi_{n-q} = G_d(s)e^{(n-q)sT}. \tag{10.30}$$

Substituting (10.30) into (10.2), we obtain

$$G_d(s)\left[a_0e^{nsT} + a_1e^{(n-1)sT} + \ldots + a_pe^{(n-p)sT}\right] = \\ = b_0e^{nsT} + b_1e^{(n-1)sT} + \ldots + b_pe^{(n-p)sT}.$$

Hence,

$$G_d(s) = \frac{b_0e^{nsT} + b_1e^{(n-1)sT} + \ldots + b_pe^{(n-p)sT}}{a_0e^{nsT} + a_1e^{(n-1)sT} + \ldots + a_pe^{(n-p)sT}}. \tag{10.31}$$

After cancellation by e^{nsT}, we obtain a formula independent of n

$$G_d(s) = \frac{b_0 + b_1e^{-sT} + \ldots + b_pe^{-psT}}{a_0 + a_1e^{-sT} + \ldots + a_pe^{-psT}} = W_d(s) \tag{10.32}$$

where $W_d(s)$ is the transfer function of the control program. Taking account of (10.32), from (10.24) and (10.27) we have

$$W_d(s,t)e^{st} = W_d(s)e^{ksT}, \quad kT < t < (k+1)T$$

i.e.,

$$W_d(s,t) = W_d(s)e^{-s(t-kT)}, \quad kT < t < (k+1)T \tag{10.33}$$

which can be represented in the equivalent form

$$W_d(s,t) = W_d(s)e^{-st}, \quad 0 < t < T; \quad W_d(s,t) = W_d(s,t+T). \tag{10.34}$$

Instead of (10.34) we can also write

$$W_d(s,t) = W_d(s)W_{a0}(s,t) \tag{10.35}$$

where $W_{a0}(s,t)$ is the PTF of a sampler with zero-order hold obtained from (9.9) for $\mu(t) = 1$

$$W_{a0}(s,t) = \mathrm{e}^{-st}, \quad 0 < t < T; \quad W_{a0}(s,t) = W_{a0}(s,t+T)\,. \tag{10.36}$$

If, in addition, the DAC generates output pulses of an arbitrary form $\mu(t)$, we have instead of (10.35)

$$W_d(s,t) = W_d(s)W_0(s,t)\mu(t-kT)\,, \quad kT < t < (k+1)T$$

which is equivalent to

$$W_d(s,t) = W_d(s)W_a(s,t) \tag{10.37}$$

where $W_a(s,t)$ is the PTF of the sampling unit (9.9).

Example 10.2 Find an expression for the parametric transfer function of the computer with the program

$$\psi_n + 3\psi_{n-1} = 2\xi_n + \xi_{n-1}$$

and $\mu(t) = 1$. Then, the transfer function of the program is given by

$$W_d(s) = \frac{2 + \mathrm{e}^{-sT}}{1 + 3\mathrm{e}^{-sT}}$$

and the associated PTF appears as

$$W_d(s,t) = \frac{2 + \mathrm{e}^{-sT}}{1 + \mathrm{e}^{-sT}} W_{a0}(s,t)$$

where $W_{a0}(s,t)$ is the PTF of a zero-order hold defined by (10.36). □

We note that, using the representation of the PTF $W_a(s,t)$ in the form of a Fourier series (9.13) and the following obvious equality,

$$W_d(s) = W_d(s + \mathrm{j}\omega)\,, \quad \omega = 2\pi/T \tag{10.38}$$

the PTF $W_d(s,t)$ (10.37) can be written as a Fourier series

$$W_d(s,t) = \frac{1}{T} \sum_{k=-\infty}^{\infty} M(s + k\mathrm{j}\omega)W_d(s + k\mathrm{j}\omega)\mathrm{e}^{k\mathrm{j}\omega t}$$

which can be represented as a DPFR (3.1)

$$W_d(s,t) = \varphi_{Q_d}(T,s,t) \tag{10.39}$$

where

$$Q_d(s) = M(s)W_d(s)\,. \tag{10.40}$$

Multiplying (10.39) by e^{st}, we obtain the function

$$D_d(s,t) = \frac{1}{T}\sum_{k=-\infty}^{\infty} M(s+kj\omega)W_d(s+kj\omega)e^{(s+kj\omega)t}$$

called the *discrete transfer function of the computer*. It can also be written as a DLT

$$D_d(s,t) = \mathcal{D}_{Q_d}(T,s,t)\,. \tag{10.41}$$

10.4 Block-diagrams of computers

Using the relations of Section 10.3, we can obtain some equivalent block-diagrams of the controlling computer used below. In what follows we regard systems with the same PTF to be identical.

1. Consider the transfer function of the program $W_d(s)$ as the transfer function of a LTI system. Let us consider the linear periodic system shown in Fig. 10.2. Its operator equations have the form

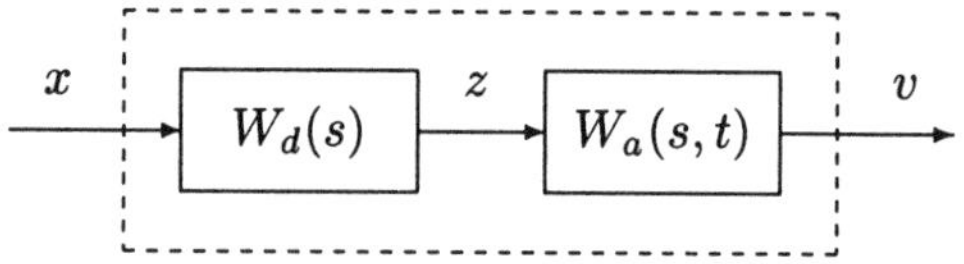

Figure 10.2: Block-diagram of a computer

$$v = W_a(s,t)z\,, \quad z = W_d(s)x\,. \tag{10.42}$$

According to the general formula (6.41), the PTF of this system is equal to

$$W(s,t) = W_a(s,t)W_d(s)$$

which coincides with (10.37). Therefore, the systems shown in Fig 10.1 and 10.2 are equivalent in the above sense.

It should be noted that the element with transfer function $W_d(s)$ has different physical meaning in the systems (10.1)–(10.3) and (10.42). In the initial scheme, the function $W_d(s)$ specifies the rule of transformation between the output and input numerical sequences inside the computer. But in (10.42) this element characterizes an element transforming a continuous input $x(t)$ into continuous output $z(t)$. The latter element can be represented as a connection of pure-delay networks, in which the delay equals the sampling period T. Indeed, let U_T be a shift element by T, so that

$$\mathsf{U}_T\,[x(t)] = x(t-T)\,. \tag{10.43}$$

The operator U_T is linear and time-invariant, and its PTF is given by

$$W_T(s) = \mathrm{e}^{-sT}\,. \tag{10.44}$$

It can be easily verified that the LTI system constructed of the elements U_T according to Fig. 10.3, has the transfer function

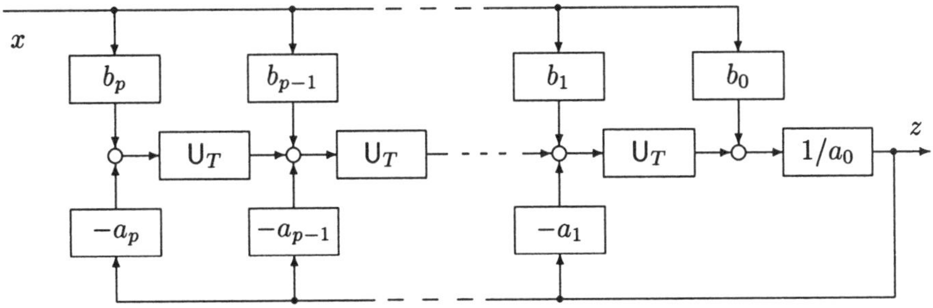

Figure 10.3: Control program as linear continuous-time system

$$W(s) = \frac{b_0 + b_1\mathrm{e}^{-sT} + \ldots + b_p\mathrm{e}^{-psT}}{a_0 + a_1\mathrm{e}^{-sT} + \ldots + a_p\mathrm{e}^{-psT}} = W_d(s)\,.$$

2. Consider the block-diagram shown in Fig. 10.4, where the element with transfer function $W_d(s)$ is described as in Fig. 10.3. By Lemma 6.1, the PTF

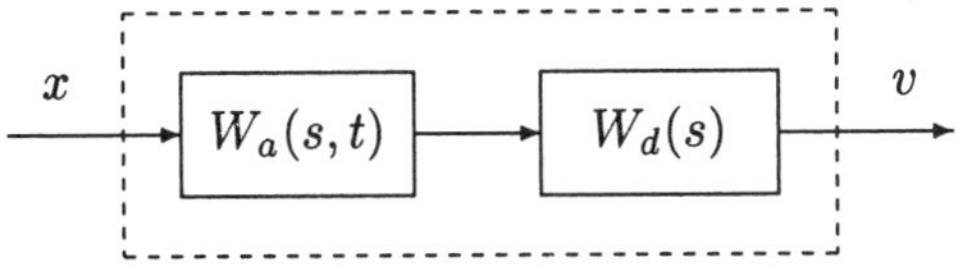

Figure 10.4: Block-diagram of computer

of this system is given by

$$W(s,t) = W_d(s)W_a(s,t) = W_p(s,t)\,.$$

Thus, the structures shown in Fig. 10.1, 10.2, and 10.4 are equivalent in the above sense. The block-diagram shown in Fig. 10.2 formally reduces the computer to an open-loop SD system (Fig. 9.4) with $F(s) = W_d(s)$. This makes it possible, using the relations of Section 9.11, to obtain the PTF of a computer with phase shift. Indeed, if the computer is working with phase shift ϕ, the ADC equation (10.1) has to be exchanged for

$$\xi_k = x(kT + \phi + 0) \tag{10.45}$$

and the DAC equation takes the form

$$v(t) = \mu(t + \phi - kT)\psi_k\,, \quad kT + \phi < t < (k+1)T + \phi\,. \tag{10.46}$$

The equation of the control program (10.2) remains valid. It can be easily understood that, with a phase shift, the sampling unit (9.1) in Fig. 10.2 is changed for the sampling unit (9.23) with phase shift ϕ. Therefore, the PTF $W_d^{\phi}(s,t)$ of a computer with phase shift is given by

$$W_d^{\phi}(s,t) = W_d(s, t-\phi) = W_d(s)\varphi_M(T, s, t-\phi)\,. \tag{10.47}$$

10.5 Transfer functions of open-loop systems with a computer

In this section we investigate the open-loop system shown in Fig. 10.5, where C denotes a computer described by (10.1)–(10.3), and $F(s)$ is the transfer function of continuous part given by (9.59). Using the block-diagram shown

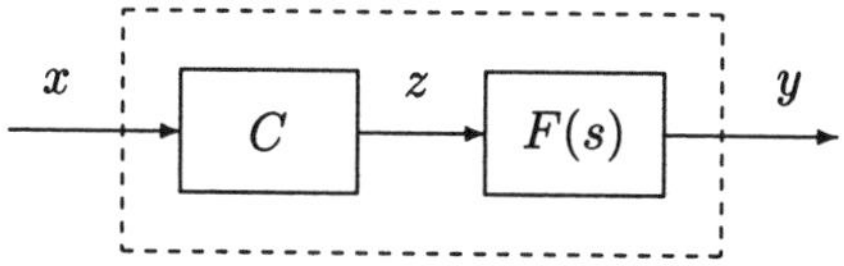

Figure 10.5: Block-diagram of an open-loop system with computer

in Fig. 10.2, we obtain the structure shown in Fig. 10.6, where $W_d(s)$ is the transfer function of the program (10.18), and U_a is an sampling unit of the form (9.1). To determine the PTF $W_{pd}(s,t)$ of this system, we take $x = \mathrm{e}^{st}$

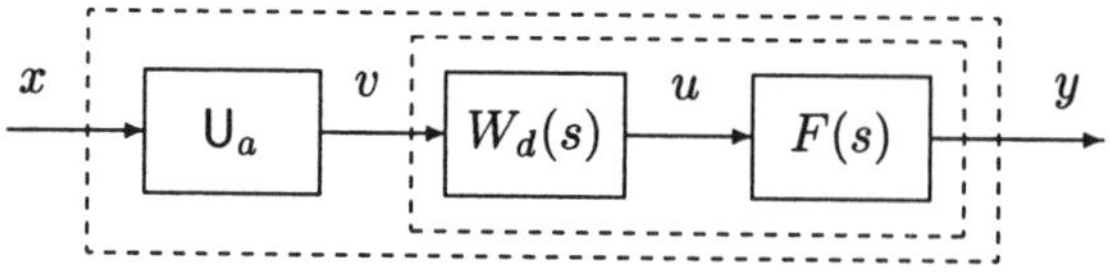

Figure 10.6: Model of a computer

and find the solution of the corresponding system of differential and difference equations having the form

$$y(s,t) = W_{pd}(s,t)\mathrm{e}^{st}, \quad W_{pd}(s,t) = W_{pd}(s, t+T)\,. \tag{10.48}$$

Using the PTF of the computer (10.37), for $x = \mathrm{e}^{st}$ we have

$$u(s,t) = W_d(s)\varphi_M(T,s,t)\mathrm{e}^{st} = W_d(s)\frac{1}{T}\sum_{k=-\infty}^{\infty} M(s+k\mathrm{j}\omega)\mathrm{e}^{(s+k\mathrm{j}\omega)t}\,. \quad (10.49)$$

The response of the continuous network to the input (10.49) has the form

$$y(s,t) = W_d(s)\frac{1}{T}\sum_{k=-\infty}^{\infty} F(s+k\mathrm{j}\omega)M(s+k\mathrm{j}\omega)\mathrm{e}^{(s+k\mathrm{j}\omega)t}\,. \quad (10.50)$$

Multiplying (10.50) by e^{-st} and using (10.48), we find the required PTF

$$W_{pd}(s,t) = W_d(s)\varphi_{FM}(T,s,t)\,. \quad (10.51)$$

The function

$$D_{pd}(s,t) \triangleq W_{pd}(s,t)\mathrm{e}^{st} \quad (10.52)$$

will be called the *discrete transfer function* (DTF) of the open-loop system under consideration. From Eqs. (10.52) and (10.51) we find

$$D_{pd}(s,t) = W_d(s)\mathcal{D}_{FM}(T,s,t)\,. \quad (10.53)$$

If the computer has a phase shift, the sampling unit (9.1) in Fig. 10.6 should be changed for an sampling unit with phase shift (9.23). In this case, similarly to (9.185), we obtain the corresponding PTF

$$W_{pd}^{\phi}(s,t) = W_d(s)\varphi_{FM}(T,s,t-\phi)\,. \quad (10.54)$$

If there is a pure delay τ in the continuous network, instead of transfer function $F(s)$ in Fig. 10.6 we have

$$F_\tau(s) = F(s)\mathrm{e}^{-s\tau}\,.$$

Then, the corresponding PTF W_{pd}^{τ} is obtained by changing $F(s)$ for $F_\tau(s)$ in (10.53), so that

$$W_{pd}^{\tau}(s,t) = W_d(s)\varphi_{F_\tau M}(T,s,t)\,. \quad (10.55)$$

If we have both, pure delay and phase shift, the corresponding PTF is given by

$$W_{pd}^{\phi\tau}(s,t) = W_d(s)\varphi_{F_\tau M}(T,s,t-\phi)\,. \quad (10.56)$$

10.6 Non-pathological conditions

Using the notation (9.146), we take

$$D_{pd}^{o}(\zeta,t) = D_{pd}(s,t)\,\big|_{\mathrm{e}^{-sT}=\zeta} \quad (10.57)$$

i.e.,

$$D_{pd}^{o}(\zeta,t) = W_d^{o}(\zeta)\mathcal{D}_{FM}^{o}(T,\zeta,t) \quad (10.58)$$

where

$$\begin{aligned} W_d^o(\zeta) &= W_d(s)\,\big|_{e^{-sT}=\zeta} \\ \mathcal{D}_{FM}^o(T,\zeta,t) &= \mathcal{D}_{FM}(T,s,t)\,\big|_{e^{-sT}=\zeta} = D_p(s,t)\,\big|_{e^{-sT}=\zeta}\,. \end{aligned} \tag{10.59}$$

We investigate some general properties of the function $D_{pd}^o(s,t)$ on the basis of Eqs. (10.58), (10.59).

1. From (10.18) and (10.19) we have

$$W_d^o(\zeta) = \frac{b^o(\zeta)}{a^o(\zeta)} \tag{10.60}$$

$$\text{with} \quad \begin{aligned} a^o(\zeta) &= a_0 + a_1\zeta + \ldots + a_p\zeta^p, \quad a_0 \neq 0 \\ b^o(\zeta) &= b_0 + b_1\zeta + \ldots + b_p\zeta^p\,. \end{aligned} \tag{10.61}$$

As follows from (10.59) and (10.61), under the given assumptions the program transfer function $W_d(s)$ is a causal rational periodic function (see Appendix A). If, in addition, $a_p \neq 0$, the control program (9.2) is called *limited.* If $a_p = 0$ and $b_p \neq 0$, the control program is called *retarded.* The real rational function (10.60) will henceforth assumed to be irreducible.
Let ζ_i $(i = 1, 2, \ldots, \rho)$ be different roots of the equation

$$a^o(\zeta) = a_0 + a_1\zeta + \ldots + a_p\zeta^p = 0 \tag{10.62}$$

with multiplicities λ_i $(i = 1, 2, \ldots, \rho)$, and let the numbers $\tilde{s}_i$ $(i = 1, 2, \ldots, \rho)$ satisfy the relations

$$e^{-\tilde{s}_i T} = \zeta_i\,, \quad -\omega/2 \leq \operatorname{Im} \tilde{s}_i < \omega/2\,. \tag{10.63}$$

As follows from the Appendix A, the numbers $\tilde{s}_i$ are prime poles of the function $W_d(s)$ and have multiplicities λ_i.

2. As was shown in Section 9.8, for any t the function $\mathcal{D}_{FM}^o(T,\zeta,t)$ is a real rational function with respect to ζ, i.e., for any t the function $\mathcal{D}_{FM}(T,s,t)$ is a r.p. function. As follows from Eqs. (9.99) and (9.106), for a fixed $t = \tilde{t}$ the set of poles of the function $\mathcal{D}_{FM}(T,s,\tilde{t}) = D_p(s,\tilde{t})$ belongs to the set of numbers

$$s_{ik} \stackrel{\triangle}{=} s_i + kj\omega\,, \quad (k = 0, \pm 1, \pm 2, \ldots) \tag{10.64}$$

where s_i are the poles of the function $F(s)$. As before, the set of different real parts of s_i will be called the set of *characteristic indices* of the transfer function $F(s)$.

3. Substituting (10.60) and (9.149) into (10.58), we obtain

$$D_{pd}^o(\zeta,t) = \frac{\beta^o(\zeta,t)}{\alpha^o(\zeta)}\,\frac{b^o(\zeta)}{a^o(\zeta)}\,. \tag{10.65}$$

For $t = 0$, using (9.160), we find

$$D^o_{pd}(\zeta, 0) = \frac{\zeta \chi^o(\zeta)}{\alpha^o(\zeta)} \frac{b^o(\zeta)}{a^o(\zeta)} \tag{10.66}$$

where χ^o, α^o, a^o and b^o are known polynomials such that $\deg \chi^o(\zeta) < n$, and $\alpha^o(\zeta)$ is given by (9.151).

Definition The open-loop system shown in Fig. 10.1 will be called *non-pathological*, if the real rational function (10.66) is irreducible.

From this definition it follows that for a system with a computer to be non-pathological it is necessary for the ratio

$$D^o_{pd}(\zeta, 0) = \mathcal{D}_{FM}(T, s, 0)\,\big|_{e^{-sT}=\zeta} \tag{10.67}$$

to be irreducible, i.e., the elementary open-loop system obtained from Fig. 10.1 with $W_d(s) = 1$ has to be non-pathological.
If the non-pathological conditions hold, the poles of the function $D^o_{pd}(\zeta, 0)$ coincide, taking account of their multiplicities, with the set of the roots of the polynomial

$$f^o(\zeta) \triangleq \alpha^o(\zeta) a^o(\zeta)\,. \tag{10.68}$$

Let f_i $(i = 1, 2, \ldots, q)$ be the roots of $f^o(\zeta)$ having multiplicities β_i. Then, the rational periodic function

$$D_{pd}(s, 0) = W_d(s)\mathcal{D}_{FM}(T, s, 0) \tag{10.69}$$

has poles of multiplicities β_i at the points

$$d_i = -\frac{1}{T} \ln f_i\,, \quad (i = 1, 2, \ldots, q)\,. \tag{10.70}$$

4. As follows from the aforesaid, the singular points of the DTF (10.53) are its poles, i.e., for any t this function is meromorphic. Obviously, for any $\tilde{t}$ the poles of the function $D_{pd}(s, \tilde{t})$ belongs to the set of values d_i (10.70).

Definition The number d_i will be called a *pole of the function* $D_{pd}(s, t)$, if it is a pole of $D_{pd}(s, \tilde{t})$ at least for one $t = \tilde{t}$. By the *multiplicity* of a pole d_i of the function $D_{pd}(s, t)$ we mean the largest multiplicity of the pole d_i of the function $D_{pd}(s, \tilde{t})$ for all possible $\tilde{t}$.

Proposition Let the system in Fig. 10.1 be non-pathological. Then, the poles of the DTF $D_{pd}(s, t)$ coincide with the numbers (10.70) and have multiplicities β_i.

Example 10.3 Examine the non-pathological condition for the system shown in Fig. 10.1 with

$$F(s) = \frac{1}{s-a}, \quad M(s) = \frac{1-e^{-sT}}{s}, \quad W_d(s) = \frac{1-e^{aT}e^{-sT}}{1+e^{aT}e^{-sT}}. \tag{10.71}$$

In this case, due to (9.115) we have

$$D_p(s,0) = \frac{1-e^{-aT}}{a} \frac{e^{(a-s)T}}{1-e^{(a-s)T}}$$

and the ratio $D_{pd}(s,0) = W_d(s)D_p(s,0)$ is reducible, i.e., if (10.71) holds, the open-loop system is pathological. □

10.7 Operator description of systems and equations for images

Using (10.51) the PTF of the open-loop system $W_{pd}(s,t)$ can be represented in the form

$$W_{pd}(s,t) = \frac{1}{T} \sum_{k=-\infty}^{\infty} Q_{pd}(s+kj\omega)e^{kj\omega t} = \varphi_{Q_{pd}}(T,s,t) \tag{10.72}$$

where

$$Q_{pd}(s) \triangleq F(s)M(s)W_d(s). \tag{10.73}$$

Taking account of the above relations, we also have

$$Q_{pd}(s) = \frac{m(s)b(s)M(s)}{d(s)a(s)}. \tag{10.74}$$

The function (10.73) will be called the *equivalent transfer function (ETF)* of the open-loop system. Below we investigate some general properties of the ETF $Q_{pd}(s)$.

Denote by $\tilde{\mu}_i$ $(i = 1, 2, \ldots, \rho)$ the different real parts of the numbers d_i (10.70) numbered in ascending order. The values $\tilde{\mu}_i$ will be called the *characteristic indices of the open-loop system.* Similarly, we can define the set of characteristic indices $\tilde{\lambda}_i$ of the function $Q_{pd}(s)$ as the set of different real parts of its poles. Since the poles of $Q_{pd}(s)$ are included in the set of roots of the equation

$$d(s)a(s) = 0 \tag{10.75}$$

the set of numbers $\tilde{\lambda}_i$ belongs to the set of characteristic indices $\tilde{\mu}_i$.

Lemma 10.1 *The sets of characteristic indices $\tilde{\lambda}_i$ and $\tilde{\mu}_i$ coincide.*

This lemma can be proved according to the scheme given in Lemma 11.1. ■

Lemma 10.2 *There exists a sufficiently small* $\epsilon > 0$*, so that for* $|s| \to \infty$ *in any stripe* $\tilde{\mu}_i + \epsilon \leq \operatorname{Re} s \leq \tilde{\mu}_{i+1} - \epsilon$, $(i = 1, 2, \ldots, \rho - 1)$ *and in the right half-plane* $\operatorname{Re} s \geq \tilde{\mu}_\rho + \epsilon$ *the following estimate holds*

$$|Q_{pd}(s)| < L|s|^{-2} \tag{10.76}$$

where L is a positive constant.

Proof Since $F(s)$ is strictly proper, in each specified stripe for $|s| \to \infty$ we have $|F(s)| < L|s|^{-1}$. Moreover, Eq. (9.11) yields $|M(s)| < K|s|^{-1}$. Besides that, in these areas the function $W_d(s)$ is uniformly bounded, as follows from Corollary 2 of Theorem A.2. ■

Theorem 10.1 *Let the open-loop system be non-pathological and* $\tilde{\mu}_i$ $(i = 1, 2, \ldots, \rho)$ *be the set of its characteristic indices. Then, in each regularity interval* $(\tilde{\mu}_i, \tilde{\mu}_{i+1})$ *, the PTF (10.72) coincides with the PTF of the LPPO*

$$\mathsf{U}_{pd_i}[x(t)] = \sum_{k=-\infty}^{\infty} \tilde{q}_i(t - kT)x(kT + 0) \tag{10.77}$$

where the pulse response $\tilde{q}_i(t)$ *is given by*

$$\tilde{q}_i(t) = \frac{1}{2\pi \mathrm{j}} \int_{c-\mathrm{j}\infty}^{c+\mathrm{j}\infty} Q_{pd}(s)\mathrm{e}^{st}\, \mathrm{d}s\,, \quad \tilde{\mu}_i < c < \tilde{\mu}_{i+1}\,. \tag{10.78}$$

Moreover, the function $\tilde{q}_i(t)$ *is continuous and* $\tilde{q}_i(t) \in \Lambda(\tilde{\mu}_i, \tilde{\mu}_{i+1})$. *We also have*

$$\tilde{q}_\rho = 0 \quad \text{for} \quad t < 0 \tag{10.79}$$

and the associated LPPO is causal.

The proof is not presented, because it reiterates the reasoning given above. ■

The set of LPPO (10.77) will henceforth be called the *family of LPPO* associated with the system involving a computer. As in Section 9.9, it can be shown that each LPPO from the family (10.77) defines, for the corresponding inputs, a particular solution of the system of equations, which describes the system under consideration. Therefore, if $x(t) \in \Lambda(\tilde{\mu}_i, \tilde{\mu}_{i+1})$, the corresponding Laplace transform of the output has the form

$$Y_i(s) = F(s)M(s)W_d(s)\mathcal{D}_x(T, s, +0)\,, \quad \tilde{\mu}_i < \operatorname{Re} s < \tilde{\mu}_{i+1} \tag{10.80}$$

and the discrete Laplace transform is given by

$$\mathcal{D}_{y_i}(T, s, t) = \mathcal{D}_{Y_i}(T, s, t) = \mathcal{D}_{FM}(T, s, t)W_d(s)\mathcal{D}_x(T, s, +0)$$
$$\tilde{\mu}_i < \operatorname{Re} s < \tilde{\mu}_{i+1}\,.$$

Next, we find the Laplace transform for $t > 0$ taking account of the initial states of the control program and the continuous element, assuming that $x(t) \in \Lambda_+(\tilde{\mu}_\rho, \infty)$. Using (9.65), we have

$$Y(s) = F(s)V(s) + \frac{m_0(s)}{d(s)} \tag{10.81}$$

where $V(s)$ is the Laplace transform of the output signal. Then, by Eq. (7.63),

$$V(s) = M(s)\psi^*(s) \tag{10.82}$$

where $\psi^*(s)$ is the DLT of the control sequence $\{\psi_k\}$ defined by (10.17). Substituting (10.17) and (10.82) into (10.81), for $\operatorname{Re} s > \tilde{\mu}_\rho$ we find

$$Y(s) = F(s)M(s)W_d(s)\mathcal{D}_x(T,s,+0) + F(s)M(s)\frac{b_0(s)}{a(s)} + \frac{m_0(s)}{d(s)}. \tag{10.83}$$

Next, we calculate the discrete Laplace transform of the function $Y(s)$. From (10.83) for any integer k we have

$$\begin{aligned} Y(s+kj\omega) &= F(s+kj\omega)M(s+kj\omega)W_d(s)\mathcal{D}_x(T,s,+0) + \\ &+ F(s+kj\omega)M(s+kj\omega)\frac{b_0(s)}{a(s)} + \frac{m_0(s+kj\omega)}{d(s+kj\omega)}. \end{aligned}$$

Therefore,

$$\begin{aligned} \mathcal{D}_Y(T,s,t) &= \frac{1}{T}\sum_{k=-\infty}^{\infty} Y(s+kj\omega)e^{(s+kj\omega)t} \\ &= \mathcal{D}_{FM}(T,s,t)W_d(s)\mathcal{D}_x(T,s,+0) + \mathcal{D}_{FM}(T,s,t)\frac{b_0(s)}{a(s)} + \mathcal{D}_{Y_0}(T,s,t) \end{aligned} \tag{10.84}$$

with

$$Y_0(s) \triangleq \frac{m_0(s)}{d(s)}, \quad m_0(s) = m_{10}s^{n-1} + m_{20}s^{n-2} + \ldots + m_{n0} \tag{10.85}$$

where m_{i0} $(i = 1, \ldots, n)$ are arbitrary constants.
Denote by $y_0(t)$ the original for the image $Y_0(s)$ in the half-plane $\operatorname{Re} s > \tilde{\mu}_\rho$. Since in the general case $\deg m_0(s) = n-1$, according to the results of Section 1.14, the function $y_0(t)$ has a discontinuity equal to m_{10} at $t = 0$. In this case, the sum of the series

$$\mathcal{D}_{y_0}(T,s,t) = \sum_{k=-\infty}^{\infty} y_0(t+kT)e^{-ksT}$$

coincides with the sum of the series $\mathcal{D}_{Y_0}(T,s,t)$ for $t \neq kT$. If $t = kT$, the function $\mathcal{D}_{y_0}(T,s,t)$ has a break, the first two terms on the right-hand side of

(10.84) are continuous with respect to t. Therefore, if $y(t)$ is the original for the image (10.83), for $t = kT$ we have

$$\mathcal{D}_y(T,s,t) = \mathcal{D}_{FM}(T,s,t)\left[W_d(s)\mathcal{D}_x(T,s,+0) + \frac{b_0(s)}{a(s)}\right] + \mathcal{D}_{y_0}(T,s,t)\,. \tag{10.86}$$

Moreover, taking the limit as $t \to +0$, we obtain

$$\mathcal{D}_y(T,s,+0) = \mathcal{D}_{FM}(T,s,0)\left[W_d(s)\mathcal{D}_x(T,s,+0) + \frac{b_0(s)}{a(s)}\right] + \mathcal{D}_{y_0}(T,s,+0)\,. \tag{10.87}$$

Example 10.4 Find a closed expression for the discrete Laplace transform $\mathcal{D}_Y(T,s,t)$ of the output of the open-loop system in Fig. 10.5 with

$$F(s) = \frac{1}{s-a}\,, \qquad M(s) = M_0(s) = \frac{1-\mathrm{e}^{-sT}}{s}\,.$$

Then,

$$Y_0(s) = \frac{m_0(s)}{s-a}$$

where m_0 is an arbitrary constant. The control program and the image $\psi^*(s)$ are assumed to be the same as in Example 10.1. Then, taking account of (4.11), for $0 < t < T$ we obtain

$$\mathcal{D}_Y(T,s,t) = \mathcal{D}_{FM_0}(T,s,t) * \tag{10.88}$$
$$* \left[\frac{\mathrm{e}^{-sT} + 5\mathrm{e}^{-2sT}}{1 + 3\mathrm{e}^{-sT} + 2\mathrm{e}^{-2sT}}\mathcal{D}_x(T,s,+0) + \frac{b_{00} + b_{01}\mathrm{e}^{-sT}}{1 + 3\mathrm{e}^{-sT} + 2\mathrm{e}^{-2sT}}\right] + \frac{m_0\mathrm{e}^{-st}}{1 - \mathrm{e}^{(a-s)T}}$$

where the function $\mathcal{D}_{FM_0}(T,s,t)$ is given by (9.115). From (10.88) it also follows that

$$\mathcal{D}_y(T,s,t) = \mathcal{D}_Y(T,s,t) \quad \text{for} \quad t \neq kT\,.$$

At the points $t_k = kT$ the image $\mathcal{D}_y(T,s,t)$ has discontinuities. Taking the limit of (10.88) as $t \to +0$, we find

$$\mathcal{D}_y(T,s,+0) = \mathcal{D}_{FM_0}(T,s,0) *$$
$$* \left[\frac{\mathrm{e}^{-sT} + 5\mathrm{e}^{-2sT}}{1 + 3\mathrm{e}^{-sT} + 2\mathrm{e}^{-2sT}}\mathcal{D}_x(T,s,+0) + \frac{b_0 + b_1\mathrm{e}^{-sT}}{1 + 3\mathrm{e}^{-sT} + 2\mathrm{e}^{-2sT}}\right] + \frac{m_0}{1 - \mathrm{e}^{(a-s)T}}$$

where the function $\mathcal{D}_{FM_0}(T,s,0)$ is defined by (9.115) with $t = 0$. □

10.8 Open-loop systems with prefilters

In this section we consider the open-loop system shown in Fig. 10.7, in which the input signal $x(t)$ is the result of the passage of a signal $a(t)$ through a LTI system with the transfer function $G(s)$. Henceforth, the function $G(s)$

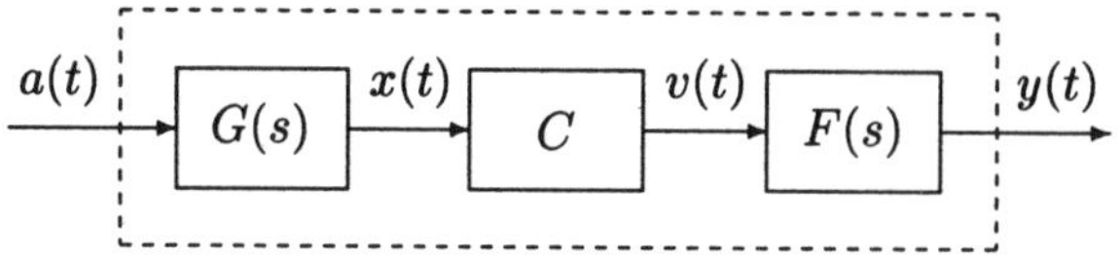

Figure 10.7: Digital system with analog prefilter

is assumed to be a strictly proper real rational function. According to the aforesaid, the PTF of the system is given by

$$\hat{W}(s,t) = G(s)\varphi_{FM}(T,s,t)W_d(s)\,. \tag{10.89}$$

Regarding the input $a(t)$, the system under consideration is not discrete, because the values of the input $x(t)$ at the sampling points $t_k = kT$ depend, in the general case, on the values of $a(t)$ for all t.
Denote by n_i the poles of the function $\hat{W}(s,t)$, and by $\hat{\mu}_i$ $(i = 1, 2, \dots, r)$ the different real parts of the poles n_i numbered in ascending order. The numbers $\hat{\mu}_i$ will be called the *characteristic indices of the open-loop system with prefilter.* Assume that $\hat{\mu}_0 = -\infty$ and $\hat{\mu}_{r+1} = \infty$. The intervals $(\hat{\mu}_i, \hat{\mu}_{i+1})$ will be called the regularity intervals of the system.

Theorem 10.2 *Under the given assumptions the following propositions hold.*
1. The following integrals converge at least in the sense of the principal value:

$$\hat{g}_i(t,u) = \frac{1}{2\pi \mathrm{j}}\int_{c-\mathrm{j}\infty}^{c+\mathrm{j}\infty} \hat{W}(s,t)\mathrm{e}^{su}\,\mathrm{d}s\,, \quad \hat{\mu}_i < c < \hat{\mu}_{i+1}\,. \tag{10.90}$$

2. The following estimations hold:

$$\left|\hat{g}_i(t,u)\mathrm{e}^{-su}\right| < K\mathrm{e}^{-\gamma|u|}\,, \quad \hat{\mu}_i < \operatorname{Re} s < \hat{\mu}_{i+1} \tag{10.91}$$

where K and γ are positive constants.

3. For $i = r$ we have

$$\hat{g}_r(t,u) = 0 \quad \text{for} \quad u < 0\,. \tag{10.92}$$

4. With the notation

$$\hat{h}_i(t,\tau) \triangleq \hat{g}_i(t,u)\,|_{u=t-\tau} \tag{10.93}$$

the integrals (operators)

$$y(t) = \int_{-\infty}^{\infty} \hat{h}_i(t,\tau)a(\tau)\,\mathrm{d}\tau \triangleq \hat{\mathrm{U}}_i[a(t)] \tag{10.94}$$

define, for $a(t) \in \Lambda(\hat{\mu}_i, \hat{\mu}_{i+1})$, particular solutions of the differential and difference equations describing the system at hand.

5. For $i = r$ the solution of (10.94) has the form

$$y(t) = \int_{-\infty}^{t} \hat{h}_r(t,\tau)a(\tau)\,\mathrm{d}\tau\,. \tag{10.95}$$

If $a(t) \in \Lambda_+(\hat{\mu}_r, \infty)$, i.e., $a(t) = 0$ for $t < 0$, then from (10.95) we obtain

$$y(t) = \int_0^t \hat{h}_r(t,\tau)a(\tau)\,d\tau\,. \tag{10.96}$$

6. For $a(t) \in \Lambda_+(\hat{\mu}_i, \hat{\mu}_{i+1})$ the solutions (10.94) can be represented in the form

$$y(t) = \frac{1}{2\pi j}\int_{c-j\infty}^{c+j\infty} \hat{W}(s,t)A(s)e^{st}\,ds\,, \quad \hat{\mu}_i < c < \hat{\mu}_{i+1}\,. \tag{10.97}$$

As a special case, the solution (10.95) is obtained from (10.97) with $c > \hat{\mu}_r$.

The proof of the theorem is not given, because it can be easily demonstrated using Theorem 7.5 with the help of techniques used in Rosenwasser (1995a). ∎

We can formulate the following remarks on the basis of this theorem.

1. From (10.90) it follows that

$$\hat{g}_i(t,u) = \hat{g}_i(t+T,u)$$

or, equivalently,

$$\hat{h}_i(t+T,\tau+T) = \hat{h}_i(t,\tau)\,.$$

Hence, Eq. (10.94) defines, with respect to the corresponding inputs, periodic integral operators $\hat{\mathrm{U}}_i$. It can be shown that Eq. (10.94) includes all linear periodic operators associated with the system under consideration. In this case, the functions $\hat{g}_i(t,u)$ and $\hat{h}_i(t,\tau)$ are the corresponding pulse responses and Green functions.

2. Equation (10.95) yields that the operator $\hat{\mathrm{U}}_r$ is causal. The remaining operators $\hat{\mathrm{U}}_i$ are not causal.

3. If there are no poles n_i on the imaginary axis, the system at hand will be called *e-dichotomic*. For an e-dichotomic system there exists a regularity interval $S_\gamma : (\hat{\mu}_\gamma, \hat{\mu}_{\gamma+1})$ such that $0 \in S_\gamma$. The Green function $\hat{h}_\gamma(t,\tau)$ associated with this interval will be called the *primary Green function*. The primary Green function satisfies the estimate

$$\left|\hat{h}_\gamma(t,\tau)\right| < Ke^{-\chi|t-\tau|} \tag{10.98}$$

where K and χ are positive constants. The corresponding operator $\hat{\mathrm{U}}_\gamma$ will also be called the *primary operator*.

4. As follows from the aforesaid, the primary operator is causal iff

$$\operatorname{Re} n_i < 0 \quad (i = 1, \dots, r) \tag{10.99}$$

i.e., all poles of the PTF (10.89) are in the left half-plane.

Equation (10.94) gives certain particular solutions of the system of equations under consideration, which are defined, in the general case, in $-\infty < t < \infty$. To determine the Laplace transform of the output under a given initial state we can use the formulae (10.83) and (10.84). Let

$$G(s) = \frac{m_G(s)}{d_G(s)}$$

where $m_G(s)$ and $d_G(s)$ are polynomials such that $\deg d_G(s) = \ell$ and $\deg m_G(s) < \ell$. Then, if $a(t) \in \Lambda_+(\hat{\mu}_r, \infty)$, using (9.65), for $\operatorname{Re} s > \hat{\mu}_r$ we have

$$X(s) = G(s)A(s) + G_0(s) \tag{10.100}$$

where

$$G_0(s) = \frac{m_{G_0}(s)}{d_G(s)} \tag{10.101}$$

with an arbitrary polynomial $m_{G_0}(s)$ such that $\deg m_{G_0}(s) < \ell$. Let $g_0(t)$ and $x(t)$ be the originals for the images (10.101) and (10.100) in the half-plane $\operatorname{Re} s > \hat{\mu}_r$. Then, similarly to the aforesaid, from (10.100) we can derive

$$\mathcal{D}_x(T, s, t) = \mathcal{D}_{GA}(T, s, t) + \mathcal{D}_{g_0}(T, s, t)\,. \tag{10.102}$$

The first term on the right-hand side of (10.102) is continuous with respect to t, while the second one may have discontinuities. Therefore, for $t \to +0$ we have

$$\mathcal{D}_x(T, s, +0) = \mathcal{D}_{GA}(T, s, 0) + \mathcal{D}_{g_0}(T, s, +0)\,. \tag{10.103}$$

Substituting (10.103) into (10.83), for $\operatorname{Re} s > \hat{\mu}_r$ we find

$$Y(s) = F(s)M(s)\left[W_d(s)\mathcal{D}_{GA}(T, s, 0) + W_d(s)\mathcal{D}_{g_0}(T, s, +0) + \frac{b_0(s)}{a(s)}\right] + \frac{m_0(s)}{d(s)}\,. \tag{10.104}$$

Similarly, from (10.84) we obtain

$$\begin{aligned} \mathcal{D}_Y(T, s, t) &= \mathcal{D}_{FM}(T, s, t) \,* \\ &* \left[W_d(s)\mathcal{D}_{GA}(T, s, 0) + W_d(s)\mathcal{D}_{g_0}(T, s, +0) + \frac{b_0(s)}{a(s)}\right] + \mathcal{D}_{Y_0}(T, s, t)\,. \end{aligned} \tag{10.105}$$

Example 10.5 Let the assumptions of Example 10.4 hold. Then for the system of Fig. 10.7 find the discrete Laplace transform of the output for

$$G(s) = \frac{1}{s - b}\,, \quad G_0(s) = \frac{g_0}{s - b}$$

where g_0 is an arbitrary constant. Then, from (10.102) we have

$$\mathcal{D}_x(T,s,+0) = \mathcal{D}_{GA}(T,s,0) + \frac{g_0}{1-\mathrm{e}^{(b-s)T}}\,. \tag{10.106}$$

Using this formula in (10.88), we find the DLT of the coresponding output. □

10.9 Multivariable open-loop systems

The theory presented above can be extended to the MIMO-case without major modifications. Let us consider the block-diagram shown in Fig. 10.7, where $\boldsymbol{a}(t)$, $\boldsymbol{x}(t)$, $\boldsymbol{v}(t)$ and $\boldsymbol{y}(t)$ are vectors, and $\boldsymbol{G}(s)$ and $\boldsymbol{F}(s)$ are real rational matrices of the corresponding dimensions. The performance of the computer is described by the following multivariable generalization of Eqs. (10.1)–(10.3).

$$\boldsymbol{\xi}_k = \boldsymbol{x}(kT+0) \tag{10.107}$$

$$\boldsymbol{a}_0\boldsymbol{\psi}_n + \boldsymbol{a}_1\boldsymbol{\psi}_{n-1} + \ldots + \boldsymbol{a}_p\boldsymbol{\psi}_{n-p} = \boldsymbol{b}_0\boldsymbol{\xi}_n + \boldsymbol{b}_1\boldsymbol{\xi}_{n-1} + \ldots + \boldsymbol{b}_p\boldsymbol{\xi}_{n-p} \tag{10.108}$$

where $\boldsymbol{a}_i$ and $\boldsymbol{b}_i$ are constant matrices of the corresponding dimensions, and $\det \boldsymbol{a}_0 \neq 0$. Moreover,

$$\boldsymbol{v}(t) = \mu(t-kT)\boldsymbol{\psi}_k\,, \quad kT < t < (k+1)T \tag{10.109}$$

where $\mu(t)$ is a scalar function, appearing in (9.1). The real rational matrices $\boldsymbol{F}(s)$ are assumed to be strictly proper, and $\boldsymbol{G}(s)$ to be proper. The PTM $\hat{\boldsymbol{W}}(s,t)$ of the system at hand has the form

$$\hat{\boldsymbol{W}}(s,t) = \varphi_{\boldsymbol{F}M}(T,s,t)\boldsymbol{W}_d(s)\boldsymbol{G}(s)\,. \tag{10.110}$$

with the notation

$$\boldsymbol{W}_d(s) \triangleq \boldsymbol{a}^{-1}(s)\boldsymbol{b}(s) \tag{10.111}$$

where

$$\begin{aligned} \boldsymbol{a}(s) &= \boldsymbol{a}_0 + \boldsymbol{a}_1\mathrm{e}^{-sT} + \ldots + \boldsymbol{a}_p\mathrm{e}^{-psT} \\ \boldsymbol{b}(s) &= \boldsymbol{b}_0 + \boldsymbol{b}_1\mathrm{e}^{-sT} + \ldots + \boldsymbol{b}_p\mathrm{e}^{-psT} \end{aligned} \tag{10.112}$$

are quasipolynomial matrices. Moreover, in (10.110) we have, similarly to the SISO-case,

$$\varphi_{\boldsymbol{F}M}(T,s,t) \triangleq \frac{1}{T}\sum_{k=-\infty}^{\infty} \boldsymbol{F}(s+kj\omega)M(s+kj\omega)\mathrm{e}^{kj\omega t}\,. \tag{10.113}$$

This matrix is the PTM of the open-loop SD system obtained with $\boldsymbol{G}(s) = \boldsymbol{I}$ and $\boldsymbol{W}_d(s) = \boldsymbol{I}$. As was shown in Rosenwasser (1996a), all main relations obtained in this section can be, without major modifiations, generalized onto the multivariable case (10.110).

10.10 Partial fraction expansion of DTF

In this section we consider the partial fraction expansion of the rational periodic function (10.53)

$$D_{pd}(s,t) = W_d(s)\mathcal{D}_{FM}(T,s,t) \tag{10.114}$$

(see Section A.4 in the Appendix). We shall assume that the function $F(s)$ can be expanded into partial fractions by (9.68). In this case we can, without loss of generality, assume that all poles s_i are primary, i.e., $-\omega/2 \le \operatorname{Im} s_i < \omega/2$, because the final results do not depend on this assumption. We shall also assume that the transfer function of the control program $W_d(s)$ has simple poles d_r, $(r = 1, 2, \ldots, m)$ with residues $\hat{c}_r = \frac{1}{T}c_r$. We further assume that all the numbers s_i and d_r are different. Since the program is causal, using (A.27) we can write

$$W_d(s) = l_\kappa \mathrm{e}^{-\kappa sT} + \ldots + l_1\mathrm{e}^{-sT} + W_{d1}(s) \tag{10.115}$$

where $W_{d1}(s)$ is a limited rational periodic function having finite limits

$$\ell_d^+ = \lim_{\operatorname{Re} s\to\infty} W_{d1}(s)\,, \quad \ell_{d1}^- = \lim_{\operatorname{Re} s\to-\infty} W_{d1}(s)\,. \tag{10.116}$$

Substituting (10.115) into (10.114), we have

$$D_{pd}(s,t) = \quad D_{pd}^{(1)}(s,t) + D_{pd}^{(2)}(s,t) \tag{10.117}$$

$$\text{with} \qquad D_{pd}^{(1)}(s,t) = \mathcal{D}_{FM}(T,s,t)\sum_{i=1}^{\kappa} l_i \mathrm{e}^{-isT} \tag{10.118}$$

$$D_{pd}^{(2)}(s,t) = W_{d1}(s)\mathcal{D}_{FM}(T,s,t)\,. \tag{10.119}$$

A partial fraction expansion of the rational periodic function (10.118) for $0 \le t \le T$ can be obtained by substituting any of the expansions (9.99) or (9.100) into (10.118). Therefore, our aim is to expand the rational periodic function (10.119) into partial fractions. For $0 \le t \le T$ this function is limited, because from Eqs. (10.116), (9.99) and (9.100) it follows that the next limits are finite

$$\ell^+\left[D_{pd}^{(2)}(s,t)\right] = \ell_d^+ h_p(t)\,, \quad \ell^-\left[D_{pd}^{(2)}(s,t)\right] = \ell_{d1}^- h_p^*(t)\,. \tag{10.120}$$

Under the given assumptions, the function $D_{pd}^{(2)}(s,t)$ has primary poles at the points $s = s_i$, $(i = 1, 2, \ldots, \ell)$ and $s = d_r$, $(r = 1, 2, \ldots, m)$. Therefore, for $0 \le t \le T$ an expansion into partial fractions of the form (A.50) is given by

$$D_{pd}^{(2)}(s,t) = T\sum_{i=1}^{\ell} \operatorname*{Res}_{q=s_i}\left[\frac{D_{pd}^{(2)}(q,t)}{1-\mathrm{e}^{(q-s)T}}\right] + T\sum_{r=1}^{m} \operatorname*{Res}_{q=d_r}\left[\frac{D_{pd}^{(2)}(q,t)}{1-\mathrm{e}^{(q-s)T}}\right] + \ell_{d1}^- h_p^*(t)\,. \tag{10.121}$$

Since the poles d_r are simple,

$$\operatorname*{Res}_{q=d_r} \left[\frac{D_{pd}^{(2)}(q,t)}{1-\mathrm{e}^{(q-s)T}} \right] = \frac{1}{T} \frac{c_r \mathcal{D}_{FM}(T,d_r,t)}{1-\mathrm{e}^{(d_r-s)T}} . \tag{10.122}$$

To calculate the residue at a pole $s = s_i$ we note that if the function $F(s)$ is analytic at $s = s_i$, then

$$\operatorname*{Res}_{s_i} \frac{F(s)}{(s-s_i)^k} = \frac{1}{(k-1)!} \left. \frac{\mathrm{d}^{k-1} F(s)}{\mathrm{d}s^{k-1}} \right|_{s=s_i} \tag{10.123}$$

or, in a compressed form,

$$\operatorname*{Res}_{s_i} \frac{F(s)}{(s-s_i)^k} = \frac{1}{(k-1)!} \frac{\mathrm{d}^{k-1} F(s_i)}{\mathrm{d}s_i^{k-1}} . \tag{10.124}$$

Since the sampled-data part is assumed to be non-pathological, from (9.164) we have

$$D_p(q,t) = \mathcal{D}_{FM}(T,s,t) = \frac{1}{T} F(q) M(q) \mathrm{e}^{qt} + C(q,t) \tag{10.125}$$

where the function $C(q,t)$ is analytic at $q = s_i$. Taking account of (9.68), Eq. (10.125) takes the form

$$\mathcal{D}_{FM}(T,q,t) = \frac{1}{T} M(q) \mathrm{e}^{qt} \sum_{i=1}^{\ell} \sum_{k=1}^{\nu_i} \frac{f_{ik}}{(q-s_i)^k} + C(q,t) . \tag{10.126}$$

Substituting (10.126) into (10.119), we find

$$D_{pd}^{(2)}(q,t) = \frac{1}{T} \sum_{i=1}^{\ell} \sum_{k=1}^{\nu_i} \frac{f_{ik}}{(q-s_i)^k} M(q) W_{d1}(q) \mathrm{e}^{qt} + \tilde{C}(q,t) \tag{10.127}$$

where the function $\tilde{C}(q,t)$, is analytic at $q = s_i$ $(i = 1, 2, \ldots, \ell)$. As a result, we have

$$\frac{D_{pd}^{(2)}(q,t)}{1-\mathrm{e}^{(q-s)T}} = \frac{1}{T} \sum_{i=1}^{\ell} \sum_{k=1}^{\nu_i} \frac{f_{ik}}{(q-s_i)^k} \frac{M(q) W_{d1}(q) \mathrm{e}^{qt}}{1-\mathrm{e}^{(q-s)T}} + \tilde{C}_1(q,t) \tag{10.128}$$

where the function $\tilde{C}_1(q,t)$ is analytic at $q = s_i$. Using (10.124), for a fixed i we obtain

$$\operatorname*{Res}_{q=s_i} \left[\frac{D_{pd}^{(2)}(q,t)}{1-\mathrm{e}^{(q-s)T}} \right] = \frac{1}{T} \sum_{i=1}^{\ell} \sum_{k=1}^{\nu_i} \frac{f_{ik}}{(k-1)!} \frac{\partial^{k-1}}{\partial {s_i}^{k-1}} \frac{M(s_i) W_{d1}(s_i) \mathrm{e}^{s_i t}}{1-\mathrm{e}^{(s_i-s)T}} . \tag{10.129}$$

Substituting (10.122) and (10.129) into (10.121), for $0 \le t \le T$ we find

$$\begin{aligned} D_{pd}^{(2)}(q,t) = \sum_{i=1}^{\ell} \sum_{k=1}^{\nu_i} \frac{f_{ik}}{(k-1)!} \frac{\partial^{k-1}}{\partial {s_i}^{k-1}} \frac{M(s_i) W_{d1}(s_i) \mathrm{e}^{s_i t}}{1-\mathrm{e}^{(s_i-s)T}} + \\ + \sum_{r=1}^{m} \frac{c_r \mathcal{D}_{FM}(T,d_r,t)}{1-\mathrm{e}^{(d_r-s)T}} + \ell_{d1}^{-} h_p^*(t) , \qquad 0 \le t \le T . \end{aligned} \tag{10.130}$$

To continue formula (10.130) over the whole t-axis, we use the relation

$$D_{pd}^{(2)}(s, t+\kappa T) = D_{pd}^{(2)}(s,t)\mathrm{e}^{\kappa s T}\,.$$

In the special case when $t = \varepsilon + \kappa T$ and $0 \leq \varepsilon < T$, from (10.130) we have

$$D_{pd}^{(2)}(s,t) = \left[\sum_{i=1}^{\ell}\sum_{k=1}^{\nu_i} \frac{f_{ik}}{(k-1)!}\frac{\partial^{k-1}}{\partial {s_i}^{k-1}}\frac{M(s_i)W_{d1}(s_i)\mathrm{e}^{s_i\varepsilon}}{1-\mathrm{e}^{(s_i-s)T}} + \right. \tag{10.131}$$

$$\left. + \sum_{r=1}^{m} \frac{c_r \mathcal{D}_{FM}(T, d_r, \varepsilon)}{1-\mathrm{e}^{(d_r-s)T}} + \ell_{d1}^{-} h_p^{*}(\varepsilon)\right]\mathrm{e}^{\kappa s T}\,.$$

In a similar way, using (A.49), we can obtain the following equivalent expansion for $t = \varepsilon + \kappa T$ and $0 \leq \varepsilon \leq T$:

$$D_{pd}^{(2)}(s,t) = \left[\sum_{i=1}^{\ell}\sum_{k=1}^{\nu_i} \frac{f_{ik}}{(k-1)!}\frac{\partial^{k-1}}{\partial {s_i}^{k-1}}\frac{M(s_i)W_{d1}(s_i)\mathrm{e}^{s_i\varepsilon}}{\mathrm{e}^{(s-s_i)T}-1} + \right. \tag{10.132}$$

$$\left. + \sum_{r=1}^{m} \frac{c_r \mathcal{D}_{FM}(T, d_r, \varepsilon)}{\mathrm{e}^{(s-d_r)T}-1} + \ell_{d}^{+} h_p(\varepsilon)\right]\mathrm{e}^{\kappa s T}\,.$$

If all poles s_i are simple, and (9.101) holds, the above formulae become simplified

$$D_{pd}^{(2)}(s,t) = \left[\sum_{i=1}^{n} \frac{f_i M(s_i)W_{d1}(s_i)\mathrm{e}^{s_i\varepsilon}}{1-\mathrm{e}^{(s_i-s)T}} + \sum_{r=1}^{m} \frac{c_r \mathcal{D}_{FM}(T, d_r, \varepsilon)}{1-\mathrm{e}^{(d_r-s)T}} + \ell_{d1}^{-} h_p^{*}(\varepsilon)\right]\mathrm{e}^{\kappa s T} \tag{10.133}$$

and

$$D_{pd}^{(2)}(s,t) = \left[\sum_{i=1}^{n} \frac{f_i M(s_i)W_{d1}(s_i)\mathrm{e}^{s_i\varepsilon}}{\mathrm{e}^{(s-s_i)T}-1} + \sum_{r=1}^{m} \frac{c_r \mathcal{D}_{FM}(T, d_r, \varepsilon)}{\mathrm{e}^{(s-d_r)T}-1} + \ell_{d}^{+} h_p(\varepsilon)\right]\mathrm{e}^{\kappa s T}. \tag{10.134}$$

Equations (9.102) and (9.103) follow from these relations as special cases.

Example 10.6 Let us consider in more detail the technique for constructing the partial fraction expansion (10.134) for the open digital system with (10.114) and

$$F(s) = \frac{1}{s-a}\,, \quad M(s) = \frac{1-\mathrm{e}^{-sT}}{s}\,, \quad W_d(s) = \frac{1+\mathrm{e}^{-sT}}{0.5-\mathrm{e}^{-sT}} \tag{10.135}$$

where $\mathrm{e}^{aT} \neq 2$. The primary pole d_1 of the transfer function $W_d(s)$ satisfies the equation $\mathrm{e}^{d_1 T} = 2$. The residue $\hat{c}_1$ at this pole equals $3/T$, i.e., $c_1 = 3$. In this case, Eqs. (10.35) and (9.115) yield

$$W_d(a) = \frac{1-\mathrm{e}^{-aT}}{0.5-\mathrm{e}^{-aT}}$$

$$\mathcal{D}_{FM}(T, d_1, t) = \frac{e^{a(t-T)} - 1}{a} + \frac{1 - e^{-aT}}{a} \frac{e^{at}}{1 - 0.5e^{aT}}, \quad 0 \le t \le T.$$

Taking into account that, in the given case,

$$\ell_{d1}^- = \ell_d^- = -1, \quad h_p^*(t) = \frac{e^{a(t-T)} - 1}{a}$$

we find the required expansion (10.134). □

Chapter 11

Closed-loop systems with a single sampling unit

11.1 General structure of single-loop systems

In this chapter we investigate the closed-loop system shown in Fig. 11.1, where $R(s)$ and $G(s)$ are the transfer functions of the plant and actuator, respectively, and C denotes a computer with the structure shown in Fig. 10.1. The system

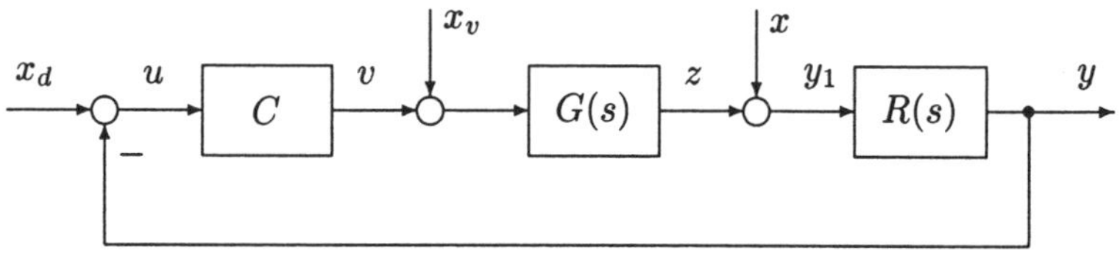

Figure 11.1: Single-loop digital system

shown in Fig. 11.1 is described by the following operator equations

$$y = R(s)(z + x)\,, \quad z = G(s)(v + x_v)\,, \quad u = x_d - y\,. \tag{11.1}$$

It is assumed that the computer is described by Eqs. (10.1)–(10.3)

$$\xi_k = u(kT + 0) \tag{11.2}$$

$$a_0\psi_n + a_1\psi_{n-1} + \ldots + a_p\psi_{n-p} = b_0\xi_n + b_1\xi_{n-1} + \ldots + b_p\xi_{n-p} \tag{11.3}$$

$$v(t) = \mu(t - kT)\psi_k\,, \quad kT < t < (k+1)T\,. \tag{11.4}$$

In Fig. 11.1, by x and x_v we denote exogenous disturbances acting on the continuous elements, while by x_d we mean a measurement error, a quantization error, or a reference signal. The system under investigation is obtained by closing the feedback for the open-loop digital system shown in Fig. 11.2.

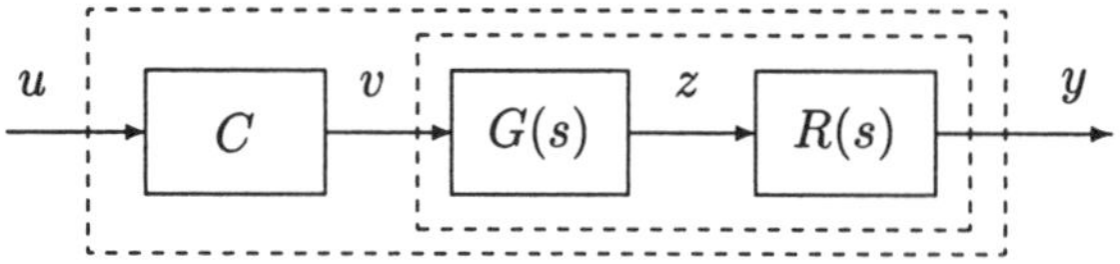

Figure 11.2: Open-loop system with computer

According to (10.51), the PTF of the system shown in Fig. 11.2 is given by

$$W_{pd}(s,t) = W_d(s)W_p(s,t) \tag{11.5}$$

where $W_d(s)$ is the transfer function of the control program (10.18), and, in the given case,

$$W_p(s,t) = \varphi_{RGM}(T,s,t) \tag{11.6}$$

where

$$\varphi_{RGM}(T,s,t) \stackrel{\triangle}{=} \frac{1}{T}\sum_{k=-\infty}^{\infty} R(s+kj\omega)G(s+kj\omega)M(s+kj\omega)e^{kj\omega t}\,. \tag{11.7}$$

Here $M(s)$ is the transfer function of the forming element (9.10).
As follows from the aforesaid, $W_p(s,t)$ is the PTF of the open-loop system obtained for $W_d(s) = 1$ (Fig. 11.3). With the notation

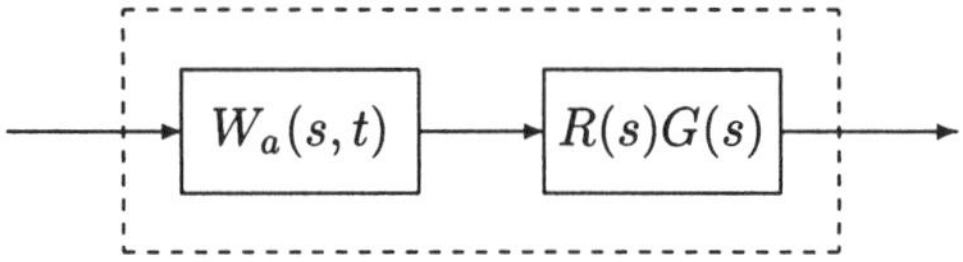

Figure 11.3: Open-loop SD system

$$F(s) \stackrel{\triangle}{=} R(s)G(s) \tag{11.8}$$

from (11.6) we obtain

$$W_p(s,t) = \varphi_{FM}(T,s,t)\,. \tag{11.9}$$

The function $F(s)$ will be henceforth called the transfer function of the continuous part of the system under investigation, while the open-loop system shown in Fig. 11.3 will be called its sampled-data part. In this case, the PTF of the open-loop system shown in Fig. 11.2 appears as

$$W_{pd}(s,t) = W_d(s)\varphi_{FM}(T,s,t) \tag{11.10}$$

and the corresponding DLT is equal to

$$D_{pd}(s,t) = W_d(s)\mathcal{D}_{FM}(T,s,t)\,. \tag{11.11}$$

The functions $R(s)$ and $G(s)$ can be either proper or strictly proper, but they should ensure that, with appropriate input, a continuous feedback signal acts at the input of the computer. Under such conditions, the real rational function $F(s)$ will always be strictly proper.

11.2 Transfer functions for sampled inputs

If $x(t) = x_v(t) = 0$, we have a closed-loop system with the input $x_d(t)$. Further we find the PTFs from the input $W_{yd}(s,t)$ and $W_{zd}(s,t)$ from the input $x_d(t)$ to the outputs $y(t)$ and $z(t)$, respectively. To derive the PTF $W_{yd}(s,t)$, according to the general technique we assume that

$$x_d(t) = \mathrm{e}^{st}, \quad y(t) = W_{yd}(s,t)\mathrm{e}^{st}, \quad W_{yd}(s,t) = W_{yd}(s,t+T). \tag{11.12}$$

Since $F(s)$ is strictly proper, the PTF $W_{yd}(s,t)$ is continuous with respect to t. Using (11.12) and Fig. 11.1, we obtain

$$u(t) = \mathrm{e}^{st} f_u(s,t) \tag{11.13}$$

where

$$f_u(s,t) \triangleq 1 - W_{yd}(s,t) = f_u(s,t+T). \tag{11.14}$$

Since the function $f_u(s,t)$ is continuous and periodic with respect to t, using the stroboscopic effect of the computer, the input of the computer can be changed equivalently to

$$\tilde{u}(t) = \mathrm{e}^{st} f_u(s,0) = \mathrm{e}^{st}\left[1 - W_{yd}(s,0)\right]. \tag{11.15}$$

Using (11.15) and the PTF (11.10), we can write

$$y(t) = \mathrm{e}^{st}\left[1 - W_{yd}(s,0)\right] W_d(s)\varphi_{FM}(T,s,t). \tag{11.16}$$

Comparing equations (11.12) and (11.16) for $y(t)$, we obtain

$$W_{yd}(s,t) = \left[1 - W_{yd}(s,0)\right] W_d(s)\varphi_{FM}(T,s,t) \tag{11.17}$$

which can be considered as a functional equation in the PTF $W_{yd}(s,t)$. Since the right-hand side of (11.17) is continuous in t, the left must have the same property. Then, taking $t = 0$ in (11.17), we obtain

$$W_{yd}(s,0) = \left[1 - W_{yd}(s,0)\right] W_d(s)\varphi_{FM}(T,s,0).$$

Hence,

$$W_{yd}(s,0) = \frac{W_d(s)\varphi_{FM}(T,s,0)}{1 + W_d(s)\varphi_{FM}(T,s,0)} \tag{11.18}$$

where

$$\varphi_{FM}(T,s,0) = \mathcal{D}_{FM}(T,s,0) = \frac{1}{T}\sum_{k=-\infty}^{\infty} F(s+k\mathrm{j}\omega)M(s+k\mathrm{j}\omega)$$

in accordance with (11.7). Substituting (11.18) into (11.17), we find

$$W_{yd}(s,t) = \frac{W_d(s)\varphi_{FM}(T,s,t)}{1 + W_d(s)\varphi_{FM}(T,s,0)}. \tag{11.19}$$

As was proved above, if Eq.(11.17) is solvable, its solution has the form (11.19). But, by direct substitution it can be shown that (11.19) satisfies Eq. (11.17), i.e., gives a solution of (11.17). Moreover, as follows from the above relations, the solution is unique. Thus, Eq. (11.19) defines the required PTF.

Then, we find the PTF $W_{zd}(s,t)$ from the input $x_d(t)$ to the output $z(t)$, assuming that the real rational function $G(s)$ is proper. As follows from (11.15) and (11.18),

$$\tilde{u}(t) = \frac{e^{st}}{1 + W_d(s)\varphi_{FM}(T,s,0)}. \tag{11.20}$$

The steady-state response of the open-loop system shown in Fig. 11.4 to this input signal equals

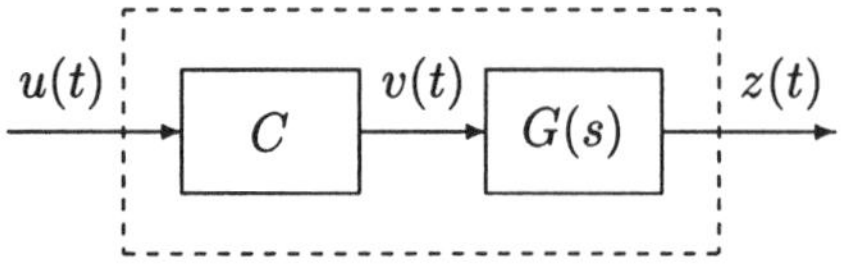

Figure 11.4: Open-loop system

$$z(t) = e^{st} \frac{W_d(s)\varphi_{GM}(T,s,t)}{1 + W_d(s)\varphi_{FM}(T,s,0)} \tag{11.21}$$

where

$$\varphi_{GM}(T,s,t) = \frac{1}{T} \sum_{k=-\infty}^{\infty} G(s + kj\omega)M(s + kj\omega)e^{kj\omega t}.$$

Multiplying (11.21) by e^{-st}, we find the desired PTF

$$W_{zd}(s,t) = \frac{W_d(s)\varphi_{GM}(T,s,t)}{1 + W_d(s)\varphi_{FM}(T,s,0)}. \tag{11.22}$$

Taking into account that

$$W_d(s) = W_d(s + j\omega), \quad \varphi_{FM}(T,s,0) = \varphi_{FM}(t,s + j\omega,0)$$

and using the Fourier series (11.7), we can represent Eq. (11.19) in the form

$$W_{yd}(s,t) = \frac{1}{T} \sum_{k=-\infty}^{\infty} Q_{yd}(s + kj\omega)e^{kj\omega t} \tag{11.23}$$

where

$$Q_{yd}(s) \triangleq F(s)M(s)\tilde{W}_d(s) = R(s)G(s)M(s)\tilde{W}_d(s) \tag{11.24}$$

$$\tilde{W}_d(s) \triangleq \frac{W_d(s)}{1 + W_d(s)\varphi_{FM}(T,s,0)}. \tag{11.25}$$

Comparing (11.23) with (10.72), we find that, as regards input $x_d(t)$ and output $y(t)$, the system at hand reduces formally to an open-loop digital system with a computer, where the transfer function of the control program is defined by (11.25). The PTF (11.22) can also be written in a form similar to (11.23), because

$$W_{zd}(s,t) = \frac{1}{T} \sum_{k=-\infty}^{\infty} Q_{zd}(s + kj\omega)e^{kj\omega t} \tag{11.26}$$

where

$$Q_{zd}(s) \triangleq G(s)M(s)\tilde{W}_d(s)\,. \tag{11.27}$$

We note that, multiplying (11.19) and (11.22) by e^{st}, we obtain the discrete transfer function

$$D_{yd}(s,t) = \mathcal{D}_{Q_{yd}}(T,s,t) = \frac{W_d(s)\mathcal{D}_{FM}(T,s,t)}{1 + W_d(s)\mathcal{D}_{FM}(T,s,0)} \tag{11.28}$$

$$D_{zd}(s,t) = \mathcal{D}_{Q_{zd}}(T,s,t) = \frac{W_d(s)\mathcal{D}_{GM}(T,s,t)}{1 + W_d(s)\mathcal{D}_{FM}(T,s,0)} \tag{11.29}$$

taking into account that $\varphi_{FM}(T,s,0) = \mathcal{D}_{FM}(t,s,0)$. For $0 \le t = \varepsilon < T$ Eqs. (11.28) and (11.29) coincide with the corresponding transfer functions in the sense of the modified discrete Laplace transformation, Zypkin (1964). For $t = 0$ Eq. (11.28) yields

$$D_{yd}(s,0) = \frac{W_d(s)\mathcal{D}_{FM}(T,s,0)}{1 + W_d(s)\mathcal{D}_{FM}(T,s,0)}\,. \tag{11.30}$$

It can be immediately verified that the function

$$D^*_{yd}(z) \triangleq D_{yd}(s,0)\,|_{e^{sT}=z} \tag{11.31}$$

coincides with the transfer function of the system in the sense of $z-$transformation, Jury (1958); Zypkin (1964).

Example 11.1 Construct the DTF of the closed-loop system shown in Fig. 11.5, where

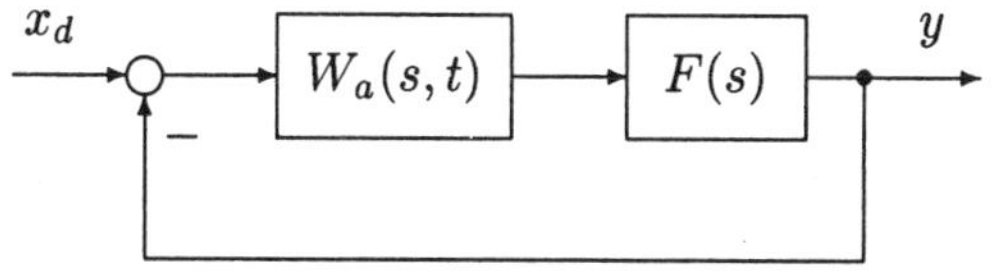

Figure 11.5: Single-loop SD system

$$F(s) = \frac{\gamma}{s-a}, \quad W_d(s) = 1\,.$$

The form of the controlling pulse is assumed to be arbitrary. Under the given assumptions, from (11.28) we have

$$D_{yd}(s,t) = \frac{\mathcal{D}_{FM}(T,s,t)}{1+\mathcal{D}_{FM}(T,s,0)}\,.$$

Due to (9.129), for $0 \le t \le T$ we have

$$\mathcal{D}_{FM}(T,s,t) = \frac{\gamma M(a)\mathrm{e}^{at}}{\mathrm{e}^{(s-a)T}-1} + h_p(t) \tag{11.32}$$

where

$$h_p(t) = \gamma \int_0^t \mathrm{e}^{a(t-\tau)}\mu(\tau)\,\mathrm{d}\tau\,. \tag{11.33}$$

For $t = 0$ Eq. (11.32) yields

$$\mathcal{D}_{FM}(T,s,0) = \frac{\gamma M(a)}{\mathrm{e}^{(s-a)T}-1}\,. \tag{11.34}$$

Substituting (11.32)–(11.34) into (5.31), we find

$$D_{yd}(s,t) = \frac{h_p(t)\left(\mathrm{e}^{sT}-\mathrm{e}^{aT}\right)+\gamma M(a)\mathrm{e}^{a(t+T)}}{\mathrm{e}^{sT}-\mathrm{e}^{aT}+\gamma\mathrm{e}^{aT}M(a)}\,. \tag{11.35}$$

The continuation of Eq. (11.35) onto the whole t–axis gives for ($mT \le t \le (m+1)T$) taking due account of (9.105),

$$D_{yd}(s,t) = \mathrm{e}^{msT}\,\frac{h_p(t-mT)\left(\mathrm{e}^{sT}-\mathrm{e}^{aT}\right)+\gamma M(a)\mathrm{e}^{at}\mathrm{e}^{-(m-1)aT}}{\mathrm{e}^{sT}-\mathrm{e}^{aT}+\gamma\mathrm{e}^{aT}M(a)}\,.$$

For $t = 0$ Eq. (11.35) yields

$$D_{yd}(s,0) = \frac{\gamma\mathrm{e}^{aT}M(a)}{\mathrm{e}^{sT}-\mathrm{e}^{aT}+\gamma\mathrm{e}^{aT}M(a)}\,.$$

□

Table 11.1 presents the PTFs and DTFs of the system at hand from the input x_d to the outputs y, z and u.

11.3 PTF for inputs of continuous-time networks

Using the stroboscopic effect of the computer, we can construct the PTFs of the system shown in Fig. 11.1 for exogenous inputs applied to continuous networks.

Table 11.1: PTF and DTF

$\diamondsuit$	$W_{\diamondsuit d}(s,t)$	$D_{\diamondsuit d}(s,t)$
y	$\dfrac{W_d(s)\varphi_{FM}(T,s,t)}{1+W_d(s)\varphi_{FM}(T,s,0)}$	$\dfrac{W_d(s)\mathcal{D}_{FM}(T,s,t)}{1+W_d(s)\mathcal{D}_{FM}(T,s,0)}$
z	$\dfrac{W_d(s)\varphi_{GM}(T,s,t)}{1+W_d(s)\varphi_{FM}(T,s,0)}$	$\dfrac{W_d(s)\mathcal{D}_{GM}(T,s,t)}{1+W_d(s)\mathcal{D}_{FM}(T,s,0)}$
u	$1-\dfrac{W_d(s)\varphi_{FM}(T,s,t)}{1+W_d(s)\varphi_{FM}(T,s,0)}$	$1-\dfrac{W_d(s)\mathcal{D}_{FM}(T,s,t)}{1+W_d(s)\mathcal{D}_{FM}(T,s,0)}$

1. First, we construct the PTF $W_{yx}(s,t)$ from the input $x(t)$ to the output $y(t)$, assuming that $x_d = x_v = 0$. With this aim in view, we take

$$x(t) = \mathrm{e}^{st}, \quad y(t) = W_{yx}(s,t)\mathrm{e}^{st}, \quad W_{yx}(s,t) = W_{yx}(s,t+T) \tag{11.36}$$

where the PTF $W_{yx}(s,t)$ is continuous with respect to t. Let Eq. (11.36) hold for a fixed s. Then, as follows from Fig. 11.1, the input of the computer has been acted upon by the signal

$$u(t) = -\mathrm{e}^{st}W_{yx}(s,t)\,. \tag{11.37}$$

Using the stroboscopic effect, the input (11.37) can be changed equivalently by

$$\tilde{u}(t) = -\mathrm{e}^{-st}W_{yx}(s,0)\,. \tag{11.38}$$

Then, as in Section 11.2, we obtain

$$y(t) = -\mathrm{e}^{st}W_d(s)\varphi_{FM}(T,s,t)W_{yx}(s,0) + R(s)\mathrm{e}^{st}\,. \tag{11.39}$$

Comparing Eqs. (11.36) and (11.39) for $y(t)$, we obtain the following functional equation with respect to the PTF $W_{yx}(s,t)$

$$W_{yx}(s,t) = -W_d(s)\varphi_{FM}(T,s,t)W_{yx}(s,0) + R(s)\,. \tag{11.40}$$

For $t = 0$ we have

$$W_{yx}(s,0) = -W_d(s)\varphi_{FM}(T,s,0)W_{yx}(s,0) + R(s)$$

whence

$$W_{yx}(s,0) = \frac{R(s)}{1+W_d(s)\varphi_{FM}(T,s,0)}\,. \tag{11.41}$$

Substituting (11.41) into (11.40), we find

$$W_{yx}(s,t) = R(s)\left[1 - \frac{W_d(s)\varphi_{FM}(T,s,t)}{1+W_d(s)\varphi_{FM}(T,s,0)}\right]. \tag{11.42}$$

2. Next, we determine the PTF $W_{zx}(s,t)$ from the input $x(t)$ to the output $z(t)$. If the input is given by (11.39), we have

$$z(t) = -e^{st} W_d(s) \varphi_{GM}(T,s,t) W_{yx}(s,0)\,.$$

Using (11.41) and multiplying it by e^{-st}, we obtain

$$W_{zx}(s,t) = -\frac{R(s)W_d(s)\varphi_{GM}(T,s,t)}{1+W_d(s)\varphi_{FM}(T,s,0)}\,. \tag{11.43}$$

3. Then, we construct the PTF $W_{yx_v}(s,t)$ from the input $x_v(t)$ to the output $y(t)$. Let

$$x_v(t) = e^{st}, \quad y(t) = W_{yx_v}(s,t)e^{st}, \quad W_{yx_v}(s,t) = W_{yx_v}(s,t+T) \tag{11.44}$$

and the PTF $W_{yx_v}(s,t)$ be continuous with respect to t. If (11.44) holds, we find from Fig. 11.1

$$u(t) = -e^{st} W_{yx_v}(s,t)$$

which can be exchanged by the equivalent input

$$\tilde{u}(t) = -e^{st} W_{yx_v}(s,0)\,. \tag{11.45}$$

Using (11.45), similarly to the aforesaid we have

$$y(t) = -e^{st} W_d(s)\varphi_{FM}(T,s,t) W_{yx_v}(s,0) + F(s)e^{st}\,. \tag{11.46}$$

Comparing Eqs.(11.44) and (11.46) for $y(t)$, we find

$$W_{yx_v}(s,t) = -W_d(s)\varphi_{FM}(T,s,t)W_{yx_v}(s,0) + F(s)\,. \tag{11.47}$$

Then, for $t=0$ we have

$$W_{yx_v}(s,0) = \frac{F(s)}{1+W_d(s)}\varphi_{FM}(T,s,0)\,. \tag{11.48}$$

Substituting this formula into (11.47), we obtain

$$W_{yx_v}(s,t) = F(s)\left[1 - \frac{W_d(s)\varphi_{FM}(T,s,t)}{1+W_d(s)\varphi_{FM}(T,s,0)}\right]. \tag{11.49}$$

4. In order to calculate the PTF $W_{zx_v}(s,t)$ from the input $x_v(t)$ to the output $z(t)$, with reference to (11.45) and (11.47) we have

$$\tilde{u}(t) = -e^{st}\frac{F(s)}{1+W_d(s)\varphi_{FM}(T,s,0)}\,. \tag{11.50}$$

From (11.50) and Fig. 11.1 it follows that

$$z(t) = -e^{st}\frac{F(s)W_d(s)\varphi_{GM}(T,s,t)}{1+W_d(s)\varphi_{FM}(T,s,0)} + G(s)e^{st} \tag{11.51}$$

which, after multiplication by e^{-st}, gives the desired PTF

$$W_{zx_v}(s,t) = G(s)\left[1 - \frac{R(s)W_d(s)\varphi_{GM}(T,s,t)}{1+W_d(s)\varphi_{FM}(T,s,0)}\right]. \tag{11.52}$$

Table 11.2 presents the PTFs of the system under investigation from the inputs x and x_v to the outputs y, z and u.

Table11.2: PTFs of the system in Fig. 11.1

$\diamondsuit$	$W_{\diamondsuit x}(s,t)$	$W_{\diamondsuit x_v}(s,t)$
y	$R(s)\left[1 - \frac{W_d(s)\varphi_{FM}(T,s,t)}{1+W_d(s)\varphi_{FM}(T,s,0)}\right]$	$F(s)\left[1 - \frac{W_d(s)\varphi_{FM}(T,s,t)}{1+W_d(s)\varphi_{FM}(T,s,0)}\right]$
z	$-\frac{R(s)W_d(s)\varphi_{GM}(T,s,t)}{1+W_d(s)\varphi_{FM}(T,s,0)}$	$G(s)\left[1 - \frac{R(s)W_d(s)\varphi_{GM}(T,s,t)}{1+W_d(s)\varphi_{FM}(T,s,0)}\right]$
u	$R(s)\left[\frac{W_d(s)\varphi_{FM}(T,s,t)}{1+W_d(s)\varphi_{FM}(T,s,0)} - 1\right]$	$F(s)\left[\frac{W_d(s)\varphi_{FM}(T,s,t)}{1+W_d(s)\varphi_{FM}(T,s,0)} - 1\right]$

Example 11.2 Let us consider the dynamics of a stabilized platform, that reduces, with some simplifications, to the block-diagram shown in Fig. 11.6, Schwarz (1981). In this case we take

$$G(s) = \frac{1}{s}, \quad R(s) = \frac{1}{s}, \quad M(s) = M_0(s) = \frac{1-e^{-sT}}{s}.$$

The PTF of the system is given by (11.52). Moreover, the sum of the series $\varphi_{GM}(T,s,t)$ is defined by (9.115), and, due to (9.137),

$$\varphi_{FM_0}(T,s,0) = \frac{T^2}{2}\frac{e^{sT}+1}{(e^{sT}-1)^2}.$$

□

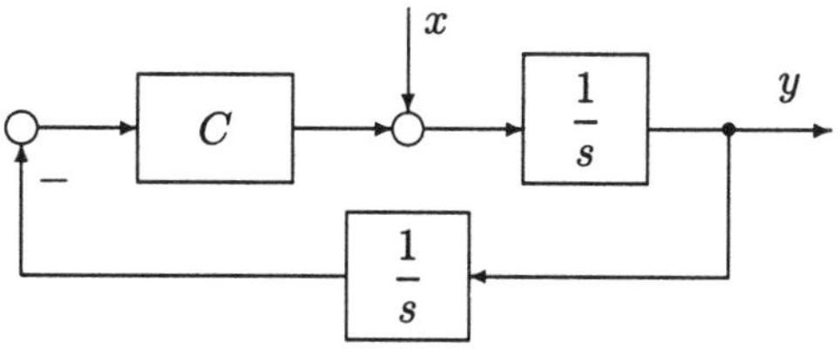

Figure 11.6: Stabilized platform

11.4 Transfer functions of time-delayed systems

In principle, the derivation of the formulae for the transfer functions in Sections 11.2 and 11.3 is independent of the form of the transfer functions $F(s)$ and $G(s)$. In fact, these formulae are applicable if the corresponding series converge and the signal at the input of the computer is continuous. These conditions hold in the special case, when the continuous networks contain pure delays, i.e.,

$$G(s) = G_1(s)\mathrm{e}^{-s\tau_1}, \quad R(s) = R_1(s)\mathrm{e}^{-s\tau_2} \tag{11.53}$$

where τ_1 and τ_2 are non-negative constants, and $G_1(s)$, $R_1(s)$ are real rational functions satisfying the conditions of Sections 11.2 and 11.3. It can be easily verified, that in this case all formulae given in these sections remain valid. To construct PTFs and DTFs of the time-delayed system we can use the formulae given in Tables 11.1 and 11.2, where $G(s)$ and $R(s)$ are given by (11.53). In this case, it is useful to remember that, under (11.53),

$$\varphi_{GM}(T,s,t) = \frac{1}{T}\sum_{k=-\infty}^{\infty} G_1(s+kj\omega)M(s+kj\omega)\mathrm{e}^{-(s+kj\omega)\tau_1}\mathrm{e}^{kj\omega t}$$

which can be written in the form

$$\varphi_{GM}(T,s,t) = \varphi_{G_1M}(t,s,t-\tau_1)\mathrm{e}^{-s\tau_1}. \tag{11.54}$$

Multiplying this equation by e^{st}, we find

$$\mathcal{D}_{GM}(T,s,t) = \mathcal{D}_{G_1M}(T,s,t-\tau_1). \tag{11.55}$$

Similarly,

$$\varphi_{FM}(T,s,t) = \varphi_{G_1R_1M}(t,s,t-\tau_1-\tau_2)\mathrm{e}^{-s(\tau_1+\tau_2)},$$

which is equivalent to

$$\mathcal{D}_{FM}(T,s,t) = \mathcal{D}_{R_1G_1M}(T,s,t-\tau_1-\tau_2). \tag{11.56}$$

For $t = 0$ Eq. (11.56) yields

$$\mathcal{D}_{FM}(T,s,0) = \varphi_{FM}(T,s,0) = \mathcal{D}_{R_1G_1M}(T,s,-\tau_1-\tau_2). \tag{11.57}$$

Let m be an integer such that

$$0 \le mT - \tau_1 - \tau_2 < T. \tag{11.58}$$

Then,

$$\tau_1 + \tau_2 = mT - \delta T, \quad 0 \le \delta < 1. \tag{11.59}$$

Using (11.59) and (11.57), we have

$$\mathcal{D}_{FM}(T,s,0) = \varphi_{FM}(T,s,0) = \mathcal{D}_{R_1G_1M}(T,s,\delta T)\mathrm{e}^{-msT}. \tag{11.60}$$

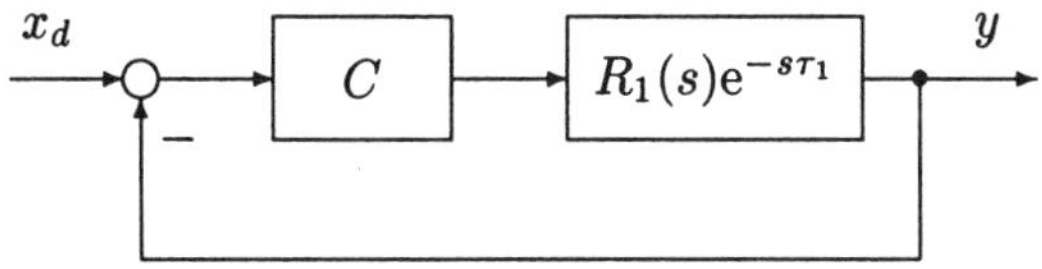

Figure 11.7: Closed-loop system with pure delay

Example 11.3 Let us construct the DTF of the closed-loop system shown in Fig. 11.7. Taking account of the above relations, the DTF $D_{yd}(s,t)$ (11.28) appears as

$$D_{yd}(s,t) = \frac{W_d(s)\mathcal{D}_{R_1M}(T,s,t-\tau_1)}{1+W_d(s)\mathcal{D}_{R_1M}(T,s,-\tau_1)}.$$

For $t=0$, taking due account of (11.60) we obtain

$$D_{yd}(s,0) = \frac{W_d(s)\mathcal{D}_{R_1M}(T,s,-\gamma T)}{e^{msT}+W_d(s)\mathcal{D}_{R_1M}(T,s,\gamma T)} \tag{11.61}$$

where the numbers m and δ are obtained from (11.59) with $\tau_2=0$. For $e^{sT}=z$ Eq. (11.61) gives the $z-$transfer function. □

Example 11.4 As a more complicated example let us construct the PTF of the closed-loop system with two arbitrary delays shown in Fig. 11.8. Using

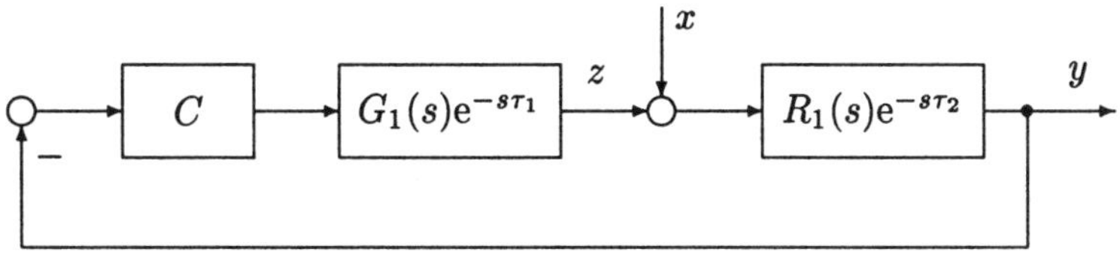

Figure 11.8: Closed-loop system with pure delays

(11.43) and the above relations, we can find the PTF $W_{zx}(s,t)$ from the input $x(t)$ to the output $z(t)$

$$W_{yx}(s,t) = -\frac{R_1(s)W_d(s)\varphi_{G_1M}(T,s,t-\tau_1)e^{-s(\tau_1+\tau_2)}}{1+W_d(s)\mathcal{D}_{R_1G_1M}(T,s,-\tau_1-\tau_2)}$$

which, taking account of (11.59) and (11.60), can be represented in the form

$$W_{yx}(s,t) = -\frac{R_1(s)W_d(s)\varphi_{G_1M}(T,s,t-\tau_1)e^{\delta sT}}{e^{msT}+W_d(s)\mathcal{D}_{R_1G_1M}(T,s,\delta T)}. \tag{11.62}$$

It should be remarked that in this example the DTF does not exist because the external excitation acts on a continuous-time element. □

11.5 Analysis of the poles of transfer functions

As will be shown below, the character of the processes in the systems is determined mostly by the poles of the DTFs and PTFs obtained above. In this section we investigate this problem under assumptions commonly satisfied in practice. Henceforth, we take

$$G(s) = \frac{m_G(s)}{d_G(s)}, \quad R(s) = \frac{m_R(s)}{d_R(s)} \tag{11.63}$$

where the numerators and denominators are co-prime polynomials. The polynomials

$$m(s) \triangleq m_G(s)m_R(s), \quad d(s) \triangleq d_G(s)d_R(s) \tag{11.64}$$

are also assumed to be co-prime, so that the ratio

$$F(s) = R(s)G(s) = \frac{m(s)}{d(s)} \tag{11.65}$$

is irreducible with the polynomial $d(s)$ of the form (9.17). The open-loop system shown in Fig. 11.3 is assumed to be non-pathological in the sense of Section 9.8. Therefore, for $t = 0$ the DTF of the open-loop system

$$D_p(s,0) = \mathcal{D}_{FM}(T,s,0)$$

is an irreducible rational periodic function. As follows from (10.60) and (10.66),

$$D_p(s,0) = \frac{e^{-sT}\chi(s)}{\alpha(s)} \tag{11.66}$$

where $\chi(s)$, $\alpha(s)$ are quasipolynomials such that $\deg\chi(s) < \deg\alpha(s)$ and, due to (9.151),

$$\alpha(s) = \left(1 - e^{s_1 T}e^{-sT}\right)^{\nu_1} \cdot \ldots \cdot \left(1 - e^{s_\ell T}e^{-sT}\right)^{\nu_\ell}. \tag{11.67}$$

With the substitution $e^{-sT} = \zeta$ we obtain from (10.53) the irreducible real rational function (10.66)

$$D^o_{pd}(\zeta,0) = \frac{\zeta\chi^o(\zeta)}{\alpha^o(\zeta)}\frac{b^o(\zeta)}{a^o(\zeta)} \tag{11.68}$$

where

$$\alpha^o(\zeta) = \left(1 - e^{s_1 T}\zeta\right)^{\nu_1} \cdot \ldots \cdot \left(1 - e^{s_\ell T}\zeta\right)^{\nu_\ell}. \tag{11.69}$$

Introduce the notation

$$f(\zeta) \triangleq \zeta\chi^o(\zeta)b^o(\zeta), \quad g(\zeta) \triangleq \alpha^o(\zeta)a^o(\zeta). \tag{11.70}$$

By the above assumptions, the polynomials $f(\zeta)$ and $g(\zeta)$ are co-prime.

Definition 1 The rational periodic function

$$\lambda(s) \stackrel{\triangle}{=} 1 + W_d(s)\mathcal{D}_{FM}(T,s,0) = 1 + \frac{e^{-sT}\chi(s)}{\alpha(s)}\frac{b(s)}{a(s)} \tag{11.71}$$

will be called the *determining function of the system* and the real rational function

$$\lambda^o(\zeta) \stackrel{\triangle}{=} \lambda(s)\,|_{e^{-sT}=\zeta} = 1 + \frac{\zeta\chi^o(\zeta)}{\alpha^o(\zeta)}\frac{b^o(\zeta)}{a^o(\zeta)} \tag{11.72}$$

will be called the *characteristic function of the system.*

Definition 2 The quasipolynomial

$$\Delta(s) \stackrel{\triangle}{=} a(s)\alpha(s) + e^{-sT}b(s)\chi(s) \tag{11.73}$$

will be called the *characteristic quasipolynomial* of the closed-loop system, and the polynomial

$$\Delta^o(\zeta) \stackrel{\triangle}{=} \Delta(s)\,|_{e^{-sT}=\zeta} = a^o(\zeta)\alpha^o(\zeta) + \zeta b^o(\zeta)\chi^o(\zeta) \tag{11.74}$$

will be called the *characteristic polynomial.*

Definition 3 The equation

$$\Delta(s) = a(s)\alpha(s) + e^{-sT}b(s)\chi(s) = 0 \tag{11.75}$$

will be called the *determining equation* of the closed-loop system, and the equation

$$\Delta^o(\zeta) = a^o(\zeta)\alpha^o(\zeta) + \zeta b^o(\zeta)\chi^o(\zeta) = 0 \tag{11.76}$$

the *characteristic equation.*

For the system shown in Fig. 11.1 we take $x_d = x_1$, $x_v = x_2$, and $x = x_3$, and correspondingly $y = y_1$, $z = y_2$, and $u = y_3$. The PTF from the input x_k to the output y_m is denoted by $W_{mk}(s,t)$. Then, the nine PTFs shown in Tables 11.1 and 11.2 taken together, constitute the matrix

$$\boldsymbol{W}(s,t) = \Big(W_{mk}(s,t)\Big), \quad (m,k = 1,2,3) \tag{11.77}$$

that is called the *parametric transfer matrix (PTM)* of the system under consideration.

Theorem 11.1 *The PTM (11.77) admits the representation*

$$\boldsymbol{W}(s,t) = \frac{\boldsymbol{N}(s,t)}{\Delta(s)} \tag{11.78}$$

where $\Delta(s)$ is the characteristic quasipolynomial (11.73), and $\boldsymbol{N}(s,t)$ is a matrix with elements $N_{mk}(s,t)$ not having poles, i.e., are entire functions.

Proof To prove the statement it suffices to show that the following equation holds for all m and k

$$W_{mk}(s,t) = \frac{N_{mk}(s,t)}{\Delta(s)} \tag{11.79}$$

where $N_{mk}(s,t)$ is an entire function in s. First we consider the PTF $W_{11}(s,t) = W_{yd}(s,t)$ given by (11.19) and the corresponding DTF (11.28). Substituting

$$\mathcal{D}_{FM}(T,s,t) = \frac{\beta(s,t)}{\alpha(s)}, \quad W_d(s) = \frac{b(s)}{a(s)} \tag{11.80}$$

into (11.28), we find

$$D_{11}(s,t) = \frac{\beta(s,t)b(s)}{a(s)\alpha(s) + \mathrm{e}^{-sT}b(s)\chi(s)} = \frac{\beta(s,t)b(s)}{\Delta(s)}$$

whence follows that

$$W_{11}(s,t) = \frac{\beta(s,t)b(s)\mathrm{e}^{-st}}{\Delta(s)}$$

i.e., for the PTF $W_{11}(s,t)$ we obtain a representation of the form (11.79), where

$$N_{11}(s,t) = \beta(s,t)b(s)\mathrm{e}^{-st}\,.$$

The proof for the remaining elements of the first column is similar.
The statement of Theorem 11.1 is proved with more efforts for the PTF of the second and third columns of the PTM $\boldsymbol{W}(s,t)$. Consider, for example, the PTF $W_{13}(s,t) = W_{yx}(s,t)$ given by (11.42). Equation (11.42) can be written in the form

$$W_{13}(s,t) = \frac{R(s) + W_d(s)R(s)\left[\varphi_{RGM}(T,s,0) - \varphi_{RGM}(T,s,t)\right]}{1 + W_d(s)\varphi_{RGM}(t,s,0)}\,.$$

With account for (11.78) and (11.66), we can represent the last relation as

$$W_{13}(s,t) = \frac{N_{13}(s,t)}{\Delta(s)}$$

where

$$N_{13}(s,t) = R(s)a(s)\alpha(s) + R(s)b(s)\alpha(s)\left[\varphi_{RGM}(T,s,0) - \varphi_{RGM}(T,s,t)\right]\,.$$

As follows from (11.67), the product $R(s)\alpha(s)$ is an entire function, i.e., all poles of $R(s)$ are cancelled by zeros of $\alpha(s)$. It remains to prove that the product

$$R(s)\alpha(s)\left[\varphi_{RGM}(T,s,0) - \varphi_{RGM}(T,s,t)\right] \tag{11.81}$$

is an entire function. We notice that

$$\begin{aligned} &\varphi_{RGM}(T,s,0) - \varphi_{RGM}(T,s,t) = \\ &\quad = \frac{1}{T}\sum_{\substack{k=-\infty \\ k\neq 0}}^{\infty} R(s+kj\omega)G(s+kj\omega)M(s+kj\omega)\left[1 - \mathrm{e}^{kj\omega t}\right]\,. \end{aligned} \tag{11.82}$$

Since the open-loop system is assumed to be non-pathological, the poles of the right side of (11.82) are

$$\tilde{s}_{ik} \stackrel{\triangle}{=} s_i + kj\omega\,, \quad k \neq 0\,.$$

with multiplicities ν_i. Therefore, the product

$$R(s)\left[\varphi_{RGM}(T,s,0) - \varphi_{RGM}(T,s,t)\right]$$

may have poles at the points $\tilde{s}_{ik}$ with multiplicities ν_i and at s_i with multiplicities not higher than ν_i. Due to (11.67), all these poles are cancelled by zeros of the function $\alpha(s)$, and the function (11.81) is entire. The remaining elements of the PTM $\boldsymbol{W}(s,t)$ can be proved in a similar way. ■

Remark A generalization of Theorem 11.1 onto the MIMO-case is given in Rosenwasser (1995a).

Definition 4 The number $\tilde{s}$ is called a *pole of the PTF* $W_{mk}(s,t)$, if, at least for a single $t = \tilde{t}$, the function $W_{mk}(s,\tilde{t})$ has a pole at $s = \tilde{s}$. With the *multiplicity of the pole* $\tilde{s}$ we mean the largest multiplicity of the pole $\tilde{s}$ of the function $W_{mk}(s,\tilde{t})$ for all possible $\tilde{t}$.

Definition 5 A number $\tilde{s}$ is called a *pole of the PTM* (11.77) with multiplicity $\tilde{\nu}$, if $\tilde{s}$ is a pole of multiplicity $\tilde{\nu}$ at least for one of the PTFs $W_{mk}(s,t)$, and none of these functions has a pole at $s = \tilde{s}$ of multiplicity higher than $\tilde{\nu}$.

Theorem 11.2 *Let the open-loop system shown in Fig. 11.8 be non-pathological, i.e., the real rational function (10.66) is irreducible. Then, each root of the determining equation (11.73) of multiplicity $\tilde{\nu}$ is a pole of the PTM $\boldsymbol{W}(s,t)$ of multiplicity $\tilde{\nu}$, and vice versa.*

Proof Denote by $\tilde{S}$ the set of roots $\tilde{s}_i$ of the determining equation (11.73), and by $\tilde{\nu}_i$ the multiplicities of these roots. Let $\tilde{D}$ be the set of the poles d_m of the PTM with multiplicities λ_m. As follows from (11.78) the PTF $\boldsymbol{W}(s,t)$ may have poles only at the points $\tilde{s}_i$, i.e., $\tilde{D} \subset \tilde{S}$, where $\lambda_i \leq \tilde{\nu}_i$. On the other hand, consider the PTF $W_{11}(s,t)$. For $t = 0$ from (11.18) we find

$$W_{11}(s,0) = \frac{\dfrac{e^{-sT}\chi(s)b(s)}{\alpha(s)a(s)}}{1 + \dfrac{e^{-sT}\chi(s)b(s)}{\alpha(s)a(s)}}$$

i.e.,

$$W_{11}(s,0) = \frac{e^{-sT}\chi(s)b(s)}{\Delta(s)}\,. \tag{11.83}$$

Assuming that $e^{-sT} = \zeta$ and using (11.72), we obtain

$$W_{11}^o(\zeta,0) = W_{11}(s,0)\,\big|_{e^{-sT}=\zeta} = \frac{\zeta\chi^o(\zeta)b^o(\zeta)}{\Delta^o(\zeta)}\,. \tag{11.84}$$

It can be easily verified that the irreducibility of the real rational function (11.68) ensures the irreducibility of the real rational function (11.84). Then, due to the relations given in Appendix A, the rational periodic function (11.83) is also irreducible, and any root d_m of its denominator with multiplicity λ_m is a pole of $W_{11}(s,0)$ of the same multiplicity. Hence, $\tilde{S} \subset \tilde{D}$, where $\lambda_i \geq \tilde{\nu}_i$. Since the inverse inclusion was proved above, the assertion is proved. ■

Corollary Let $\tilde{\zeta}_1, \ldots, \tilde{\zeta}_r$ be the roots of the characteristic equation (11.76) of multiplicities $\tilde{\kappa}_1, \ldots, \tilde{\kappa}_r$, respectively, and let the numbers $\hat{s}_1, \ldots, \hat{s}_r$ be defined by the relations

$$e^{-\hat{s}_i T} = \tilde{\zeta}_i\,, \quad -\omega/2 \leq \operatorname{Im}\,\hat{s}_i < \omega/2\,. \tag{11.85}$$

Then, the set of the poles of the PTM $\boldsymbol{W}(s,t)$ consists of the numbers

$$\tilde{s}_{ik} = \hat{s}_i + kj\omega\,, \quad (k = 0, \pm 1, \ldots) \tag{11.86}$$

of multiplicities $\tilde{\kappa}_i$.

Corollary It can be shown, that if the continuous networks include pure delays, i.e., in the case of (11.53), the propositions of Theorem 11.1 and 11.2 remain valid if they hold for $\tau_1 = 0$, $\tau_2 = 0$. In this case, the associated determining function $\Delta_\tau(s)$ is given by

$$\Delta_\tau(s) = 1 + W_d(s)\mathcal{D}_{R_1G_1M}(T,s,\delta T)e^{-msT} \tag{11.87}$$

where m and δ are defined from (11.59). With the substitution $\zeta = e^{-sT}$, we obtain the characteristic function

$$\Delta_\tau^o(\zeta) = \Delta_\tau(s)\,\big|_{e^{-sT}=\zeta} = 1 + W_d^o(\zeta)\mathcal{D}_{R_1G_1M}^o(T,\zeta,\delta T)\zeta^m\,. \tag{11.88}$$

Using the last relations, we can easily find the determining quasipolynomial and the characteristic polynomial for a time-delayed system.

11.6 Deadbeat systems

In principle, by an appropriate choice of the control program (11.3), we can reach

$$\Delta^o(\zeta) = \Delta_0 = \text{const.} \tag{11.89}$$

i.e., the characteristic polynomial reduces to a constant. As follows from Corollary 1 to Theorem 11.2, under the condition (11.89) the PTM $\boldsymbol{W}(s,t)$ is an entire function in s. The case (11.89) is very important for applications. Henceforth, a closed-loop system will be called a *deadbeat system*, if Eq. (11.89) holds.

1. Consider Eq. (11.86) in detail. It is equivalent, taking account of (11.74), to

$$\alpha^o(\zeta)a^o(\zeta) + \zeta\chi^o(\zeta)b^o(\zeta) = \Delta_0 . \tag{11.90}$$

Assuming that $\zeta = 0$ and taking (11.69) and (10.61) into account, we have

$$\Delta_0 = a^o(0) = a_0 .$$

Since the rational periodic function $W_d(s)$ is assumed to be causal, we have $a_0 \neq 0$, and it is possible to take $a_0 = 1$. As a result, Eq. (11.90) takes the form

$$\alpha^o(\zeta)a^o(\zeta) + \zeta\chi^o(\zeta)b^o(\zeta) = 1 . \tag{11.91}$$

The characteristic function (11.72) appears as

$$\lambda^o(\zeta) = \frac{1}{\alpha^o(\zeta)a^o(\zeta)} . \tag{11.92}$$

The determining function (11.71) gets the form

$$\lambda(s) = 1 + W_d(s)\varphi_{FM}(T, s, 0) = \frac{1}{\alpha(s)a(s)} . \tag{11.93}$$

Substituting (11.93) in the formulae of Tables 11.1 and 11.2, we obtain simplified formulae for the PTFs and DTFs. For instance, from (11.93) and (11.80) we have

$$D_{11}(s,t) = \beta(s,t)b(s) . \tag{11.94}$$

For $0 \leq t \leq T$, the right-hand side of (11.94) is a quasipolynomial in $\zeta = \mathrm{e}^{-sT}$.

2. For known transfer functions $G(s)$ and $R(s)$ and the form of the pulse $\mu(t)$, i.e., for known polynomials $\alpha^o(\zeta)$ and $\chi^o(\zeta)$, we can consider Eq. (11.91) as a *Diophantine equation* with respect to the unknown polynomials $a^o(\zeta)$ and $b^o(\zeta)$, which determine the control program As is known from the general theory of such equations, Volgin (1986); Kučera (1979), Eq.(11.91) is solvable iff the polynomials $\alpha^o(\zeta)$ and $\zeta\chi^o(\zeta)$ are co-prime, i.e., the ratio

$$W_p^o(\zeta, 0) = \mathcal{D}_{FM}^o(T, \zeta, 0) = \frac{\zeta\chi^o(\zeta)}{\alpha^o(\zeta)}$$

is irreducible.

3. Using the aforesaid, we propose that Eq. (11.91) is solvable if the SD part is non-pathological. The general solution of Eq. (11.91) takes the form

$$a^o(\zeta) = a_m^o(\zeta) + a_0(\zeta) , \quad b^o(\zeta) = b_m^o(\zeta) + b_0(\zeta) \tag{11.95}$$

where $a_m^o(\zeta)$, $b_m^o(\zeta)$ is a pair forming a particular solution of Eq. (11.91), and the polynomials $a_0(\zeta$ and $b_0(\zeta)$ constitute an arbitrary solution of the homogeneous equation

$$\alpha^o(\zeta)a_0(\zeta) + \zeta\chi^o(\zeta)b_0(\zeta) = 0\,. \tag{11.96}$$

Moreover, as follows from (11.91), $a^o(0) \neq 0$, i.e., any program providing for (11.91) is causal.
In this case a particular solution can be constructed directly. Indeed, consider the partial fraction expansion

$$\frac{1}{\zeta\alpha^o(\zeta)\chi^o(\zeta)} = \frac{\tilde{a}(\zeta)}{\zeta\chi^o(\zeta)} + \frac{\tilde{b}(\zeta)}{\alpha^o(\zeta)} \tag{11.97}$$

where $\tilde{a}(\zeta)$ and $\tilde{b}(\zeta)$ are known polynomials, such that

$$\deg\tilde{a}(\zeta) < \deg\zeta\chi^o(\zeta)\,, \quad \deg\tilde{b}(\zeta) < \deg\alpha^o(\zeta)\,. \tag{11.98}$$

From (11.97) we obtain

$$1 = \alpha^o(\zeta)\tilde{a}(\zeta) + \zeta\chi^o(\zeta)\tilde{b}(\zeta)$$

i.e., we can take

$$a_m^o(\zeta) = \tilde{a}(\zeta)\,, \quad b_m^o(\zeta) = \tilde{b}(\zeta)\,. \tag{11.99}$$

To define the general solution of the homogenuous equation (11.96) we rewrite it as

$$\frac{b_0(\zeta)}{a_0(\zeta)} = -\frac{\alpha^o(\zeta)}{\zeta\chi^o(\zeta)}\,. \tag{11.100}$$

Since the ratio at the right-hand side is irreducible, Eq. (11.99) can be valid only with

$$a_0(\zeta) = -c(\zeta)\zeta\chi^o(\zeta)\,, \quad b_0(\zeta) = c(\zeta)\alpha^o(\zeta) \tag{11.101}$$

where $c(\zeta)$ is an arbitrary polynomial. Thus, the set of polynomials $a^o(\zeta), b^o(\zeta)$ satisfying Eq.(11.91) takes the form

$$\begin{aligned} a^o(\zeta) &= \tilde{a}(\zeta) - c(\zeta)\zeta\chi^o(\zeta) \\ b^o(\zeta) &= \tilde{b}(\zeta) + c(\zeta)\alpha^o(\zeta)\,. \end{aligned} \tag{11.102}$$

As follows from (11.102) for $c(\zeta) \neq 0$,

$$\deg a^o(\zeta) \geq \deg\zeta\chi^o(\zeta)\,, \quad \deg b^o(\zeta) \geq \deg\alpha^o(\zeta) \tag{11.103}$$

is valid.

4. Comparing (11.98) and (11.103), we find that the polynomials $a^o(\zeta)$ and $b^o(\zeta)$ have minimal degrees in the solution (11.99) obtained from the expansion (11.97). The solution (11.99) will henceforth be called the *mimimal solution.* Taking some additional information into account, we can obtain more detailed information on the degrees of the polynomials $a^o(\zeta)$ and $b^o(\zeta)$. First of all, from (11.91) it follows that

$$\deg[\alpha^o(\zeta)a^o(\zeta)] = \deg[\zeta\chi^o(\zeta)b^o(\zeta)]$$

i.e.,

$$\deg\alpha^o(\zeta) + \deg a^o(\zeta) = \deg\zeta\chi^o(\zeta) + \deg b^o(\zeta)$$

or, equivalently, taking account of (11.69),

$$n + \deg a^o(\zeta) = \deg\zeta\chi^o(\zeta) + \deg b^o(\zeta)\,. \qquad (11.104)$$

There are two cases.

1.
$$\deg\zeta\chi^o(\zeta) = \deg\alpha^o(\zeta) = n\,. \qquad (11.105)$$
Then, from (11.104) we obtain
$$\deg a^o(\zeta) = \deg b^o(\zeta) \qquad (11.106)$$
i.e., the desired control program
$$W_d(s) = \frac{b(s)}{a(s)}$$
is limited.

2. If
$$\deg\zeta\chi^o(\zeta) < \deg\alpha^o(\zeta) = n \qquad (11.107)$$
we have
$$\deg a^o(\zeta) < \deg b^o(\zeta)$$
and the desired control program $W_d(s)$ is retarded. As was shown in Rosenwasser (1994c), the case (11.107) has no practical meaning. Therefore, henceforth we consider only the case of Eq. (11.106).

Example 11.5 Let us construct the set of all deadbeat controllers for the system of Fig. 11.5 with

$$F(s) = \frac{\gamma}{s-a}$$

and arbitrary form of the pulse $\mu(t)$. Then, taking account of (11.34), we have

$$\alpha^o(\zeta) = 1 - e^{aT}\zeta\,, \quad \chi^o(\zeta) = \gamma M(a)e^{aT}\,.$$

Moreover, $M(a) \neq 0$, because the SD part is assumed to be non-pathological. It can be immediately verified that

$$\frac{1}{\zeta\gamma M(a)e^{aT}\left(1-e^{aT}\zeta\right)} = \frac{1}{\zeta\gamma M(a)e^{aT}} + \frac{1}{\gamma M(a)}\,\frac{1}{1-e^{aT}\zeta}\,.$$

The minimal solution has the form

$$\tilde{a}(\zeta) = 1\,, \quad \tilde{b}(\zeta) = \frac{1}{\gamma M(a)}$$

and the general solution (11.102) appears as

$$a^o(\zeta) = 1 - c(\zeta)\zeta\gamma M(a)e^{aT}$$
$$b^o(\zeta) = \frac{1}{\gamma M(a)} + c(\zeta)\left(1 - e^{aT}\zeta\right)\,.$$

□

5. We note in conclusion that all results can be extended to time-delayed systems. In that case we have (11.53), and (11.91) must be replaced by

$$\alpha^o(\zeta)a(\zeta) + \zeta^m \beta^o(\zeta, \delta T)b^o(\zeta) = 1 . \tag{11.108}$$

11.7 Operator description of systems and equations for images

First we consider the PTF $W_{11}(s,t) = W_{yd}(st)$ described by Eqs. (11.23)–(11.25). Taking into account that

$$W_d(s) = \frac{b(s)}{a(s)}, \quad \varphi_{FM}(T, s, 0) = \frac{\mathrm{e}^{-sT}\chi(s)}{\alpha(s)}$$

these relations can be rewritten in the form

$$W_{11}(s,t) = \frac{1}{T} \sum_{k=-\infty}^{\infty} Q_{yd}(s + kj\omega)\mathrm{e}^{kj\omega t} \tag{11.109}$$

where

$$Q_{yd}(s) = \frac{R(s)G(s)M(s)b(s)\alpha(s)}{\alpha(s)a(s) + \mathrm{e}^{-sT}\chi(s)b(s)} = \frac{R(s)G(s)M(s)b(s)\alpha(s)}{\Delta(s)} . \tag{11.110}$$

The numerator and the denominator of the ratio (11.110) are entire functions in s, because the poles of the product $F(s) = R(s)G(s)$ are cancelled by zeros of the function $\alpha(s)$. Therefore, the set of poles of the function $Q_{yd}(s)$ belongs to the set of numbers (11.86). Denote by $\tilde{\mu}_1, \ldots, \tilde{\mu}_r$ the different real parts of the values (11.86) numbered in ascending order, and by $\tilde{\nu}_1, \ldots, \tilde{\nu}_m$ we denote different real parts of the poles of the function $Q_{yd}(s)$ numbered in the same way.

Lemma 11.1 *The sets of the numbers $\tilde{\mu}_i$ and $\tilde{\nu}_i$ coincide, i.e., $r = m$ and $\tilde{\mu}_i = \tilde{\nu}_i$.*

Proof To prove the proposition it suffices to show that for a characteristic index $\tilde{\mu}_i$ of the function (11.84) there exists a characteristic index $\tilde{\nu}_i = \tilde{\mu}_i$ of the function $Q_{yd}(s)$. Assume the inverse, i.e., that a number $\tilde{\mu}_\gamma$ is a characteristic index for $W_{11}(s,0)$ and is not one of $Q_{yd}(s)$. Then, the function $Q_{yd}(s)$ is analytic in an interval $\tilde{\alpha} < \tilde{\mu}_\gamma < \tilde{\beta}$. Then, by Theorem 3.3 and Eq. (11.109), the function $W_{11}(s,0)$ is analytic for $\tilde{\alpha} < \operatorname{Re} s < \tilde{\beta}$, which contradicts the initial assumption that $\tilde{\mu}_\gamma$ is a characteristic index. Therefore, the number $\tilde{\mu}_\gamma$ is also a characteristic index of $Q_{yd}(s)$. ∎

Lemma 11.2 *The rational periodic function (11.25)*

$$\tilde{W}_d(s) = \frac{W_d(s)}{1 + W_d(s)\varphi_{FM}(T, s, 0)}$$

is causal.

Proof Taking account of (10.19) and (9.143), we have

$$\ell^{+}\left[\tilde{W}_{d}(s)\right] = \lim_{\mathrm{Re}\, s\to\infty} \frac{W_d(s)}{1+W_d(s)\varphi_{FM}(T,s,0)} = \frac{b_0}{a_0}$$

whence follows the proposition. ■

Thus, the PTF (11.109) is associated with an open-loop system with a computer and the equivalent transfer function

$$Q_{yd}(s) = R(s)G(s)M(s)\tilde{W}_d(s)\,. \tag{11.111}$$

Moreover, the rational periodic function $\tilde{W}_d(s)$ is causal. The numbers $\tilde{\mu}_i$ will be called the *characteristic indices* of the closed-loop system. Then, taking into account that the function $F(s) = R(s)G(s)$ is strictly proper, and using the results of Section 9.9, we can propose that the PTF (11.19) defines a family of LPPO U_{d_i} of the form (6.97), which corresponds to the regularity intervals $(\tilde{\mu}_i, \tilde{\mu}_{i+1})$ with pulse responses

$$\tilde{q}_i(t) = \frac{1}{2\pi \mathrm{j}} \int_{c-\mathrm{j}\infty}^{c+\mathrm{j}\infty} R(s)G(s)M(s)\tilde{W}_d(s)\mathrm{e}^{st}\,\mathrm{d}s\,, \quad \tilde{\mu}_i < c < \tilde{\mu}_{i+1}\,. \tag{11.112}$$

Discretizing the integral (11.112), we obtain the equivalent expression

$$\tilde{q}_i(t) = \frac{T}{2\pi \mathrm{j}} \int_{c-\mathrm{j}\omega/2}^{c+\mathrm{j}\omega/2} \mathcal{D}_{RGM}(T,s,t)\tilde{W}_d(s)\,\mathrm{d}s = \frac{T}{2\pi \mathrm{j}} \int_{c-\mathrm{j}\omega/2}^{c+\mathrm{j}\omega/2} D_{11}(s,t)\,\mathrm{d}s$$

$$\tilde{\mu}_i < c < \tilde{\mu}_{i+1}\,. \tag{11.113}$$

In this case, for $x_d(t) \in \Lambda(\tilde{\mu}_i, \tilde{\mu}_{i+1})$ the Laplace transform of the output $y(t) = \mathsf{U}_{d_i}[x_d(t)]$ is given by

$$Y(s) = R(s)G(s)M(s)\tilde{W}_d(s)\varphi_{x_d}(T,s,+0)\,, \quad \tilde{\mu}_i < \mathrm{Re}\, s < \tilde{\mu}_{i+1}\,. \tag{11.114}$$

Similarly, for the corresponding discrete Laplace transform of the output we have

$$\mathcal{D}_y(T,s,t) = \mathcal{D}_Y(T,s,t) = \mathcal{D}_{RGM}(T,s,t)\tilde{W}_d(s)\mathcal{D}_{x_d}(T,s,+0)\,. \tag{11.115}$$

In a similar way we can perform an operator description and construct equations in images for the input $x_d(t)$ and outputs $z(t)$ and $u(t)$. As regards the inputs $x_v(t)$ and $x(t)$, the above relations are inapplicable, because the system is not discrete with respect to these inputs. Nevertheless, in each special case all problems can be easily solved using the technique of Chapter 7. Consider, for instance, the PTF

$$W_{13}(s,t) = W_{yx}(s,t) = R(s)\left[1 - \frac{W_d(s)\varphi_{FM}(T,s,t)}{1+W_d(s)\varphi_{FM}(T,s,0)}\right] \tag{11.116}$$

assuming that the function $R(s)$ is strictly proper. Equation (11.116) can be written in the form

$$W_{13}(s,t) = R(s) - R(s)W_{11}(s,t) \tag{11.117}$$

where $W_{11}(s,t)$ is given by (11.19). The first term in (11.117) is associated with a LTI system with transfer function $R(s)$. The second summand is associated with an open-loop system with a prefilter considered in Section 6.7. Then, taking into account that the poles of the PTF $W_{13}(s,t)$ (11.42) are the numbers (11.86), we can state:

1. The integrals

$$\tilde{g}_i(t,u) = \frac{1}{2\pi \mathrm{j}} \int_{c-\mathrm{j}\infty}^{c+\mathrm{j}\infty} W_{13}(s,t)\mathrm{e}^{su}\,\mathrm{d}s\,, \quad \tilde{\mu}_i < c < \tilde{\mu}_{i+1} \tag{11.118}$$

converge at least in the sense of the principal value. In this case, for any $t = \tilde{t}$ we have $\tilde{g}_i(\tilde{t},u) \in \Lambda(\tilde{\mu}_i, \tilde{\mu}_{i+1})$ uniformly with respect to t.

2. As regards the input $x(t)$ and output $y(t)$, the closed-loop system defines a family of linear periodic operators

$$y(t) \triangleq \mathsf{U}_{0i}[x(t)] = \int_{-\infty}^{\infty} \tilde{h}_i(t,\tau)x(\tau)\,\mathrm{d}\tau \tag{11.119}$$

where

$$\tilde{h}_i(t,\tau) = \tilde{g}_i(t,u)\,|_{u=t-\tau}\ . \tag{11.120}$$

and the operator U_{0r} is causal.

3. For $x(t) \in \Lambda(\tilde{\mu}_i, \tilde{\mu}_{i+1})$ the solutions of (11.119) can be represented in the form

$$y(t) = \frac{1}{2\pi \mathrm{j}} \int_{c-\mathrm{j}\infty}^{c+\mathrm{j}\infty} W_{13}(s,t)X(s)\mathrm{e}^{st}\,\mathrm{d}s\,, \quad \tilde{\mu}_i < c < \tilde{\mu}_{i+1} \tag{11.121}$$

in this case the conditions of applicability of Eq. (7.37) hold, and the DTF of the solution (11.119) takes the form

$$\mathcal{D}_y(T,s,t) = \frac{1}{T}\sum_{k=-\infty}^{\infty} W_{13}(s+k\mathrm{j}\omega,t)X(s+k\mathrm{j}\omega,t)\mathrm{e}^{(s+k\mathrm{j}\omega)t}\,.$$

Substituting (11.116) in the last equation and taking into account that for any integer k

$$\varphi_{FM}(T,s+k\mathrm{j}\omega,t) = \varphi_{FM}(T,s,t)\mathrm{e}^{-k\mathrm{j}\omega t}$$

we obtain

$$\mathcal{D}_y(T,s,t) = \mathcal{D}_{RX}(T,s,t) - \frac{\mathcal{D}_{RGM}(T,s,t)W_d(s)}{1+W_d(s)\varphi_{RGM}(T,s,0)}\mathcal{D}_{RX}(T,s,0)\,. \tag{11.122}$$

Using (4.35), from (11.122) we find the Laplace transform

$$Y(s) = R(s)X(s) - \frac{R(s)G(s)M(s)W_d(s)}{1+W_d(s)\varphi_{RGM}(T,s,0)}\mathcal{D}_{RX}(T,s,0)\,. \tag{11.123}$$

11.8 Equations of images for transients

If we take $x_d = x_v = x = 0$, the system shown in Fig. 11.1 reduces to the block-diagram shown in Fig. 11.9. Using (10.86) for $\operatorname{Re} s > \alpha$, where α is a

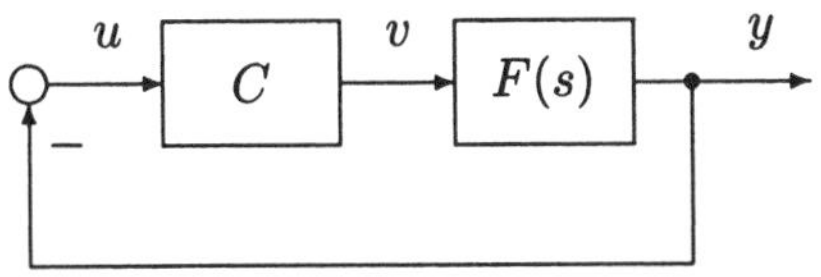

Figure 11.9: Autonomous SD control system

sufficiently large number, and taking into account the feedback equation

$$u(t) = -y(t) \tag{11.124}$$

we obtain

$$\mathcal{D}_y(T,s,t) = \mathcal{D}_{FM}(T,s,t)\left[-W_d(s)\mathcal{D}_y(T,s,+0) + \frac{b_0(s)}{a(s)}\right] + \mathcal{D}_{y_0}(T,s,t)\,. \tag{11.125}$$

Taking the limit for $t \to +0$, similarly to (10.87) we find

$$\mathcal{D}_y(T,s,+0) = \mathcal{D}_{FM}(T,s,0)\left[-W_d(s)\mathcal{D}_y(T,s,+0) + \frac{b_0(s)}{a(s)}\right] + \mathcal{D}_{y_0}(T,s,+0)\,.$$

Then,

$$\mathcal{D}_y(T,s,+0) = \frac{\mathcal{D}_{FM}(T,s,0)\,\frac{b_0(s)}{a(s)}}{1 + W_d(s)\mathcal{D}_{FM}(T,s,0)} + \frac{\mathcal{D}_{y_0}(T,s,+0)}{1 + W_d(s)\mathcal{D}_{FM}(T,s,0)}\,. \tag{11.126}$$

Substituting this equation into (11.125), we obtain

$$\mathcal{D}_y(T,s,t) = \mathcal{D}_{dy}(T,s,t) + \mathcal{D}_{0y}(T,s,t) \tag{11.127}$$

with

$$\begin{aligned}
\mathcal{D}_{dy}(T,s,t) &= \frac{\mathcal{D}_{FM}(T,s,t)}{1 + W_d(s)\mathcal{D}_{FM}(T,s,0)}\,\frac{b_0(s)}{a(s)} \\
\mathcal{D}_{0y}(T,s,t) &= -\frac{W_d(s)\mathcal{D}_{FM}(T,s,t)}{1 + W_d(s)\mathcal{D}_{FM}(T,s,0)}\mathcal{D}_{y_0}(T,s,+0) + \mathcal{D}_{y_0}(T,s,t)\,.
\end{aligned} \tag{11.128}$$

The image $\mathcal{D}_{dy}(T,s,t)$ defines the term of the transient dependent on the initial conditions of the control program. The image $\mathcal{D}_{0y}(T,s,t)$ is associated with the term caused by the initial conditions of the continuous part. Using the above technique, it can be shown that

$$\mathcal{D}_{dy}(T,s,t) = \frac{N_{yd}(s,t)}{\Delta(s)}\,, \quad \mathcal{D}_{0y}(T,s,t) = \frac{N_{0d}(s,t)}{\Delta(s)}$$

where the numerators of both ratios are, for any s and $0 \leq t \leq T$, polynomials in e^{-sT}. The associated Laplace transforms appear as

$$
\begin{aligned}
Y_d(s) &= \frac{F(s)M(s)}{1 + W_d(s)\mathcal{D}_{FM}(T, s, 0)} \frac{b_0(s)}{a(s)} \\
\tilde{Y}_0(s) &= -\frac{F(s)M(s)W_d(s)}{1 + W_d(s)\mathcal{D}_{FM}(T, s, 0)} \mathcal{D}_{y_0}(T, s, +0) + Y_0(s) .
\end{aligned}
\qquad (11.129)
$$

The above relations define processes in the system without exogenous disturbances. Such processes will be called the natural motions. Superimposing these images and those obtained in Section 11.7, we receive the general solution with exogenous disturbances and non-zero initial conditions.

11.9 Standard MIMO systems

1. In this section we construct the PTM for a generalized SD system with the block-diagram shown in Fig. 11.10, where $\boldsymbol{P}$ is a generalized continuous net-

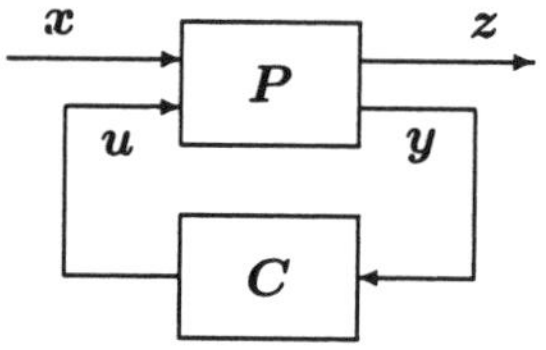

Figure 11.10: Generalized SD system

work with the real rational transfer function (matrix) $\boldsymbol{P}(s)$, $\boldsymbol{C}$ is a computer, and $\boldsymbol{u}$, $\boldsymbol{x}$, $\boldsymbol{y}$ and $\boldsymbol{z}$ are vectors of dimensions $k \times 1$, $m \times 1$, $\ell \times 1$, and $n \times 1$, respectively, Chen and Francis (1995). In an expanded form the connection between the inputs and outputs of the continuous element shown in Fig. 11.10 is given by

$$
\begin{aligned}
\boldsymbol{z} &= \boldsymbol{K}(s)\boldsymbol{x} + \boldsymbol{L}(s)\boldsymbol{u} \\
\boldsymbol{y} &= \boldsymbol{R}(s)\boldsymbol{x} + \boldsymbol{N}(s)\boldsymbol{u}
\end{aligned}
\qquad (11.130)
$$

where $\boldsymbol{K}(s)$, $\boldsymbol{L}(s)$, $\boldsymbol{R}(s)$, and $\boldsymbol{N}(s)$ are real rational matrices of dimensions $n \times m$, $n \times k$, $\ell \times m$, and $\ell \times k$, respectively. As follows from (11.130),

$$
\boldsymbol{P}(s) = \begin{array}{c|cc|} & m & k \\ n & \boldsymbol{K} & \boldsymbol{L} \\ \ell & \boldsymbol{R} & \boldsymbol{N} \end{array} \qquad (11.131)
$$

where the letters outside the matrix denote the sizes of the corresponding blocks. It is assumed that the computer is described by the relations given in Section 10.9.

2. Then, we find an explicit expression for the transfer matrix of the system from the input $\boldsymbol{x}(t)$ to the output $\boldsymbol{z}(t)$.

Theorem 11.3 *Let the matrix $\boldsymbol{L}(s)$ in (11.131) be proper and the matrix $\boldsymbol{N}(s)$ be strictly proper. Then, there exists uniquely the PTM of the standard system*

$$\boldsymbol{W}_c(s,t) = \boldsymbol{K}(s) + \varphi_{\boldsymbol{L}M}(T,s,t)\tilde{\boldsymbol{W}}_{\boldsymbol{N}}(s)\boldsymbol{R}(s) \tag{11.132}$$

where

$$\tilde{\boldsymbol{W}}_{\boldsymbol{N}}(s) = \boldsymbol{W}_d(s)\left[\boldsymbol{I}_\ell - \varphi_{\boldsymbol{N}M}(T,s,0)\boldsymbol{W}_d(s)\right]^{-1}. \tag{11.133}$$

In Eqs. (11.132) and (11.133) we used the notation

$$\varphi_{\boldsymbol{L}M}(T,s,t) = \frac{1}{T}\sum_{k=-\infty}^{\infty} \boldsymbol{L}(s+kj\omega)M(s+kj\omega)e^{kj\omega t} \tag{11.134}$$

$$\varphi_{\boldsymbol{N}M}(T,s,t) = \frac{1}{T}\sum_{k=-\infty}^{\infty} \boldsymbol{N}(s+kj\omega)M(s+kj\omega)e^{kj\omega t}. \tag{11.135}$$

Moreover, $M(s)$ denotes the transfer function of the forming element (9.10) and $\boldsymbol{W}_d(s)$ is the transfer matrix of the control program (10.111).

The proof follows directly from the more general result presented in Section 12.10. ∎

Corollary 1 The restrictions imposed on the matrices $\boldsymbol{L}(s)$ and $\boldsymbol{N}(s)$ are meaningful. The properness of the real rational matrix $\boldsymbol{L}(s)$ is necessary for the convergence of the series (11.134), while the strict properness of $\boldsymbol{N}(s)$ ensures the continuity of the matrix $\varphi_{\boldsymbol{N}M}(T,s,t)$ with respect to t, which is necessary for Eq. (11.133) to be meaningful.

Corollary 2 Equations (11.132) and (11.133) remain valid for the case of time-delayed systems.

3. It should be noted that the standard structure shown in Fig. 11.10 is fairly general, and an arbitrary single-rate system with a single DAC and ADC can be reduced to it. Systems encountered in applications do not, as a rule, have standard structure, but can be reduced to the latter by structural transformations. Nevertheless, such transformations are not necessary, because the PTM can always be obtained directly by an initial model of the system at hand. Moreover, having constructed the PTM for the initial system, we can easily define the matrix $\boldsymbol{P}(s)$ for the corresponding standard structure.

Example 11.6 Let us examine the possibility of constructing the standard form of a system by means of its PTM. Consider the single-loop multivariable system shown in Fig. 11.11, which has not the standard structure. As was

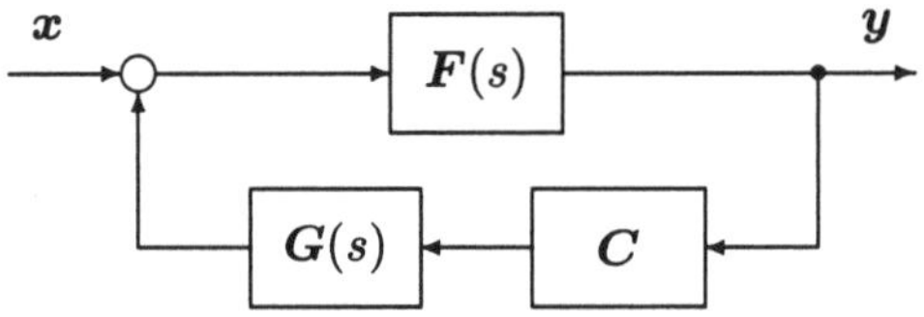

Figure 11.11: Control system of non-standard structure

shown in Rosenwasser (1995a), the PTM of this system from the input $\boldsymbol{x}$ to the output $\boldsymbol{y}$ takes the form

$$\boldsymbol{W}(s,t) = \boldsymbol{F}(s) + \varphi_{\boldsymbol{FGM}}(T,s,t)\boldsymbol{W}_d(s)\left[\boldsymbol{I} - \varphi_{\boldsymbol{FGM}}(T,s,0)\boldsymbol{W}_d(s)\right]^{-1}\boldsymbol{F}(s)\,.$$

Comparing this equation with (11.132) and (11.133), we find that the system shown in Fig. 11.11 can be associated with the standard structure, where

$$\boldsymbol{P}(s) = \begin{bmatrix} \boldsymbol{F}(s) & \boldsymbol{F}(s)\boldsymbol{G}(s) \\ \boldsymbol{F}(s) & \boldsymbol{F}(s)\boldsymbol{G}(s) \end{bmatrix}$$

□

4. For $\boldsymbol{R}(s) = 0$ the standard system, regarding the input $\boldsymbol{x}(t)$ and the output $\boldsymbol{y}(t)$, transforms into a LTI system with the transfer function $\boldsymbol{K}(s)$. Equation (11.132) gives an adequate result as well

$$\boldsymbol{W}_c(s,t) = \boldsymbol{K}(s)\,.$$

It can also be shown that for the case when $\boldsymbol{K}(s) = 0$, and $\boldsymbol{R}(s) = \boldsymbol{I}_m$ the PTM $\boldsymbol{W}_c(s,t)$ coincides with the transfer matrix of the corresponding discrete-time system in the sense of the modified Laplace transform up to an exponential multiplier. Hence, the PTM $\boldsymbol{W}_c(s,t)$ gives an universal frequency domain description of multivariable SD systems, from which standard frequency domain descriptions for LTI continuous-time and discrete-time systems are obtained as special cases.

5. For $s = \mathrm{j}\nu$, where $\mathrm{j} = \sqrt{-1}$ and ν is a real variable, from (11.132) and (11.133) we obtain the expression for the parametric frequency response of the standard system

$$\boldsymbol{\Phi}_c(\mathrm{j}\nu,t) \triangleq \boldsymbol{W}_c(s,t)\,|_{s=\mathrm{j}\nu} =$$

$$= \boldsymbol{K}(\mathrm{j}\nu) + \varphi_{\boldsymbol{LM}}(T,\mathrm{j}\nu,t)\boldsymbol{W}_d(\mathrm{j}\nu)\left[\boldsymbol{I}_\ell - \varphi_{\boldsymbol{NM}}(T,\mathrm{j}\nu,0)\boldsymbol{W}_d(\mathrm{j}\nu)\right]^{-1}\boldsymbol{R}(s)$$

which can be considered as a result of an analytic continuation of the PTM $\boldsymbol{W}_c(s,t)$ on the whole imaginary axis. In such an approach, the notion of frequency response is applicable for unstable systems as well.

Chapter 12

Systems with several sampling units

12.1 General reasoning

In the applications considered above for a SD system with a single sampling unit there exists a PTF $W(s,t) = W(s,t+T)$ defined on the whole complex plane, except for the set of its poles located on a finite number of vertical lines $\operatorname{Re} s = \mu_i$, $(i = 1, 2, \ldots .r)$, where μ_i are the corresponding characteristic indices. In the associated regularity intervals $\mu_i < \operatorname{Re} s < \mu_{i+1}$ and in the half-planes $\operatorname{Re} s < \mu_1$ and $\operatorname{Re} s > \mu_r$ the PTF is analytic. By a known PTF $W(s,t)$ we can determine linear periodic operators U_i of the system under consideration. Moreover, the PTF of each of the operators U_i coincides with the PTF $W(s,t)$ of the system in the corresponding stripes of convergence. To find the transfer functions of parallel and series connections of systems possessing these properties we can employ formal rules given in Section 6.3. In this case the set of characteristic indices of the complex system is the union of the corresponding sets of its elements.

If connected systems have different, but multiple periods, the system will be periodic with a period equal to the least common multiple of the periods of the connected elements. In practice, the transfer function of a complex system with an input $x(t)$ and output $y(t)$ can be found on the basis of Eq. (6.19).

12.2 Open-loop synchronous systems

A system with several sampling units will be called *synchronous*, if all sampling units have the same sampling period. For concreteness, in this section we consider the series connection shown in Fig. 12.1, where

$$W_{pd}^{(i)}(s,t) = \varphi_{F_i M_i W_{di}}(T,s,t) = W_{di}(s)\varphi_{F_i M_i}(T,s,t)\,, \quad (i = 1, 2) \qquad (12.1)$$

are parametric transfer functions (10.51) of the open-loop systems considered in Chapter 10, and ϕ_1 and ϕ_2 are the corresponding phase shifts. Moreover, the transfer functions $F_1(s)$, $F_2(s)$ are assumed to be strictly proper.

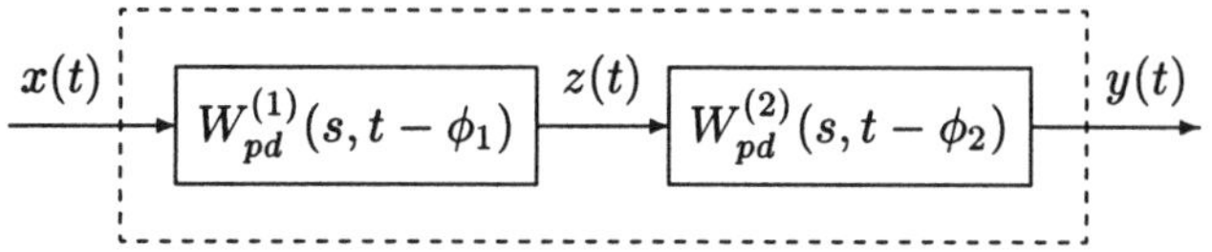

Figure 12.1: Synchronous sampled-data system

Using the general technique, we take $x(t) = \mathrm{e}^{st}$ and assume that

$$y(t) = W_c(s,t)\,, \quad W_c(s,t) = W_c(s,t+T) \tag{12.2}$$

where $W_c(s,t)$ is the desired PTF. But, by definition of the PTF, for $x = \mathrm{e}^{st}$ we have

$$z(t) = W_{pd}^{(1)}(s, t-\phi_1)\mathrm{e}^{st}\,. \tag{12.3}$$

Due to the stroboscopic effect, the input of the second element (12.3) can be changed equivalently by

$$\tilde{z}(t) = W_{pd}^{(1)}(s, \phi_2-\phi_1)\mathrm{e}^{st}\,. \tag{12.4}$$

Hence,

$$y(t) = W_{pd}^{(2)}(s, t-\phi_2)W_{pd}^{(1)}(s, \phi_2-\phi_1)\mathrm{e}^{st}\,. \tag{12.5}$$

Comparing (12.2) with (12.5), we find the required PTF of the system

$$W_c(s,t) = W_{pd}^{(2)}(s, t-\phi_2)W_{pd}^{(1)}(s, \phi_2-\phi_1)\,. \tag{12.6}$$

Taking into account that

$$\varphi_{F_iM_i}(T,s,t) = \frac{1}{T}\sum_{k=-\infty}^{\infty} F_i(s+k\mathrm{j}\omega)M_i(s+k\mathrm{j}\omega)\mathrm{e}^{k\mathrm{j}\omega t}\,, \quad \omega = 2\pi/T$$

and

$$W_{di}(s) = W_{di}(s+\mathrm{j}\omega)$$

we can represent Eq. (12.6) in the form

$$W_c(s,t) = \frac{1}{T}\sum_{k=-\infty}^{\infty} Q_c(s+k\mathrm{j}\omega)\mathrm{e}^{k\mathrm{j}\omega(t-\phi_1)} = \varphi_{Q_c}(T,s,t-\phi_1) \tag{12.7}$$

where

$$Q_c(s) \triangleq F_2(s)M_2(s)W_{d2}(s)W_{d1}(s)\varphi_{F_1M_1}(T,s,\phi_2-\phi_1)\,. \tag{12.8}$$

Comparing (12.7) with (6.115), we find that the system shown in Fig. 12.1 can be associated with an open-loop SD system with phase shift ϕ_1 and the equivalent transfer function (ETF) (12.8). Let $\mu_1, \ldots, \mu_r$ be the set of characteristic indices of $Q_c(s)$, which is the union of the corresponding sets of the functions $F_1(s)$, $F_2(s)$, $W_{d1}(s)$, and $W_{d2}(s)$. Then, the original for the image (12.8) takes the form

$$q_{ci}(t) = \frac{1}{2\pi \mathrm{j}} \int_{c-\mathrm{j}\infty}^{c+\mathrm{j}\infty} Q_c(s)\mathrm{e}^{st}\,\mathrm{d}s\,, \quad \mu_i < c < \mu_{i+1}\,. \tag{12.9}$$

Discretization of the integral (12.9) taking account of (12.8) yields

$$q_{ci}(t) = \frac{T}{2\pi \mathrm{j}} \int_{c-\mathrm{j}\omega/2}^{c+\mathrm{j}\omega/2} D_0(T,s,t)\,\mathrm{d}s\,, \quad \mu_i < c < \mu_{i+1} \tag{12.10}$$

where

$$D_0(T,s,t) \triangleq \mathcal{D}_{F_2M_2}(T,s,t-\phi_1)\mathcal{D}_{F_1M_1}(T,s,\phi_2-\phi_1)W_{d1}(s)W_{d2}(s)\,. \tag{12.11}$$

If $x(t) \in \Lambda(\mu_i, \mu_{i+1})$, by (7.115) and (12.4) the Laplace transform of the output $Y(s)$ has the form

$$Y(s) = F_2(s)M_2(s)W_{d1}(s)W_{d2}(s)\,\varphi_{F_1M_1}(T,s,\phi_2-\phi_1)\varphi_x(T,s,\phi_1+0) \tag{12.12}$$

while the discrete Laplace transform of the output is given by

$$\begin{gathered}\mathcal{D}_y(T,s,t) = \\ \mathcal{D}_{F_2M_2}(T,s,t-\phi_2)\mathcal{D}_{F_1M_1}(T,s,\phi_2-\phi_1)W_{d1}(s)W_{d2}(s)\mathcal{D}_x(T,s,\phi_1+0)\,.\end{gathered} \tag{12.13}$$

12.3 PTF decomposition for limited programs

To investigate a SD system with several sampling units which have different periods, it is necessary to use for the open loop with a computer which has the period T an equivalent representation in the form of a parallel connection of open-loop systems with period $T_N = NT$, where $N > 1$ is an integer. This representation is based on Eqs. (3.123), (3.129), (4.111) and (4.112) and will be referred to as polyphase representation, Crochiere and Rabiner (1983).
Let us have an open-loop system with a computer, the PTF of which is given by

$$W_{pd}(s,t) = W_d(s)\varphi_{FM}(T,s,t) \tag{12.14}$$

and the DTF (10.53) is

$$D_{pd}(s,t) = W_d(s)\mathcal{D}_{FM}(T,s,t)\,. \tag{12.15}$$

Equation (12.15) can be written as a discrete Laplace transform

$$D_{pd}^{T}(s,t) = \frac{1}{T}\sum_{k=-\infty}^{\infty} Q_{pd}(s+kj\omega)e^{kj\omega t} = \mathcal{D}_{Q_{pd}}(T,s,t) \tag{12.16}$$

where

$$Q_{pd}(s) = F(s)M(s)W_d(s) \tag{12.17}$$

is the corresponding ETF. The letter 'T' on the left-hand side of (12.16) means that this DTF is constructed for period T. According to (4.112), for any integer $N > 1$ we have

$$\breve{D}_{pd}^{NT}(s,t) = \frac{1}{N}\sum_{q=0}^{N-1} D_{pd}^{T}(s+\frac{qj\omega}{N},t)\,. \tag{12.18}$$

Using the relations of Section 10.10, we can obtain closed expressions for the function $\breve{D}_{pd}^{NT}(s,t)$ for a given N. First, we assume that the program $W_d(s)$ is limited. Then, under the conditions of Section 10.10, for $0 \le t \le T$ we have

$$\begin{aligned} D_{pd}^{T}(s,t) &= \sum_{r=1}^{m} \frac{c_r \mathcal{D}_{FM}(T,d_r,t)}{1-e^{(d_r-s)T}} + \\ &+ \sum_{i=1}^{\ell}\sum_{k=1}^{\nu_i} \frac{f_{ik}}{(k-1)!}\frac{\partial^{k-1}}{\partial {s_i}^{k-1}}\frac{W_d(s_i)M(s_i)e^{s_i t}}{1-e^{(s_i-s)T}} + \ell_d^{-} h_p^{*}(t)\,. \end{aligned} \tag{12.19}$$

Hence,

$$\begin{aligned} D_{pd}^{T}(s+\frac{qj\omega}{N},t) &= \sum_{r=1}^{m} \frac{c_r \mathcal{D}_{FM}(T,d_r,t)}{1-e^{(d_r-s)T}e^{-qj\omega T/N}} + \\ &+ \sum_{i=1}^{\ell}\sum_{k=1}^{\nu_i} \frac{f_{ik}}{(k-1)!}\frac{\partial^{k-1}}{\partial {s_i}^{k-1}}\frac{W_d(s_i)M(s_i)e^{s_i t}}{1-e^{(s_i-s)T}e^{-qj\omega T/N}} + \ell_d^{-} h_p^{*}(t)\,. \end{aligned} \tag{12.20}$$

Substituting (12.20) into (12.18), we have

$$\begin{aligned} \breve{D}_{pd}^{NT}(s,t) &= \frac{1}{N}\Bigg[\sum_{r=1}^{m} c_r \mathcal{D}_{FM}(T,d_r,t)\psi_{N0}(s,d_r) + \\ &+ \sum_{i=1}^{\ell}\sum_{k=1}^{\nu_i} \frac{f_{ik}}{(k-1)!}\frac{\partial^{k-1}}{\partial {s_i}^{k-1}} W_d(s_i)M(s_i)e^{s_i t}\psi_{N0}(s,s_i)\Bigg] + \ell_d^{-} h_p^{*}(t) \end{aligned} \tag{12.21}$$

with the notation

$$\psi_{N0}(s,a) \triangleq \sum_{q=0}^{N-1} \frac{1}{1-e^{(a-s)T}e^{-q2\pi j/N}}\,. \tag{12.22}$$

For further exposition we use the following auxiliary relations.

1. Denote

$$d_{kN} \triangleq \mathrm{e}^{\frac{k2\pi\mathrm{j}}{N}} \tag{12.23}$$

where k is an integer. Let

$$k = \mu N + \rho \tag{12.24}$$

where μ and ρ are integers, and $0 \leq \rho < N$. Then,

$$d_{kN} = d_{\rho N} . \tag{12.25}$$

2. Denote, as in (3.126),

$$\sigma_{\lambda N} = \sum_{k=0}^{N-1} d_{kN}^{\lambda} . \tag{12.26}$$

An explicit expression for the sum (12.26) is given by Eq. (3.127). Then, for any integer ϑ,

$$\sigma_{\lambda+\vartheta N,N} = \sigma_{\lambda N} \tag{12.27}$$

because, by definition,

$$\sigma_{\lambda+\vartheta N,N} = \sum_{k=0}^{N-1} d_{kN}^{\lambda+\vartheta N} = \sum_{k=0}^{N-1} d_{kN}^{\lambda} = \sigma_{\lambda N} .$$

3. Denote

$$\gamma_{N,r}(z) \triangleq \sum_{k=0}^{N-1} \frac{d_{kN}^{r}}{1 - z d_{kN}^{-1}} \tag{12.28}$$

where r is an integer. Then, for any integer p, we have

$$\gamma_{N,r+pN}(z) = \gamma_{Nr}(z) . \tag{12.29}$$

Moreover, for any integer r such that $0 \leq r \leq N-1$ the following formula holds

$$\gamma_{Nr}(z) = \frac{Nz^r}{1 - z^N} . \tag{12.30}$$

Proof Equation (12.29) follows immediately from (12.25). Next, we prove Eq. (12.30). Consider the ratio

$$f(z) \triangleq \frac{Nz^r}{1 - z^N} . \tag{12.31}$$

This function has simple poles at the points $z_k = d_{kN}$ $(k = 0, 1, \ldots, N-1)$. Therefore, the following partial fraction expansion holds

$$f(z) = -\sum_{k=0}^{N-1} \frac{d_{kN}^{r}}{d_{kN}^{N-1}} \frac{1}{z - d_{kN}} = \sum_{k=0}^{N-1} \frac{d_{kN}^{r}}{1 - z d_{kN}^{N-1}} \tag{12.32}$$

which is equivalent to (12.30). ■

It follows from (12.30) and (12.22) with $z = \mathrm{e}^{(a-s)T}$

$$\psi_{N0}(s,a) = \frac{N}{1-\mathrm{e}^{(a-s)NT}}\,. \tag{12.33}$$

Taking due account of (12.33), from Eq. (12.21) we obtain

$$\begin{aligned} \breve{D}_{pd}^{NT}(s,t) = \sum_{r=1}^{m} \frac{c_r \mathcal{D}_{FM}(T,d_r,t)}{1-\mathrm{e}^{(d_r-s)NT}} + \\ + \sum_{i=1}^{\ell}\sum_{k=1}^{\nu_i} \frac{f_{ik}}{(k-1)!}\frac{\partial^{k-1}}{\partial {s_i}^{k-1}}\frac{W_d(s_i)M(s_i)\mathrm{e}^{s_i t}}{1-\mathrm{e}^{(s_i-s)NT}} + \ell_d^- h_p^*(t)\,, \quad 0 \le t \le T\,. \end{aligned} \tag{12.34}$$

Formula (12.34) defines $\breve{D}_{pd}^{NT}$ in the interval $0 \le t \le T$. To continue this formula into the interval $T \le t < NT$, we take

$$t = \varepsilon + \lambda T\,, \quad 0 \le \varepsilon < T\,, \quad 1 \le \lambda \le N-1\,. \tag{12.35}$$

Since the expansion holds for all t, substituting (12.35) into (12.18), we obtain

$$\breve{D}_{pd}^{NT}(s,\varepsilon+\lambda T) = \frac{1}{N}\sum_{q=0}^{N-1} D_{pd}^{T}(s+\frac{qj\omega}{N},\varepsilon+\lambda T)\,. \tag{12.36}$$

Since λ is an integer, we have

$$D_{pd}^{T}(s+\frac{qj\omega}{N},\varepsilon+\lambda T) = D_{pd}^{T}(s+\frac{qj\omega}{N},\varepsilon)\mathrm{e}^{\lambda(s+\frac{qj\omega}{N})T}\,. \tag{12.37}$$

In this case $\mathrm{e}^{\frac{\lambda qj\omega}{N}} = d_{qN}^{\lambda}$, and the preceding relation takes the form

$$D_{pd}^{T}(s+\frac{qj\omega}{N},\varepsilon+\lambda T) = \mathrm{e}^{\lambda sT} D_{pd}^{T}(s+\frac{qj\omega}{N},\varepsilon)d_{qN}^{\lambda}\,. \tag{12.38}$$

Substituting (12.20) and (12.38) into (12.36), we find

$$\begin{aligned} \breve{D}_{pd}^{NT}(s,\varepsilon+\lambda T) = \frac{\mathrm{e}^{\lambda sT}}{N}\Bigg[\sum_{r=1}^{m} c_r\mathcal{D}_{FM}(T,d_r,\varepsilon)\psi_{N\lambda}(s,d_r) + \\ + \sum_{i=1}^{\ell}\sum_{k=1}^{\nu_i}\frac{f_{ik}}{(k-1)!}\frac{\partial^{k-1}}{\partial {s_i}^{k-1}} W_d(s_i)M(s_i)\mathrm{e}^{s_i t}\psi_{N\lambda}(s,s_i) + \ell_d^- h_p^*(t)\sigma_{\lambda N}\Bigg] \end{aligned} \tag{12.39}$$

where

$$\psi_{N\lambda}(s,a) \triangleq \sum_{q=0}^{N-1}\frac{d_{qN}^{\lambda}}{1-\mathrm{e}^{(a-s)T}d_{qN}^{-1}}\,. \tag{12.40}$$

As follows from (12.30),

$$\psi_{N\lambda}(s,a) = \frac{N\mathrm{e}^{\lambda(a-s)T}}{1-\mathrm{e}^{(a-s)NT}}\,. \tag{12.41}$$

Then, taking account of (12.41) and (3.127), Eq. (12.39) yields

$$\begin{aligned}\breve{D}_{pd}^{NT}(s,\varepsilon+\lambda T) = \sum_{r=1}^{m} \frac{c_r \mathcal{D}_{FM}(T,d_r,\varepsilon)\mathrm{e}^{\lambda d_r T}}{1-\mathrm{e}^{(d_r-s)NT}} + \\ + \sum_{i=1}^{\ell}\sum_{k=1}^{\nu_i} \frac{f_{ik}}{(k-1)!}\frac{\partial^{k-1}}{\partial {s_i}^{k-1}} \frac{W_d(s_i)M(s_i)\mathrm{e}^{s_i(\varepsilon+\lambda T)}}{1-\mathrm{e}^{(s_i-s)NT}}\,.\end{aligned} \tag{12.42}$$

Using (12.35) once again, and combining Eqs. (12.34) and (12.42), for $0 \le t \le T$ we can write

$$\begin{aligned}\breve{D}_{pd}^{NT}(s,t) = \sum_{r=1}^{m} \frac{c_r \mathcal{D}_{FM}(T,d_r,t)}{1-\mathrm{e}^{(d_r-s)NT}} + \\ + \sum_{i=1}^{\ell}\sum_{k=1}^{\nu_i} \frac{f_{ik}}{(k-1)!}\frac{\partial^{k-1}}{\partial {s_i}^{k-1}} \frac{W_d(s_i)M(s_i)\mathrm{e}^{s_i t}}{1-\mathrm{e}^{(s_i-s)NT}} + R_N(t)\ell_d^- h_p^*(t)\end{aligned} \tag{12.43}$$

where

$$R_N(t) = \begin{cases} 1 & 0 < t < T \\ 0 & T < t < NT\,. \end{cases} \tag{12.44}$$

Formula (12.43) can be continued onto the whole $t-$axis by

$$\breve{D}_{pd}^{NT}(s,t+NT) = \breve{D}_{pd}^{NT}(s,t)\mathrm{e}^{NsT}\,. \tag{12.45}$$

If $F(s)$ has only simple poles and the expansion (3.102) holds, Eq. (12.43) takes the simpler form

$$\breve{D}_{pd}^{NT}(s,t) = \sum_{r=1}^{m} \frac{c_r \mathcal{D}_{FM}(T,d_r,t)}{1-\mathrm{e}^{(d_r-s)NT}} + \sum_{i=1}^{n} \frac{f_i W_d(s_i)M(s_i)\mathrm{e}^{s_i t}}{1-\mathrm{e}^{(s_i-s)NT}} + R_N(t)\ell_d^- h_p^*(t)\,. \tag{12.46}$$

Example 12.1 Perform the PTF decomposition (9.129) for $N = 2$. Using (12.46) and (9.129), we have

$$\tilde{D}_p^{2T}(t,s) = \begin{cases} \dfrac{M(a)\mathrm{e}^{at}}{1-\mathrm{e}^{2(a-s)T}} - \displaystyle\int_t^T \mathrm{e}^{a(t-\tau)}\mu(\tau)\,\mathrm{d}\tau & 0 \le t \le T \\ \dfrac{M(a)\mathrm{e}^{at}}{1-\mathrm{e}^{2(a-s)T}} & T \le t \le 2T\,. \end{cases} \tag{12.47}$$

□

Finally, we notice that, by multiplying the above relations by e^{-st}, we obtain formulae for the corresponding PTFs.

12.4 PTF decomposition in the general case

If the control algorithm is not limited, it can be represented in the form (A.27)

$$W_d(s) = l_\kappa \mathrm{e}^{-\kappa sT} + \ldots + l_1 \mathrm{e}^{-sT} + \tilde{W}_{d1}(s) \tag{12.48}$$

where $\tilde{W}_{d1}(s)$ is limited. Then, Eq. (12.15) yields

$$D_{pd}(s,t) = \mathcal{D}_{FM}(T,s,t) \sum_{\mu=1}^{\kappa} l_\mu \mathrm{e}^{-\mu sT} + \tilde{W}_{d1}(s)\mathcal{D}_{FM}(T,s,t)\,. \tag{12.49}$$

The product $\tilde{W}_{d1}(s)D_{FM}(T,s,t)$ can be decomposed on the basis of the relations of Section 12.3. Therefore, to decompose the DTF D_{pd}^T in the general case it remains to decompose a function of the form

$$D_{\mu p}^T(s,t) \triangleq \mathrm{e}^{-\mu sT}\mathcal{D}_{FM}(T,s,t) \tag{12.50}$$

where μ is a positive integer. By analogy with (12.18), we have

$$\breve{D}_{\mu p}^{NT}(s,t) = \frac{1}{N}\sum_{q=0}^{N} D_{\mu p}^T(s + \frac{q\mathrm{j}\omega}{N}, t) \tag{12.51}$$

or, in expanded form,

$$\breve{D}_{\mu p}^{NT}(s,t) = \frac{\mathrm{e}^{-\mu sT}}{N}\sum_{q=0}^{N-1} \mathcal{D}_{FM}(T, s + \frac{q\mathrm{j}\omega}{N}, t)\, d_{qN}^{-\mu} \tag{12.52}$$

where the numbers d_{qN} are obtained from (12.23). Relation (12.52) holds for all t. Since, due to (9.100), in the interval $0 \le t \le T$ we have

$$\mathcal{D}_{FM}(T,s,t) = \sum_{i=1}^{\ell}\sum_{k=1}^{\nu_i} \frac{f_{ik}}{(k-1)!}\frac{\partial^{k-1}}{\partial {s_i}^{k-1}}\frac{M(s_i)\mathrm{e}^{s_i t}}{1-\mathrm{e}^{(s_i-s)T}} + h_p^*(t)\,. \tag{12.53}$$

Hence,

$$\mathcal{D}_{FM}(T, s + \frac{q\mathrm{j}\omega}{N}, t) = \sum_{i=1}^{\ell}\sum_{k=1}^{\nu_i} \frac{f_{ik}}{(k-1)!}\frac{\partial^{k-1}}{\partial {s_i}^{k-1}}\frac{M(s_i)\mathrm{e}^{s_i t}}{1-\mathrm{e}^{(s_i-s)T}d_{qN}^{-1}} + h_p^*(t)\,. \tag{12.54}$$

Substituting (12.54) into (12.52), we find

$$\breve{D}_{\mu p}^{NT}(s,t) = \qquad (0 \le t \le T) \tag{12.55}$$

$$\frac{\mathrm{e}^{-\mu sT}}{N}\left[\sum_{i=1}^{\ell}\sum_{k=1}^{\nu_i} \frac{f_{ik}}{(k-1)!}\frac{\partial^{k-1}}{\partial {s_i}^{k-1}} M(s_i)\mathrm{e}^{s_i t}\psi_{N,-\mu}(s,s_i) + \sigma_{-\mu,N} h_p^*(t)\right]$$

with

$$\psi_{-\mu,N}(s,a) = \sum_{q=0}^{N-1} \frac{d_{qN}^{-\mu}}{1 - e^{(a-s)T} d_{qN}^{-1}}, \qquad \sigma_{-\mu,N} = \sum_{q=0}^{N-1} d_{qN}^{-\mu}. \tag{12.56}$$

Let

$$-\mu = c + \vartheta N \tag{12.57}$$

where c and ϑ are integers such that $0 \le c < N$. Then, from (12.56) we have

$$\sigma_{-\mu,N} = \sigma_{cN} \tag{12.58}$$

and, taking due accout of (12.31),

$$\psi_{N,-\mu}(s,a) = \psi_{Nc}(s,a) = \frac{N e^{c(a-s)T}}{1 - e^{(a-s)NT}}. \tag{12.59}$$

Taking account of (A.58) and (12.59), Eq. (12.55) takes the form

$$\breve{D}_{\mu p}^{NT}(s,t) = e^{\vartheta N s T} \sum_{i=1}^{\ell} \sum_{k=1}^{\nu_i} \frac{f_{ik}}{(k-1)!} \frac{\partial^{k-1}}{\partial {s_i}^{k-1}} \frac{M(s_i) e^{s_i(t+cT)}}{1 - e^{(s_i-s)NT}} + \frac{e^{-\mu s T}}{N} \sigma_{cN} h_p^*(t)$$
$$0 \le t \le T. \tag{12.60}$$

In order to continue formula (12.60) on the interval $T \le t \le NT$, we assume that (12.35) holds. Then, taking into account that $0 \le \varepsilon < T$ and $1 \le \lambda \le N-1$, and using Eq. (12.52), we have

$$\begin{aligned} \breve{D}_{\mu p}^{NT}(s,\varepsilon+\lambda T) &= \frac{e^{-\mu s T}}{N} \sum_{q=0}^{N-1} \mathcal{D}_{FM}(T, s + \frac{qj\omega}{N}, \varepsilon + \lambda T)\, d_{qN}^{-\mu} \\ &= \frac{e^{-\mu s T}}{N} \sum_{q=0}^{N-1} \mathcal{D}_{FM}(T, s + \frac{qj\omega}{N}, \varepsilon) e^{\lambda(s+\frac{qj\omega}{N})T}\, d_{qN}^{-\mu} \\ &= \frac{e^{(\lambda-\mu)sT}}{N} \sum_{q=0}^{N-1} \mathcal{D}_{FM}(T, s + \frac{qj\omega}{N}, \varepsilon)\, d_{qN}^{\lambda-\mu}. \end{aligned} \tag{12.61}$$

The right-hand side of this equation is obtained from the right-hand side of (12.52) by substitution of $\lambda - \mu$ for μ. Then, defining the integers c_λ and ϑ_λ from

$$\lambda - \mu = c_\lambda + \vartheta_\lambda N \tag{12.62}$$

from (12.60) we obtain

$$\breve{D}_{\mu p}^{NT}(s,t) = e^{\vartheta_\lambda N s T} \sum_{i=1}^{\ell} \sum_{k=1}^{\nu_i} \frac{f_{ik}}{(k-1)!} \frac{\partial^{k-1}}{\partial {s_i}^{k-1}} \frac{M(s_i) e^{s_i(\varepsilon+c_\lambda T)}}{1 - e^{(s_i-s)NT}} +$$
$$+ \frac{e^{(\lambda-\mu)sT}}{N} \sigma_{c_\lambda N} h_p^*(\varepsilon), \qquad 0 \le \varepsilon < T, \quad 1 \le \lambda \le N-1. \tag{12.63}$$

Example 12.2 Perform the decomposition with $N=2$ for the DTF

$$D_{\mu p}^{T}(s,t)=\mathrm{e}^{-sT}\mathcal{D}_{FM}(T,s,t) \tag{12.64}$$

where

$$F(s)=\frac{1}{s-a}$$

and the form of the pulse is not specified. According to (9.129), we have

$$D_{p}^{T}(s,t)=\frac{M(a)\mathrm{e}^{at}}{1-\mathrm{e}^{(a-s)T}}+h_{p}^{*}(t)\,,\quad 0\le t\le T\,.$$

Since in this case $\mu=1$, taking account of (12.57) we take $c=1$ and $\vartheta=-1$. Then, $\sigma_{cN}=0$, and Eq. (12.60) yields

$$\breve{D}_{\mu p}^{2T}(s,t)=\mathrm{e}^{-2sT}\,\frac{M(a)\mathrm{e}^{a(t+T)}}{1-\mathrm{e}^{2(a-s)T}}\,,\quad 0\le t\le T\,.$$

For the interval $T\le t\le 2T$ we have $t=\varepsilon+T$ with $0\le\varepsilon\le T$, i.e., $\lambda=1$. Since $\lambda-\mu=0$, we should take $c_\lambda=0$ and $\vartheta_\lambda=0$. Hence, $\sigma_{c_\lambda N}=2$, and Eq.(12.63) yields

$$\breve{D}_{\mu p}^{2T}(s,t)=\frac{M(a)\mathrm{e}^{a(t-T)}}{1-\mathrm{e}^{2(a-s)T}}+h_{p}^{*}(t-T)\,,\quad T\le t\le 2T\,.$$

□

12.5 Ascending SD systems

By an *ascending SD system* we mean a series connection of two open-loop SD systems, L_{NT} and L_T, in those cases when the sampling period of the first system equals N times the period of the second one, with an integer N (Fig. 12.2). The operator equations of the systems have the form

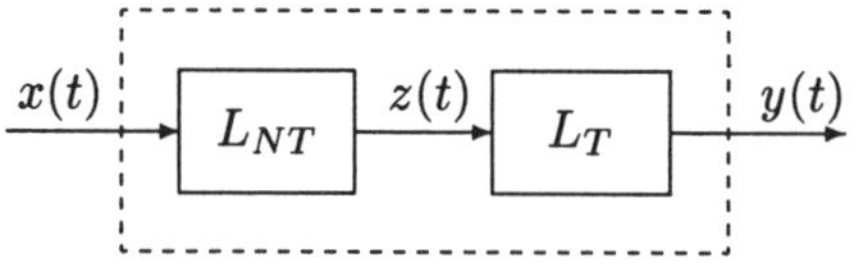

Figure 12.2: Ascending SD system

$$y=W_{d2}(s)\varphi_{F_2M_2}(T,s,t)z\,,\quad z=W_{d1}(s)\varphi_{F_1M_1}(NT,s,t)x \tag{12.65}$$

where, by construction,

$$\begin{aligned}\varphi_{F_1M_1}(NT,s,t)&=\frac{1}{NT}\sum_{k=-\infty}^{\infty}F_1\Big(s+\frac{kj\omega}{N}\Big)M_1\Big(s+\frac{kj\omega}{N}\Big)\,\mathrm{e}^{\frac{kj\omega}{N}t}\\ \varphi_{F_2M_2}(T,s,t)&=\frac{1}{T}\sum_{k=-\infty}^{\infty}F_2(s+kj\omega)M_2(s+kj\omega)\,\mathrm{e}^{kj\omega}\end{aligned} \tag{12.66}$$

with $\omega = 2\pi/T$ and

$$W_{d1}(s) = W_{d1}(s + \frac{\mathrm{j}\omega}{N}), \quad W_{d2}(s) = W_{d2}(s + \mathrm{j}\omega). \tag{12.67}$$

The transfer functions $F_1(s)$, $F_2(s)$ are assumed to be strictly proper. Decomposing the second system, we can write the polyphase representation in the form

$$y(t) = \sum_{\lambda=0}^{N-1} y_\lambda(t) \tag{12.68}$$

where the variable $y_\lambda(t)$ is defined by the operator equations

$$y_\lambda = \breve{W}_2(NT, s, t - \lambda T)z, \quad z = W_{d1}(s)\varphi_{F_1 M_1}(NT, s, t)x \tag{12.69}$$

where the PTF $\breve{W}_2(NT, s, t)$ is defined by the following relations based on (3.129)

$$\breve{W}_2(NT, s, t) = \frac{1}{NT} \sum_{k=-\infty}^{\infty} F_2(s + \frac{k\mathrm{j}\omega}{N})M_2(s + \frac{k\mathrm{j}\omega}{N})W_{d2}(s + \frac{k\mathrm{j}\omega}{N})\,\mathrm{e}^{\frac{k\mathrm{j}\omega}{N}t}. \tag{12.70}$$

Equation (12.70) coincides with the PTF of an open-loop system with period NT and equivalent transfer function

$$Q_2(s) = F_2(s)M_2(s)W_{d2}(s). \tag{12.71}$$

Then, Eq. (12.69) describes a series connection of two open-loop systems with the period NT. Therefore, the system (12.69) also has the period NT. Assuming that $x(t) = \mathrm{e}^{st}$ and using the stroboscopic effect, we immediately obtain

$$y_\lambda(t) = \breve{W}_2(NT, s, t - \lambda T)W_{d1}(s)\varphi_{F_1 M_1}(NT, s, \lambda T)\mathrm{e}^{st}.$$

Hence, the PTF of the system (12.69) appears as

$$W_{v\lambda}(s, t) = \breve{W}_2(NT, s, t - \lambda T)W_{d1}(s)\varphi_{F_1 M_1}(NT, s, \lambda T). \tag{12.72}$$

This PTF can be written in the form

$$W_{v\lambda}(s, t) = \frac{1}{NT} \sum_{k=-\infty}^{\infty} Q_{v\lambda}(s + \frac{k\mathrm{j}\omega}{N})\mathrm{e}^{\frac{k\mathrm{j}\omega}{N}t} \tag{12.73}$$

where

$$Q_{v\lambda}(s) = F_2(s)M_2(s)W_{d2}(s)W_{d1}(s)\varphi_{F_1 M_1}(NT, s, \lambda T). \tag{12.74}$$

As follows from (12.73), Eq. (12.69) is associated with an polyphase open-loop system with period NT, which has the PTF (12.73) and the ETF (12.74). As follows from (12.68) and (12.73), the PTF $W_v(s, t)$ of the system under consideration can be represented in the form

$$W_v(s,t) = \frac{1}{NT} \sum_{k=-\infty}^{\infty} Q_v(s + \frac{kj\omega}{N}) e^{\frac{kj\omega}{N} t} \tag{12.75}$$

where

$$Q_v(s) \triangleq F_2(s) M_2(s) W_{d2}(s) W_{d1}(s) \sum_{\lambda=0}^{N-1} \varphi_{F_1 M_1}(NT, s, \lambda T) \tag{12.76}$$

which corresponds to an open-loop SD system with the ETF (12.76) without phase shift.
As follows from the above formulae, the set of characteristic indices μ_i ($i = 1, \dots, r$) of the system at hand is the union of the corresponding sets of the functions $F_1(s)$, $F_2(s)$, $W_{d1}(s)$, and $W_{d2}(s)$. Moreover, in any regularity stripe $|Q_v(s)|$ decreases as $|s|^{-2}$ for $|s| \to \infty$. Therefore, for $x(t) \in \Lambda(\mu_i, \mu_{i+1})$ the Laplace transform of the output is given by

$$Y(s) = F_2(s) M_2(s) W_{d2}(s) W_{d1}(s) \sum_{\lambda=0}^{N-1} \varphi_{F_1 M_1}(NT, s, \lambda T) \varphi_x(NT, s, +0) . \tag{12.77}$$

Using (4.27) and

$$\varphi_{F_1 M_1}(NT, s + \frac{kj\omega}{N}, \lambda T) = \varphi_{F_1 M_1}(NT, s, \lambda T) e^{-\frac{kj\omega}{N} \lambda T}$$

from (12.77) we obtain the discrete Laplace transform of the output

$$\begin{aligned} \mathcal{D}_Y(NT, s, t) &= \mathcal{D}_y(NT, s, t) = \\ &= \sum_{\lambda=0}^{N-1} \breve{\mathcal{D}}_{F_2 M_2 W_{d2}}(NT, s, t - \lambda T) \mathcal{D}_{F_1 M_1}(NT, s, \lambda T) W_{d1}(s) \mathcal{D}_x(NT, s, +0) . \end{aligned} \tag{12.78}$$

The formula

$$D_v(s,t) = W_{d1}(s) \sum_{\lambda=0}^{N-1} \breve{\mathcal{D}}_{F_2 M_2 W_{d2}}(NT, s, t - \lambda T) \mathcal{D}_{F_1 M_1}(NT, s, \lambda T) \tag{12.79}$$

gives the DTF of the ascending system.

Example 12.3 Find the DTF of an ascending system with $N = 2$ and

$$F_1(s) = \frac{1}{s-b}, \quad F_2(s) = \frac{1}{s-a}, \quad W_{d1}(s) = W_{d2}(s) = 1 \tag{12.80}$$

with an arbitrary form of the pulses. Then, Eq. (12.79) yields

$$D_v(s,t) = \breve{\mathcal{D}}_{F_2 M_2}(2T, s, t) \mathcal{D}_{F_1 M_1}(2T, s, 0) + \breve{\mathcal{D}}_{F_2 M_2}(2T, s, t-T) \mathcal{D}_{F_1 M_1}(2T, s, T) . \tag{12.81}$$

Using (9.129), we have

$$\mathcal{D}_{F_1M_1}(2T,s,t) = \frac{M_1(b)\mathrm{e}^{bt}}{1-\mathrm{e}^{2(b-s)T}} - \int_t^{2T} \mathrm{e}^{b(t-\tau)}\mu_1(\tau)\,\mathrm{d}\tau\,, \quad 0 \le t \le 2T\,. \tag{12.82}$$

Moreover, from (12.47) it follows that

$$\breve{\mathcal{D}}_{F_2M_2}(2T,s,t) = \begin{cases} \dfrac{M_2(a)\mathrm{e}^{at}}{1-\mathrm{e}^{2(a-s)T}} - \displaystyle\int_t^T \mathrm{e}^{a(t-\tau)}\mu_2(\tau)\,\mathrm{d}\tau & 0 \le t \le T \\ \dfrac{M_2(a)\mathrm{e}^{at}}{1-\mathrm{e}^{2(a-s)T}} & T \le t \le 2T\,. \end{cases} \tag{12.83}$$

In the above relations

$$-\int_t^{2T} \mathrm{e}^{b(t-\tau)}\mu_1(\tau)\,\mathrm{d}\tau = h_{p1}^*(t)\,, \quad -\int_t^T \mathrm{e}^{a(t-\tau)}\mu_2(\tau)\,\mathrm{d}\tau = h_{p2}^*(t)\,. \tag{12.84}$$

These formulae should be added by the relations

$$\begin{aligned} \mathcal{D}_{F_1M_1}(2T,s,t+2T) &= \mathcal{D}_{F_1M_1}(2T,s,t)\mathrm{e}^{2bsT} \\ \breve{\mathcal{D}}_{F_2M_2}(2T,s,t+2T) &= \breve{\mathcal{D}}_{F_2M_2}(2T,s,t)\mathrm{e}^{2asT} \end{aligned} \tag{12.85}$$

Substituting (12.82)–(12.85) into (12.81), we can obtain a closed expressions for the desired PTF. □

12.6 Descending SD systems

By a *descending SD system* we mean a series connection of two elementary SD systems in cases, when the sampling period of the second system equals N times the period of the first one, with an integer N (Fig. 12.3), Crochiere and Rabiner (1983); Spring and Unger (1994). By analogy with (12.65) the

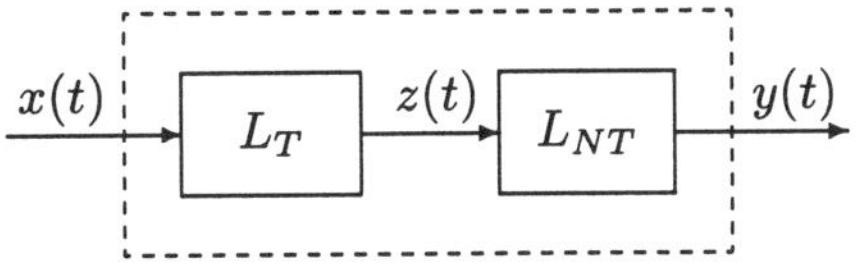

Figure 12.3: Descending SD system

equations of the descending system can be written in the form

$$y = W_{d2}(s)\varphi_{F_2M_2}(NT,s,t)z\,, \quad z = W_{d1}(s)\varphi_{F_1M_1}(T,s,t)x \tag{12.86}$$

where

$$W_{d1}(s) = W_{d1}(s+\mathrm{j}\omega)\,, \quad W_{d1}(s) = W_{d1}(s+\frac{\mathrm{j}\omega}{N})\,. \tag{12.87}$$

Taking due account of (12.87), we can write

$$
\begin{aligned}
W_{d1}(s)\varphi_{F_1M_1}(T,s,t) = \varphi_{F_1M_1W_{d1}}(T,s,t) &\stackrel{\triangle}{=} W_1(T,s,t) \\
W_{d2}(s)\varphi_{F_2M_2}(NT,s,t) = \varphi_{F_2M_2W_{d2}}(NT,s,t) &\stackrel{\triangle}{=} W_2(NT,s,t)\,.
\end{aligned}
\tag{12.88}
$$

Decomposing the PTF $W_1(T,s,t)$, we get the polyphase structure

$$
W_1(T,s,t) = \sum_{\lambda=0}^{N-1} \breve{\varphi}_{F_1M_1W_{d1}}(NT,s,t-\lambda T)\,. \tag{12.89}
$$

From (12.86) and (12.89) we obtain

$$
y(t) = \sum_{\lambda=0}^{N-1} y_\lambda(t)\,, \tag{12.90}
$$

where the functions $y_\lambda(t)$ are given by the operator equations

$$
y_\lambda = \varphi_{F_2M_2W_{d2}}(NT,s,t)z_\lambda\,, \quad z_\lambda = \breve{\varphi}_{F_1M_1W_{d1}}(NT,s,-\lambda T)x\,. \tag{12.91}
$$

By the stroboscopic effect, the PTF of the series connection (12.91) takes the form

$$
W_a(s,t) = \varphi_{F_2M_2W_{d2}}(NT,s,t)\breve{\varphi}_{F_1M_1W_{d1}}(NT,s,-\lambda T)\,. \tag{12.92}
$$

Summing up Eq. (12.92) for different λ, we obtain the PTF of the descending system

$$
W_a(s,t) = \varphi_{F_2M_2W_{d2}}(NT,s,t)\sum_{\lambda=0}^{N-1} \breve{\varphi}_{F_1M_1W_{d1}}(NT,s,-\lambda T)\,. \tag{12.93}
$$

But, as follows from (3.123),

$$
\sum_{\lambda=0}^{N-1} \breve{\varphi}_{F_1M_1W_{d1}}(NT,s,-\lambda T) = \varphi_{F_1M_1W_{d1}}(T,s,0) = W_{d1}(s)\varphi_{F_1M_1}(T,s,0)
$$

and Eq. (12.93) takes the form

$$
W_a(s,t) = \varphi_{F_2M_2}(NT,s,t)\varphi_{F_1M_1}(T,s,0)W_{d2}(s)W_{d1}(s)\,. \tag{12.94}
$$

To find the transforms of the output, we notice that the PTF (12.92) can be represented in the form

$$
W_{\lambda a}(s,t) = \frac{1}{NT}\sum_{k=-\infty}^{\infty} Q_{\lambda a}(s+\frac{kj\omega}{N})e^{\frac{kj\omega}{N}(t-\lambda T)} \tag{12.95}
$$

with

$$
Q_{\lambda a}(s) = F_2(s)M_2(s)W_{d2}(s)\breve{\varphi}_{F_1M_1W_{d1}}(NT,s,-\lambda T)\,. \tag{12.96}
$$

Equation (12.95) is associated with an open-loop system with ETF (12.96) and phase shift λT. The set of characteristic indices μ_i, $(i = 1, \dots, r)$ of the system is the union of the corresponding sets of the functions $F_1(s)$, $F_2(s)$, $W_{d1}(s)$, and $W_{d2}(s)$. For $x(t) \in \Lambda(\mu_i, \mu_{i+1})$ the Laplace transform $Y_\lambda(s)$ of the function $y_\lambda(t)$ (12.91) appears as

$$Y_\lambda(s) = F_2(s)M_2(s)W_{d2}(s)\breve{\varphi}_{F_1M_1W_{d1}}(NT, s, -\lambda T)\varphi_x(NT, s, \lambda T+0)\,. \quad (12.97)$$

Summing up Eq. (12.97) for different λ, we obtain the Laplace transform of the output

$$Y(s) = F_2(s)M_2(s)W_{d2}(s)\sum_{\lambda=0}^{N-1}\breve{\varphi}_{F_1M_1W_{d1}}(NT, s, -\lambda T)\varphi_x(NT, s, \lambda T + 0)\,. \quad (12.98)$$

Example 12.4 Find the PTF of the descending SD system with $N = 2$ and (12.80) for an arbitrary form of the pulse. In this case, due to (12.94), the PTF of the system is given by

$$W_a(s,t) = \varphi_{F_2M_2}(2T, s, t)\varphi_{F_1M_1}(T, s, 0)\,.$$

Moreover,

$$\varphi_{F_2M_2}(2T, s, t) = \mathrm{e}^{-st}\left[\frac{M_2(a)\mathrm{e}^{at}}{1-\mathrm{e}^{2(a-s)T}} - \int_t^T \mathrm{e}^{a(t-\tau)}\mu_2(\tau)\,\mathrm{d}\tau\right], \quad 0 \le t \le 2T$$

$$\varphi_{F_1M_1}(T, s, 0) = \frac{M(b)}{\mathrm{e}^{(s-b)T} - 1}\,. \qquad \square$$

12.7 Feedback synchronous systems

To simplify the derivation, we consider only the system shown in Fig. 12.4, where the open-loop system has the structure shown in Fig. 12.1. The system

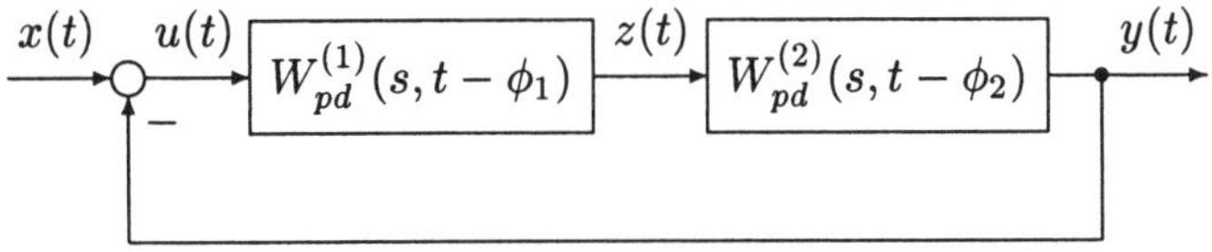

Figure 12.4: Feedback synchronous system

can be decribed by an operator equation of the form

$$y = \varphi_{F_2M_2W_{d2}}(T, s, t-\phi_2)z\,, \quad z = \varphi_{F_1M_1W_{d1}}(T, s, t-\phi_1)u\,, \quad u = x - y\,. \quad (12.99)$$

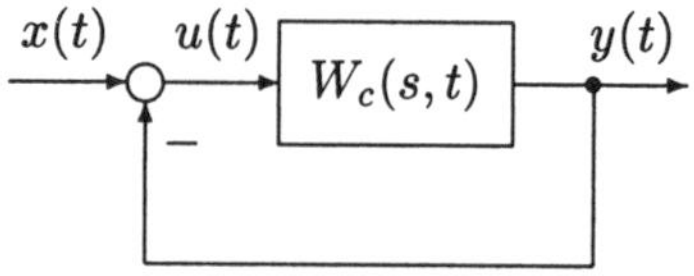

Figure 12.5: Equivalent block-diagram to Fig. 12.4

Using the formulae obtained in Section 12.2, we can replace the block-diagram shown in Fig. 12.4 by an equivalent one presented in Fig. 12.5, where the PTF $W_c(s,t)$ is given by (12.7). To construct the PTF $W_{cl}(s,t)$ of the closed-loop system we put up, as usual,

$$x(t) = \mathrm{e}^{st}\,, \quad y(t) = W_{cl}(s,t)\mathrm{e}^{st}\,, \quad W_{cl}(s,t) = W_{cl}(s,t+T)\,. \tag{12.100}$$

Let Eq. (12.100) hold. Then, from Fig. 12.5 it follows that

$$u(t) = \mathrm{e}^{st} f_u(t) \tag{12.101}$$

where

$$f_u(t) \stackrel{\triangle}{=} 1 - W_{cl}(s,t) = f_u(t+T)\,. \tag{12.102}$$

Taking account of the stroboscopic effect, the input (12.101) can be replaced equivalently by

$$\tilde{u}(t) = \mathrm{e}^{st} f_u(\phi_1) = \mathrm{e}^{st}\left[1 - W_{cl}(s,\phi_1)\right]\,. \tag{12.103}$$

From (12.103) and Fig. 12.5 we have

$$y(t) = \mathrm{e}^{st} W_c(s,t)\left[1 - W_{cl}(s,\phi_1)\right]\,. \tag{12.104}$$

Comparing Eqs. (12.100) and (12.104) for $y(t)$, we obtain the functional equation

$$W_{cl}(s,t) = W_c(s,t)\left[1 - W_{cl}(s,\phi_1)\right]\,. \tag{12.105}$$

Substituting $t = \phi_1$ at both sides of Eq. (12.105), we find

$$W_{cl}(s,\phi_1) = \frac{W_c(s,\phi_1)}{1 + W_c(s,\phi_1)}\,. \tag{12.106}$$

Substituting (12.106) into (12.105), we obtain

$$W_{cl}(s,t) = \frac{W_c(s,t)}{1 + W_c(s,\phi_1)}\,. \tag{12.107}$$

As follows from (12.7),

$$W_c(s,\phi_1) = \frac{1}{T}\sum_{k=-\infty}^{\infty} Q_c(s + kj\omega) = \mathcal{D}_{Q_c}(T,s,0) \tag{12.108}$$

where, by (12.8),

$$\mathcal{D}_{Q_c}(T,s,0) = W_{d2}(s)W_{d1}(s) *$$
$$* \frac{1}{T}\sum_{k=-\infty}^{\infty} F_2(s+kj\omega)M_2(s+kj\omega)\varphi_{F_1M_1}(T,s+kj\omega,\phi_2-\phi_1)\,.$$

Since

$$\varphi_{F_1M_1}(T,s+kj\omega,\phi_2-\phi_1) = \varphi_{F_1M_1}(T,s,\phi_2-\phi_1)\mathrm{e}^{kj\omega(\phi_1-\phi_2)}$$

from (12.108) we obtain

$$\begin{aligned} W_c(s,\phi_1) &= \mathcal{D}_{Q_c}(T,s,0) \\ &= W_{d2}(s)W_{d1}(s)\varphi_{F_2M_2}(T,s,\phi_1-\phi_2)\varphi_{F_1M_1}(T,s,\phi_2-\phi_1) \\ &= W_{d2}(s)W_{d1}(s)\mathcal{D}_{F_2M_2}(T,s,\phi_1-\phi_2)\mathcal{D}_{F_1M_1}(T,s,\phi_2-\phi_1) \end{aligned}$$

and formula (12.107) takes the form

$$W_{cl}(s,t) = \frac{W_c(s,t)}{1+W_{d1}(s)W_{d2}(s)\mathcal{D}_{F_1M_1}(T,s,\phi_2-\phi_1)\mathcal{D}_{F_2M_2}(T,s,\phi_1-\phi_2)}\,. \tag{12.109}$$

The denominator of the latter equation is periodic with respect to s. Therefore, using Eq. (12.7), we can represent Eq. (12.109) in the form

$$W_{cl}(s,t) = \frac{1}{T}\sum_{k=-\infty}^{\infty} Q_{cl}(s+kj\omega)\mathrm{e}^{kj\omega(t-\phi_1)} \tag{12.110}$$

where

$$Q_{cl}(s) = \frac{F_2(s)M_2(s)W_{d1}(s)W_{d2}(s)\varphi_{F_1M_1}(T,s,\phi_2-\phi_1)}{1+W_{d1}(s)W_{d2}(s)\mathcal{D}_{F_1M_1}(T,s,\phi_2-\phi_1)\mathcal{D}_{F_2M_2}(T,s,\phi_1-\phi_2)}\,. \tag{12.111}$$

The relations (12.110) and (12.111) are associated with an open-loop SD system with ETF (12.111) and phase shift ϕ_1. Therefore, if μ_i, $(i=1,\dots,r)$ are the characteristic indices of the function (12.111), for $x(t)\in\Lambda(\mu_i,\mu_{i+1})$ the Laplace transform of the output $Y(s)$ is given by

$$Y(s) = \tag{12.112}$$
$$\frac{F_2(s)M_2(s)W_{d1}(s)W_{d2}(s)\varphi_{F_1M_1}(T,s,\phi_2-\phi_1)}{1+W_{d1}(s)W_{d2}(s)\mathcal{D}_{F_1M_1}(T,s,\phi_2-\phi_1)\mathcal{D}_{F_2M_2}(T,s,\phi_1-\phi_2)}\varphi_x(T,s,\phi_1+0).$$

Using (12.112) and (4.27), we can easily find the discrete Laplace transform of the output.

Example 12.5 Construct the PTF of the synchronous SD system shown in Fig. 12.6, Ackermann (1988), where the sampling unit I operates at the points $t_{1k}=kT+\phi$ with $0\le\phi<T$, and the unit II operates at $t_{2k}=kT$. In this case $W_{d1}(s)=W_{d2}(s)=1$, $F_1(s)=s^{-1}$, and $F_2(s)=(1+s)^{-1}$. As a result,

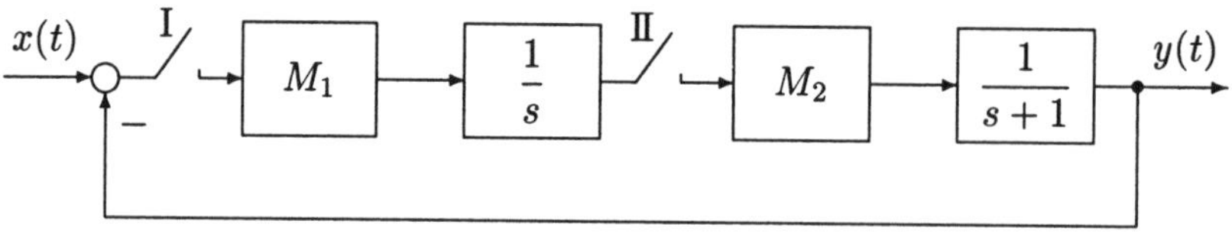

Figure 12.6: Synchronous SD system

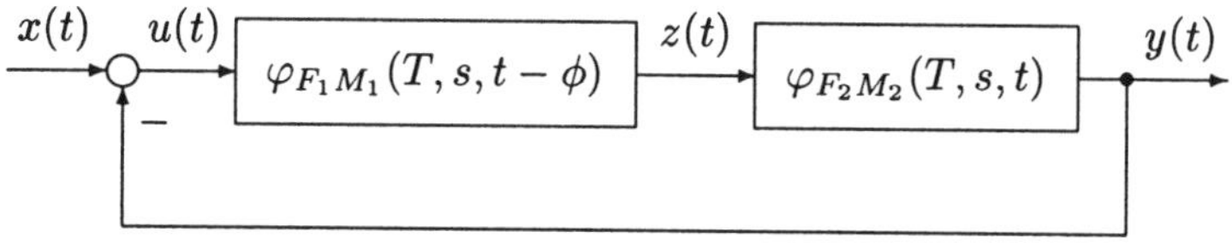

Figure 12.7: Equivalent block-diagram for Fig. 12.6

the block-diagram shown in Fig. 12.6 can be represented as shown in Fig. 12.7. For this example Eq. (12.109) yields

$$W_{cl}(s,t) = \frac{\varphi_{F_2M_2}(T,s,t)\varphi_{F_1M_1}(T,s,-\phi)}{1+\mathcal{D}_{F_1M_1}(T,s,-\phi)\mathcal{D}_{F_2M_2}(T,s,\phi)}\,. \tag{12.113}$$

Moreover, by (9.110) for the interval $0 \le t \le T$ we have

$$\begin{aligned}\varphi_{F_1M_1}(T,s,t) &= \mathrm{e}^{-st}\left[\frac{M_1(0)}{\mathrm{e}^{sT}-1}+\int_0^t \mu_1(\tau)\,\mathrm{d}\tau\right]\\ \varphi_{F_2M_2}(T,s,t) &= \mathrm{e}^{-st}\left[\frac{M_2(-1)\mathrm{e}^{-t}}{\mathrm{e}^{(s+1)T}-1}+\int_0^t \mathrm{e}^{-(t-\tau)}\mu_2(\tau)\,\mathrm{d}\tau\right]\,.\end{aligned} \tag{12.114}$$

With the help of Eq. (12.114) it is easy to obtain closed expression for the PTF (12.113). □

12.8 Ascending closed-loop systems

The block-diagram of this system is given in Fig. 12.8, where the open-loop

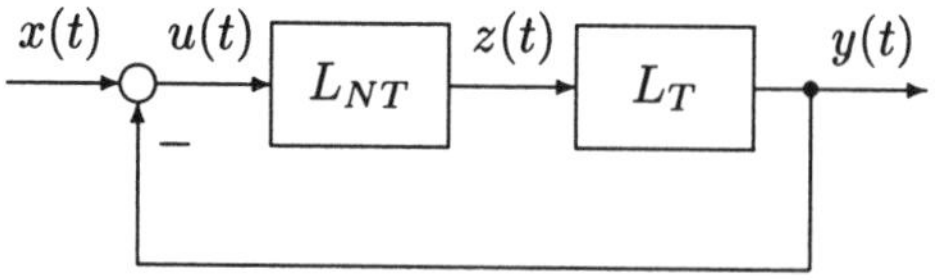

Figure 12.8: Ascending closed-loop system

system satisfies the conditions of Section 12.5. Using (12.75), we can obain the equivalent block-diagram shown in Fig. 12.9, where the open-loop system is a SD system with ETF (12.76) and without phase shift.
Let, as usual,

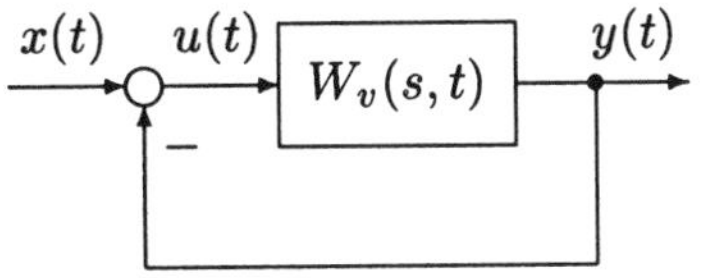

Figure 12.9: Equivalent block-diagram to Fig. 12.8

$$x(t) = e^{st}, \quad y(t) = W_{vl}(s,t)e^{st}, \quad W_{vl}(s,t) = W_{vl}(s,t+T). \tag{12.115}$$

Then,

$$u(t) = e^{st} f_u(t) \tag{12.116}$$

where

$$f_u(t) \triangleq 1 - W_{vl}(s,t) = f_u(t+NT). \tag{12.117}$$

Using the stroboscopic effect, the signal (12.116) can be replaced by an equivalent one

$$\tilde{u}(t) = e^{st} f_u(0) = e^{st}[1 - W_{vl}(s,0)]. \tag{12.118}$$

Using (12.118) and Fig. 12.9, we find

$$y(t) = W_v(s,t)[1 - W_{vl}(s,0)]\, e^{st}. \tag{12.119}$$

Comparing (12.115) with (12.119), we obtain

$$W_{vl}(s,t) = W_v(s,t)[1 - W_{vl}(s,0)]$$

whence

$$W_{vl}(s,0) = \frac{W_v(s,0)}{1 + W_v(s,0)}$$

and

$$W_{vl}(s,t) = \frac{W_v(s,t)}{1 + W_v(s,0)}. \tag{12.120}$$

As follows from (12.75),

$$W_v(s,0) = \frac{1}{NT} \sum_{k=-\infty}^{\infty} Q_v(s + \frac{kj\omega}{N}) = W_v(s + \frac{kj\omega}{N}, 0). \tag{12.121}$$

Therefore, taking account of (12.75), the PTF (9.120) can be written in the form

$$W_{vl}(s,t) = \frac{1}{NT} \sum_{k=-\infty}^{\infty} Q_{vl}(s + \frac{kj\omega}{N}) e^{\frac{kj\omega}{N} t} \tag{12.122}$$

where

$$Q_{vl}(s) = \frac{Q_v(s)}{1 + W_v(s,0)} \tag{12.123}$$

and $Q_v(s)$ is given by (12.76). Moreover, from (12.121) and (12.75) we have

$$
\begin{aligned}
W_v(s,0) &= \sum_{\lambda=0}^{N-1} \breve{\varphi}_{F_2M_2W_{d2}}(NT,s,-\lambda T)\varphi_{F_1M_1}(NT,s,\lambda T)W_{d1}(s) \\
&= \sum_{\lambda=0}^{N-1} \breve{\mathcal{D}}_{F_2M_2W_{d2}}(NT,s,-\lambda T)\mathcal{D}_{F_1M_1}(NT,s,\lambda T)W_{d1}(s)\,.
\end{aligned}
\tag{12.124}
$$

As follows from (12.122), the system at hand can be considered as an open-loop system with ETF (12.123) and without phase shift. Therefore, if we introduce the set of characteristic indices μ_i as usual, for $x(t) \in \Lambda(\mu_i, \mu_{i+1})$ the Laplace transform of the output takes the form

$$
Y(s) = \frac{F_2(s)M_2(s)W_{d2}(s)W_{d1}(s)\sum\limits_{\lambda=0}^{N-1} \varphi_{F_1M_1}(NT,s,\lambda T)}{1 + W_{d1}(s)\sum\limits_{\lambda=0}^{N-1} \breve{\mathcal{D}}_{F_2M_2W_{d2}}(NT,s,-\lambda T)\mathcal{D}_{F_1M_1}(NT,s,\lambda T)}\,\varphi_x(NT,s,+0)\,. \tag{12.125}
$$

12.9 Descending closed-loop systems

The block-diagram of the systems under consideration is shown in Fig. 12.10, where the open-loop system satisfies the conditions of Section 12.6. The system

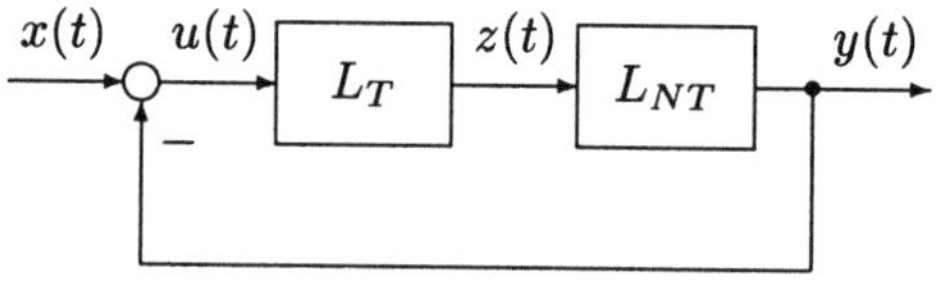

Figure 12.10: Descending closed-loop system

can be represented with operator equations of the form

$$
y = W_{d2}(s)\varphi_{F_2M_2}(NT,s,t)z\,, \quad z = W_{d1}(s)\varphi_{F_1M_1}(T,s,t)z\,, \quad u = x - y\,. \tag{12.126}
$$

Using the decomposition (12.89), the block-diagram shown in Fig. 12.10 can be transformed into the polyphase structure of Fig. 12.11.
Let $W_{al}(s,t) = W_{al}(s,t+NT)$ be the desired PTF of the closed-loop system and

$$
x(t) = \mathrm{e}^{st}\,, \quad y(t) = W_{al}(s,t)\mathrm{e}^{st}\,. \tag{12.127}
$$

Then, as follows from Fig. 12.11,

$$
u(t) = \mathrm{e}^{st} f_u(t) \tag{12.128}
$$

where

$$
f_u(t) = 1 - W_{al}(s,t) = f_u(t+T)\,.
$$

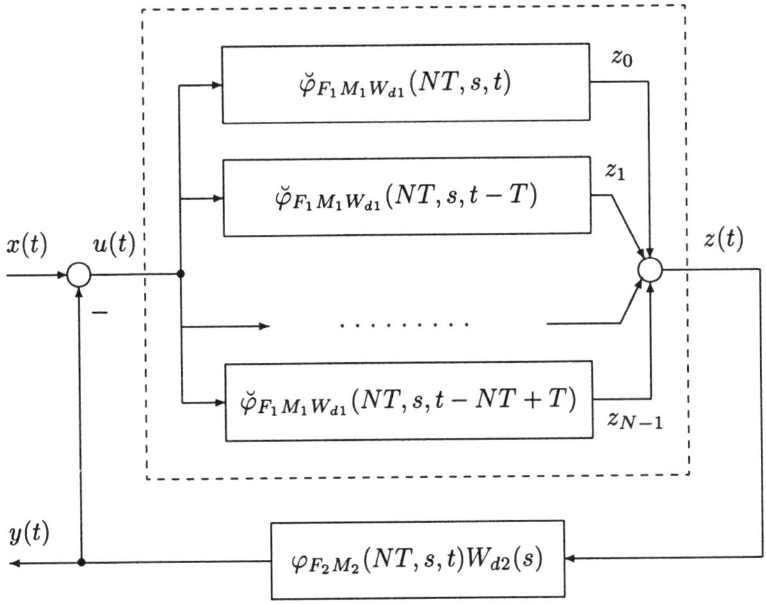

Figure 12.11: Polyphase structure for Fig. 12.10

Using Eq. (12.128) and taking the stroboscopic effect into account, from Fig. 12.11 we have

$$z_\lambda(t) = \mathrm{e}^{st}\breve{\varphi}_{F_1M_1W_{d1}}(NT,s,t-\lambda T)\,[1-W_{al}(s,\lambda T)]\;.$$

Hence,

$$y(t) = W_{d2}(s)\varphi_{F_2M_2}(NT,s,t)\sum_{\lambda=0}^{N-1}\breve{\varphi}_{F_1M_1W_{d1}}(NT,s,-\lambda T)\,[1-W_{al}(s,\lambda T)]\,\mathrm{e}^{st}.$$

Taking account of (3.123), the last relation appears as

$$y(t) = W_{d2}(s)\varphi_{F_2M_2}(NT,s,t)\mathrm{e}^{st} \, * \qquad (12.129)$$
$$* \left[W_{d1}(s)\varphi_{F_1M_1}(T,s,0) - \sum_{\lambda=0}^{N-1}\breve{\varphi}_{F_1M_1W_{d1}}(NT,s,-\lambda T)W_{al}(s,\lambda T)\right]\;.$$

From (12.127) and (12.129) we find a functional equation for the required PTF

$$W_{al}(s,t) = W_{d2}(s)\varphi_{F_2M_2}(NT,s,t) \, * \qquad (12.130)$$
$$* \left[W_{d1}(s)\varphi_{F_1M_1}(T,s,0) - \sum_{\lambda=0}^{N-1}\breve{\varphi}_{F_1M_1W_{d1}}(NT,s,-\lambda T)W_{al}(s,\lambda T)\right]\;.$$

Denote

$$A(s) \triangleq W_{d1}(s)\varphi_{F_1M_1}(T,s,0) - \sum_{\lambda=0}^{N-1} \breve{\varphi}_{F_1M_1W_{d1}}(NT,s,-\lambda T)W_{al}(s,\lambda T). \tag{12.131}$$

Then, Eq. (12.130) takes the form

$$W_{al}(s,t) = W_{d2}(s)\varphi_{F_2M_2}(NT,s,t)A(s). \tag{12.132}$$

Substituting (12.132) into the right-hand side of (12.131), we find

$$A(s) = W_{d1}(s)\varphi_{F_1M_1}(T,s,0) - \\ - A(s)W_{d2}(s)\sum_{\lambda=0}^{N-1} \breve{\varphi}_{F_1M_1W_{d1}}(NT,s,-\lambda T)\varphi_{F_2M_2}(NT,s,\lambda T).$$

Then,

$$A(s) = \frac{W_{d1}(s)\varphi_{F_1M_1}(T,s,0)}{1 + W_{d2}(s)\sum\limits_{\lambda=0}^{N-1} \breve{\varphi}_{F_1M_1W_{d1}}(NT,s,-\lambda T)\varphi_{F_2M_2}(NT,s,\lambda T)}. \tag{12.133}$$

Substituting (12.133) into (12.132), we find

$$W_{al}(s,t) = \frac{W_{d1}(s)W_{d2}(s)\varphi_{F_1M_1}(T,s,0)\varphi_{F_2M_2}(NT,s,t)}{1 + W_{d2}(s)\sum\limits_{\lambda=0}^{N-1} \breve{\varphi}_{F_1M_1W_{d1}}(NT,s,-\lambda T)\varphi_{F_2M_2}(NT,s,\lambda T)}. \tag{12.134}$$

Next, we determine the Laplace transform of the output. We notice that the denominator in (12.134) is periodic with respect to s with period $\mathrm{j}\omega/N$, because of $W_{d2}(s) = W_{d2}(s+\mathrm{j}\omega/N)$ and

$$\sum_{\lambda=0}^{N-1} \breve{\varphi}_{F_1M_1W_{d1}}(NT,s,-\lambda T)\varphi_{F_2M_2}(NT,s,\lambda T) = \\ = \sum_{\lambda=0}^{N-1} \breve{\mathcal{D}}_{F_1M_1W_{d1}}(NT,s,-\lambda T)\mathcal{D}_{F_2M_2}(NT,s,\lambda T).$$

Therefore, using (12.92) and (12.93), we obtain

$$W_{al}(s,t) = \sum_{\lambda=0}^{N-1} \breve{W}_{\lambda a}(s,t) \tag{12.135}$$

where

$$\breve{W}_{\lambda a}(s,t) = \frac{1}{NT}\sum_{k=-\infty}^{\infty} \breve{Q}_{\lambda a}\left(s+\frac{k\mathrm{j}\omega}{N}\right)\mathrm{e}^{\frac{k\mathrm{j}\omega}{N}(t-\lambda T)} \tag{12.136}$$

and

$$\breve{Q}_{\lambda a}(s) = \frac{F_2(s)M_2(s)W_{d2}(s)\breve{\varphi}_{F_1M_1W_{d1}}(NT,s,-\lambda T)}{1+W_{d2}(s)\sum\limits_{\lambda=0}^{N-1}\breve{\varphi}_{F_1M_1W_{d1}}(NT,s,-\lambda T)\varphi_{F_2M_2}(NT,s,\lambda T)}\,. \tag{12.137}$$

Relations (12.136) and (12.137) are equivalent to an open-loop SD system with ETF (12.137) and phase shift λT. Let μ_i, $(i = 1,\dots,r)$ be the set of characteristic indices of the function $Q_{\lambda a}(s)$. Then, for $x(t) \in \Lambda(\mu_i, \mu_{i+1})$ the Laplace transform $Y_\lambda(s)$ of the output of the system with PTF (12.136) takes the form

$$Y_\lambda(s) = \breve{Q}_{\lambda a}(s)\varphi_x(NT,s,\lambda T+0)\,. \tag{12.138}$$

Summing up Eq. (12.138) for different λ, we obtain the Laplace transform of the output

$$Y(s) = \sum_{\lambda=0}^{N-1}\breve{Q}_{\lambda a}(s)\varphi_x(NT,s,\lambda T+0)\,. \tag{12.139}$$

12.10 Standard multirate systems

In this section we construct the PTF for a generalized structure shown in Fig. 12.12, Lampe and Rosenwasser (1997a). Here $\boldsymbol{P}$ is a generalized plant

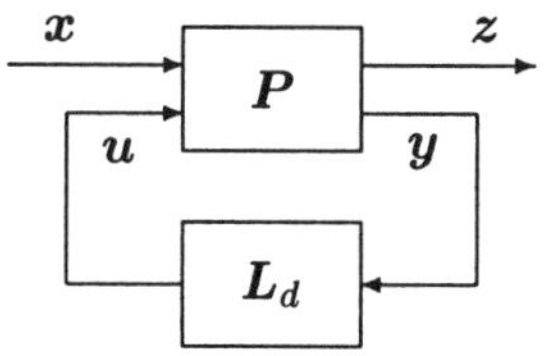

Figure 12.12: Standard multirate system

with the transfer matrix (11.131), and $\boldsymbol{L}_d$ is a generalized discrete controller having the structure shown in Fig. 12.13, where $\varphi_{\boldsymbol{Q}_i}(T,s,t)$ are the PTMs of

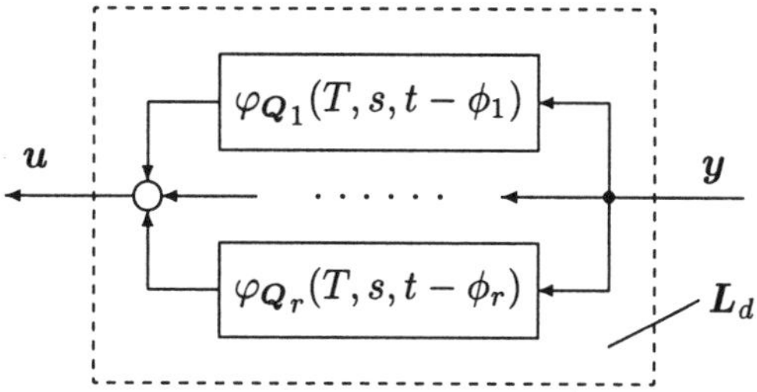

Figure 12.13: Structure of generalized multirate controller

SD systems with period T, having equivalent transfer functions (matrices) $\boldsymbol{Q}_i$ and phase shifts ϕ_i, $(i = 1,\dots,r)$. By polyphase decomposition we are able to reduce any system with periodic discrete controllers, having commensurate sampling periods T_κ. In this case, the period T of the system equals the least common multiple of the periods T_κ. The detailed structure associated with Fig. 12.12 and and 12.13 is shown in Fig. 12.14.

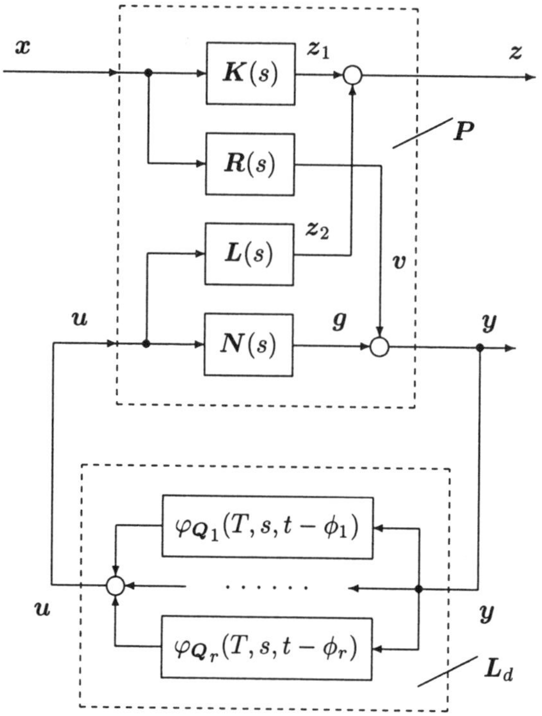

Figure 12.14: Detailed structure of generalized multirate system

The following theorem determines the PTM of the system shown in Fig. 12.14.

Theorem 12.1 *Let the series $\varphi_{\boldsymbol{LQ}_i}(T,s,t)$ and $\varphi_{\boldsymbol{NQ}_i}(T,s,t)$ converge and the matrices $\varphi_{\boldsymbol{NQ}_i}(T,s,t)$ be continuous with respect to t. Let the matrix system of equations*

$$\boldsymbol{W}_\mu(s) = I + \sum_{\lambda=1}^{r} \varphi_{\boldsymbol{NQ}_\lambda}(T,s,\phi_\mu - \phi_\lambda)\boldsymbol{W}_\lambda(s) \tag{12.140}$$

have a unique solution. Then, there exists the PTM $\boldsymbol{W}_{\boldsymbol{zx}}(s,t)$ of the system from the input $\boldsymbol{x}(t)$ to the output $\boldsymbol{z}(t)$, which is unique and given by

$$\boldsymbol{W}_{\boldsymbol{zx}}(s,t) = \boldsymbol{K}(s) + \sum_{\lambda=1}^{r} \varphi_{\boldsymbol{LQ}_\lambda}(T,s,t-\phi_\lambda)\boldsymbol{W}_\lambda(s)\boldsymbol{R}(s)\,. \tag{12.141}$$

Proof 1. Let

$$v = \boldsymbol{I}\mathrm{e}^{st}, \quad \boldsymbol{y} = \boldsymbol{W}_{\boldsymbol{yv}}(s,t)\mathrm{e}^{st}, \quad \boldsymbol{W}_{\boldsymbol{yv}}(s,t) = \boldsymbol{W}_{\boldsymbol{yv}}(s,t+T). \tag{12.142}$$

By the above assumptions, the matrix $\boldsymbol{W}_{\boldsymbol{yv}}(s,t)$ is continuous with respect to t. Therefore, using the stroboscopic effect, we have

$$\boldsymbol{u}(t) = \sum_{\lambda=1}^{r} \varphi_{\boldsymbol{Q}_\lambda}(T,s,t-\phi_\lambda)\boldsymbol{W}_{\boldsymbol{yv}}(s,\phi_\lambda)\mathrm{e}^{st}. \tag{12.143}$$

Hence,

$$\boldsymbol{g}(t) = \sum_{\lambda=1}^{r} \varphi_{\boldsymbol{NQ}_\lambda}(T,s,t-\phi_\lambda)\boldsymbol{W}_{\boldsymbol{yv}}(s,\phi_\lambda)\mathrm{e}^{st}. \tag{12.144}$$

From (12.144) and Fig. 12.14 we find

$$\boldsymbol{y}(t) = \left[\sum_{\lambda=1}^{r} \varphi_{\boldsymbol{NQ}_\lambda}(T,s,t-\phi_\lambda)\boldsymbol{W}_{\boldsymbol{yv}}(s,\phi_\lambda) + \boldsymbol{I}\right]\mathrm{e}^{st}. \tag{12.145}$$

Comparing Eqs. (12.142) and (12.145) for $\boldsymbol{y}(t)$, we obtain

$$\boldsymbol{W}_{\boldsymbol{yv}}(s,t) = \sum_{\lambda=1}^{r} \varphi_{\boldsymbol{NQ}_\lambda}(T,s,t-\phi_\lambda)\boldsymbol{W}_{\boldsymbol{yv}}(s,\phi_\lambda) + \boldsymbol{I}. \tag{12.146}$$

2. Substituting $t = \phi_\mu$ $(\mu = 1,\dots,r)$ into (12.146), we obtain a system of the form (12.140)

$$\boldsymbol{W}_{\boldsymbol{yv}}(s,\phi_\mu) = I + \sum_{\lambda=1}^{r} \varphi_{\boldsymbol{NQ}_\lambda}(T,s,\phi_\mu-\phi_\lambda)\boldsymbol{W}_{\boldsymbol{yv}}(s,\phi_\lambda). \tag{12.147}$$

By the conditions of the theorem, the system (12.147) has a unique solution

$$\boldsymbol{W}_\mu(s) \triangleq \boldsymbol{W}_{\boldsymbol{yv}}(s,\phi_\mu), \quad (\mu = 1,\dots,r). \tag{12.148}$$

Substituting (12.148) into (12.143), we have

$$\boldsymbol{u}(t) = \sum_{\lambda=1}^{r} \varphi_{\boldsymbol{Q}_\lambda}(T,s,t-\phi_\lambda)\boldsymbol{W}_\lambda(s)\mathrm{e}^{st}. \tag{12.149}$$

3. From (12.149) and Fig. 12.14 it follows that, under the condition (12.142),

$$z_2(t) = \sum_{\lambda=1}^{r} \varphi_{\boldsymbol{LQ}_\lambda}(T,s,t-\phi_\lambda)\boldsymbol{W}_\lambda(s)\mathrm{e}^{st}. \tag{12.150}$$

Hence, with reference to Fig. 12.14, for $\boldsymbol{x}(t) = \boldsymbol{I}\mathrm{e}^{st}$ we find

$$z_2(t) = \sum_{\lambda=1}^{r} \varphi_{\boldsymbol{LQ}_\lambda}(T, s, t - \phi_\lambda) \boldsymbol{W}_\lambda(s) \boldsymbol{R}(s) \mathrm{e}^{st} .$$

Therefore, for $x(t) = \boldsymbol{I}\mathrm{e}^{st}$ we have

$$z(t) = z_1(t) + z_2(t) = \left[\sum_{\lambda=1}^{r} \varphi_{\boldsymbol{LQ}_\lambda}(T, s, t - \phi_\lambda) \boldsymbol{W}_\lambda(s) \boldsymbol{R}(s) + \boldsymbol{K}(s) \right] \mathrm{e}^{st}$$

Multiplying the last equation by e^{-st}, we obtain (12.141). ∎

Corollary If all components of the discrete controller operate without a phase shift, we have

$$\boldsymbol{u} = \varphi_{\boldsymbol{Q}}(T, s, t) \boldsymbol{y} .$$

In this case, the system (12.140) reduces to the single equation

$$\boldsymbol{W}_1(s) = \boldsymbol{I} + \varphi_{\boldsymbol{NQ}}(T, s, 0) \boldsymbol{W}_1(s)$$

which yields

$$\boldsymbol{W}_1(s) = \left[\boldsymbol{I} - \varphi_{\boldsymbol{NQ}}(T, s, 0) \right]^{-1} .$$

Then, the PTM of the system is given by the expression

$$\boldsymbol{W}_{\boldsymbol{zx}}(s, t) = \boldsymbol{K}(s) + \varphi_{\boldsymbol{LQ}}(T, s, t) \left[\boldsymbol{I} - \varphi_{\boldsymbol{NQ}}(T, s, 0) \right]^{-1} \boldsymbol{R}(s)$$

which gives Eqs. (11.132) and (11.133) as special cases.

Part IV

Analysis of SD systems in continuous time

Introduction In this part we consider a calculation technique for the responses of linear periodic sampled-data systems to deterministic and stochastic disturbances, investigating steady-state and transient processes. In all cases the problem reduces to the determination of the inverse Laplace transform for a discrete Laplace transform in continuous-time. For this purpose we use the formulas obtained in Chapter 4. The practical calculation of the inversion integrals is performed, using theorems of Sections 4.4 and 4.5, by contour integration in the $s-$plane. This technique allows us to obtain closed expressions for the output of the system in continuous-time.

Chapter 13

Analysis of open-loop SD systems under deterministic disturbances

13.1 Transient processes in open-loop systems

As one of the simplest examples of the application of the proposed method, we consider the elementary open-loop sampled-data system shown in Fig. 13.1, where U_a is the sampling unit (sample and hold) (9.6) and $F(s)$ is the transfer

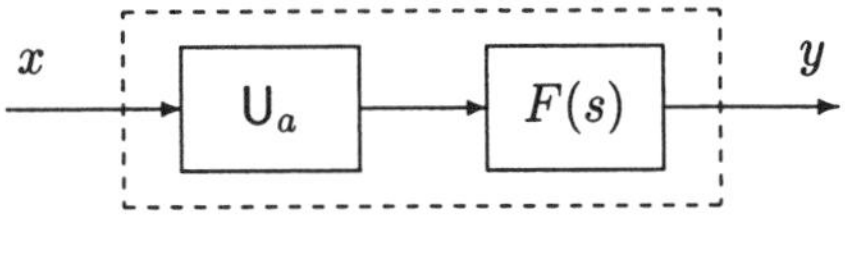

Figure 13.1: Elementary open-loop SD system

function of the continuous-time network, which possesses the form

$$F(s) = \sum_{i=1}^{\ell} \sum_{k=1}^{\nu_i} \frac{f_{ik}}{(s - s_i)^k} \,. \tag{13.1}$$

We shall assume that the exogenous disturbence takes the form

$$x(t) \triangleq x_m(t) = \begin{cases} \frac{1}{(m-1)!} t^{m-1} e^{bt} & t > 0 \\ 0 & t < 0 \end{cases} \tag{13.2}$$

where $m \geq 1$ is an integer and b is a real or complex constant. In practice, almost any exogenous signal encountered in applications can be approximated by a sum of signals of the form (13.2).

The Laplace transform for the signal (13.2) has the form

$$X(s) \triangleq X_m(s) = \frac{1}{(s-b)^m}, \quad \operatorname{Re} s > b. \tag{13.3}$$

The DPFR of the image (13.3) is given by the Fourier series

$$\varphi_{X_m}(T,s,t) = \frac{1}{T}\sum_{k=-\infty}^{\infty} \frac{e^{kj\omega t}}{(s+kj\omega-b)^m}, \quad \omega = \frac{2\pi}{T}. \tag{13.4}$$

For $\operatorname{Re} s > b$, the series (13.4) coincides with the Fourier series for the DPFR

$$\varphi_{x_m}(T,s,t) = \sum_{k=-\infty}^{\infty} x_m(t+kT)e^{-s(t+kT)} \tag{13.5}$$

and, due to (3.90), we have

$$\varphi_{x_m}(T,s,t) = \frac{1}{(m-1)!}\frac{\partial^{m-1}}{\partial b^{m-1}}\frac{e^{(b-s)t}}{1-e^{(b-s)T}}, \quad 0 < t < T. \tag{13.6}$$

For $m > 1$, the function $\varphi_{x_m}(T,s,t)$ is continuous on the interval $-\infty < t < \infty$. For $m = 1$ it has discontinuities at $t = qT$, where q is an integer. Moreover, for any $m \geq 1$, we have

$$\varphi_{x_m}(T,s,+0) = \frac{1}{(m-1)!}\frac{\partial^{m-1}}{\partial b^{m-1}}\frac{1}{1-e^{(b-s)T}} = \mathcal{D}_{x_m}(T,s,+0). \tag{13.7}$$

Let μ_r be the largest characteristic index of the function $F(s)$. Then, for $\operatorname{Re} s > \max\{\mu_r, \operatorname{Re} b\} \triangleq \lambda$ the Laplace transform of the output is given by Eq. (9.181), where $m_0(t)$ is an arbitrary polynomial that is determined by the initial conditions. For $m_0(s) = 0$, we have processes with zero initial energy with the image

$$Y(s) = F(s)M(s)\mathcal{D}_{x_m}(T,s,+0) = F(s)M(s)\varphi_{x_m}(T,s,+0). \tag{13.8}$$

As was proved in the Appendix, the rational periodic function $\mathcal{D}_{x_m}(T,s,+0)$ is uniformly bounded in the half-plane $\operatorname{Re} s > \lambda$. Therefore, in the half-plane of convergence $|Y(s)|$ decreases as $|s|^{-2}$ for $|\operatorname{Im} s| \to \infty$. Therefore, due to Theorem 1.4,

$$y(t) = \frac{1}{2\pi j}\int_{c-j\infty}^{c+j\infty} F(s)M(s)\mathcal{D}_{x_m}(T,s,+0)e^{st}\,ds, \quad c > \lambda \tag{13.9}$$

where the integral converges absolutely. The function $y(t)$ is continuous, and the analyticity of $Y(s)$ in the half-plane $\operatorname{Re} s > \lambda$ ensures that $y(t) = 0$ for $t < 0$. Discretizing the integral (13.9) with step $j\omega$ (this is correct due to its absolute convergence), we have

$$y(t) = \frac{T}{2\pi \mathrm{j}} \int_{c-\mathrm{j}\omega/2}^{c+\mathrm{j}\omega/2} \mathcal{D}_{FM}(T,s,t)\mathcal{D}_{x_m}(T,s,+0)\,\mathrm{d}s \tag{13.10}$$

where

$$\mathcal{D}_{FM}(T,s,t) = \frac{1}{T} \sum_{k=-\infty}^{\infty} F(s+k\mathrm{j}\omega)M(s+k\mathrm{j}\omega)\mathrm{e}^{(s+k\mathrm{j}\omega)t}\,. \tag{13.11}$$

Closed expressions for the sum of the series (13.11) are given in Section 9.6. For further transformation of the integral (13.10) we use Eq. (9.132)

$$\mathcal{D}_{FM}(T,s,t) = D_p(s,t) = \sum_{i=1}^{\ell} \sum_{k=1}^{\nu_i} \frac{f_{ik}}{(k-1)!} \frac{\partial^{k-1}}{\partial s_i^{k-1}} D_{p1}(s,s_i,t) \tag{13.12}$$

where $D_{p1}(s,a,t)$ is the sum of the series

$$D_{p1}(s,a,t) = \frac{1}{T} \sum_{k=-\infty}^{\infty} \frac{M(s+k\mathrm{j}\omega)}{s+k\mathrm{j}\omega - a}\, \mathrm{e}^{(s+k\mathrm{j}\omega)t} \tag{13.13}$$

which is given, for $0 \le t \le T$, by the equivalent formulae (9.129)

$$\begin{aligned} D_{p1}(s,a,t) &= \frac{M(a)\mathrm{e}^{at}}{\mathrm{e}^{(s-a)T}-1} + \int_0^t \mathrm{e}^{a(t-\tau)}\mu(\tau)\,\mathrm{d}\tau \\ D_{p1}(s,a,t) &= \frac{M(a)\mathrm{e}^{at}}{1-\mathrm{e}^{(s-a)T}} - \int_t^T \mathrm{e}^{a(t-\tau)}\mu(\tau)\,\mathrm{d}\tau\,. \end{aligned} \tag{13.14}$$

Substituting (13.12) into (13.10) and changing the order of integration and differentiation (this is correct in the given case), we obtain

$$y(t) = \sum_{i=1}^{\ell} \sum_{k=1}^{\nu_i} \frac{f_{ik}}{(k-1)!} \frac{\partial^{k-1}}{\partial s_i^{k-1}} y_{1m}(t,s_i,b) \tag{13.15}$$

with

$$y_{1m}(t,a,b) \triangleq \frac{T}{2\pi \mathrm{j}} \int_{c-\mathrm{j}\omega/2}^{c+\mathrm{j}\omega/2} D_{p1}(s,a,t)\mathcal{D}_{x_m}(T,s,+0)\,\mathrm{d}s\,, \quad c > \lambda\,. \tag{13.16}$$

From (13.15) it follows that for the calculation of $y(t)$ it suffices to calculate the integrals (13.16). This problem, in its turn, can be reduced to the calculation of a still simpler integral. With this aim in view, we substitute (13.7) into (13.16), so that

$$y_{1m}(t,a,b) = \frac{1}{(m-1)!} \frac{\partial^{m-1}}{\partial b^{m-1}} y_{11}(t,a,b) \tag{13.17}$$

with

$$y_{11}(t,a,b) \triangleq \frac{T}{2\pi \mathrm{j}} \int_{c-\mathrm{j}\omega/2}^{c+\mathrm{j}\omega/2} \frac{D_{p1}(s,a,t)}{1-\mathrm{e}^{(b-s)T}}\,\mathrm{d}s\,, \quad c > \max\{\operatorname{Re} a, \operatorname{Re} b\}\,. \tag{13.18}$$

Taking account of the above relations, the general formula (13.15) can be represented in the form

$$y(t) = \sum_{i=1}^{\ell} \sum_{k=1}^{\nu_i} \frac{f_{ik}}{(k-1)!(m-1)!} \frac{\partial^{k-1}}{\partial s_i^{k-1}} \frac{\partial^{m-1}}{\partial b^{m-1}} y_{11}(t, s_i, b)\,. \tag{13.19}$$

Thus, the problem of calculation of the output by Eq. (13.10) is reduced to the calculation of a standard integral of the form (13.18). From (13.14), for $0 \le t \le T$ we have

$$\begin{aligned} \ell^{+}[D_{p1}(s,a,t)] &= \int_0^t \mathrm{e}^{a(t-\tau)} \mu(\tau)\,\mathrm{d}\tau \\ \ell^{-}[D_{p1}(s,a,t)] &= -\int_t^T \mathrm{e}^{a(t-\tau)} \mu(\tau)\,\mathrm{d}\tau\,. \end{aligned} \tag{13.20}$$

With the notation

$$f(s,t,a,b) \triangleq \frac{D_{p1}(s,a,t)}{1-\mathrm{e}^{(b-s)T}}\,. \tag{13.21}$$

we get, for $0 \le t \le T$,

$$\ell^{+}[f(s,t,a,b)] = \int_0^t \mathrm{e}^{a(t-\tau)} \mu(\tau)\,\mathrm{d}\tau\,, \qquad \ell^{-}[f(s,t,a,b)] = 0\,. \tag{13.22}$$

Since $f(s,t+T,a,b) = f(s,t,a,b)\mathrm{e}^{sT}$, we also have

$$\ell^{-}[f(s,t,a,b)] = 0\,, \quad t \ge 0\,. \tag{13.23}$$

Therefore, using (A.83), from (13.18) we have

$$y_{11}(t,a,b) = T \sum_i \operatorname*{Res}_{\tilde{s}_i} f(s,t,a,b) \tag{13.24}$$

where the $\tilde{s}_i$ are the primary poles of the function (13.21). Since the function (13.21) is periodic with respect to s and b with period $\mathrm{j}\omega$, we can, without loss of generality, assume that the function $f(s,t,a,b)$ has primary poles at $s = a$ and $s = b$. The final result remains valid irrespectively of this assumption.

For practical calculation of the integral (13.18) we have to consider two variants. If $\mathrm{e}^{aT} \ne \mathrm{e}^{bT}$ in (13.21), we shall say that we have the non-resonant case. In this case the function (13.21) has two primary poles $s = a$ and $s = b$. Assume that the non-pathological condition $M(a) \ne 0$ holds. Then, as follows from (13.13)

$$\operatorname*{Res}_{a} f(s,t,a,b) = \frac{1}{T} \frac{M(a)\mathrm{e}^{at}}{1-\mathrm{e}^{(b-a)T}} \ne 0\,. \tag{13.25}$$

The residue at the pole $s = b$ is given by

$$\operatorname*{Res}_{b} f(s,t,a,b) = \frac{1}{T} D_{p1}(b,a,t) \,. \tag{13.26}$$

Substituting (13.25) and (13.26) into (13.24), we find a closed expression for the integral (13.18)

$$y_{11}(t,a,b) = \frac{M(a)e^{at}}{1 - e^{(b-a)T}} + D_{p1}(b,a,t) \,, \quad t \geq 0 \tag{13.27}$$

where, by virtue of (13.14), for $0 \leq t \leq T$ we have

$$D_{p1}(b,a,t) = \frac{M(a)e^{at}}{e^{(b-a)T} - 1} + \int_0^t e^{a(t-\tau)} \mu(\tau) \, d\tau$$

$$D_{p1}(b,a,t) = \frac{M(a)e^{at}}{1 - e^{(b-a)T}} - \int_t^T e^{a(t-\tau)} \mu(\tau) \, d\tau \,.$$

Example 13.1 Let us consider the transient process for vanishing initial energy in the system of Fig. 13.1 with

$$F(s) = \frac{1}{s-a} \,, \quad M(s) = \frac{1 - e^{-sT}}{s} \,, \quad a < 0 \tag{13.28}$$

and the input signal $x(t)$ has the form (1.42), so that $b = 0$ and $X(s) = s^{-1}$. From (13.27) for $b = 0$ and $t \geq 0$ we find

$$y(t) = \frac{M(a)e^{at}}{1 - e^{-aT}} + D_{p1}(0,a,t) = \frac{e^{at}}{a} + D_{p1}(0,a,t) \,. \tag{13.29}$$

From (13.13) for $s = 0$ we have

$$D_{p1}(0,a,t) = \frac{1}{T} \sum_{k=-\infty}^{\infty} \frac{M(kj\omega)}{kj\omega - a} e^{kj\omega t} \,.$$

But, as follows from (13.28),

$$M(kj\omega) = 0 \,, \quad (k \neq 0) \,; \quad M(0) = T \,. \tag{13.30}$$

Therefore, $D_{p1}(0,a,t) = a^{-1}$, and from (13.29) we find

$$y(t) = \frac{e^{at} - 1}{a} \,. \tag{13.31}$$

This result has a simple physical meaning: If the input of a zero-order hold is acted upon by a signal of the form (1.42), and the input values after the break are fixed, the input of the continuous-time part is affected, for $t \geq 0$, by the same signal (1.42). The corresponding response of the continuous-time network with zero initial energy has the form (13.31). □

Example 13.2 Consider the transient process for vanishing initial conditions in the system of Example 13.1 with

$$x(t) = \begin{cases} t & t > 0 \\ 0 & t < 0 \end{cases} \tag{13.32}$$

i.e., $m = 2$ and $X(s) = X_2(s) = s^{-2}$. Then, using (13.19), we have

$$y(t) = \lim_{b \to 0} \frac{\partial y_{11}(t, a, b)}{\partial b}. \tag{13.33}$$

But, by Eqs. (13.27) and (9.115), for $qT \le t \le (q+1)T$ we have

$$y_{11}(t, a, b) = \frac{\left(1 - e^{-aT}\right) e^{at}}{a\left[1 - e^{(b-a)T}\right]} + \left[\frac{1 - e^{-aT}}{a} \frac{e^{a(t-qT)}}{e^{(b-a)T} - 1} + \frac{e^{a(t-qT)} - 1}{a}\right] e^{qbT}. \tag{13.34}$$

From (13.33) and (13.34) it follows that

$$y(t) = \frac{qT}{a} + \frac{Te^{a(t-T)}\left(1 - e^{-qaT}\right)}{a\left(1 - e^{-aT}\right)}, \quad qT \le t \le (q+1)T. \tag{13.35}$$

Figure 13.2 presents the processes computed by this formula for $T = 1$, $a = -1$ and $a = -2$. □

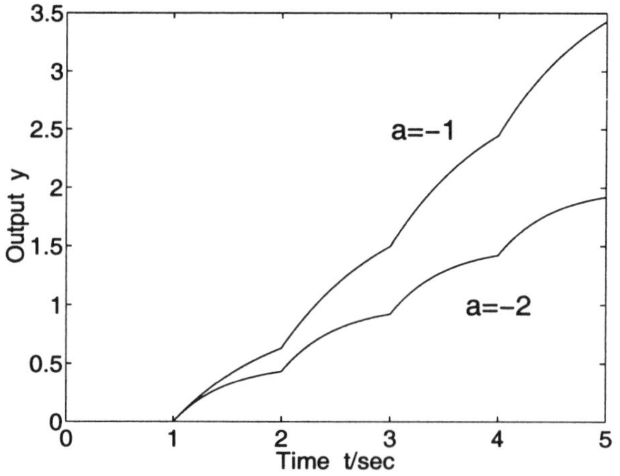

Figure 13.2: Processes for Example 13.2

13.2 Processes under resonance

We shall say that there is *resonance* in the system shown in Fig. 13.1, if $e^{aT} = e^{bT}$. In this case we can assume that the integrand in (13.18) has a multiple primary pole at the point $s = a$, and all relations of Section (13.1) lose their sense. Therefore, to construct the processes under resonance, we have to pass to the limit in formulae of Section 13.1. This technique will be demonstrated for the function

$$y_{nm}(t) = \lim_{e^{bT} \to e^{aT}} \frac{1}{(n-1)!} \frac{1}{(m-1)!} \frac{\partial^{n-1}}{\partial a^{n-1}} \frac{\partial^{m-1}}{\partial b^{m-1}} y_{11}(t, a, b) . \tag{13.36}$$

We note, that for $qT \le t \le (q+1)T$ the first equation in (13.14) yields

$$D_{p1}(s, a, t) = \left[-\frac{M(a) e^{a(t-qT)}}{1 - e^{(s-a)T}} + e^{-qaT} \int_0^{t-qT} e^{a(t-\tau)} \mu(\tau) \, d\tau \right] e^{qsT} . \tag{13.37}$$

Hence, for $s = b$ we find

$$D_{p1}(b, a, t) = \left[-\frac{M(a) e^{a(t-qT)}}{1 - e^{(b-a)T}} + e^{-qaT} \int_0^{t-qT} e^{a(t-\tau)} \mu(\tau) \, d\tau \right] e^{qbT} . \tag{13.38}$$

Substituting (13.38) into (13.27), we have

$$y_{11}(t, a, b) = \frac{M(a) e^{at} \left[1 - e^{q(b-a)T} \right]}{1 - e^{(b-a)T}} + e^{q(b-a)T} \int_0^{t-qT} e^{a(t-\tau)} \mu(\tau) \, d\tau . \tag{13.39}$$

Taking into account that for $q > 0$ we have

$$\frac{1 - e^{q(b-a)T}}{1 - e^{(b-a)T}} = \sum_{k=0}^{q-1} e^{k(b-a)T} \tag{13.40}$$

we find from Eq. (13.39)

$$y_{11}(t, a, b) = M(a) e^{at} \sum_{k=0}^{q-1} e^{k(b-a)T} + e^{q(b-a)T} \int_0^{t-qT} e^{a(t-\tau)} \mu(\tau) \, d\tau . \tag{13.41}$$

Substituting this relation into (13.36), we find that under resonance for $qT \le t \le (q+1)T$

$$y_{nm}(t) = \frac{1}{(n-1)!} \frac{1}{(m-1)!} \frac{\partial^{n-1}}{\partial a^{n-1}} \frac{\partial^{m-1}}{\partial a^{m-1}} \tag{13.42}$$

$$\left. \left[M(a) e^{at} \sum_{k=0}^{q-1} e^{k(b-a)T} + e^{q(b-a)T} \int_0^{t-qT} e^{a(t-\tau)} \mu(\tau) \, d\tau \right] \right|_{e^{bT} = e^{aT}} .$$

The substitution $e^{bT} = e^{aT}$ is possible, because the expression inside the brackets is analytic for all a and b.

Example 13.3 Let us consider the transient process for vanishing initial conditions in the open-loop system with

$$F(s) = \frac{1}{s-a}, \quad X(s) = \frac{1}{s-a}$$

where the form of the control pulses is arbitrary. In this case we have $m = 1$, $n = 1$, and Eq. (13.41) yields

$$y_{11}(t,a,b)\,|_{e^{bT}=e^{aT}} = M(a)e^{at}q + \int\limits_0^{t-qT} e^{a(t-\tau)}\mu(\tau)\,d\tau\,, \quad qT \le t \le (q+1)T\,.$$

In the special case, when $a = b = 0$, i.e., $F(s) = s^{-1}$ and $X(s) = s^{-1}$, we obtain

$$y_{11}(t,a,b)\,|_{a=b=0} = M(0)q + \int_0^{t-qT} \mu(\tau)\,d\tau\,, \quad qT \le t \le (q+1)T\,. \qquad (13.43)$$

Figure 13.3 presents the process computed by (13.43) for $T = 1$ and

$$\mu(t) = \begin{cases} 1 & 0 < t < 0.5T \\ 0 & 0.5T < t < T\,. \end{cases}$$

Then,

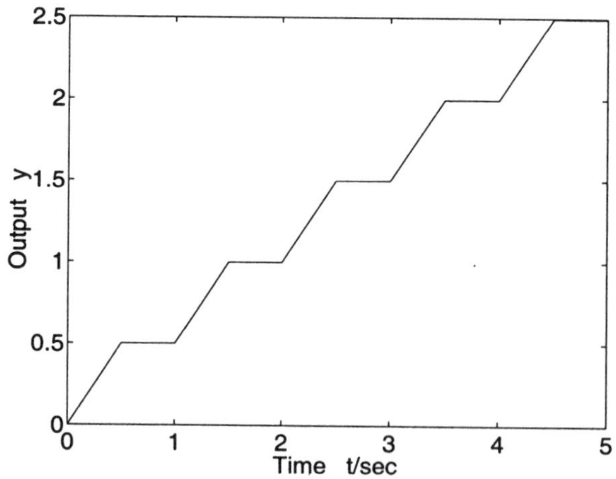

Figure 13.3: Process under resonance

$$M(s) = \frac{1 - e^{-0.5sT}}{s}, \quad M(0) = 0.5T\,.$$

As is evident from Fig. 13.3, the output process increases infinitely for a bounded disturbance. This fact characterizes the resonant case. □

13.3 Transients in open-loop SD systems

In this section we consider processes with zero initial energy in the open-loop digital system shown in Fig. 10.5. We consider the processes for $t > 0$, assuming that the input signal has the form (13.2). Due to the choice of initial conditions, for $b_0(s) = 0$ and $y_0(t) = 0$ Eq. (10.86) yields the discrete Laplace transform of the output

$$\mathcal{D}_y(T, s, t) = \mathcal{D}_{FM}(T, s, t)W_d(s)\mathcal{D}_{x_m}(T, s, +0) \stackrel{\triangle}{=} f_d(s, t) \,. \tag{13.44}$$

The output process is determined by the inversion formula

$$y(t) = \frac{T}{2\pi \mathrm{j}} \int_{c-\mathrm{j}\omega/2}^{c+\mathrm{j}\omega/2} f_d(s, t) \, \mathrm{d}s \,, \quad c > \max\{\mu_r, \operatorname{Re} b\} \tag{13.45}$$

where μ_r is the largest characteristic index of the system. If the transfer function of the continuous-time part has the form (13.1), by analogy with (13.19) it can be shown that

$$y(t) = \frac{1}{(m-1)!} \frac{\partial^{m-1}}{\partial b^{m-1}} \sum_{i=1}^{\ell} \sum_{k=1}^{\nu_i} \frac{f_{ik}}{(k-1)!} \frac{\partial^{k-1}}{\partial s_i^{k-1}} \frac{T}{2\pi \mathrm{j}} \int_{c-\mathrm{j}\omega/2}^{c+\mathrm{j}\omega/2} \frac{D_{p1}(s, s_i, t)W_d(s)}{1 - \mathrm{e}^{(b-s)T}} \, \mathrm{d}s \tag{13.46}$$

where the function $D_{p1}(s, a, t)$ is given by (13.14). Hence, it is necessary to consider the calculation of integrals of the form

$$\tilde{y}_{11}(t, a, b) \stackrel{\triangle}{=} \frac{T}{2\pi \mathrm{j}} \int_{c-\mathrm{j}\omega/2}^{c+\mathrm{j}\omega/2} \frac{D_{p1}(s, a, t)W_d(s)}{1 - \mathrm{e}^{(b-s)T}} \, \mathrm{d}s \tag{13.47}$$

where all poles of the integrand are located in the half-plane $\operatorname{Re} s < c$. Assuming that the control program is causal and using (A.27), we can find

$$W_d(s) = l_\kappa \mathrm{e}^{-\kappa s T} + \ldots + l_1 \mathrm{e}^{-sT} + l_0 + \tilde{W}_{d1}(s) \tag{13.48}$$

where l_i $(i = 0, 1, \ldots, \kappa)$ are constants, and the function $\tilde{W}_{d1}(s)$ is limited, so that

$$\ell^- \left[\tilde{W}_{d1}(s)\right] = 0 \,. \tag{13.49}$$

Substituting (13.48) into (13.47), we find

$$\tilde{y}_{11}(t, a, b) = \sum_{\mu=1}^{\kappa} l_\mu y_\mu(t, a, b) + \tilde{y}_d(t, a, b) \tag{13.50}$$

where

$$y_\mu(t, a, b) \stackrel{\triangle}{=} \frac{T}{2\pi \mathrm{j}} \int_{c-\mathrm{j}\omega/2}^{c+\mathrm{j}\omega/2} \frac{D_{p1}(s, a, t)\mathrm{e}^{-\mu s T}}{1 - \mathrm{e}^{(b-s)T}} \, \mathrm{d}s \,, \tag{13.51}$$

$$\tilde{y}_d(t,a,b) \triangleq \frac{T}{2\pi \mathrm{j}} \int_{c-\mathrm{j}\omega/2}^{c+\mathrm{j}\omega/2} \frac{D_{p1}(s,a,t)\tilde{W}_{d1}(s)}{1-\mathrm{e}^{(b-s)T}}\,\mathrm{d}s\,. \tag{13.52}$$

As a result, the problem reduces to the calculation of integrals of the form (13.51) and (13.52). The functions $y_\mu(t,a,b)$ appearing in (13.51) can be obtained directly by the formulae of Sections 13.1. Indeed, since

$$D_{p1}(s,a,t)\mathrm{e}^{-\mu sT} = D_{p1}(s,a,t-\mu T)$$

we have

$$y_\mu(t,a,b) = \frac{T}{2\pi \mathrm{j}} \int_{c-\mathrm{j}\omega/2}^{c+\mathrm{j}\omega/2} \frac{D_{p1}(s,a,t-\mu T)}{1-\mathrm{e}^{(b-s)T}}\,ds = y_{11}(t-\mu T,a,b) \tag{13.53}$$

where the function $y_{11}(t,a,b)$ is defined by (13.27) and (13.39).
Then, we calculate the integral (13.52). To simplify the derivation, we assume that all poles of the function $\tilde{W}_d(s)$ are simple. Then, using (13.49) and (A.60), we have

$$\tilde{W}_{d1}(s) = T\sum_{i=1}^{\rho} \frac{c_i}{1-\mathrm{e}^{(r_i-s)T}} \tag{13.54}$$

where the r_i $(i=1,\dots,\rho)$ are the primary poles of the transfer function $W_d(s)$, and c_i are the residues at these poles. Substituting (13.54) into (13.52), we obtain that the calculation of the function $\tilde{y}_d(t,a,b)$ reduces to the calculation of integrals of the form

$$y_{11}(t,a,b,d) \triangleq \frac{T}{2\pi \mathrm{j}} \int_{c-\mathrm{j}\omega/2}^{c+\mathrm{j}\omega/2} f(s,t,a,b,d)\,\mathrm{d}s \tag{13.55}$$

with

$$f(s,t,a,b,d) \triangleq \frac{D_{p1}(s,a,t)}{\left[1-\mathrm{e}^{(b-s)T}\right]\left[1-\mathrm{e}^{(d-s)T}\right]} \tag{13.56}$$

where a, b and d are known complex numbers lying in the half-plane $\operatorname{Re} s < c$. In this case

$$\ell^-[f(s,t,a,b,d)] = 0\,, \quad t \geq 0\,. \tag{13.57}$$

Then, using Theorem A.3, for $t \geq 0$ we find

$$y_{11}(t,a,b,d) = T\sum_i \operatorname*{Res}_{\tilde{s}_i} f(s,t,a,b,d) \tag{13.58}$$

where the sum is extended on all primary poles $\tilde{s}_i$ of the function $f(s,t,a,b,d)$. Since $f(s,t,a,b,d)$ is periodic with respect to s, b and d, we can assume, without loss of generality, that the numbers a, b and d are primary poles of the function (13.56). Let

$$\mathrm{e}^{aT} \neq \mathrm{e}^{bT}\,, \quad \mathrm{e}^{aT} \neq \mathrm{e}^{dT}\,, \quad \mathrm{e}^{bT} \neq \mathrm{e}^{dT}\,. \tag{13.59}$$

This case will be called *non-resonance.* In this case, under the given assumptions the function (13.56) has simple poles at the points $s = a$, $s = b$, and $s = d$. Then,

$$\begin{aligned}
\operatorname*{Res}_{a} f(s,t,a,b,d) &= \frac{1}{T}\,\frac{M(a)\mathrm{e}^{at}}{\left[1-\mathrm{e}^{(b-a)T}\right]\left[1-\mathrm{e}^{(d-a)T}\right]}\\
\operatorname*{Res}_{b} f(s,t,a,b,d) &= \frac{1}{T}\,\frac{D_{p1}(b,a,t)\mathrm{e}^{bT}}{\mathrm{e}^{bT}-\mathrm{e}^{dT}}\\
\operatorname*{Res}_{d} f(s,t,a,b,d) &= \frac{1}{T}\,\frac{D_{p1}(d,a,t)\mathrm{e}^{dT}}{\mathrm{e}^{dT}-\mathrm{e}^{bT}}\,.
\end{aligned} \tag{13.60}$$

As a result, Eq. (13.58) yields

$$y_{11}(t,a,b,d) = \frac{M(a)\mathrm{e}^{at}\mathrm{e}^{(b+d)T}}{\left(\mathrm{e}^{aT}-\mathrm{e}^{bT}\right)\left(\mathrm{e}^{aT}-\mathrm{e}^{dT}\right)} + \frac{D_{p1}(b,a,t)\mathrm{e}^{bT}-D_{p1}(d,a,t)\mathrm{e}^{dT}}{\mathrm{e}^{bT}-\mathrm{e}^{dT}}\,. \tag{13.61}$$

Using (13.53), (13.61), and (13.46), we can construct the transients without resonance.

Example 13.4 Let us consider the transient process in the open-loop SD system with

$$F(s) = \frac{1}{s-a}\,, \quad X(s) = \frac{1}{s} \quad W_d(s) = \frac{1}{\mathrm{e}^{sT}-0.5}\,, \quad M(s) = \frac{1-\mathrm{e}^{-sT}}{s}$$

and vanishing initial conditions for the continuous-time element and the digital filter program. Moreover, let $a > 0$ be a real number such that $\mathrm{e}^{aT} \neq 0.5$. In this case $b = 0$ and $\mathrm{e}^{r_1 T} = 0.5$. The integral (13.47) has the form

$$y(t) = \frac{T}{2\pi \mathrm{j}} \int_{c-\mathrm{j}\omega/2}^{c+\mathrm{j}\omega/2} \frac{D_{p1}(s,a,t)}{\left(\mathrm{e}^{sT}-0.5\right)\left(1-\mathrm{e}^{-sT}\right)}\,\mathrm{d}s\,.$$

Direct calculation by the residue theorem gives

$$y(t) = \frac{\mathrm{e}^{at}}{a\left(\mathrm{e}^{aT}-0.5\right)} + 2\left[D_{p1}(0,a,t) - D_{p1}(r_1,a,t)\right]$$

which, after some transformations, reduces to

$$y(t) = \frac{\mathrm{e}^{at}\left(1-0.5^q\mathrm{e}^{-aqT}\right)}{a\left(\mathrm{e}^{aT}-0.5\right)} + \frac{2}{a}\left(0.5^q-1\right)\,, \qquad qT \leq t \leq (q+1)T\,.$$

Figure 13.4 shows these curves for $T = 1$ with $a = -1$ and $a = -2$. □

If any of the conditions (13.59) do not hold, there is *resonance* in the open-loop system. To find the transient for the resonant case, it is necessary to pass to the limit similarly to Section 13.2. The corresponding calculations are given in detail in Rosenwasser (1994c).

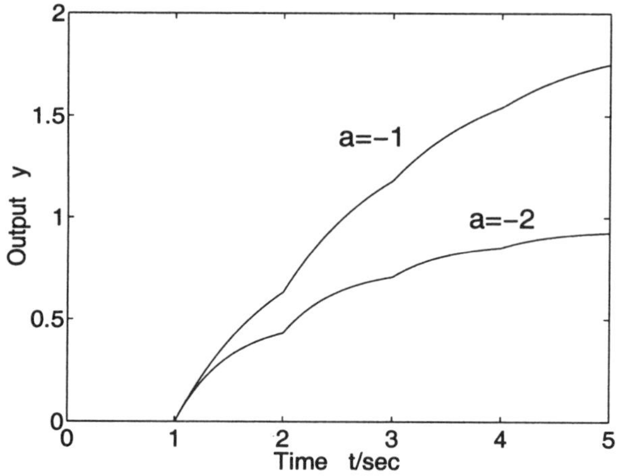

Figure 13.4: Processes for Example 13.4

13.4 Steady-state processes in open-loop SD systems

Consider again the integral (13.45), which, taking account of (13.44), can be represented in the form

$$y(t) = \frac{1}{(m-1)!} \frac{\partial^{m-1}}{\partial b^{m-1}} y_1(t,b) \tag{13.62}$$

where

$$y_1(t,b) \triangleq \frac{T}{2\pi \mathrm{j}} \int_{c-\mathrm{j}\omega/2}^{c+\mathrm{j}\omega/2} \frac{\mathcal{D}_{FM}(T,s,t)W_d(s)}{1-\mathrm{e}^{(b-s)T}} \, \mathrm{d}s \tag{13.63}$$

where $W_d(s)$ is the transfer function of the control program, and the function $\mathcal{D}_{FM}(T,s,t) = D_p(s,t)$ is defined by Eqs. (9.106) and (9.107). We assume that $s = b$ is not a pole for $\mathcal{D}_{FM}(T,s,t)$ or $W_d(s)$. As follows from Section A.8, the function $W_d(s)$ is not limited, so the residue theorem is not, generally speaking, applicable for the integral (13.63), because the integrand in (13.63) does not always tend to zero for $\operatorname{Re} s \to -\infty$. Nevertheless, it can be easily verified that the integral (13.63) satisfies all the conditions of Section A.8. Hence, there is an integer $k \geq 0$ such that for $t > kT$ the residue theorem is applicable, and we have

$$y_1(t,b) = T \sum_i \operatorname*{Res}_{\tilde{s}_i} \left[\frac{\mathcal{D}_{FM}(T,s,t)W_d(s)}{1-\mathrm{e}^{(b-s)T}} \right] \tag{13.64}$$

where $\tilde{s}_i$ are the primary poles of the expression in the brackets. We denote all the primary poles of the product $D_{pd}(s,t) = \mathcal{D}_{FM}(T,s,t)W_d(s)$ by $\tilde{d}_i$. Then, since

$$T \operatorname*{Res}_{b} \left[\frac{\mathcal{D}_{FM}(T,s,t)W_d(s)}{1-\mathrm{e}^{(b-s)T}} \right] = D_{pd}(b,t)$$

from Eq. (13.64) we have

$$y_1(t,b) = \quad y_{10}(t,b) + y_{1\infty}(t,b) \tag{13.65}$$

$$\text{with} \qquad y_{10}(t,b) = T \sum \operatorname*{Res}_{\tilde{d}_i} \left[\frac{\mathcal{D}_{FM}(T,s,t)W_d(s)}{1-\mathrm{e}^{(b-s)T}} \right]$$

$$y_{1\infty}(t,b) = \mathcal{D}_{FM}(T,b,t)W_d(b) = D_{pd}(b,t)\,. \tag{13.66}$$

Substituting the last equation into (13.62), we have

$$y(t) = \quad y_0(t) + y_\infty(t) \tag{13.67}$$

$$\text{with} \quad y_0(t) = \frac{1}{(m-1)!}\frac{\partial^{m-1}}{\partial b^{m-1}} y_{10}(t,b) \tag{13.68}$$

$$y_\infty(t) = \frac{1}{(m-1)!}\frac{\partial^{m-1}}{\partial s^{m-1}} \left[\mathcal{D}_{FM}(T,s,t)W_d(s)\right]\Big|_{s=b}\,. \tag{13.69}$$

Assume that all poles of the product $\mathcal{D}_{FM}(T,s,t)W_d(s)$ are located in the left half-plane, i.e., $\operatorname{Re}\tilde{d}_i < 0$. Then, taking the corresponding residues,we can show that

$$\lim_{t\to\infty} y_0(t) = 0\,.$$

Therefore, there exists the asymptotic limit

$$\lim_{t\to\infty} y(t) = y_\infty(t) \tag{13.70}$$

henceforth called the *steady-state process.* The function $y_0(t)$ will be called the *transient process.* Thus, the steady-state process is the process at the output of the system after the transient process dies down. Using (10.52)

$$D_{pd}(s,t) = W_{pd}(s,t)\mathrm{e}^{st}$$

where

$$W_{pd}(s,t) = \varphi_{FM}(T,s,t)W_d(s)$$

is the PTF of the open-loop SD system, we can write Eq. (13.69) in the form

$$y_\infty(t) = \frac{1}{(m-1)!}\frac{\partial^{m-1}}{\partial s^{m-1}} \left[W_{pd}(s,t)\mathrm{e}^{st}\right]\Big|_{s=b} \tag{13.71}$$

which is a special case of the general formula (7.18). Applying differentiation in (13.71) and taking into account that

$$\frac{\partial^r W_{pd}(s,t)}{\partial s^r} = \frac{\partial^r W_{pd}(s,t+T)}{\partial s^r}$$

we can obtain an expression of the form

$$y_\infty(t) = e^{bt}\left[f_0(t) + f_1(t)t + \ldots + f_{m-1}(t)t^{m-1}\right] \tag{13.72}$$

where $f_i(t) = f_i(t+T)$ $(i = 0, 1, \ldots, m-1)$ are known periodic functions. Formula (13.72) defines the general form of the output of an open-loop digital system under an input of the form (13.2). In the special case, when $m = 1$, i.e., $x(t) = e^{bt}$ for $t > 0$ and $x(t) = 0$ for $t < 0$, from (13.71) we find $y_\infty(t) = W_{pd}(b,t)e^{bt}$ in accordance with the general definition of the PTF.

Next, we consider an important special case of Eq. (13.71).
Let the input signal be a unit step $\mathbb{1}(t)$ given by (1.42) with $m = 1$ and $b = 0$. Then, from (13.71) we find

$$y_\infty(t) = \varphi_{FM}(T, 0, t)W_d(0)\,. \tag{13.73}$$

Since $\varphi_{FM}(T, s, t) = \varphi_{FM}(T, s, t+T)$, the steady-state process is in this case a periodic oscillatory process. Using (9.94) and (9.96), we can easily obtain closed expressions for $y_\infty(t)$. Indeed, assuming $s = 0$ in the aforesaid equations, for $0 \le t \le T$ we have the equivalent expressions

$$\begin{aligned}
\varphi_{FM}(T,0,t) &= \sum_{i=1}^{\ell}\sum_{k=1}^{\nu_i} \frac{f_{ik}}{(k-1)!}\frac{\partial^{k-1}}{\partial s_i^{k-1}}\frac{M(s_i)e^{s_i t}}{e^{-s_i T}-1} + \int_0^t h_p(t-\tau)\mu(\tau)\,d\tau \\
\varphi_{FM}(T,0,t) &= \sum_{i=1}^{\ell}\sum_{k=1}^{\nu_i} \frac{f_{ik}}{(k-1)!}\frac{\partial^{k-1}}{\partial s_i^{k-1}}\frac{M(s_i)e^{s_i t}}{1-e^{s_i T}} - \int_t^T h_p(t-\tau)\mu(\tau)\,d\tau\,.
\end{aligned} \tag{13.74}$$

If the transfer function $F(s)$ has simple poles, these relations become simplified, and under (9.101) we have

$$\begin{aligned}
\varphi_{FM}(T,0,t) &= \sum_{i=1}^{n} \frac{f_i M(s_i)e^{s_i t}}{e^{-s_i T}-1} + \int_0^t h_p(t-\tau)\mu(\tau)\,d\tau \\
\varphi_{FM}(T,0,t) &= \sum_{i=1}^{n} \frac{f_i M(s_i)e^{s_i t}}{1-e^{s_i T}} - \int_t^T h_p(t-\tau)\mu(\tau)\,d\tau\,.
\end{aligned} \tag{13.75}$$

To calculate the function $\varphi_{FM}(T, 0, t)$ we can also use the relation

$$\varphi_{FM}(t,0,t) = \frac{1}{T}\sum_{k=-\infty}^{\infty} F(kj\omega)M(kj\omega)e^{kj\omega t}\,, \quad \omega = \frac{2\pi}{T} \tag{13.76}$$

that helps to construct the Fourier expansion of the output steady-state process.
In order to estimate the system accuracy with a constant input, it is often useful to employ the average output oscillation defined by

$$\overline{y}_\infty \stackrel{\triangle}{=} \frac{1}{T}\int_0^T y_\infty(t)\,dt\,. \tag{13.77}$$

From (13.76) and (13.77) we have

$$\overline{y}_\infty = \frac{1}{T} F(0)M(0)W_d(0)\,. \tag{13.78}$$

Example 13.5 Find the quasi-stationary process in the open-loop system with

$$F(s) = \frac{1}{s-a}\,, \quad h_p(t) = \mathrm{e}^{at}\,, \quad W_d(s) = 1$$

and the form of pulse given by

$$\mu(t) = \begin{cases} 1 & 0 < t < \gamma T \\ 0 & \gamma T < t < T \end{cases}$$

The system is driven by the step input $x(t) = \mathbb{1}(t)$. It follows from this assumptions

$$M(s) = \frac{1 - \mathrm{e}^{-s\gamma T}}{s}\,, \quad M(0) = \gamma T$$

and

$$h_p^*(t) = -\int_t^T \mathrm{e}^{a(t-\tau)}\mu(\tau)\,\mathrm{d}\tau = \begin{cases} \frac{1}{a}\left[\mathrm{e}^{a(t-\gamma T)} - 1\right] & 0 \le t \le \gamma T \\ 0 & \gamma T \le t \le T\,. \end{cases}$$

From the second relation in (13.75) we find

$$y_\infty(t) = \begin{cases} \dfrac{1 - \mathrm{e}^{-\gamma a T}}{a} \dfrac{\mathrm{e}^{at}}{1 - \mathrm{e}^{aT}} + \dfrac{\mathrm{e}^{a(t-\gamma T)} - 1}{a} & 0 \le t \le \gamma T \\ \dfrac{1 - \mathrm{e}^{-\gamma a T}}{a} \dfrac{\mathrm{e}^{at}}{1 - \mathrm{e}^{aT}} & \gamma T \le t \le T\,. \end{cases} \tag{13.79}$$

Figure 13.5 shows the steady-state process computed by Eq. (13.79) for $T = 1$ and $\gamma = 0.5$ with $a = -1$ and $a = -2$. Since $M(0) = 0.5$ in the given case, the average output oscillation (13.78) equals

$$\overline{y}_\infty = -\frac{1}{2a}\,.$$

□

For the most important practical case, when the sampling element is a zero-order hold and $\mu(t) = 1$, the above relations can be simplified. In that case

$$M(0) = T\,, \quad M(k\mathrm{j}\omega) = 0\,, \; (k \neq 0)$$

i.e.,

$$\varphi_{FM}(T,0,t) = \varphi_{FM}(T,0,0) = F(0) \tag{13.80}$$

and, from (13.73) and (13.76) we obtain

$$y_\infty(t) = F(0)W_d(0) = \text{const.} \tag{13.81}$$

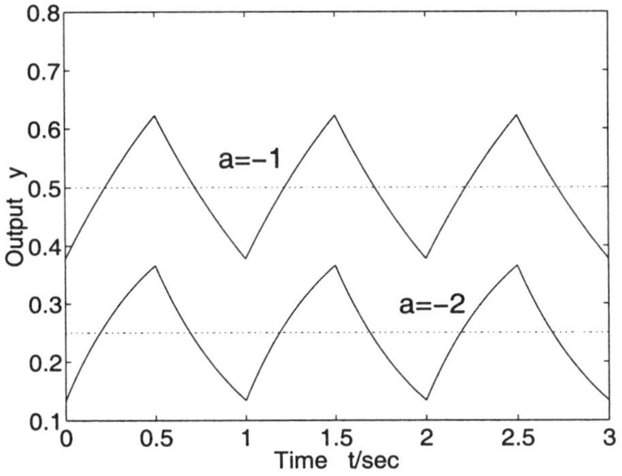

Figure 13.5: Steady-state process for Example 13.5

Now we consider the calculation of the steady-state response to a polynomial excitation

$$x(t) = \begin{cases} \frac{t^{m-1}}{(m-1)!} & t > 0 \\ 0 & t < 0 . \end{cases}$$

The general expression for the output process can be obtained from (13.71)

$$y_\infty(t) = \frac{1}{(m-1)!} \frac{\partial^{m-1}}{\partial s^{m-1}} \left[\varphi_{FM}(T,s,t) W_d(s) e^{st} \right]_{| s=0} . \tag{13.82}$$

Example 13.6 Find the steady-state output process under the conditions of Example 13.2. In this case Eq. (13.82) yields

$$y_\infty(t) = \left[t e^{st} \varphi_{FM}(T,s,t) + \frac{\partial \varphi_{FM}(T,s,t)}{\partial s} e^{st} \right]_{| s=0} .$$

Taking account of (9.115), under the conditions of Example 13.2 for $0 \le t \le T$ we have

$$\varphi_{FM}(T,s,t) = e^{st} \left[\frac{1 - e^{-aT}}{a} \frac{e^{at}}{e^{(s-a)T} - 1} + \frac{e^{at} - 1}{a} \right] .$$

Hence,

$$\varphi_{FM}(T,0,t) = -\frac{1}{a}$$

$$\frac{\partial \varphi_{FM}(T,s,t)}{\partial s}_{| s=0} = \frac{t}{a} + \frac{T e^{at}}{a(1 - e^{aT})} , \quad 0 < t < T$$

Therefore,

$$y_\infty(t) = -\frac{t}{a} + \left[\frac{t}{a} + \frac{Te^{at}}{a(1-e^{aT})}\right], \quad 0 \le t \le T.$$

The last formula can be continued over the whole $t-$axis by

$$y_\infty(t) = -\frac{t}{a} + \frac{t-qT}{a} + \frac{Te^{a(t-qT)}}{a(1-e^{aT})}, \quad qT \le t \le (q+1)T.$$

Figure 13.6 shows the steady-state process for $T = 1$ with $a = -1$ and

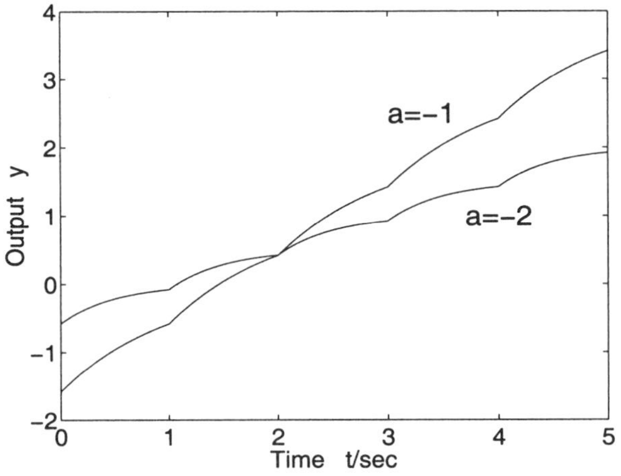

Figure 13.6: Steady-state process for Example 13.6

$a = -2$. □

13.5 Steady-state processes under harmonic excitation

If

$$x(t) = \begin{cases} e^{j\nu t} & t > 0 \\ 0 & t < 0 \end{cases}$$

where ν is a real parameter, a formula for the output steady-state process is obtained from (13.71) with $b = j\nu$ and $m = 1$

$$y_\infty(t) = W_{pd}(j\nu, t)e^{j\nu t} = \varphi_{FM}(T, j\nu, t)W_d(j\nu)e^{j\nu t}.$$

Using (6.57), we can represent the last relation in the form

$$y_\infty(t) = \Phi_{pd}(j\nu, t)e^{j\nu t} \tag{13.83}$$

where

$$\Phi_{pd}(\mathrm{j}\nu, t) = W_{pd}(s, t)\,|_{s=\mathrm{j}\nu} \tag{13.84}$$

is the parametric frequency response of the system.
Using the relations obtained in Section 6.4, we can find the steady-state response of the system to the inputs $x(t) = \cos\nu t$ and $x(t) = \sin\nu t$, denoting the corresponding outputs by $y_{c\infty}(t)$ and $y_{s\infty}(t)$, respectively. Using (6.62), we immediately obtain

$$\begin{aligned} y_{c\infty}(t) &= P_{pd}(\nu, t)\cos\nu t - Q_{pd}(\nu, t)\sin\nu t \\ y_{s\infty}(t) &= P_{pd}(\nu, t)\sin\nu t + Q_{pd}(\nu, t)\cos\nu t \end{aligned} \tag{13.85}$$

where, in accordance with the general theory,

$$\begin{aligned} P_{pd}(\nu, t) &= \mathrm{Re}[\varphi_{FM}(T, \mathrm{j}\nu, t)W_d(\mathrm{j}\nu)] \\ Q_{pd}(\nu, t) &= \mathrm{Im}[\varphi_{FM}(T, \mathrm{j}\nu, t)W_d(\mathrm{j}\nu)]\,. \end{aligned} \tag{13.86}$$

By construction, we have

$$P_{pd}(\nu, t) = P_{pd}(\nu, t+T)\,, \quad Q_{pd}(\nu, t) = Q_{pd}(\nu, t+T)\,. \tag{13.87}$$

We note some general properties of the functions (13.85), which follow from the relations proved above:

1. If the sampling frequency $\omega = 2\pi/T$ and the frequency of the exogenous disturbance ν are not commensurate, the functions $y_{c\infty}(t)$ and $y_{s\infty}(t)$ are not periodic in t. In this case they are 'almost periodic' (or, rigorously, 'doubly periodic') functions in t. The same can be said about the steady-state process (13.83).

2. If the frequencies ω and ν are *commensurate*, i.e.,

$$\frac{\omega}{\nu} = \frac{m}{n}$$

with co-prime positive integers m and n, the output steady-state process $y_\infty(t)$ is periodic with respect to t with period T_a, where

$$T_a \triangleq nT_\nu = mT\,, \quad T_\nu = \frac{2\pi}{\nu}\,, \quad T = \frac{2\pi}{\omega}\,.$$

Moreover, it can be immediately verified that

$$y_{c\infty}(t) = y_{c\infty}(t + T_a)\,, \quad y_{s\infty}(t) = y_{s\infty}(t + T_a)\,.$$

3. If we consider the input values only at discrete time points, it is possible to obtain a number of additional important relations. A discrete sequence of the form

$$f_k \triangleq A\mathrm{e}^{kj\overline{\omega}}, \quad (k = 0, \pm 1, \dots) \tag{13.88}$$

where $A > 0$ and $\overline{\omega} > 0$ are real numbers, will be called a *harmonic sequence* with the *relative frequency* $\overline{\omega}$ and the *relative amplitude* A. The sequence

$$x_k = A\mathrm{e}^{kj\overline{\omega}+j\phi} \tag{13.89}$$

with a real ϕ will be called the *phase shifted sequence* by ϕ with respect to the sequence (13.88). An input signal $x(t) = \mathrm{e}^{j\nu t}$ at discrete points $t_k = \varepsilon + kT$, $0 \leq \varepsilon < T$, is associated with the input sequence

$$x_k = \mathrm{e}^{j(k\nu T + \nu\varepsilon)} \tag{13.90}$$

with the relative frequency

$$\overline{\omega} = \nu T = \frac{2\pi\nu}{\omega}$$

and the phase shift $\nu\varepsilon$. The input sequence (13.90) for the points t_k is associated, taking account of (13.83) with the output sequence

$$y_k \triangleq y_\infty(t_k) = W_{pd}(j\nu, \varepsilon)\mathrm{e}^{j(k\nu T+\nu\varepsilon)} \,. \tag{13.91}$$

Let us consider, for a fixed ε, with the use of Eq. (6.60),

$$\Phi_{pd}(j\nu, t) = A_{pd}(\nu, \varepsilon)\mathrm{e}^{j\phi_{pd}(\nu,\varepsilon)} \,.$$

Then, from (13.91) we obtain

$$y_k = A_{pd}(\nu, \varepsilon)\mathrm{e}^{j[k\nu T+\nu\varepsilon+\phi_{pd}(\nu,\varepsilon)]} \tag{13.92}$$

where the values $A_{pd}(\nu, \varepsilon)$ and $\phi_{pd}(\nu, \varepsilon)$ are independent of k. From (13.92) it follows that, for any fixed ε and the input sequence given by (13.90), the sampling of the steady-state output at the points $t_k = \varepsilon + kT$ is a discrete harmonic sequence of the same relative frequency, but having different amplitude and phase. Extracting the real and imaginary parts in (13.90), it can be shown that a similar conclusion is valid for the sequences

$$x_{ck} \triangleq \cos(k\nu T + \nu\varepsilon) \,, \quad x_{sk} \triangleq \sin(k\nu T + \nu\varepsilon) \,.$$

The corresponding output steady-state sequences have the form

$$\begin{aligned} y_{ck} &= A(\nu, \varepsilon)[\cos(k\nu T + \nu\varepsilon + \phi_{pd}(\nu, \varepsilon)] \\ y_{sk} &= A(\nu, \varepsilon)[\sin(k\nu T + \nu\varepsilon + \phi_{pd}(\nu, \varepsilon)] \,. \end{aligned} \tag{13.93}$$

As follows from the above relations, the character of the steady-state process in SD systems considered in continuous time is qualitatively different in comparison with the discrete one. The principal differences are listed below.

A) The steady-state continuous-time response to a harmonic input is not a harmonic signal of the same frequency. Moreover, if the frequency of the input signal is not commensurate with the sampling frequency, the steady-state output process is not periodic at all.

B) The response of the system to sine and cosine input signals satisfies relations similar to (13.93) and are complicated functions of the time.

C) The amplitude and phase frequency responses $A(\nu, t)$ and $\phi(\nu, t)$ considered in continuous time do not possess properties following from (13.92).

Then, we investigate yet other properties of parametric frequency responses following from the above relations.

1. First of all, using (13.11), we have

$$\Phi_{pd}(\mathrm{j}\nu, t) = W_d(\mathrm{j}\nu)\frac{1}{T}\sum_{k=-\infty}^{\infty} F(\mathrm{j}\nu + k\mathrm{j}\omega)M(\mathrm{j}\nu + k\mathrm{j}\omega)\mathrm{e}^{k\mathrm{j}\omega t}. \tag{13.94}$$

Hence, $\Phi_{pd}(\mathrm{j}\nu, t)$ is completely defined by the frequency response of the continuous part $F(\mathrm{j}\nu)$, the forming element $M(\mathrm{j}\nu)$ and the control program $W_d(\mathrm{j}\nu)$.

2. From (13.94) it immediately follows that

$$\Phi_{pd}[\mathrm{j}(\nu + \omega), t] = \Phi_{pd}(\mathrm{j}\nu, t)\mathrm{e}^{-\mathrm{j}\omega t} \tag{13.95}$$

i.e., the values of $\Phi_{pd}(\mathrm{j}\nu, t)$ for $-\infty < t < \infty$ are completely defined by its values for $0 \le \nu \le \omega$ and $0 \le t \le T$.

3. Extracting the real and imaginary parts in (13.95), taking account of (13.86) we find

$$\begin{aligned} P(\nu + \omega, t) &= P(\nu, t)\cos\omega t + Q(\nu, t)\sin\omega t \\ Q(\nu + \omega, t) &= -P(\nu, t)\sin\omega t + Q(\nu, t)\cos\omega t \end{aligned} \tag{13.96}$$

i.e., the frequency responses $P(\nu, t)$ and $Q(\nu, t)$ are completely defined by their values in the frequency range $0 \le \nu \le \omega$.

Example 13.7 Find the parametric frequency response of the open-loop system with

$$F(s) = \frac{1}{s-a}, \quad W_d(s) = 1, \quad M(s) = \frac{1 - \mathrm{e}^{-sT}}{s}.$$

Then, for $0 \le t \le T$ we have

$$\Phi_{pd}(\mathrm{j}\nu, t) = \mathrm{e}^{-\mathrm{j}\nu t}\left[\frac{1 - \mathrm{e}^{-aT}}{a}\,\frac{\mathrm{e}^{at}}{\mathrm{e}^{(\mathrm{j}\nu - a)T} - 1} + \frac{\mathrm{e}^{at} - 1}{a}\right].$$

Extracting the real and imaginary parts, we obtain the real and imaginary frequency response, $P_{pd}(\nu, t)$ and $Q_{pd}(\nu, t)$. Figure 13.7 shows these functions for $a = -1$, $T = 0.44$ and $\nu = 0.6\omega$. □

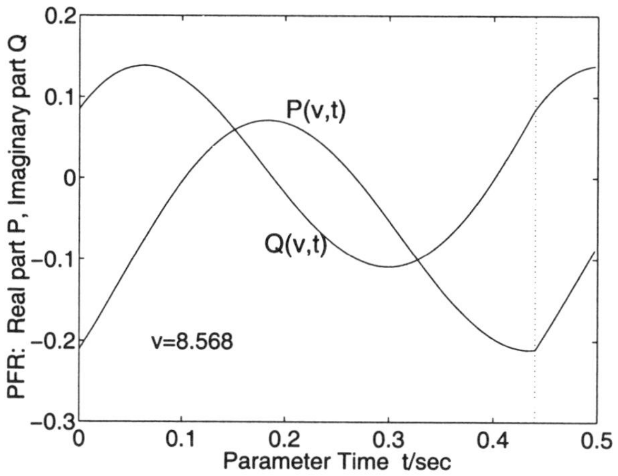

Figure 13.7: Real and imaginary part of a parametric frequency response

13.6 Experimental determination of the parametric frequency response

Using the general relations of the Sections 6.4 und 13.5 we are able to find general principles for the experimental determination of the parametric frequency response of stable linear periodic systems, including SD systems. The idea is as follows, Volovodov et al. (1991), Sommer et al. (1994), Lampe and Rosenwasser (1995):
Let a stable linear periodic system with the PTF $W(s,t)$ with respect to Fig. 5.1 be given. Then, for harmonic excitation $x(t) = \mathrm{e}^{\mathrm{j}\nu t}$ the quasi-stationary process at the output is determined by (6.58)

$$y_{\infty}(t) = W(\mathrm{j}\nu, t)\mathrm{e}^{\mathrm{j}\nu t}. \tag{13.97}$$

Therefore, the quasi-stationary output for the input $x_s(t) = \sin \nu t$ has the form

$$\begin{aligned} y_{s\infty}(t) = \quad & P(\nu,t)\sin\nu t + Q(\nu,t)\cos\nu t && (13.98)\\ \text{with} \quad & P(\nu,t) = \operatorname{Re} W(\mathrm{j}\nu,t)\,, \quad Q(\nu,t) = \operatorname{Im} W(\mathrm{j}\nu,t) && (13.99)\\ & P(\nu,t) = P(\nu,t+T)\,, \quad Q(\nu,t) = Q(\nu,t+T)\,. && (13.100) \end{aligned}$$

By analogy the quasi-stationary output for the excitation $x_c(t) = \cos(\nu t)$ is given by

$$y_{c\infty}(t) = P(\nu,t)\cos\nu t - Q(\nu,t)\sin\nu t\,. \tag{13.101}$$

From the mathematical point of view (13.98), (13.101) are particular solutions of the equations, that connect the input $x(t)$ with the output $y(t)$ for certain initial conditions. If the system for vanishing initial conditions is excited by $x_s(t)$ and $x_c(t)$, then at the output

$$y_s(t) = y_{sn}(t) + y_{s\infty}(t)$$
$$y_c(t) = y_{cn}(t) + y_{c\infty}(t)$$

is observed, where $y_{sn}(t)$ and $y_{cn}(t)$ describe the transients of the output signal. Because of the supposed stability of the system the transient processes are extinguished after a certain duration of time. After this time interval only the quasi-stationary motions (13.98) and (13.101) are measured. Though, the direct measuring and analysis of this processes is connected with great technological problems because $y_{s\infty}(t)$ and $y_{c\infty}(t)$ are in general not periodic functions of the time. However, these difficulties can easily be surmounted, because we get by (13.98) and (13.101)

$$\begin{aligned} P(\nu,t) &= y_{s\infty}(t)\sin\nu t + y_{c\infty}(t)\cos\nu t \\ Q(\nu,t) &= y_{s\infty}(t)\cos\nu t - y_{c\infty}(t)\sin\nu t\,. \end{aligned} \qquad (13.102)$$

These equations constitute one possibility for the pointwise experimental determination of the PFR by means of the measured steady-state reactions $y_{s\infty}(t)$ and $y_{c\infty}(t)$ for a fixed frequency ν of the input signals, respectively. For that purpose after a transient time of N periods of the input signal the parameter t is to be selected in the range $N\tau < t < N\tau + T$, $\tau = 2\pi/\nu$. The duration of the transient depends on the system behaviour, and it can be seen as finished, if the processes

$$P'(\nu,t) = y_s(t)\sin\nu t + y_c(t)\cos\nu t$$
$$Q'(\nu,t) = y_s(t)\cos\nu t - y_c(t)\sin\nu t$$

look periodical.

If the sampling frequency and the exciting frequency are in rational relation with

$$\frac{\omega}{\nu} = \frac{\tau}{T} = \frac{r}{m}, \quad \frac{2m}{r} \text{ not integer} \qquad (13.103)$$

and natural numbers r, m, then there exists a further possibility for the experimental determination of the PFR, managing with only one measurement for each frequency ν. This method also starts from the representation (13.98). With respect to (13.100) these components are developed into Fourier series. From the practical viewpoint they break off after a limited number of terms, which generates

$$\begin{aligned} P(\nu,t) &\approx \sum_{k=0}^{n} A_k(\nu)\sin k\omega t + B_k(\nu)\cos k\omega t \\ Q(\nu,t) &\approx \sum_{k=0}^{n} C_k(\nu)\sin k\omega t + D_k(\nu)\cos k\omega t\,. \end{aligned} \qquad (13.104)$$

Substituting this in (13.93), we get

$$\begin{aligned} y_s(t) = &\sum_{k=0}^{n} \Big(A_k(\nu) \sin k\omega t + B_k(\nu) \cos k\omega t \Big) \sin \nu t \\ &+ \sum_{k=0}^{n} \Big(C_k(\nu) \sin k\omega t + D_k(\nu) \cos k\omega t \Big) \cos \nu t \,. \end{aligned}$$

Applying the addition theorem yields

$$\begin{aligned} y_s(t) = \ &\frac{1}{2} \sum_{k=0}^{n} \Big[A_k(\nu) \Big(\cos(k\omega - \nu)t - \cos(k\omega + \nu)t \Big) \\ &+ B_k(\nu) \Big(-\sin(k\omega - \nu)t + \sin(k\omega + \nu)t \Big) \\ &+ C_k(\nu) \Big(\sin(k\omega - \nu)t + \sin(k\omega + \nu)t \Big) \\ &+ D_k(\nu) \Big(\cos(k\omega - \nu)t - \cos(k\omega + \nu)t \Big) \Big] \,. \end{aligned} \tag{13.105}$$

Summing up the terms with equal frequencies, multiplying it by the harmonics of the frequencies, and integration over the (least) common period $m\tau = rT$, we find

$$\begin{aligned} D_k(\nu) + A_k(\nu) &= \frac{1}{m\tau} \int_0^{m\tau} y_s(t) \cos(k\omega - \nu)t \, dt = J_{1k} \\ C_k(\nu) - B_k(\nu) &= \frac{1}{m\tau} \int_0^{m\tau} y_s(t) \sin(k\omega - \nu)t \, dt = J_{2k} \\ D_k(\nu) - A_k(\nu) &= \frac{1}{m\tau} \int_0^{m\tau} y_s(t) \cos(k\omega + \nu)t \, dt = J_{3k} \\ B_k(\nu) + C_k(\nu) &= \frac{1}{m\tau} \int_0^{m\tau} y_s(t) \sin(k\omega + \nu)t \, dt = J_{4k} \,. \end{aligned} \tag{13.106}$$

After evaluation of the integrals in (13.106) for $k = 0, 1, \ldots, n$ we obtain for each fixed ν a system of linear equations for the unknown quantities A_k, B_k, C_k, D_k. Hence follows that $P(\nu, t)$ and $Q(\nu, t)$ are calculated by (13.104). Because of its special form the system of equations is easy to solve for every k:

$$A_k = \frac{J_{1k} - J_{3k}}{2}\,, \ B_k = \frac{J_{4k} - J_{2k}}{2}\,, \ C_k = \frac{J_{2k} + J_{4k}}{2}\,, \ D_k = \frac{J_{1k} + J_{3k}}{2}\,. \tag{13.107}$$

If during the experiment the harmonic excitations act on the sampling unit, which is always the case if the computer itself generates the input, it then follows from (13.96), that it is sufficient to restrict the measurement to the frequency range $0 \leq \nu < \omega$.

Chapter 14

Analysis of feedback SD systems under deterministic disturbances

14.1 Process equations with zero initial energy

In this chapter we investigate the construction of transient and steady-state processes in the system shown in Fig. 11.1. For concreteness, we consider the signal $y(t)$ as an output and $x(t)$ as an input. In this case, the operator equation of the system has the form

$$y = W(s,t)x \tag{14.1}$$

where, by (11.42),

$$W(s,t) \triangleq W_{yx}(s,t) = R(s)\left[1 - \frac{W_d(s)\varphi_{FM}(T,s,t)}{1 + W_d(s)\varphi_{FM}(T,s,0)}\right]. \tag{14.2}$$

Henceforth, it is assumed that the functions $F(s) = G(s)R(s)$ and $R(s)$ are strictly proper, and $F(s)$ has the form (13.1). In this case,

$$W_d(s) = \frac{b(s)}{a(s)}, \quad \varphi_{FM}(T,s,0) = \mathcal{D}_{FM}(T,s,0) = \frac{e^{-sT}\chi(s)}{\alpha(s)} \tag{14.3}$$

where $\alpha(s)$, $\chi(s)$, $b(s)$ and $a(s)$ are quasi-polynomials in $\zeta = e^{-sT}$. We shall assume that the non-pathological conditions (9.161) and (9.162) hold, and therefore the rational periodic function $\varphi_{FM}(T,s,0)$ is irreducible.
Under the given assumptions, by Theorem 11.2, the PTF (14.2) has the form

$$W(s,t) = \frac{n(s,t)}{\Delta(s)} \tag{14.4}$$

where

$$\Delta(s) = \alpha(s)a(s) + \mathrm{e}^{-sT}\chi(s)b(s) \tag{14.5}$$

is the characteristic quasi-polynomial of the system, and

$$n(s,t) = R(s)\alpha(s)a(s) + R(s)\alpha(s)b(s)\left[\varphi_{FM}(t,s,0) - \varphi_{FM}(T,s,t)\right] \tag{14.6}$$

is an entire function in s for all t. As follows from (14.4), if the determining equation

$$\Delta(s) = 0 \tag{14.7}$$

has the primary roots $\tilde{s}_i$ $(i = 1,\dots,r)$ of multiplicities $\nu_1,\dots,\nu_r$, for all t the function $W(s,t)$ may have poles only at $s_{im} = \tilde{s}_i + m\mathrm{j}\omega$, where m is an arbitrary integer. Moreover, the multiplicity of a pole s_{im} does not exceed ν_i. Henceforth, we shall assume that all the primary roots $\tilde{s}_i$ lie in the left half-plane, i.e., $\operatorname{Re}\tilde{s}_i < 0$, $(i = 1,\dots,r)$.

We also assume that the input has the form (13.2) and its Laplace image is given by (13.3). As in Section 11.7, by $\tilde{\mu}_1,\dots,\tilde{\mu}_\kappa$ we denote the characteristic indices of the system, i.e., the different real parts of the primary roots $\tilde{s}_i$ numbered in ascending order. Let c be a real constant such that $c > \tilde{\mu}_\kappa$. Then, the integral

$$\tilde{h}_\kappa(t,\tau) = \frac{1}{2\pi\mathrm{j}}\int_{c-\mathrm{j}\infty}^{c+\mathrm{j}\infty} W(s,t)\mathrm{e}^{s(t-\tau)}\,\mathrm{d}s \tag{14.8}$$

gives the Green function of the causal operator

$$y(t) = \int_{-\infty}^{t} \tilde{h}_\kappa(t,\tau)x(\tau)\,\mathrm{d}\tau\,. \tag{14.9}$$

If this integral converges, it determines the output of the system. Under the given assumptions the solution (14.9) exists and can be represented as an inversion integral of the form (11.120)

$$y(t) = \frac{1}{2\pi\mathrm{j}}\int_{c_1-\mathrm{j}\infty}^{c_1+\mathrm{j}\infty} W(s,t)X(s)\mathrm{e}^{st}\,\mathrm{d}s\,, \quad c_1 > \lambda \stackrel{\triangle}{=} \max\{\mu_\kappa, \operatorname{Re} b\}\,. \tag{14.10}$$

Since $x(t) = 0$ for $t < 0$, from (14.9) it follows that the solution (14.10) is equivalent to the integral

$$y(t) = \int_{0}^{t} \tilde{h}_\kappa(t,\tau)x(\tau)\,\mathrm{d}\tau \tag{14.11}$$

which will be called the *solution with zero initial energy*. Discretization of the integral (14.10) gives

$$y(t) = \frac{T}{2\pi\mathrm{j}}\int_{c_1-\mathrm{j}\omega/2}^{c_1+\mathrm{j}\omega/2} \mathcal{D}_y(T,s,t)\,\mathrm{d}s \tag{14.12}$$

where $\mathcal{D}_y(T,s,t)$ is the discrete Laplace transform of the output defined by (11.122), so that

$$\mathcal{D}_y(T,s,t) = \mathcal{D}_{RX}(T,s,t) - \frac{W_d(s)\mathcal{D}_{RGM}(T,s,t)}{1 + W_d(s)\mathcal{D}_{RGM}(T,s,0)}\mathcal{D}_{RX}(T,s,0)\,. \tag{14.13}$$

Remark We note that, in principle, formula (14.13) can be extended to the case when the function $R(s)$ is proper, but not strictly proper. Then, if $m > 1$ in (13.2), formula (14.13) remains its form. If $m = 1$, we have to substitute $\mathcal{D}_{RX}(T, s, +0)$ instead of $\mathcal{D}_{RX}(T, s, 0)$ on the right-hand side of (14.13).

14.2 Calculation of processes with zero initial energy

It can be easily verified, that for $0 \le t \le T$ the rational periodic function (14.13) has a finite limit

$$\ell^{+}[\mathcal{D}_y(T,s,t)] = \lim_{\mathrm{Re}\, s\to\infty} \mathcal{D}_y(T,s,t) \tag{14.14}$$

i.e., the function (14.13) is causal. Moreover, all the assumptions of Section (A.8) are satisfied. Therefore, to calculate the integral (14.12) we can use the general formulae (A.116). First of all, we must investigate the behaviour of the rational periodic function (14.13) for $\mathrm{Re}\, s \to -\infty$ and find the set of its poles.

Theorem 14.1 *Let the control program $W_d(s)$ be limited and*

$$1 + \ell^{-}[W_d(s)]\,\ell^{-}[\mathcal{D}_{FM}(T,s,0)] \ne 0\,. \tag{14.15}$$

Then,

$$\ell^{-}[\mathcal{D}_y(T,s,t)] = 0\,, \quad t \ge 0\,. \tag{14.16}$$

Proof Taking the limit of (14.13) as $\mathrm{Re}\, s \to -\infty$, we have

$$\ell^{-}[\mathcal{D}_y(T,s,t)] =$$

$$= \ell^{-}[\mathcal{D}_{RX}(T,s,t)] - \frac{\ell^{-}[W_d(s)]\ell^{-}[\mathcal{D}_{RGM}(T,s,t)]}{1+\ell^{-}[W_d(s)]\ell^{-}[\mathcal{D}_{RGM}(T,s,0)]}\,\ell^{-}[\mathcal{D}_{RX}(T,s,0)]\,.$$

Since the rational periodic function $R(s)X(s)$ is strictly proper, from (4.67) we have

$$\ell^{-}[\mathcal{D}_{RX}(T,s,t)] = 0\,, \quad t \ge +0\,.$$

Moreover, since $F(s)$ is strictly proper, Eqs. (9.140) and (9.142) yield

$$\ell^{-}[\mathcal{D}_{FM}(T,s,t)] = h_p^*(t)\,, \quad 0 \le t \le T; \quad \ell^{-}[\mathcal{D}_{FM}(T,s,t)] = 0\,, \quad t \ge T\,.$$

The above relations immediately yield (14.16). ■

Theorem 14.2 *Let the control program $W_d(s)$ be not limited, so that the representation (A.27) holds*

$$W_d(s) = l_\kappa \mathrm{e}^{-\kappa s T} + \ldots + l_1 \mathrm{e}^{-sT} + W_{d1}(s) \tag{14.17}$$

where $W_{d1}(s)$ is a limited function and $l_\kappa \ne 0$. Let also

$$\ell^{-}[\mathcal{D}_{FM}(T,s,0)] = h_p^*(0) \ne 0\,. \tag{14.18}$$

Then, Eq. (14.16) holds.

Proof Taking account of (14.17), we represent (14.13) in the form

$$\mathcal{D}_y(T,s,t) = \mathcal{D}_{RX}(T,s,t) - \frac{e^{\kappa sT} W_d(s)\mathcal{D}_{RGM}(T,s,t)}{e^{\kappa sT} + e^{\kappa sT} W_d(s)\mathcal{D}_{RGM}(T,s,0)}\mathcal{D}_{RX}(T,s,0)\,.$$

Taking the limit as $\operatorname{Re} s \to -\infty$, we have

$$\ell^-[\mathcal{D}_y(T,s,t)] = \ell^-[\mathcal{D}_{RX}(T,s,t)] -$$

$$-\frac{\ell^-[e^{\kappa sT} W_d(s)]\,\ell^-[\mathcal{D}_{RGM}(T,s,t)]}{\ell^-[e^{\kappa sT}] + \ell^-[e^{\kappa sT} W_d(s)]\,\ell^-[\mathcal{D}_{RGM}(T,s,0)]}\,\ell^-[\mathcal{D}_{RX}(T,s,0)]\,.$$

From (14.17) it follows that $\ell^-[e^{\kappa sT} W_d(s)] = l_\kappa \neq 0$, therefore, as in Theorem 14.1, the transition to the limit for s $t > 0$ yields (14.16). ■

Henceforth we assume that the above conditions of applicability of Eq. (14.16) are satisfied. Then, using (A.83), for $t > 0$ we have

$$y(t) = \sum_r \operatorname*{Res}_{\tilde{q}_r} \mathcal{D}_y(T,s,t) \tag{14.19}$$

where $\tilde{q}_r$ are the primary poles of the function (14.13). The set of these poles are given by the following theorem.

Theorem 14.3 *Let the non-pathological conditions (9.161) and (9.162) hold, i.e., the quasi-polynomials* $e^{-sT}\chi(s)$ *and* $\alpha(s)$ *are co-prime. Let* $\tilde{s}_i$ $(i = 1,\dots,r)$ *be the primary roots of the determining equation (14.7) and the following conditions hold*

$$e^{\tilde{s}_i T} \neq e^{bT}\,, \quad (i = 1,\dots,r)\,. \tag{14.20}$$

Then, for any t the set of poles of the function (14.13) consists of the numbers $s_{im} = \tilde{s}_i + mj\omega$ *and* $b_m = b + mj\omega$*, where m is an arbitrary integer.*

Proof Consider the product

$$Y(s,t) \triangleq W(s,t)X(s) = \left[R(s) - \frac{W_d(s)\varphi_{FM}(T,s,t)R(s)}{1 + W_d(s)\varphi_{FM}(t,s,0)}\right]\frac{1}{(s-b)^m}\,. \tag{14.21}$$

Due to (14.4), the function in the square brackets has poles at s_{im} of multiplicities not higher than ν_i. Therefore, taking account of (14.20), we find that the function $Y(s,t)$ has poles at $s = s_{im}$ and at $s = b$. From (11.121) it follows that

$$\mathcal{D}_y(T,s,t) = \frac{1}{T}\sum_{k=-\infty}^{\infty} Y(s+kj\omega,t)e^{(s+kj\omega)t} = \frac{1}{T}\sum_{k=-\infty}^{\infty}\frac{W(s+kj\omega,t)}{(s+kj\omega-b)^m}\,e^{(s+kj\omega)t}. \tag{14.22}$$

In any region of the $s-$plane that does not contain the points s_{im} and b_m, the series (14.22) converges absolutely and uniformly with respect to s and t. Therefore, the poles of $\mathcal{D}_y(T,s,t)$ are located only at the points s_{im} and b_m. ■

Remark Since $\mathcal{D}_y(T,s,t)$ is periodic with respect to s, we can assume, without loss of generality, that the numbers $s=\tilde{s}_i$ and $s=b$ define the set of primary poles of $\mathcal{D}_y(T,s,t)$.

Next, we calculate $y(t)$ directly by (14.19). To simplify the calculation, we assume that all poles of the determining equation (14.7) are simple and none of the roots $\tilde{s}_i$ is a pole of $F(s)$. Then, representing (14.13) in the form

$$\mathcal{D}_y(T,s,t) = \mathcal{D}_{RX}(T,s,t) - \frac{W_d(s)a(s)\alpha(s)\mathcal{D}_{FM}(T,s,t)}{\Delta(s)}\mathcal{D}_{RX}(T,s,0)$$

we immediately obtain

$$\operatorname*{Res}_{\tilde{s}_i} \mathcal{D}_y(T,s,t) = A_i \mathcal{D}_{FM}(T,\tilde{s}_i,t) \tag{14.23}$$

where

$$A_i \triangleq -\frac{W_d(\tilde{s}_i)a(\tilde{s}_i)\alpha(\tilde{s}_i)\mathcal{D}_{RX}(T,\tilde{s}_i,0)}{\Delta'(\tilde{s}_i)} \tag{14.24}$$

and $\Delta'(s) = \mathrm{d}\Delta(s)/\,\mathrm{d}s$.
To calculate the residue at the primary pole $s=b$, we note that from the given assumptions and (14.22) it follows that

$$\operatorname*{Res}_{b} \mathcal{D}_y(T,s,t) = \frac{1}{T}\operatorname*{Res}_{b}\left[\frac{W(s,t)\mathrm{e}^{st}}{(s-b)^m}\right].$$

Since, by assumption, the product $W(s,t)\mathrm{e}^{st}$ is analytic at $s=b$, we have

$$\operatorname*{Res}_{b} \mathcal{D}_y(T,s,t) = \frac{1}{T}\frac{1}{(m-1)!}\frac{\partial^{m-1}}{\partial s^{m-1}}\left[W(s,t)\mathrm{e}^{st}\right]_{|\,s=b}. \tag{14.25}$$

Substituting (14.23) and (14.25) into (14.19), we find a closed expression for the output process with zero initial energy

$$y(t) = T\sum_i A_i\mathcal{D}_{FM}(T,\tilde{s}_i,t) + \frac{1}{(m-1)!}\frac{\partial^{m-1}}{\partial s^{m-1}}\left[W(s,t)\mathrm{e}^{st}\right]_{|\,s=b}. \tag{14.26}$$

Example 14.1 Consider the process with zero initial energy in the SD system shown in Fig. 14.1, where U_a is the sampling unit and

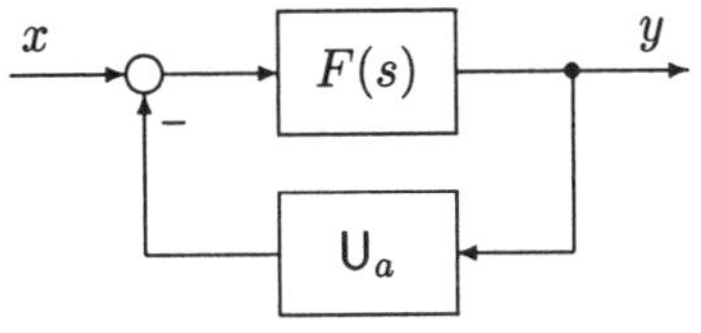

Figure 14.1: SD system for Example 14.1

$$F(s) = \frac{\gamma}{s-a}, \quad M(s) = \frac{1-\mathrm{e}^{-sT}}{s}$$

where a, γ ($a \neq \gamma$) are real numbers. For the input disturbance we have $X(s) = s^{-1}$, so that $m = 1$ and $b = 0$.
First of all, we show that, for this example, Eq. (14.13) can be written in a simpler form. Indeed, we have

$$M(s) = \left(1 - \mathrm{e}^{-sT}\right) X(s) .$$

Hence,

$$\mathcal{D}_{FM}(T,s,t) = \left(1 - \mathrm{e}^{-sT}\right) \mathcal{D}_{RX}(T,s,t) .$$

Therefore, for $F(s) = R(s)$ from (14.13) we obtain

$$\mathcal{D}_y(T,s,t) = \frac{\mathcal{D}_{FX}(T,s,t)}{1+\mathcal{D}_{FM}(T,s,0)} . \tag{14.27}$$

Then, Eq.(13.14) yields

$$\mathcal{D}_{FM}(T,s,t) = \frac{\gamma M(a)\mathrm{e}^{aT}}{\mathrm{e}^{sT} - \mathrm{e}^{aT}} . \tag{14.28}$$

Therefore,

$$\ell^{-}\left[\mathcal{D}_{FM}(T,s,0)\right] = -\gamma M(a)$$

so that the condition (14.15) takes the form

$$1 - \gamma M(a) \neq 0 .$$

If the last inequality holds, to calculate the process $y(t)$ we may use formula (14.19). Since

$$F(s)X(s) = \frac{\gamma}{a}\frac{1}{s-a} - \frac{\gamma}{a}\frac{1}{s}$$

Eq. (4.42) yields

$$\mathcal{D}_{FX}(T,s,t) = \frac{\gamma \mathrm{e}^{sT}}{a}\left[\frac{\mathrm{e}^{a(t-qT)}}{\mathrm{e}^{sT}-\mathrm{e}^{aT}} - \frac{1}{\mathrm{e}^{sT}-1}\right]\mathrm{e}^{qsT}, \quad qT \leq t \leq (q+1)T . \tag{14.29}$$

Using (14.28) and (14.29), from (14.27) we obtain

$$\mathcal{D}_y(T,s,t) = \frac{m(s,t)}{d(s)} \tag{14.30}$$

where

$$m(s,t) \triangleq \frac{\gamma \mathrm{e}^{sT}}{a}\left[\mathrm{e}^{a(t-qT)} - \frac{\mathrm{e}^{sT}-\mathrm{e}^{aT}}{\mathrm{e}^{sT}-1}\right]\mathrm{e}^{qsT}, \quad qT \leq t \leq (q+1)T \tag{14.31}$$

$$d(s) \triangleq \mathrm{e}^{sT} - \mathrm{e}^{aT} + \gamma \mathrm{e}^{aT} M(a) . \tag{14.32}$$

Let $\tilde{s}$ be the primary root of the equation $d(s) = 0$. Then,

$$e^{\tilde{s}T} = e^{aT}[1 - \gamma M(a)]. \tag{14.33}$$

Since $d'(s) = Te^{sT}$, the residue of $\mathcal{D}_y(T, s, t)$ at the pole $\tilde{s}$ equals

$$\operatorname*{Res}_{\tilde{s}} \mathcal{D}_y(T, s, t) = \frac{\gamma}{aT}\left[e^{a(t-qT)} - \frac{e^{\tilde{s}T} - e^{aT}}{e^{\tilde{s}T} - 1}\right] e^{q\tilde{s}T}, \quad qT \le t \le (q+1)T.$$

Besides the primary pole $\tilde{s}$, the function $\mathcal{D}_y(T, s, t)$ has another primary pole $s = 0$. The residue at this pole equals

$$\operatorname*{Res}_{s=0} \mathcal{D}_y(T, s, t) = -\frac{\gamma}{aT} \frac{1 - e^{aT}}{1 - e^{aT} - \gamma M(a)e^{aT}}.$$

Since

$$M(a) = \frac{1 - e^{-aT}}{a}$$

taking due account of (14.32) and (14.33) we obtain

$$\frac{e^{\tilde{s}T} - e^{aT}}{e^{\tilde{s}T} - 1} = \frac{\gamma}{\gamma - a}, \quad \frac{1 - e^{aT}}{1 - e^{aT} - \gamma M(a)e^{aT}} = \frac{a}{q - \gamma}.$$

As a result, the output process is given by

$$y(t) = \frac{\gamma}{a}\left[e^{a(t-qT)} - \frac{\gamma}{\gamma - a}\right] e^{q\tilde{s}T} + \frac{\gamma}{\gamma - a}, \quad qT \le t \le (q+1)T.$$

The processes for $T = 1$ and $\gamma = 1$ with $a = -1$ and $a = -2$ are presented in Fig. 14.2. □

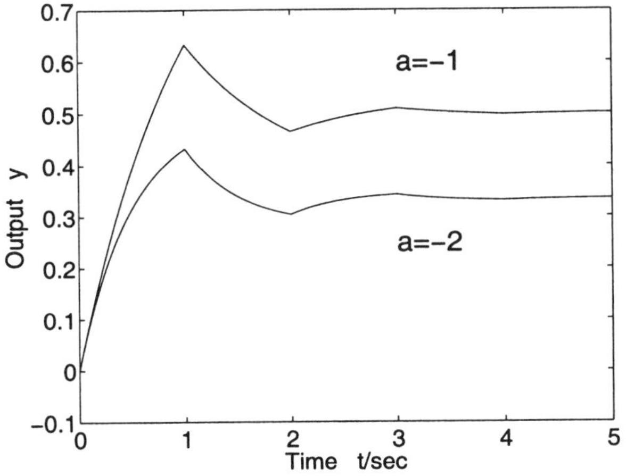

Figure 14.2: Output processes for Example 14.1

14.3 Steady-state processes

If (14.20) holds, Eq. (14.19) yields, in the general case,

$$y(t) = T \sum_i \operatorname*{Res}_{\tilde{s}_i} \mathcal{D}_y(T,s,t) + T \operatorname*{Res}_{b} \mathcal{D}_y(T,s,t)\,. \tag{14.34}$$

As it was shown in detail by Rosenwasser (1994c) that if a primary pole $\tilde{s}_i$ has the multiplicity ν_i,

$$\operatorname*{Res}_{\tilde{s}_i} \mathcal{D}_y(T,s,t) = \mathrm{e}^{\tilde{s}_i t} \sum_{k=0}^{\nu_i - 1} t^k g_{ik}(t) \tag{14.35}$$

where $g_{ik}(t) = g_{ik}(t+T)$ are bounded periodic functions. As follows from (14.35), if for all primary poles we have $\operatorname{Re} \tilde{s}_i < 0$, then, for $t \to \infty$ the solution (14.34) tends asymptotically to a steady-state process

$$y_\infty(t) = T \operatorname*{Res}_{b} \mathcal{D}_y(T,s,t)\,. \tag{14.36}$$

Using (14.25), from (14.36) we obtain

$$y_\infty(t) = \frac{1}{(m-1)!} \frac{\partial^{m-1}}{\partial s^{m-1}} \left[W(s,t) \mathrm{e}^{st} \right]_{|\, s=b} \tag{14.37}$$

which coincides with (7.18). Similar relations can be obtained for different choices of the input and output. Next, we consider an important special case of formula (14.37).

Let $X(s) = s^{-1}$, i.e., the input is a unit step (1.42). In this case, $b = 0$ and $m = 1$, and (14.37) yields

$$y_\infty(t) = W(0,t)\,. \tag{14.38}$$

Since

$$W(s,t) = W(s,t+T)$$

under the given assumption the steady-state output process $y_\infty(t)$ is a periodic oscillatory process with a period equal to the sampling period T. Taking due account of (14.2), we have

$$y_\infty(t) = R(0) \left[1 - \frac{W_d(0) \varphi_{FM}(T,0,t)}{1 + W_d(0) \varphi_{FM}(T,0,0)} \right]\,. \tag{14.39}$$

If the functions $F(s)$ or $W_d(s)$ have poles at $s = 0$, it is necessary to pass to the limit in (14.39).

The control error in the steady-state mode is given by

$$e_\infty(t) = x(t) - y_\infty(t) = 1 - y_\infty(t)$$

and is a periodic function of time. Therefore, in the general case $e_\infty(t) \neq 0$, and the notion of static correctness loses its meaning. Nevertheless, we can use this concept with respect to the average value of the steady-state process, if

$$\overline{e}_\infty = \frac{1}{T}\int_0^T e_\infty(t)\,\mathrm{d}t = 1 - \frac{1}{T}\int_0^T y_\infty(t)\,\mathrm{d}t = 0\,.$$

Moreover, there is an important exception, when $\mu(t) = 1$ and (9.12) holds. Then,

$$\varphi_{FM}(T,0,t) = \varphi_{FM}(T,0,0) = F(0)$$

and Eq. (14.39) takes the form

$$y_\infty(t) = \frac{R(0)}{1 + W_d(0)F(0)} = \text{const.} \tag{14.40}$$

In this case we have a constant steady-state error

$$e_\infty = 1 - y_\infty = 1 - \frac{R(0)}{1 + W_d(0)F(0)}$$

and the parameters of the system can be chosen so that the value e_∞ vanishes. This corresponds to the static correctness with respect to a constant input signal.

Remark Consider the continuous-time system shown in Fig. 14.3, which is obtained from Fig. 11.1 by changing the computer C for a linear continuous-time element with transfer function $W_d(s)$. For $x(t) = 1$ the steady-state output of the system is defined by (14.40). It can be shown that this technique of calculation of the steady-state process under a constant input is correct for any choice of input and output, if Eq. (9.12) is valid.

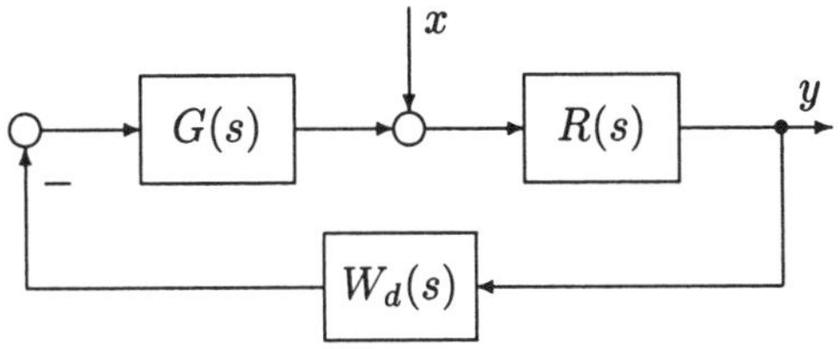

Figure 14.3: Static equivalent continuous-time system

For a polynomial input $X(s) = s^{-m}$, Eq. (14.37) yields

$$y_\infty(t) = \frac{1}{(m-1)!}\left\{\frac{\partial^{m-1}}{\partial s^{m-1}}\left[W(s,t)\mathrm{e}^{st}\right]\right\}_{\big| s=0} \tag{14.41}$$

which can be, after some transformations, presented in the form

$$y_\infty(t) = f_0(t) + f_1(t)t + \ldots + f_{m-1}(t)t^{m-1} \tag{14.42}$$

where $f_i(t)$ are known bounded periodic functions. As follows from (14.42), for $m > 1$ the notion of continuous-time static correctness possesses no meaning for closed-loop SD systems.

Considering the output processes in discrete-time, we have, generally speaking, a different situation. For instance, for $t_k = kT$, where k is an arbitrary integer, we have

$$y_\infty(kT) = f_0(kT) + f_1(kT)kT + \ldots + f_{m-1}(kT)(kT)^{m-1}. \tag{14.43}$$

Therefore, if $f_0(0) = 1$ and $f_i(0) = 0$ $(i = 1, \ldots, m-1)$, we have $e_\infty(kT) = 0$ for all integers k.

Next, we investigate steady-state processes under a harmonic excitation, when $m = 1$ and $b = \mathrm{j}\nu$ with a real parameter (frequency) ν. In this case, from (14.37) we obtain

$$y_\infty(t) = \Phi(\mathrm{j}\nu, t)\mathrm{e}^{\mathrm{j}\nu t} \tag{14.44}$$

where

$$\Phi(\mathrm{j}\nu, t) = W(s,t)\,|_{s=\mathrm{j}\nu}$$

is the parametric frequency response of the system. If the periods $T = 2\pi/\omega$ and $T_\nu = 2\pi/\nu$ are not commensurate, the output (14.44) is a doubly periodic function in t and is not a harmonic continuous function. Nevertheless, if we consider discrete samplings of the input, there appears a situation similar to that for open-loop systems. Indeed, assuming $t_k = \varepsilon + kT$ with $0 \le \varepsilon \le T$, we have a harmonic sequence (13.90). Then, since $W(\mathrm{j}\nu, t_k) = W(\mathrm{j}\nu, \varepsilon)$, we have

$$y_\infty(t_k) = \Phi(\mathrm{j}\nu, \varepsilon)\mathrm{e}^{\mathrm{j}(k2\pi\nu/\omega+\varepsilon)}$$

which is a harmonic sequence with the relative frequency $\overline{\omega} = 2\pi\nu/\omega$. Moreover, Eqs. (13.92) and (13.93) are valid, and all the differences between the behaviour of continuous- and discrete-time processes stated in Section 13.5 remain. In reality, these differences are caused by the fact, that a SD system considered in continuous time is a time-variable system with periodically varying parameters.

14.4 Analysis of deadbeat systems

As it was shown in Section 11.6, for a definite choice of the control program the PTFs of a feedback system has no poles and is an entire function in s. For such systems, the integrals defining processes with zero initial energy, cannot be calculated by the residue theorem, and other methods should be used. For concreteness, we consider the system shown in Fig. 11.1 with $x_v(t) = x(t) = 0$ with respect to the output $y(t)$. Then, the operator equation has the form

$$y = W(s,t)x_d \tag{14.45}$$

where

$$W(s,t) \triangleq W_{11}(s,t) = \frac{W_d(s)\varphi_{FM}(T,s,t)}{1 + W_d(s)\varphi_{FM}(T,s,0)}. \tag{14.46}$$

The corresponding discrete transfer function $D(s,t)$ will be written in the form

$$D(s,t) \triangleq W_{11}(s,t)e^{st} = \frac{W_d(s)\mathcal{D}_{FM}(T,s,t)}{1+W_d(s)\mathcal{D}_{FM}(T,s,0)} \tag{14.47}$$

taking into account that $\varphi_{FM}(T,s,0) = \mathcal{D}_{FM}(T,s,0)$. The discrete Laplace transform of the output is given by (11.115), which can be written in the form

$$\mathcal{D}_y(T,s,t) = \frac{W_d(s)\mathcal{D}_{FM}(T,s,t)}{1+W_d(s)\mathcal{D}_{FM}(T,s,0)}\mathcal{D}_{x_d}(T,s,+0)\,. \tag{14.48}$$

Henceforth we assume that all the conditions set in Section 14.1 hold, and the input $x_d(t)$ has the form (13.2). Then, from (13.7) we have

$$\mathcal{D}_{x_d}(T,s,+0) = \frac{1}{(m-1)!}\frac{\partial^{m-1}}{\partial b^{m-1}}\frac{e^{sT}}{e^{sT}-e^{bT}}\,. \tag{14.49}$$

The inversion integral for the output process takes the form

$$y(t) = \frac{T}{2\pi j}\int_{c-j\omega/2}^{c+j\omega/2}\frac{W_d(s)\mathcal{D}_{FM}(T,s,t)}{1+W_d(s)\mathcal{D}_{FM}(T,s,0)}\frac{1}{(m-1)!}\frac{\partial^{m-1}}{\partial b^{m-1}}\frac{e^{sT}}{e^{sT}-e^{bT}}\,ds \tag{14.50}$$

where c is a constant such that all poles of the integrand lie in the half-plane $\operatorname{Re} s < c$. Equation (14.50) can be represented in the form

$$y(t) = \frac{1}{(m-1)!}\frac{\partial^{m-1}}{\partial b^{m-1}}y_1(t,b) \tag{14.51}$$

where

$$y_1(t,b) \triangleq \frac{T}{2\pi j}\int_{c-j\omega/2}^{c+j\omega/2}\frac{W_d(s)\mathcal{D}_{FM}(T,s,t)}{1+W_d(s)\mathcal{D}_{FM}(T,s,0)}\frac{e^{sT}}{e^{sT}-e^{bT}}\,ds\,. \tag{14.52}$$

Thus, the problem is reduced to the calculation of the integral (14.52). For further transformations, we note that changing the variable ζ for e^{-sT} in (9.149)–(9.151), we have

$$\mathcal{D}_{FM}(T,s,t) = \frac{\beta(s,t)}{\alpha(s)}\,,\quad 0 \le t \le T \tag{14.53}$$

where

$$\begin{gathered}\beta(s,t) = \beta_0(t) + \beta_1(t)e^{-sT} + \ldots + \beta_n(t)e^{-nsT}\\ \alpha(s) = \prod_{i=1}^{\ell}\left(1-e^{s_iT}\right)^{\nu_i}\,,\quad \sum_{i=1}^{\ell}\nu_i = n.\end{gathered} \tag{14.54}$$

Moreover, as follows from (9.160) for $\zeta = e^{-sT}$,

$$\mathcal{D}_{FM}(T,s,0) = \frac{e^{-sT}\chi(s)}{\alpha(s)} \tag{14.55}$$

where

$$\chi(s) = \beta_1(0) + \beta_2(0)e^{-sT} + \ldots + \beta_n(0)e^{-(n-1)sT} . \tag{14.56}$$

Let the transfer function $W_d(s)$ have the form (10.18) and (10.19) with $a_0 = 1$. Then, the transients have a finite duration if (11.93) holds. In this case, Eq. (11.94) yields

$$D(s,t) = \beta(s,t)b(s) , \quad 0 \le t \le T . \tag{14.57}$$

Henceforth we assume that

$$a(s) = \tilde{a}(\zeta)\,|_{\zeta=e^{-sT}} , \quad b(s) = \tilde{b}(\zeta)\,|_{\zeta=e^{-sT}} \tag{14.58}$$

where the polynomials $\tilde{a}(\zeta)$ and $\tilde{b}(\zeta)$ give the minimal solution of Eq. (11.91) satisfying the condition (11.98). Then, $\deg \tilde{b}(\zeta) \le n-1$, and, taking account of (14.54), we can write

$$D(s,t) = \sum_{k=0}^{2n-1} d_k(t)e^{-ksT} , \quad 0 \le t \le T \tag{14.59}$$

where $d_k(t)$ are known functions. Since

$$D(s, t+T) = D(s,t)e^{sT} \tag{14.60}$$

from (14.59) for an arbitrary integer $q \ge 0$ we have

$$D(s,t) = \sum_{k=0}^{2n-1} d_k(t-qT)e^{(q-k)sT} , \quad qT \le t \le (q+1)T . \tag{14.61}$$

Substituting (14.61) into (14.52), for $qT \le t \le (q+1)T$ we find

$$y_1(t,b) = \frac{T}{2\pi j} \sum_{k=0}^{2n-1} d_k(t-qT) \int_{c-j\omega/2}^{c+j\omega/2} \frac{e^{(q-k)sT}e^{sT}}{e^{sT} - e^{bT}} \, ds , \quad c > \operatorname{Re} b . \tag{14.62}$$

The residue theorem is not, in the general case, applicable for calculating the integrals in (14.62), because the integrand does not vanish as $\operatorname{Re} s \to -\infty$. Nevertheless, the function $y_1(t,b)$ can be calculated directly. With this aim in view, we write (14.62) in the form

$$y_1(t,b) = \sum_{k=0}^{2n-1} d_k(t-qT)J_{q-k} , \quad qT \le t \le (q+1)T \tag{14.63}$$

where

$$J_m = \frac{T}{2\pi j} \int_{c-j\omega/2}^{c+j\omega/2} \frac{e^{(m+1)sT}}{e^{sT} - e^{bT}} \, ds , \quad \operatorname{Re} s > \operatorname{Re} b . \tag{14.64}$$

Lemma 14.1 *The following equations hold:*

$$J_m = \begin{cases} e^{mbT} & m \ge 0 \\ 0 & m < 0 . \end{cases} \tag{14.65}$$

Proof For $m \geq 0$ we have

$$\ell^{-}\left[\frac{e^{(m+1)sT}}{e^{sT}-e^{bT}}\right] = 0\,.$$

Therefore,

$$\operatorname*{Res}_{b}\left[\frac{e^{(m+1)sT}}{e^{sT}-e^{bT}}\right] = \frac{1}{T}e^{mbT}$$

and the first relation in (14.65) follows from (A.83). If $m < 0$,

$$\ell^{+}\left[\frac{e^{(m+1)sT}}{e^{sT}-e^{bT}}\right] = 0$$

so that the integrand in (14.64) is analytic in $\operatorname{Re} s \geq c$, and the second equation in (14.65) follows from (A.82). ■

Using Lemma 14.1, we obtain an explicit expression for the output process. With this aim in view, we rewrite (14.63) for $0 \leq q < 2n-1$ in the form

$$y_1(t,b) = \sum_{k=0}^{q} d_k(t-qT)J_{q-k} + \sum_{k=q+1}^{2n-1} d_k(t-qT)J_{q-k} \quad qT \leq t \leq (q+1)T\,.$$

Due to (14.65), the second sum on the right-hand side equals zero, and we have

$$y_1(t,b) = \sum_{k=0}^{q} d_k(t-qT)J_{q-k}\,, \quad qT \leq t \leq (q+1)T$$

or, equivalently,

$$y_1(t,b) = \sum_{k=0}^{q} d_k(t-qT)e^{(q-k)bT}\,, \quad qT \leq t \leq (q+1)T\,. \tag{14.66}$$

Equations (14.66) can be written in an expanded form as

$$y_1(t,b) = \begin{cases} d_0(t) & 0 \leq t \leq T \\ d_0(t-T)e^{bT} + d_1(t-T) & T \leq t \leq 2T \\ \vdots & \vdots \\ d_0(t-qT)e^{qbT} + \ldots + d_q(t-qT) & qT \leq t \leq (q+1)T\,. \end{cases} \tag{14.67}$$

For $q \geq 2n-1$ Eq. (14.63) yields

$$y_1(t,b) = \sum_{k=0}^{2n-1} d_k(t-qT)e^{(q-k)bT}\,, \quad qT \leq t \leq (q+1)T\,. \tag{14.68}$$

Comparing (14.61) with (14.68), we obtain

$$y_1(t,b) = D(b,t) = W(b,t)e^{bt}$$

which is the steady-state process $y_{1\infty}(t)$. Then, the output process for any m is obtained from (14.51).

The function

$$\tilde{y}_1(t,b) \triangleq \begin{cases} y_1(t,b) - D(b,t) & t \le (2n-1)T \\ 0 & t \ge (2n-1)T \end{cases} \tag{14.69}$$

presents the transient of $y_1(t,b)$. From (14.69) it follows that, if we construct the control program on the basis of the minimal solution of the Diophantine equation (11.91), the transient in the system will not need more than $2n-1$ periods. In Rosenwasser (1994c) it is shown that this property remains valid for arbitrary initial conditions as well. If we use a solution of (11.91) that is not minimal, by (11.103) we have $\deg b(s) \ge n$ and the degree of the quasi-polynomial $\beta(s,t)b(s)$ increases. Accordingly, the duration of the transient increases also. Hence, a controller associated with the minimal solution of the polynomial equation (11.91) ensures time-optimal transient processes.

Example 14.2 Consider the time-transient behaviour of the feedback digital system shown in Fig. 14.4, where

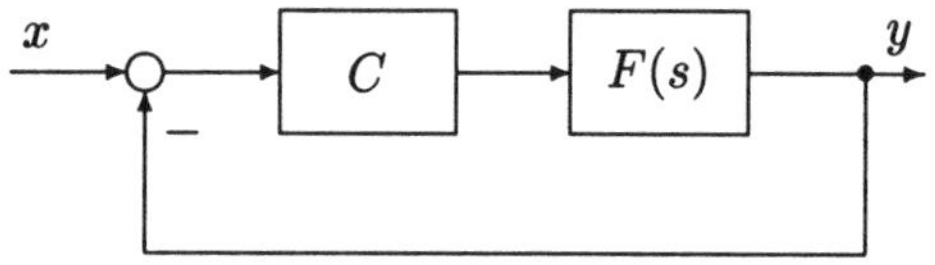

Figure 14.4: Digital control system for Example 14.2

$$F(s) = \frac{\gamma}{s-a}, \quad X(s) = \frac{1}{s-b}$$

with an arbitrary pulse form. For $0 \le t \le T$ from the second equation of (9.129) we have

$$\mathcal{D}_{FM}(T,s,t) = \gamma \frac{M(a)e^{at} - \left[1 - e^{(a-s)T}\right] \int_t^T e^{a(t-\tau)}\mu(\tau)\,d\tau}{1 - e^{(a-s)T}}$$

which can be represented in the form

$$\mathcal{D}_{FM}(T,s,t) = \frac{h_p(t) - e^{-sT}h_p^*(t)e^{aT}}{1 - e^{(a-s)T}}$$

with

$$h_p(t) = \gamma \int_0^t e^{a(t-\tau)}\mu(\tau)\,d\tau\,, \quad h_p^*(t) = -\gamma \int_t^T e^{a(t-\tau)}\mu(\tau)\,d\tau\,.$$

Hence, taking account of (9.149)–(9.151), we have

$$\alpha(s) = 1 - \mathrm{e}^{aT}\mathrm{e}^{-sT}, \quad \beta_0(t) = h_p(t), \quad \beta_1(t) = -\mathrm{e}^{aT} h_p^*(t).$$

Moreover, from the results of Example 11.5 it follows that the minimal solution of the Diophantine equation (11.91) is associated with the transfer function $W_d(s)$ with

$$a(s) = 1, \quad b(s) = \frac{1}{\gamma M(a)}.$$

Therefore, taking account of (14.59), for $0 \le t \le T$ we have

$$D(s,t) = \beta(s,t)b(s) = d_0(t) + d_1(t)\mathrm{e}^{-sT}$$

where

$$d_0(t) = \frac{1}{M(a)} \int_0^t \mathrm{e}^{a(t-\tau)}\mu(\tau)\,\mathrm{d}\tau, \quad d_1(t) = \frac{\mathrm{e}^{aT}}{M(a)} \int_t^T \mathrm{e}^{a(t-\tau)}\mu(\tau)\,\mathrm{d}\tau.$$

Using (14.67), we find

$$y_1(t,b) = \begin{cases} d_0(t) & 0 \le t \le T \\ d_0(t-T)\mathrm{e}^{bT} + d_1(t-T) & T \le t \le 2T \end{cases}$$

whence follows that the transient vanishes for $t \ge T$, i.e.,

$$y(t) = y_\infty(t), \quad t \ge T$$

and the steady-state process has the form

$$y_\infty(t) = W(b,t)\mathrm{e}^{bt} \tag{14.70}$$

where for $0 \le t \le T$

$$W(b,t) = \frac{\mathrm{e}^{-bt}}{M(a)} \left[\int_0^t \mathrm{e}^{a(t-\tau)}\mu(\tau)\,\mathrm{d}\tau + \mathrm{e}^{(a-b)T} \int_t^T \mathrm{e}^{a(t-\tau)}\mu(\tau)\,\mathrm{d}\tau \right] \tag{14.71}$$

and $W(b,t) = W(b,t+T)$. For $b = 0$ the solution (14.70) gives the steady-state periodic oscillatory process

$$y_\infty(t) = \frac{1}{M(a)} \left[\int_0^t \mathrm{e}^{a(t-\tau)}\mu(\tau)\,\mathrm{d}\tau + \mathrm{e}^{aT} \int_t^T \mathrm{e}^{a(t-\tau)}\mu(\tau)\,\mathrm{d}\tau \right], \quad 0 \le t \le T. \tag{14.72}$$

This result can also be obtained from a different viewpoint. For $a(s) = 1$ and $b(s) = [\gamma M(a)]^{-1}$ the system shown in Fig. 14.4 reduces to the SD system shown in Fig. 14.5, which, for $x(t) = 1$, is described by the equations

$$\frac{\mathrm{d}y}{\mathrm{d}t} - ay = \frac{\mu(t-qT)}{M(a)}[-y(qT) + 1], \quad 0 \le qT \le t \le (q+1)T.$$

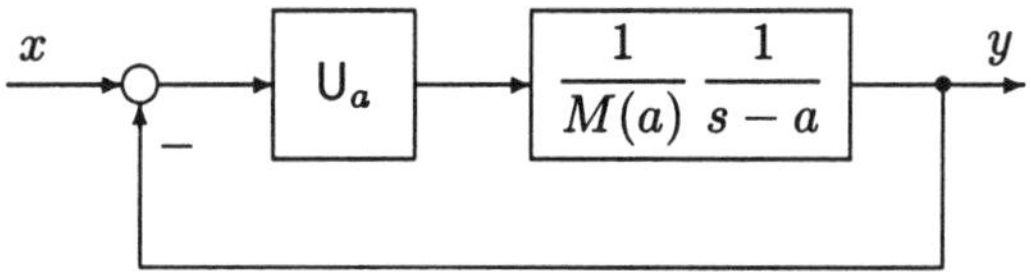

Figure 14.5: Special system representation of Example 14.2

The solution with zero initial energy, i.e., for $y(0) = 0$, has the form

$$y(t) = \frac{1}{M(a)} \int_0^t e^{a(t-\tau)} \mu(\tau)\, d\tau\,, \quad 0 \le t \le T$$

whence

$$y(T) = \frac{e^{aT}}{M(a)} \int_0^T e^{-a\tau} \mu(\tau)\, d\tau = e^{aT}\,.$$

On the interval $T \le t \le 2T$ we obtain

$$\frac{dy}{dt} - ay = \frac{\mu(t-T)}{M(a)} \left(1 - e^{aT}\right)$$

which must be solved under the condition $y(T) = e^{aT}$. This solution can be written in the form

$$y(t) = e^{at} + \frac{1 - e^{aT}}{M(a)} \int_0^{t-T} e^{a(t-u-T)} \mu(u)\, du\,, \quad T \le t \le 2T\,. \tag{14.73}$$

It can be immediately verified, that $y(2T) = e^{aT}$. Hence, for $t \ge T$, the steady-state process in the system is a periodic process. For $0 \le t \le T$ the solution (14.73) has the form

$$y(t) = e^{a(t+T)} + \frac{1 - e^{aT}}{M(a)} \int_0^t e^{a(t-\tau)} \mu(\tau)\, d\tau\,. \tag{14.74}$$

Taking into account that

$$\int_0^t e^{a(t-\tau)} \mu(\tau)\, d\tau = e^{at} M(a) - \int_t^T e^{a(t-\tau)} \mu(\tau)\, d\tau$$

we find that Eqs. (14.72) and (14.74) coincide. □

14.5 Natural oscillations and stability

In this section we consider the feedback system shown in 14.6, where the transfer function of the continuous-time element $F(s)$ has the form (13.1), and the relation between the input $y(t)$ and the output $v(t)$ of the computer C is defined by Eqs. (10.1)–(10.3) with $R = 1$. This system will be called *stable*, if without exogenous disturbance and with arbitrary initial conditions of the continuous-time element and control program for $t > 0$ the estimates

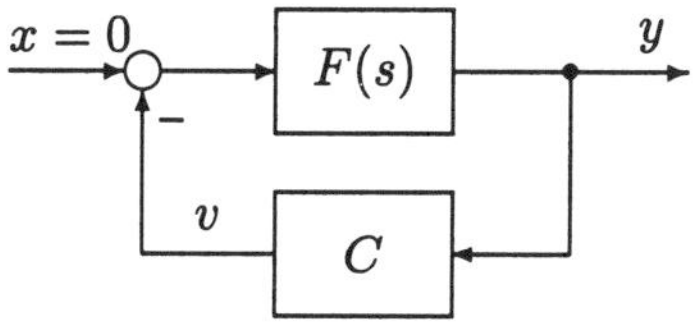

Figure 14.6: Feedback SD system

$$|y(t)| < c_1 e^{-\lambda t}, \quad |v(t)| < c_2 e^{-\lambda t} \tag{14.75}$$

hold, where c_1, c_2, and λ are positive constants, and λ is independent of the initial conditions. This definition is, in essense, equivalent to Lyapunov's definition of stability.

Theorem 14.4 *Let the non-pathological conditions (9.161) and (9.162) hold. Then, for the system to be stable it is necessary and sufficient that the determining equation*

$$\Delta(s) = \alpha(s)a(s) - e^{-sT}\chi(s)b(s) = 0 \tag{14.76}$$

is free of roots in the closed right half-plane.

Proof Consider the discrete Laplace transforms of the output $y(t)$ given by (11.127) and (11.128). In this case, the corresponding components of the output are given by the inversion integrals

$$\begin{aligned} y_d(t) &\triangleq \frac{T}{2\pi j}\int_{c-j\omega/2}^{c+j\omega/2} \frac{N_{dy}(s,t)}{\Delta(s)}\,ds \\ \tilde{y}_0(t) &\triangleq \frac{T}{2\pi j}\int_{c-j\omega/2}^{c+j\omega/2} \frac{N_{0\tilde{y}}(s,t)}{\Delta(s)}\,ds \end{aligned} \tag{14.77}$$

where c is a real constant such that all roots of Eq. (14.76) lie in the half-plane $\operatorname{Re} s < c$. The integrals (14.77) satisfy all the conditions, under which (A.116) holds. In this case, there are two possibilities. If

$$\Delta(s) = \text{const.} \neq 0 \tag{14.78}$$

the integrals (14.77) for any arbitrary conditions define deadbeat processes that decay faster than any exponent.

Consider the second variant, when $\Delta(s) \neq \text{const.}$ In this case, the integrand in (14.77) has a finite number of primary poles $\tilde{s}_i$, that are roots of Eq. (14.76). Then, using (A.116), for sufficiently large $t > 0$ we have

$$\begin{aligned} y_d(t) &= T\sum_i \operatorname*{Res}_{\tilde{s}_i} \frac{N_{dy}(s,t)}{\Delta(s)} \\ \tilde{y}_0(t) &= T\sum_i \operatorname*{Res}_{\tilde{s}_i} \frac{N_{0\tilde{y}}(s,t)}{\Delta(s)}\,. \end{aligned}$$

The corresponding residues can be represented in the form (14.35). Then, if all roots $\tilde{s}_i$ lie in the half-plane $\operatorname{Re} s \leq -a$ with $a > 0$, for sufficiently large $t > 0$ we have

$$|y(t)| < c\mathrm{e}^{-(a-\epsilon)t}$$

where ϵ is a sufficiently small positive number, i.e., $y(t)$ satisfies the condition (14.75). Thus, the sufficiency of the conditions of the theorem is proved.

Let the determining equation (14.76) have a primary root $\tilde{s}_0$ in the half-plane $\operatorname{Re} s \geq 0$. Since the quasi-polynomial $b(s)$ in (11.127) is arbitrary, it can always be chosen in such a way that the image $\mathcal{D}_{y_d}(T, s, t)$ has a pole $s = \tilde{s}_0$. Then, using the residue theorem, we obtain that the component of the output associated with the pole $\tilde{s}_0$, is not a damped function and the system is not stable. Similar reasoning can be used with respect to the variable $v(t)$. ∎

Corollary 1 Substituting ζ for e^{-sT} in (14.76), we get an algebraic characteristic equation

$$\Delta^o(\zeta) = \alpha^o(\zeta)a^o(\zeta) + \zeta\chi^o(\zeta)b^o(\zeta) = 0\,. \tag{14.79}$$

In this case, for the system to be stable it is necessary and sufficient that Eq. (14.79) is free of poles inside and on the unit circle.

Corollary 2 Let $\deg \Delta^o(\zeta) = m$. Denote

$$\tilde{\Delta}(z) = \Delta^o(z^{-1})z^m\,. \tag{14.80}$$

Then, under the given assumptions, for the stability it is necessary and sufficient that all roots of the polynomial $\tilde{\Delta}(z)$ lie inside the unit circle.

Example 14.3 Consider the stability behaviour of the feedback system shown in Fig. 14.1, where

$$F(s) = \frac{\gamma}{s-a}\,, \quad W_d(s) = 1\,, \quad M(a) \neq 0$$

and the pulse form $\mu(t)$ is arbitrary. In this case,

$$\alpha(s) = 1 - \mathrm{e}^{aT}\mathrm{e}^{-sT}\,, \quad \chi(s) = \gamma M(a)\mathrm{e}^{aT}$$

and the determining equation (14.76) takes the form

$$1 - \mathrm{e}^{aT}\mathrm{e}^{-sT} + \gamma M(a)\mathrm{e}^{aT}\mathrm{e}^{-sT} = 0\,.$$

The characteristic equation (14.79) appears as

$$1 + \zeta\,[\,\gamma M(a) - 1\,]\,\mathrm{e}^{aT} = 0\,.$$

Equation (14.80) has the form

$$z - e^{aT} + \gamma M(a)e^{aT} = 0\,.$$

In this case, the feedback system is stable iff

$$e^{aT}|1 - \gamma M(a)| < 1\,.$$

□

Example 14.4 Let us consider a special case in the stability analysis of the feedback system shown in Fig. 14.1 with

$$F(s) = \frac{\gamma}{s-a}\,, \quad W_d(s) = \frac{1 - e^{aT}e^{-sT}}{1 - 0.5e^{-sT}}$$

where a is a real number such that $e^{aT} \neq 0$. The pulse form is assumed to be arbitrary and $M(a) \neq 0$. Then,

$$a(s) = 1 - 0.5e^{-sT}\,, \quad b(s) = 1 - e^{aT}e^{-sT}$$

and the determining equation (14.76) takes the form

$$\Delta(s) = \left(1 - e^{aT}e^{-sT}\right)\left[1 - 0.5e^{-sT} + \gamma M(a)e^{aT}e^{-sT}\right] = 0\,.$$

The characteristic equation (14.79) has the form

$$\Delta^o(\zeta) = \left(1 - e^{aT}\zeta\right)\left[1 - 0.5\zeta + \gamma M(a)e^{aT}\zeta\right] = 0\,.$$

This equation has for arbitrary a and independently of the pulse form one root $\zeta = e^{-aT}$. Therefore, for $a \geq 0$ the system is unstable. We note that the ratio

$$\mathcal{D}_{FM}(T, s, 0)W_d(s) = \frac{e^{-sT}\gamma M(a)e^{sT}}{1 - e^{aT}e^{-sT}}\,\frac{1 - e^{aT}e^{-sT}}{1 - 0.5e^{-sT}}$$

is reducable. Nevertheless, cancellation of a common multiplier on the right-hand side can lead to an incorrect conclusion as regards the stability of the feedback system. □

Remark We note without a proof that all results of this section are extended, without major modifications, to time-delayed systems. If

$$F(s) = F_1(s)e^{-s\tau} \tag{14.81}$$

where $\tau > 0$ is a constant and $F_1(s)$ is a real rational function of the form (13.1). In this case, the determining equation has the form

$$\Delta_\tau(s) \triangleq \alpha(s)a(s) + e^{-msT}\beta(s, \delta T)b(s) = 0 \tag{14.82}$$

where $m \geq 0$ is an integer such that $\tau = mT - \delta T$ with $0 < \delta < 1$. The corresponding characteristic equation is obtained by substituting ζ for e^{-sT}.

14.6 Stabilization of feedback systems

Under the given assumptions the system is stable iff the characteristic polynomial $\Delta^o(\zeta)$ is free of roots inside and on the unit circle. Such polynomials will henceforth be called *stable*. The set of all stable polynomials is henceforth denoted by Γ_+. This condition can be written as

$$\Delta^o(\zeta) \in \Gamma_+ . \tag{14.83}$$

It is noteworthy that any nonzero constant a belongs to Γ_+. An arbitrary element of the set Γ_+ will be denoted by $\Delta_+(\zeta)$. Then, taking account of (14.79) the condition (14.83) can be written in the form

$$\alpha^o(\zeta)a^o(\zeta) + \zeta\chi^o(\zeta)b^o(\zeta) = \Delta_+(\zeta) . \tag{14.84}$$

For known polynomials $\alpha^o(\zeta)$, $\chi^o(\zeta)$ and $\Delta_+(\zeta)$, Eq. (14.84) can be considered as a polynomial equation with respect to the unknown polynomials $a^o(\zeta)$ and $b^o(\zeta)$ that define the control program (10.60). If we consider Eq. (14.84) for all stable polynomials $\Delta_+(\zeta)$, it defines the set of all polynomial pairs $\{a^o(\zeta), b^o(\zeta)\}$, that stabilize the feedback system. Such pairs will be henceforth called *stabilizing*. Any stabilizing pair $\{a^o(\zeta), b^o(\zeta)\}$ can be associated with a real rational function

$$W_d^o(\zeta) = \frac{b^o(\zeta)}{a^o(\zeta)} \tag{14.85}$$

which will be called the transfer function of the pair $\{a^o(\zeta), b^o(\zeta)\}$. In this section we find all stabilizing pairs and corresponding transfer functions.
Since the constant $1 \in \Gamma_+$, any solution $\{\tilde{a}^o(\zeta), \tilde{b}^o(\zeta)\}$ of the Diophantine equation

$$\alpha^o(\zeta)\tilde{a}^o(\zeta) + \zeta\chi^o(\zeta)\tilde{b}^o(\zeta) = 1 \tag{14.86}$$

is a stabilizing pair. Henceforth, any pair $\{\tilde{a}^o(\zeta), \tilde{b}^o(\zeta)\}$ satisfying (14.86) will be called a *basic pair* or a *basis*. The basic pair $\{\tilde{a}(\zeta), \tilde{b}(\zeta)\}$ satisfying (11.98) will be called *minimal*. As is known, Volgin (1986), Kučera (1979), the general solution of Eq. (14.84) for all possible $\Delta_+(\zeta)$ can be represented in the form

$$\begin{aligned} a^o(\zeta) &= \tilde{a}^o(\zeta)\Delta_+(\zeta) - \zeta c(\zeta)\chi^o(\zeta) \\ b^o(\zeta) &= \tilde{b}^o(\zeta)\Delta_+(\zeta) + c(\zeta)\alpha^o(\zeta) \end{aligned} \tag{14.87}$$

where $\{\tilde{a}^o(\zeta), \tilde{b}^o(\zeta)\}$ is a basic pair and $c(\zeta)$ is an arbitrary polynomial (it can also be a constant).

Example 14.5 Find the set of stabilizing pairs for the feedback system under the conditions of Example 14.2. According to the results of Example 11.5, the minimal stabilizing pair for this example is given by

$$\tilde{a}(\zeta) = 1 , \quad \tilde{b}(\zeta) = \frac{1}{\gamma M(a)} .$$

Then, using (14.87), we can represent the set of all stabilizing pairs in the form

$$a^o(\zeta) = \Delta_+(\zeta) - \zeta c(\zeta)\gamma M(a)e^{aT}$$
$$b^o(\zeta) = \frac{1}{\gamma M(a)}\Delta_+(\zeta) + c(\zeta)\left(1 - e^{aT}\zeta\right)$$

□

Let $\{a^o(\zeta), b^o(\zeta)\}$ be an arbitrary stabilizing pair and $\{\tilde{a}^o(\zeta), \tilde{b}^o(\zeta)\}$ be a basis. Then, the representation (14.87) will be called a *basis representation* of the pair $\{a^o(\zeta), b^o(\zeta)\}$ (in the basis $\{\tilde{a}^o(\zeta), \tilde{b}^o(\zeta)\}$). We note some important properties of the basis representation (14.87).

1. For a given basis $\{\tilde{a}^o(\zeta), \tilde{b}^o(\zeta)\}$ the representation (14.87) is unique in the sense that if, together with (14.87), we have

$$\begin{aligned} a^o(\zeta) &= \tilde{a}^o(\zeta)\Delta_{+1}(\zeta) - \zeta c_1(\zeta)\chi^o(\zeta) \\ b^o(\zeta) &= \tilde{b}^o(\zeta)\Delta_{+1}(\zeta) + c_1(\zeta)\alpha^o(\zeta) \end{aligned} \tag{14.88}$$

where $\Delta_{+1}(\zeta)$ and $c_1(\zeta)$ are polynomials, then

$$\Delta_+(\zeta) = \Delta_{+1}(\zeta), \quad c(\zeta) = c_1(\zeta).$$

Proof Subtracting (14.88) from (14.87), we obtain

$$[\Delta_+(\zeta) - \Delta_{+1}(\zeta)]\,\tilde{a}^o(\zeta) - \zeta\,[c(\zeta) - c_1(\zeta)]\,\chi^o(\zeta) = 0$$
$$[\Delta_+(\zeta) - \Delta_{+1}(\zeta)]\,\tilde{b}^o(\zeta) - [c(\zeta) + c_1(\zeta)]\,\alpha^o(\zeta) = 0.$$

But, due to the choice of $\{\tilde{a}^o(\zeta), \tilde{b}^o(\zeta)\}$, we have

$$\det\begin{bmatrix} \tilde{a}^o(\zeta) & -\zeta\chi^o(\zeta) \\ \tilde{b}^o(\zeta) & \alpha^o(\zeta) \end{bmatrix} = 1. \tag{14.89}$$

Hence, $\Delta_+(\zeta) = \Delta_{+1}(\zeta)$ and $c(\zeta) = c_1(\zeta)$. ■

2. The polynomials $a^o(\zeta)$ and $b^o(\zeta)$ appearing in (14.87) have a common root iff the polynomials $\Delta_+(\zeta)$ and $c(\zeta)$ have one.

Proof Let $\zeta = \zeta_0$ with $a^o(\zeta_0) = b^o(\zeta_0) = 0$. Then,

$$\tilde{a}^o(\zeta_0)\Delta_+(\zeta_0) - \zeta_0 c(\zeta_0)\chi^o(\zeta_0) = 0$$
$$\tilde{b}^o(\zeta_0)\Delta_+(\zeta_0) + c(\zeta_0)\alpha^o(\zeta_0) = 0$$

and, taking account of (14.89), we obtain $\Delta_+(\zeta_0) = 0$ and $c(\zeta_0) = 0$. Conversely, if $\Delta_+(\zeta_0) = c(\zeta_0) = 0$, from (14.87) it follows that $a^o(\zeta_0) = b^o(\zeta_0) = 0$. ■

3. Since a stable polynomial does not possess poles inside the unit circle, we always have $\Delta_+(0) \neq 0$. Moreover, $\alpha^o(0) = 1$ due to (11.69). Then, for $\zeta = 0$ from (14.87) we have

$$a^o(0) \neq 0 \tag{14.90}$$

for any stabilizing pair $\{a^o(\zeta), b^o(\zeta)\}$.

4. The transfer function $W_d^o(\zeta)$ (14.85) will be called *stabilizing* if its numerator and denominator form a stabilizing pair. The set of all stabilizing functions is given by the following theorem.

Theorem 14.5 *Let $\{\tilde{a}^o(\zeta), \tilde{b}^o(\zeta)\}$ be a basis. Then, the set of all irreducible transfer functions can be defined by*

$$W_d^o(\zeta) = \frac{\tilde{b}^o(\zeta) + \phi(\zeta)\alpha^o(\zeta)}{\tilde{a}^o(\zeta) - \zeta\phi(\zeta)\chi^o(\zeta)} \tag{14.91}$$

where

$$\phi(\zeta) = \frac{c(\zeta)}{\Delta_+(\zeta)} \tag{14.92}$$

is an irreducible ratio with an arbitrary polynomial $c(\zeta)$ and an arbitrary stable polynomial (or a nonzero constant) $\Delta_+(\zeta)$.

Proof Denote by $\mathcal{W}_d$ the set of all irreducible stabilizing transfer functions, and by $\mathcal{W}_0$ the set of all ratios of the form (14.91) and (14.92). Let the polynomials $\Delta_+(\zeta)$ and $c(\zeta)$ be co-prime. Substituting (14.92) in (14.91), we obtain the real rational function

$$W_d^o(\zeta) = \frac{\tilde{b}^o(\zeta)\Delta_+(\zeta) + c(\zeta)\alpha^o(\zeta)}{\tilde{a}^o(\zeta)\Delta_+(\zeta) - \zeta c(\zeta)\chi^o(\zeta)} \tag{14.93}$$

which is stabilizing. Due to Property 2 and the irreducibility of the ratio $\phi(\zeta)$, the ratio (14.93) is irreducible. Thus, $\mathcal{W}_0 \subset \mathcal{W}_d$. On the other hand, let the ratio (14.85) be stabilizing and irreducible. Then, the polynomials $a^o(\zeta)$ and $b^o(\zeta)$ can be represented in the form (14.87), and we can write $W_d^o(\zeta)$ in the form (14.91) and (14.92), i.e., $\mathcal{W}_d \subset \mathcal{W}_0$. Therefore, the sets $\mathcal{W}_d$ and $\mathcal{W}_0$ coincide. ∎

5. Taking into account that $\Delta_+(0) \neq 0$ and $\tilde{a}^o(0) \neq 0$, from (14.91) we find that the value $W_d^o(0)$ is bounded, i.e., any stabilizing transfer function is analytic for $\zeta = 0$. Hence, using the definition of Section 2, we arrive at the conclusion that any stabilizing transfer function is causal.

6. The representation of a stabilizing transfer function in the form (14.91)–(14.92) will be called its *basis representation.*

Theorem 14.6 *For a given basis* $\{\tilde{a}^o(\zeta), \tilde{b}^o(\zeta)\}$ *the representation (14.91)–(14.92) is unique, i.e., if we have at the same time*

$$W_d^o(\zeta) = \frac{\tilde{b}^o(\zeta) + \phi_1(\zeta)\alpha^o(\zeta)}{\tilde{a}^o(\zeta) - \zeta\phi_1(\zeta)\chi^o(\zeta)}$$

where

$$\phi_1(\zeta) = \frac{c_1(\zeta)}{\Delta_{+1}(\zeta)}$$

then $\phi_1(\zeta) = \phi(\zeta)$.

Proof Under the given assumptions we simultaneously have

$$W_d^o(\zeta) = \frac{\tilde{b}^o(\zeta)\Delta_+(\zeta) + c(\zeta)\alpha^o(\zeta)}{\tilde{a}^o(\zeta)\Delta_+(\zeta) - \zeta c(\zeta)\chi^o(\zeta)}$$
$$W_d^o(\zeta) = \frac{\tilde{b}^o(\zeta)\Delta_{+1}(\zeta) + c_1(\zeta)\alpha^o(\zeta)}{\tilde{a}^o(\zeta)\Delta_{+1}(\zeta) - \zeta c_1(\zeta)\chi^o(\zeta)}.$$

Since the upper ratio is irreducible by assumption, we have

$$\tilde{a}^o(\zeta)\Delta_{+1}(\zeta) - \zeta c_1(\zeta)\chi^o(\zeta) = d(\zeta)\left[\tilde{a}^o(\zeta)\Delta_+(\zeta) - \zeta c(\zeta)\chi^o(\zeta)\right]$$
$$\tilde{b}^o(\zeta)\Delta_{+1}(\zeta) + c_1(\zeta)\alpha^o(\zeta) = d(\zeta)\left[\tilde{b}^o(\zeta)\Delta_+(\zeta) + c(\zeta)\alpha^o(\zeta)\right]$$

where $d(\zeta)$ is a polynomial. The last equation can be represented as

$$\tilde{a}^o(\zeta)\left[\Delta_{+1}(\zeta) - d(\zeta)\Delta_+(\zeta)\right] - \zeta\chi^o(\zeta)\left[c_1(\zeta) - d(\zeta)c(\zeta)\right] = 0$$
$$\tilde{b}^o(\zeta)\left[\Delta_{+1}(\zeta) - d(\zeta)\Delta_+(\zeta)\right] + \alpha^o(\zeta)\left[c_1(\zeta) - d(\zeta)c(\zeta)\right] = 0$$

whence, taking account of (14.89), we have $\Delta_{+1}(\zeta) = d(\zeta)\Delta_+(\zeta)$ and $c_1(\zeta) = d(\zeta)c(\zeta)$ that completes the proof. ∎

7. In principle, other parametrization of the set of stabilizing transfer functions can be used. For instance, let $\{a_0(\zeta), b_0(\zeta)\}$ be a stabilizing pair satisfying the Diophantine equation

$$\alpha^o(\zeta)a_0(\zeta) + \zeta\chi^o(\zeta)b_0(\zeta) = \Delta_{+0}(\zeta). \tag{14.94}$$

Then, the transfer function

$$W_{d0}(\zeta) = \frac{\tilde{b}_0(\zeta) + \phi(\zeta)\alpha^o(\zeta)}{\tilde{a}_0(\zeta) - \zeta\phi(\zeta)\chi^o(\zeta)} \tag{14.95}$$

where $\phi(\zeta)$ is an irreducible arbitrary real rational function of the form (14.92), is also stabilizing. Indeed, using (14.95) and (14.92), we can represent $W_{d0}(\zeta)$ in the form (14.85) with

$$\begin{aligned} b^o(\zeta) &= b^o(\zeta)\Delta_+(\zeta) + c(\zeta)\alpha^o(\zeta) \\ a^o(\zeta) &= a^o(\zeta)\Delta_+(\zeta) - \zeta c(\zeta)\chi^o(\zeta)\,. \end{aligned} \tag{14.96}$$

If (14.96) holds, the characteristic polynomial of the feedback system has the form

$$\begin{aligned} \Delta^o(\zeta) &= \alpha^o(\zeta)\left[a_0(\zeta)\Delta_+(\zeta) - \zeta c(\zeta)\chi^o(\zeta)\right] + \zeta\chi^o(\zeta)\left[b_0(\zeta)\Delta_+(\zeta) + c(\zeta)\alpha^o(\zeta)\right] \\ &= \Delta_+(\zeta)\left[\alpha^o(\zeta)a_0(\zeta) + \zeta\chi^o(\zeta)b_0(\zeta)\right] = \Delta_+(\zeta)\Delta_{+0}(\zeta)\,. \end{aligned} \tag{14.97}$$

The polynomial on the right-hand side is stable, because the pair $\{a_0(\zeta), b_0(\zeta)\}$ is stabilizing. From (14.97) it follows that the characteristic polynomial of the closed loop is divisible by the polynomial $\Delta_{+0}(\zeta)$ irrespective of the choice of the irreducible parameter $\phi(\zeta)$ that leads to unjustified increase of the orders of the control program and the system in general. Nevertheless, the parametrization (14.95) is often used in applications when there is a possibility of immediately obtaining the first stabilizing pair $\{a_0(\zeta), b_0(\zeta)\}$.
As a special case, if the open-loop system is stable, i.e., all poles of the transfer function $F(s)$ are located in the left half-plane, the polynomial $\alpha^o(\zeta)$ is stable. In this case Eq. (14.94) is satisfied with

$$a_0(\zeta) = 1\,, \quad b_0(\zeta) = 0\,, \quad \Delta_{+0}(\zeta) = \alpha^o(\zeta) \tag{14.98}$$

and from (14.95) we obtain

$$W_{d0}(\zeta) = \frac{\phi(\zeta)\alpha^o(\zeta)}{1 - \zeta\phi(\zeta)\chi^o(\zeta)} = \frac{c(\zeta)\alpha^o(\zeta)}{\Delta_+(\zeta) - \zeta c(\zeta)\chi^o(\zeta)} \tag{14.99}$$

where $c(\zeta)$ is an arbitrary polynomial, and $\Delta_+(\zeta)$ is an arbitrary stable polynomial. This gives the stabilizing pair

$$\begin{aligned} a^o(\zeta) &= \Delta_+(\zeta) - \zeta c(\zeta)\chi^o(\zeta) \\ b^o(\zeta) &= c(\zeta)\alpha^o(\zeta)\,. \end{aligned}$$

As follows from the aforesaid, with the use of stabilizing pairs of this form the characteristic polynomial of the feedback system is given by

$$\Delta^o(\zeta) = \Delta_+(\zeta)\alpha^o(\zeta)\,.$$

Example 14.6 Build the set of stabilizing pairs under the conditions of Example 14.5 where additionally it is assumed that the continuous-time part of the system is stable, i.e. $a < 0$. Then, the above set of stabilizing pairs takes the form

$$a^o(\zeta) = \Delta_+(\zeta) - \zeta c(\zeta)\gamma M(a)\mathrm{e}^{aT}$$
$$b^o(\zeta) = c(\zeta)\left(1 - \mathrm{e}^{aT}\zeta\right) .$$

The corresponding set of transfer functions can be defined as

$$W_{d0}(\zeta) = \frac{\phi(\zeta)\left(1 - \mathrm{e}^{aT}\zeta\right)}{1 - \zeta\phi(\zeta)\gamma M(a)\mathrm{e}^{aT}}$$

□

8. The theory presented above can be, without major modifications, extended to time-delayed systems, when Eq.(14.81) holds. In this case, taking account of (11.88), the Diophantine equation (14.84) is replaced by

$$\alpha^o(\zeta)a^o(\zeta) + \zeta^m\beta^o(\zeta,\delta T)b^o(\zeta) = \Delta_+(\zeta) . \qquad (14.100)$$

If the non-pathological conditions (9.161) and (9.162) hold, from Corollary 1 of Theorem (9.1) it follows that the ratio

$$\mathcal{D}^o_{FM}(T,\zeta,t) = \frac{\beta^o(\zeta,t)}{\alpha^o(\zeta)}$$

is irreducible for $0 \le t \le T$. Therefore, the polynomials $\alpha^o(\zeta)$ and $\zeta^m\beta^o(\zeta,t)$ have no common roots, because $\alpha^o(0) \neq 0$. Therefore, if Eqs. (9.161) and (9.162) hold, the equation (14.101) is solvable for any right-hand side. Any pair $\{\tilde{a}^o(\zeta),\tilde{b}^o(\zeta)\}$ satisfying the equation

$$\alpha^o(\zeta)\tilde{a}^o(\zeta) + \zeta^m\beta^o(\zeta,\delta T)\tilde{b}^o(\zeta) = 1 \qquad (14.101)$$

will be called a *basis*.
Then, it can be shown that for any known basis $\{\tilde{a}^o(\zeta),\tilde{b}^o(\zeta)\}$ the set of all stabilizing pairs is given by

$$\begin{aligned} b^o(\zeta) &= b^o(\zeta)\Delta_+(\zeta) + \alpha^o(\zeta)c(\zeta) \\ a^o(\zeta) &= a^o(\zeta)\Delta_+(\zeta) - \zeta^m\beta^o(\zeta,\delta T)c(\zeta) \end{aligned} \qquad (14.102)$$

where $\Delta_+(\zeta)$ is an arbitrary stable polynomial, and $c(\zeta)$ is an arbitrary polynomial. Moreover, the polynomials $a^o(\zeta)$ and $b^o(\zeta)$ have no common roots if the polynomials $\Delta_+(\zeta)$ and $c(\zeta)$ are co-prime. As follows from the aforesaid, for the control program (14.102) the characteristic polynomial of the feedback system equals $\Delta_+(\zeta)$. The set of stabilizing transfer functions, corresponding to Eq. (14.102), has the form

$$W^o_d(\zeta) = \frac{\tilde{b}^o(\zeta) + \phi(\zeta)\alpha^o(\zeta)}{\tilde{a}^o(\zeta) - \zeta^m\phi(\zeta)\beta^o(\zeta,\delta T)} . \qquad (14.103)$$

Remark 1 From (14.103) and (14.90) it follows that the function $W^o_d(\zeta)$ is analytic in $\zeta = 0$, i.e., any stabilizing control program for the process with time-delay is causal.

Remark 2 A generalization of the results of this section for the MIMO case is given in Lampe and Rosenwasser (1996).

14.7 Insensitivity of stabilization

During the calculation of the polynomial solution of the stabilization problem for continuous-time systems there are some unexpected difficulties. Let us consider a linear continuous-time system shown in Fig. (14.7), where $F(s)$ is the

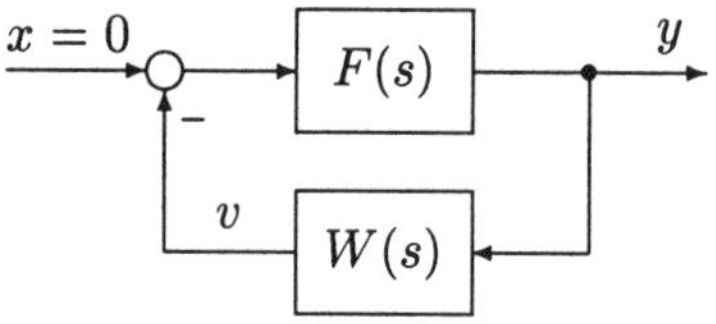

Figure 14.7: Continuous-time system

transfer function of the plant and $W(s)$ is that of the controller. Let also

$$F(s) = \frac{\beta(s)}{\alpha(s)}, \quad W(s) = \frac{b(s)}{a(s)}$$

where $\{\alpha(s), \beta(s)\}$ and $\{a(s), b(s)\}$ are known polynomial pairs, such that $\deg \alpha(s) > \deg \beta(s)$ and the ratio $F(s)$ is irreducible. The system will be called stable, if with any initial conditions of the plant and controller for $t > 0$ we have

$$|y(t)| < c_1 e^{-\lambda t}, \quad |v(t)| < c_2 e^{-\lambda t} \tag{14.104}$$

where c_1,c_2 and λ are positive constants, and λ is independent of the initial conditions. Under the given assumptions, the system at hand is stable, if all roots of the characteristic polynomial

$$\alpha(s)a(s) + \beta(s)b(s) = d(s) \tag{14.105}$$

lie in the left half-plane. With given polynomials $\alpha(s)$, $\beta(s)$, and $d(s)$, Eq.(14.105) can be considered as a polynomial equation defining stabilizing pairs $\{a(s), b(s)\}$. Since the ratio $F(s)$ is assumed to be irreducible, Eq. (14.105) is solvable for an arbitrary $d(s)$. Therefore, if we fix a polynomial $d^*(s)$ having all its roots in the left half-plane and find a solution $\{a^*(s), b^*(s)\}$ of Eq. (14.105) for $d(s) = d^*(s)$, we obtain the transfer function of a stabilizing controller

$$W^*(s) = \frac{b^*(s)}{a^*(s)}. \tag{14.106}$$

With the controller (14.106), the characteristic polynomial of the feedback system equals $d^*(s)$. Nevertheless, it can be shown, that in some cases this solution of by the stabilization problem is only formal and the system with the controller (14.106) can be practically unstable. Indeed, due to uncertainty of the parameters of an actual plant the numerator and denominator of the transfer function $F(s)$ differ from the modelled ones, so that the system will be stable in practice if this property remains under small parameter variation. In this case we shall say that the controller (14.106) gives an *insensitive* solution of the stabilization problem.

We shall show that not every solution of the stabilization problem obtained by solving Eq. (14.105) is insensitive. Assume that we actually have

$$F(s) = \frac{\beta(s) + \beta_1(s)}{\alpha(s) + \alpha_1(s)}$$

where $\alpha_1(s)$ and $\beta_1(s)$ are polynomials having sufficiently small absolute values of coefficients, such that $\deg \beta_1(s) = \deg \beta(s)$ and $\deg \alpha_1(s) = \deg \alpha(s)$. Then, with the controller (14.106) we have the characteristic polynomial

$$\tilde{d}(s) = d^*(s) + d_1(s) \tag{14.107}$$

where

$$d_1(s) = \beta_1(s)b^*(s) + \alpha_1(s)a^*(s)\,. \tag{14.108}$$

As follows from (14.107) and (14.108), if

$$\deg d_1(s) > \deg d^*(s) \tag{14.109}$$

the polynomial $\tilde{d}(s)$ may have roots in the right half-plane for arbitrarily small coefficients of $\alpha_1(s)$ and $\beta_1(s)$. This is caused by the fact, that after summation of the polynomials $d_1(s)$ and $d^*(s)$, the latter may have a higher degree and can give roots of the characteristic equation for arbitrarily small coefficients of $d_1(s)$. Thus, this solution of the stabilization problem in the case of (14.109) is sensitive and, therefore, is not useful. As was shown in Petrov (1987), the situation (14.109) is often encountered.

If, instead of (14.109), for small parameter variation of the plant we have

$$\deg d_1(s) \leq \deg d^*(s) \tag{14.110}$$

the obtained solution of the stabilization problem is insensitive.

The stabilization method for SD systems considered in Section 14.6 is also connected with solving a polynomial equation. Therefore, it can be presumed, that similar problems can arise for SD systems as well. Nevertheless, the following theorem claims that a solution of the stabilization problem for SD system is always insensitive.

Theorem 14.7 *Let the non-pathological conditions (9.161) and (9.162) hold, and*

$$\mathcal{D}^o_{FM}(T,\zeta,0) = \frac{\zeta\chi^o(\zeta)}{\alpha^o(\zeta)}\,. \tag{14.111}$$

Let us, instead of (14.111), for small variations of the parameter vectors θ of the system have

$$\mathcal{D}_{FM}^{o}(T,\zeta,0)=\frac{\zeta\left[\chi^{o}(\zeta)+\chi_1(\zeta,\theta)\right]}{\alpha^{o}(\zeta)+\alpha_1(\zeta,\theta)} \tag{14.112}$$

where $\chi_1(\zeta,\theta)$ *and* $\alpha_1(\zeta,\theta)$ *are polynomials such that*

$$\chi_1(\zeta,0)=0\,,\quad \alpha_1(\zeta,0)=0\,. \tag{14.113}$$

It is assumed that the coefficients of the polynomials $\chi_1(\zeta,\theta)$ *and* $\alpha_1(\zeta,\theta)$ *depend continuously on the components of the vector* θ. *Then, if the irreducible transfer function*

$$W_d^{o}(\zeta)=\frac{b^{o}(\zeta)}{a^{o}(\zeta)}$$

is stabilizing, i.e., the polynomial

$$\Delta^{o}(\zeta)=\alpha^{o}(\zeta)a^{o}(\zeta)+\zeta\chi^{o}(\zeta)b^{o}(\zeta)$$

is stable, the feedback system remains stable under sufficiently small absolute variation of the components of the vector θ.

Proof Let us have (14.112). Then, the characteristic polynomial of the feedback system has the form

$$\tilde{\Delta}(\zeta)=\Delta^{o}(\zeta)+\Delta_1(\zeta,\theta)\,,$$

where

$$\Delta_1(\zeta,\theta)=\alpha_1(\zeta,\theta)a^{o}(\zeta)+\zeta\chi_1(\zeta,\theta)b^{o}(\zeta)\,.$$

Denote

$$\delta\triangleq\min_{|\zeta|=1}|\Delta^{o}(\zeta)|\,. \tag{14.114}$$

Obviously, $\delta>0$, because the polynomial $\Delta^{o}(\zeta)$ has no roots on the unit circle. Since the coefficients of the polynomial $\Delta_1(\zeta,\theta)$ depend continuously on θ and (14.113) holds, there is a number $\epsilon>0$ such that for $\|\theta\|<\epsilon$ we have

$$\max_{|\zeta|=1}|\Delta_1(\zeta,\theta|<\delta\,. \tag{14.115}$$

Comparing (14.114) with (14.115), we find that for $\|\theta\|<\epsilon$ at any point of the unit circle

$$|\Delta_1(\zeta,\theta)|<\Delta^{o}(\zeta)\,. \tag{14.116}$$

By the theorem of Rouché, Titchmarsh (1932), from (14.116) it follows that the polynomials $\Delta^{o}(\zeta)$ and $\tilde{\Delta}(\zeta)=\Delta^{o}(\zeta)+\Delta_1(\zeta,\theta)$ have the same number of roots inside the unit circle. Therefore, if the polynomial $\Delta^{o}(\zeta)$ is stable, then for $\|\theta\|<\epsilon$ the polynomial $\tilde{\Delta}(\zeta)$ is also stable. ∎

Corollary 1 Theorem 14.7 claims that there exists a number $\epsilon>0$ such that for $\|\theta\|<\epsilon$ the feedback system is stable with a given controller $W_d^{o}(\zeta)$, but provides no constructive estimate of the value ϵ. Such an estimation should be performed during the investigation of the designed system.

Corollary 2 The proof of Theorem (14.7) could be directly extended to the case of time-delayed systems.

Corollary 3 A generalization of Theorem (14.7) to the MIMO case is given in Rosenwasser (1994a).

14.8 Stabilization with given static gain

For a given basis $\{\tilde{a}^o(\zeta), \tilde{b}^o(\zeta)\}$ Eqs. (14.91) and (14.92) gives the set $\mathcal{W}_d$ of all irreducible stabilizing transfer functions $W_d^o(\zeta)$. For some design problems it is necessary to separate out a subset $\tilde{\mathcal{W}}_d$ of the set $\mathcal{W}_d$, which ensures some special properties of the feedback system in addition to the stability. In this section we consider the construction of a subset $\tilde{W}_d^o(\zeta)$ that provides for stability and a given static gain of the feedback system with respect to chosen input and output.

Example 14.7 Let us consider the stationary output behaviour of the stable continuous-time system shown in Fig. 14.8, and of the stable SD system of Fig. 14.9, provided that the input signal is the unit step $x(t) = 1\!\!1(t)$. The terms

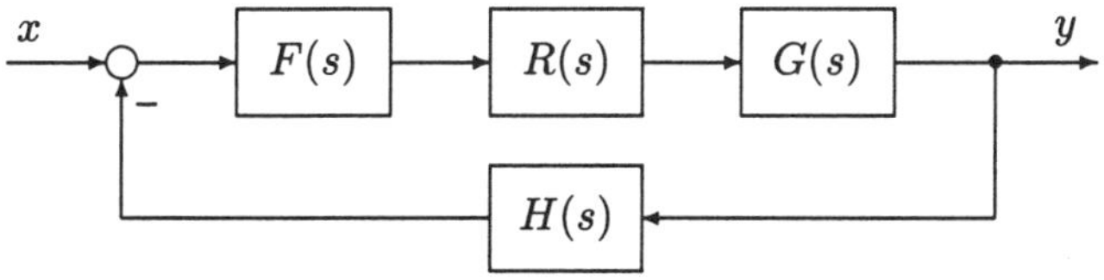

Figure 14.8: Continuous-time control system for Example 14.7

$F(s)$, $R(s)$, $G(s)$, and $H(s)$ are known real rational functions. The transfer function of the feedback system from the input $x(t)$ to the output $y(t)$ is equal to

$$W(s) = \frac{F(s)R(s)G(s)}{1 + F(s)R(s)G(s)H(s)}.$$

If $x(t) = 1\!\!1(t)$, after the transient dies down, the steady-state output is

$$y_\infty = W(0) = \frac{F(0)R(0)G(0)}{1 + F(0)R(0)G(0)H(0)}. \tag{14.117}$$

Assume that we also have a stable SD system shown in Fig. 14.9, where $F(s)$, $G(s)$, and $H(s)$ are the same as in Fig. 14.8, and C is a computer described by Eqs. (10.1)–(10.3). The transfer function of the forming element $M(s)$ is assumed to be (9.12). In accordance with the rule of Section 14.3, under a unit step we have

$$y_{d\infty} = \frac{F(0)G(0)W_d(0)}{1 + F(0)G(0)H(0)W_d(0)}. \tag{14.118}$$

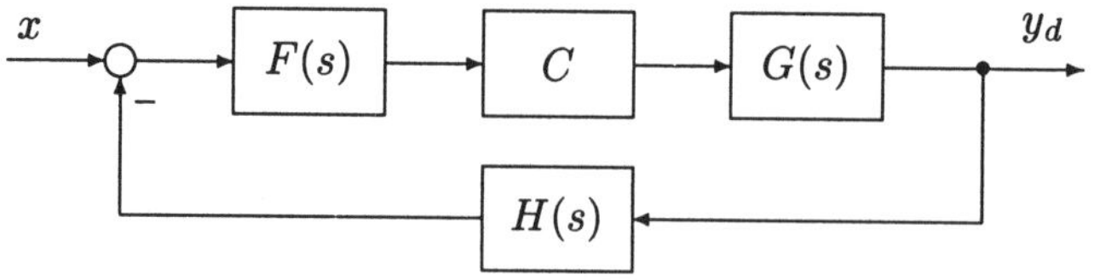

Figure 14.9: Corresponding SD system for Fig. 14.8

Assume that $y_\infty \neq 0$, $y_{d\infty} \neq 0$, and $R(0) \neq 0$. Then, using (14.117) and (14.118), it can be shown that to meet the demand for equal static gains for both system, i.e.,

$$y_\infty = y_{d\infty}$$

it is necessary and sufficient that

$$W_d(0) = R(0) \tag{14.119}$$

or, equivalently,

$$W_d^o(\zeta)\,|_{\zeta=1} = R(0)\,. \tag{14.120}$$

Thus, if it is required to stabilize the system and ensure equal static gains of the systems considered, we have to separate out a subset of the set of stabilizing controllers $W_d^o(\zeta)$ such that Eq. (14.120) is satisfied. □

The situation considered in Example 14.7 is quite typical. If we have a stable feedback SD system with a zero-order hold and PTF $W(s,t)$ and its input is acting upon by the unit step $x(t) = \mathbb{1}(t)$, we have

$$y_\infty = W(0,t) = \text{const.} = a\,. \tag{14.121}$$

Equation (14.121) is equivalent, except for some special cases, to the equality

$$W_d(0) = N_d \tag{14.122}$$

where N_d is a known constant, or, equivalently,

$$W_d^o(1) = \frac{b^o(1)}{a^o(1)} = N_d\,. \tag{14.123}$$

Using (14.91), we have

$$\frac{\tilde{b}^o(1) + \phi(1)\alpha^o(1)}{\tilde{a}^o(1) - \phi(1)\chi^o(1)} = N_d\,,$$

whence follows

$$\phi(1) = \frac{N_d\tilde{a}^o(1) - \tilde{b}^o(1)}{N_d\chi^o(1) + \alpha^o(1)} \triangleq K_d\,. \tag{14.124}$$

We note that the value K_d in (14.124) is finite under the given assumptions. Indeed, it may appear to be infinite if

$$\alpha^o(1) + N_d\chi^o(1) = 0\,. \tag{14.125}$$

But in this case from (14.123) and (14.125) we have

$$\Delta^o(1) = \alpha^o(1)a^o(1) + \chi^o(1)b^o(1) = 0$$

which contradicts the assumption on the stability of the system.
Using (14.124) we have

$$\phi(\zeta) = K_d + (\zeta - 1)\phi_1(\zeta)\,. \tag{14.126}$$

Substituting this expression into (14.91), we obtain

$$W_d^o(\zeta) = \frac{\tilde{b}^o(\zeta) + K_d\alpha^o(\zeta) + (\zeta - 1)\phi_1(\zeta)\alpha^o(\zeta)}{\tilde{a}^o(\zeta) - \zeta K_d\chi^o(\zeta) - \zeta(\zeta - 1)\phi_1(\zeta)\chi^o(\zeta)} \tag{14.127}$$

where $\phi_1(\zeta)$ is an arbitrary real rational function free of poles inside and on the unit circle (or a polynomial). Formula (14.127) gives the parametrization of the subset of stabilizing controllers with transfer functions satisfying Eq. (14.122).

Example 14.8 Under the conditions of Example 14.5 the set of stabilizing pairs $\{a^o(\zeta), b^o(\zeta)\}$ satisfying the condition (14.122) is given by

$$\begin{aligned} a^o(\zeta) &= \Delta_+(\zeta) - \zeta K_d\gamma M(a)e^{aT} - \zeta(\zeta - 1)c(\zeta)\gamma M(a)e^{aT} \\ b^o(\zeta) &= \frac{1}{\gamma M(a)}\Delta_+(\zeta) + K_d\left(1 - e^{aT}\zeta\right) + (\zeta - 1)c(\zeta)\left(1 - e^{aT}\zeta\right) \end{aligned}$$

where $c(\zeta)$ is an arbitrary polynomial, and $\Delta_+(\zeta)$ is an arbitrary stable polynomial. □

14.9 Characteristic equations of finite-dimensional SD systems

Stability is one of the most important requirements for control systems. This problem is especially relevant for SD systems, because sampling units, in general, worsen processes in the system and often lead to instability. In this section we present a general approach to the stability investigation of a general class of finite-dimensional SD systems.

Definition A periodic SD system of an arbitrary structure with an arbitrary number of sampling units will be called *finite-dimensional*, if, for any initial conditions of its elements and without exogenous disturbances, any generalized coordinate $y_i(t)$ has the discrete Laplace transform $\mathcal{D}_{y_i}(T, s, t)$ that is for $0 \le t \le T$ a causal rational periodic function in s.

It can be shown that a system is finite-dimensional if it possesses the following properties.

1. The system consists of an arbitrary finite number of continuous-time elements with real rational transfer functions and an arbitrary finite number of discrete-time elements (pulse elements, sample and hold circuits, computers, ...), the periods of the latter being commensurate so that the system as a whole has the period T.

2. Besides the above elements, the system may include pure delay elements. Each such element must be included in a closed loop containing at least one sampling unit.

Henceforth, it is assumed that the structure of the system is such that without exogenous disturbances for any initial condition at the starting moment $t_0 = 0$, for all $t > 0$ the inputs of the discrete-time elements are affected by continuous-time signals.

Assume that without exogenous disturbances a finite-dimensional system for $t > 0$ performs natural oscillations caused by the initial conditions of some elements for $t_0 = 0$. Then, an arbitrary generalized coordinate $y_i(t)$ can be expressed by the inversion integral

$$y_i(t) = \frac{T}{2\pi \mathrm{j}} \int_{c-\mathrm{j}\omega/2}^{c+\mathrm{j}\omega/2} \mathcal{D}_{y_i}(T,s,t)\, \mathrm{d}s \tag{14.128}$$

where c is a real constant such that all poles of the function $\mathcal{D}_{y_i}(T,s,t)$ lie in the half-plane $\operatorname{Re} s < c$. Because of the above assumptions, for $0 \le t \le T$ the image $\mathcal{D}_{y_i}(T,s,t)$ can be represented in the form

$$\mathcal{D}_{y_i}(T,s,t) = \frac{B(s,t)}{\Delta(s)} \tag{14.129}$$

where

$$\begin{aligned} B(s,t) &= b_0(t) + b_1(t)\mathrm{e}^{-sT} + \ldots + b_r(t)\mathrm{e}^{-rT} \\ \Delta(s) &= a_0 + a_1\mathrm{e}^{-sT} + \ldots + a_r\mathrm{e}^{-rT} \end{aligned} \tag{14.130}$$

are quasi-polynomials, $b_i(t)$, $(i = 0, 1, \ldots, r)$ are known functions depending on the initial conditions, and a_i $(i = 0, 1, \ldots, r)$ are known constants independent of the initial conditions. Moreover, it is assumed that there are such initial conditions that the ratio (14.129) is irreducible for some t.

Definition The system under investigation will be called stable with respect to an output $y(t)$, if without exogenous disturbances for any initial conditions at the moment $t_0 = 0$ for $t > 0$ we have

$$|y(t)| < c\mathrm{e}^{-\lambda t} \tag{14.131}$$

where c and λ are positive constants, such that λ is independent of the initial conditions.

The following theorem gives necessary and sufficient conditions of the stability of the system with respect to a given output.

Theorem 14.8 *Under the above assumptions, the system is stable with respect to an output $y(t)$ iff the equation*

$$\Delta(s) = 0 \tag{14.132}$$

has not roots in the closed right half-plane.

Proof *Sufficiency:* If $\Delta(s) = \text{const} \neq 0$ is valid, the image $\mathcal{D}_y(T, s, t)$ for $0 \leq t \leq T$ is a polynomial in $\zeta = \mathrm{e}^{-sT}$. In this case the transients die down in a finite time interval and the system is stable. If the function $\Delta(s)$ is not reduced to a constant, as follows from Section A.8, for sufficiently large $t > 0$ we have

$$y(t) = T \sum_i \operatorname*{Res}_{\tilde{s}_i} \mathcal{D}_y(T, s, t) \tag{14.133}$$

where $\tilde{s}_i$, $(i = 1, \ldots, n)$ are primary poles of the function $\mathcal{D}_y(T, s, t)$, which are roots of Eq. (14.132). Using (14.129) and (14.130), it can be shown, Rosenwasser (1994c), that the solution (14.133) can be represented in the form

$$y(t) = \sum_i \sum_k t^k \mathrm{e}^{\tilde{s}_i t} g_{ki}(t) \tag{14.134}$$

where $g_{ki}(t) = g_{ki}(t + T)$ are bounded functions, and the sets of indices i and k are finite. From (14.134) it follows that for $\operatorname{Re} \tilde{s}_i < 0$, $(i = 1, \ldots, n)$ we have (14.131), i.e., the sufficiency of the conditions of the theorem is proved.

Necessity: Let the function $\Delta(s)$ have a primary root $\tilde{s}_0$ such that $\operatorname{Re} s_0 \geq 0$. Then, by assumption, for some initial conditions the value $\tilde{s}_0$ is a primary pole of $\mathcal{D}_y(T, s, t)$. In this case, the expansion (14.134) contains non-damped terms, therefore, Eq. (14.131) is not valid. ■

Using Theorem 14.8, we can propose a general method of constructing Eq. (14.132) for SD systems, satisfying the above assumptions.
Since functions of the form $f(t) = t^k \mathrm{e}^{s_i t} g(t)$ with $g(t) = g(t + T)$ are independent for different k and the system is linear, each term in Eq. (14.134) defines, for some initial conditions, natural system dynamics. Therefore, if $\tilde{s}_i$ is a pole of the image $\mathcal{D}_y(T, s, t)$ given by (14.129), for some initial conditions without exogenous disturbances the following process is possible:

$$y(t) = \mathrm{e}^{\tilde{s}_i t} g(t)\,, \quad g(t + T) = g(t)\,. \tag{14.135}$$

It can be shown that, conversely, if for a number $\tilde{s}_i$ it is possible, without exogenous disturbances, to obtain an output process of the form (14.135), then the number $\tilde{s}_i$ is a pole of the image $\mathcal{D}_y(T, s, t)$.
Let $y_\mu(t)$, $(\mu = 1, \ldots, \rho)$ be the aggregate of all considered generalized coordinates of the system, such that all transforms $\mathcal{D}_y(T, s, t)$ satisfy the above

conditions. Then, the images of all outputs of the system can be represented in a form similar to (14.129)

$$\mathcal{D}_{y_\mu}(T,s,t) = \frac{B_\mu(s,t)}{\Delta(s)} \tag{14.136}$$

where the $B_\mu(s,t)$ are quasi-polynomials depending on μ, and $\Delta(s)$ is a fixed quasi-polynomial having constant coefficients independent of μ. Following the above reasonings, it can be shown that the number $\tilde{s}_i$ is a pole of the images (14.136) iff the natural motion

$$y_{\mu_i}(t) = \mathrm{e}^{\tilde{s}_i t} g_{\mu_i}(t)\,, \quad g_{\mu_i}(t+T) = g_{\mu_i}(t) \tag{14.137}$$

is possible in the system. The assumption on the existence of a natural oscillation of the form (14.137) leads to an equation with respect to the numbers $s = \tilde{s}_i$, which will be called the *characteristic numbers* of the system. Changing the variable for $\zeta = \mathrm{e}^{-sT}$ or $z = \mathrm{e}^{sT}$, we obtain an algebraic equation called the *characteristic equation* of the system.
As follows from the aforesaid, the system is stable within the terms of the above definition with respect to all considered outputs, if it has no characteristic numbers in the closed right half-plane.

Example 14.9 We apply the considerations of the above section to construct the characteristic equation of the closed-loop system shown in Fig. 14.6, where the transfer function of the continuous-time part has the form (14.81) and the shape of the controlling pulses are arbitrary. Moreover, the transfer function $F_1(s)$ is assumed to be strictly proper. In this case we may assume that

$$y(t) = \mathrm{e}^{\lambda t} f(t)\,, \quad f(t) = f(t{+}T)\,; \quad v(t) = \mathrm{e}^{\lambda t} g(t)\,, \quad g(t) = g(t{+}T) \tag{14.138}$$

where $f(t)$ and $g(t)$ are continuous functions. Then, using the stroboscopic effect and Eqs. (9.13) and (10.37), we immediately obtain

$$u(t) = -\mathrm{e}^{\lambda t} W_d(\lambda) \varphi_M(T,\lambda,t) f(0) \tag{14.139}$$

where

$$\varphi_M(T,s,t) = \frac{1}{T}\sum_{k=-\infty}^{\infty} M(s+kj\omega)\mathrm{e}^{kj\omega t}\,.$$

The steady-state response of the continuous element to the input signal (14.139) is given by

$$y(t) = -\mathrm{e}^{\lambda t} W_d(\lambda) \varphi_{F_1 M}(T,\lambda,t-\tau)\mathrm{e}^{-\lambda\tau} f(0)\,. \tag{14.140}$$

Comparing Eqs. (14.138) and (14.140), we find

$$f(t) = -W_d(\lambda) \varphi_{F_1 M}(T,\lambda,t-\tau)\mathrm{e}^{-\lambda\tau} f(0) \tag{14.141}$$

where the right-hand side is continuous with respect to t. Therefore, substituting s for λ, for $t=0$ and $f(0) \neq 0$ we find

$$1 + \mathcal{D}_{F_1 M}(T, s, -\tau) W_d(s) = 0\,. \tag{14.142}$$

Assuming, as before,

$$W_d(s) = \frac{b(s)}{a(s)}\,, \quad \mathcal{D}_{F_1 M}(T, s, -\tau) = \frac{\beta(s, \delta T)}{\alpha(s)}\, \mathrm{e}^{-msT} \quad \tau = mT - \delta T$$

we obtain the equation with respect to the characteristic numbers

$$\lambda_\tau(s) = 1 + \frac{b(s)}{a(s)} \frac{\beta(s, \delta T)}{\alpha(s)}\, \mathrm{e}^{-msT} = 0\,. \tag{14.143}$$

If the ratio at the left-hand side is irreducible, the latter equation is equivalent to Eq. (14.82), from which, assuming that $\zeta = \mathrm{e}^{-sT}$, we obtain the corresponding characteristic equation.

In some special cases the ratio at the left side of (14.143) may be reducible, but it cannot be cancelled during the construction of the characteristic equation. This results from the fact that the cancellation is related to a compensation in the open chain that corresponds to uncontrollable or unobservable parts in the systems. These parts do not appear in the transfer function. Their excitation by initial values, disturbances or rounding errors certainly happens in practice so that they have to be involved into the stability considerations. From practical points of view under any small parameter variations of the process or control program the reducibility does not take place anyway. Hence, the cancellation of the ratio in (14.143) may lead to incorrect conclusions on the stability of the system, as is shown in Example 14.4. □

14.10 Hybrid feedback systems

Consider the hybrid feedback system shown in Fig. 14.10, where $F(s)$ and $G(s)$

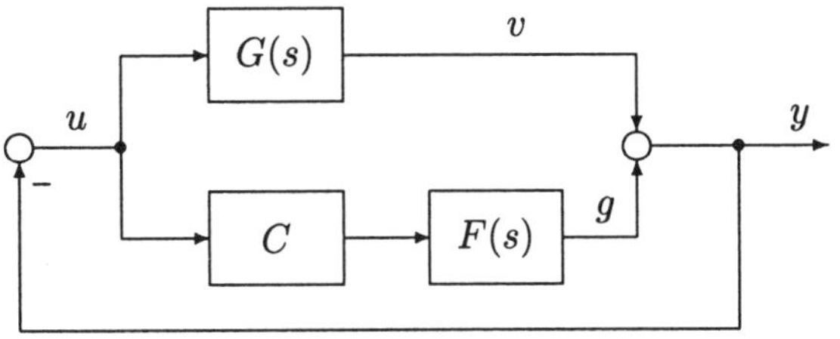

Figure 14.10: Hybrid feedback system

are strictly proper real rational functions, and C is a computer described by Eqs. (10.1)–(10.3). Since the period of the system equals T, we assume that

$$y(t) = \mathrm{e}^{\lambda t} f(t)\,, \quad f(t) = f(t + T) \tag{14.144}$$

where $f(t)$ is a continuous function. Let

$$f(t) = \sum_{k=-\infty}^{\infty} f_k e^{kj\omega t}, \quad f_k = \frac{1}{T}\int_0^T f(\tau)e^{-kj\omega\tau}\,d\tau. \tag{14.145}$$

Then, from Fig. 14.10 we have

$$v(t) = -\sum_{k=-\infty}^{\infty} G(\lambda + kj\omega) f_k e^{(\lambda+kj\omega)t}$$

$$g(t) = -W_d(\lambda)\varphi_{FM}(T,\lambda,t)e^{\lambda t} f(0).$$

Hence,

$$y(t) = -\sum_{k=-\infty}^{\infty} G(\lambda + kj\omega) f_k e^{(\lambda+kj\omega)t} - W_d(\lambda)\varphi_{FM}(T,\lambda,t)e^{\lambda t} f(0). \tag{14.146}$$

Comparing Eqs. (14.144) and (14.146) for $y(t)$, we have

$$f(t) = -\sum_{k=-\infty}^{\infty} G(\lambda + kj\omega) f_k e^{kj\omega t} - W_d(\lambda)\varphi_{FM}(T,\lambda,t) f(0).$$

Comparing the Fourier coefficients on the left and right-hand sides, we obtain

$$f_k = -\frac{1}{T} H(\lambda + kj\omega) M(\lambda + kj\omega) W_d(\lambda) f(0) \tag{14.147}$$

where

$$H(\lambda) \triangleq \frac{F(\lambda)}{1 + G(\lambda)}. \tag{14.148}$$

If $\lim_{\lambda\to\infty} G(\lambda) \neq -1$. Then, $\lim_{\lambda\to\infty} H(\lambda) = 0$ and the real rational function $H(\lambda)$ is strictly proper. Substituting (14.147) into (14.145), we obtain

$$f(t) = -W_d(\lambda)\varphi_{HM}(T,\lambda,t) f(0).$$

For $t = 0$ we obtain the compatibility condition

$$1 + W_d(s)\mathcal{D}_{HM}(T,s,0) = 0. \tag{14.149}$$

Example 14.10 Let us consider in more detail the technique of constructing the characteristic equation for the given hybrid system with

$$F(s) = \frac{\gamma_1}{s-a}, \quad G(s) = \frac{\gamma_2}{s-b}, \quad W_d(s) = 1$$

where a, b, γ_1, and γ_2 are real constants, and the form of pulse is arbitrary. From (14.148) we have

$$H(s) = \frac{\gamma_1(s-b)}{(s-a)(s-b+\gamma_2)}. \tag{14.150}$$

First, we consider the case, when

$$a \neq b, \quad a - b + \gamma_2 \neq 0.$$

Then,

$$H(s) = \frac{\alpha_1}{s-a} + \frac{\alpha_2}{s-b+\gamma_2}$$

where

$$\alpha_1 = \frac{\gamma_1(a-b)}{a-b+\gamma_2}, \quad \alpha_2 = \frac{\gamma_1\gamma_2}{a-b+\gamma_2}.$$

In this case, from (9.102) we find

$$\mathcal{D}_{HM}(T,s,0) = \frac{\alpha_1 M(a)}{e^{(s-a)T}-1} + \frac{\alpha_2 M(b-\gamma_2)}{e^{(s-b+\gamma_2)T}-1} =$$

$$= \frac{\gamma_1}{a-b+\gamma_2} \frac{(a-b)M(a)\left[e^{(s-b+\gamma_2)T}-1\right] + \gamma_2 M(b-\gamma_2)\left[e^{(s-a)T}-1\right]}{\left[e^{(s-a)T}-1\right]\left[e^{(s-b+\gamma_2)T}-1\right]}$$

and Eq. (14.149), after the substitution $e^{sT} = z$, has the form of a quadratic equation

$$\begin{gathered}(a-b+\gamma_2)\left(ze^{-aT}-1\right)\left[ze^{(-b+\gamma_2)T}-1\right] + \\ +\gamma_1(a-b)M(a)\left[ze^{(-b+\gamma_2)T}-1\right] + \gamma_1\gamma_2 M(b-\gamma_2)\left(ze^{-aT}-1\right) = 0.\end{gathered} \tag{14.151}$$

Under the given assumptions the system is stable if all roots of Eq. (14.151) are inside the unit circle.

Next, we consider the special case, when

$$a = b, \quad \gamma_2 \neq 0. \tag{14.152}$$

Then, Eq. (14.151) takes the form

$$\gamma_2\left(ze^{-aT}-1\right)\left[ze^{(-a+\gamma_2)T} - 1 + \gamma_1 M(a-\gamma_2)\right] = 0. \tag{14.153}$$

Equation (14.153) also has the degree two and has a root $z = e^{aT}$ irrespective of γ_1, γ_2 and $M(s)$. We note, that in the case of (14.152) the ratio (14.150) is reducible. Then, performing the cancellation in (14.150) and passing to Eq. (14.149), we finally obtain a first-order equation in z. The root $z = e^{aT}$ will be ignored, which may lead to an incorrect conclusion concerning the system stability.

Similarly, we can consider a special case, when

$$a \neq b, \quad a = b - \gamma_2$$

where the transfer function (14.148) appears as

$$H(s) = \frac{\gamma_1(s-b)}{(s-a)^2}$$

and has a multiple pole at $s = a$. Using (9.99), we obtain a second-order equation in z. Taking $a = b$ in this equation, we obtain a second-order equation again. □

Remark The above example demonstrates the fact that intermediate cancellations are not, generally speaking, admissible when constructing the characteristic equation by the present technique.

14.11 Characteristic equations for synchronous systems with phase shift

Consider the system shown in Fig. 14.11, where $W_{pd}^{(i)}(s,t)$, $(i = 1,2)$ are the

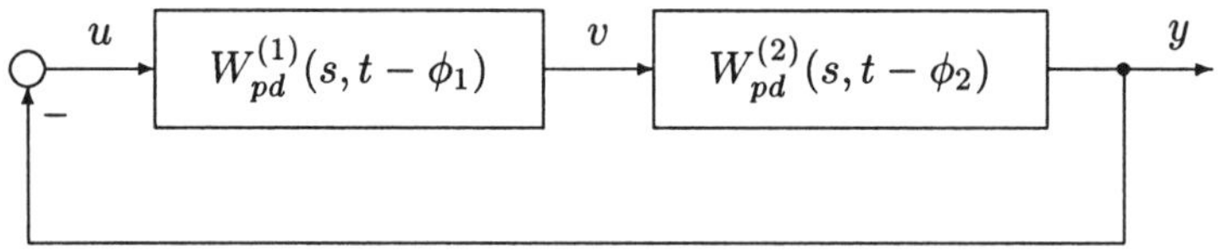

Figure 14.11: Synchronous feedback system with phase shift

PTFs of open-loop SD systems with period T of the form

$$W_{pd}^{(i)}(s,t) = W_{di}(s)\varphi_{F_i M_i}(T,s,t)$$

and ϕ_i, $(0 \le \phi_i < T)$ are the corresponding phase shifts. The transfer function of the continuous-time elements F_i are assumed to be strictly proper, so that the PTFs are continuous in t. To construct the characteristic equation, we take

$$y(t) = \mathrm{e}^{\lambda t} f(t)\,, \;\; f(t) = f(t+T)\,, \qquad v(t) = \mathrm{e}^{\lambda t} g(t)\,, \;\; g(t) = g(t+T) \tag{14.154}$$

with continuous functions $f(t)$ and $g(t)$. Then, using the stroboscopic effect,

$$v(t) = -\mathrm{e}^{\lambda t} W_{pd}^{(1)}(\lambda, t-\phi_1) f(\phi_1)\,.$$

Repeated use of the stroboscopic effect yields

$$y(t) = -\mathrm{e}^{\lambda t} W_{pd}^{(2)}(\lambda, t-\phi_2) W_{pd}^{(1)}(\lambda, \phi_2-\phi_1) f(\phi_1)\,. \tag{14.155}$$

Comparing Eqs.(14.154) and (14.155) for $y(t)$, we find

$$f(t) = -W_{pd}^{(2)}(\lambda, t-\phi_2) W_{pd}^{(1)}(\lambda, \phi_2-\phi_1) f(\phi_1)\,.$$

Taking $t = \phi_1$ and assuming that $f(\phi_1) \ne 0$, we obtain the equation

$$1 + W_{pd}^{(2)}(s, \phi_1-\phi_2) W_{pd}^{(1)}(s, \phi_2-\phi_1) = 0 \tag{14.156}$$

where the variable λ is substituted by s. Taking due account of (10.52) and (10.53), we can represent Eq. (14.156) in the form

$$1 + W_{d1}(s) W_{d2}(s) \mathcal{D}_{F_1 M_1}(T, s, \phi_1-\phi_2) \mathcal{D}_{F_2 M_2}(T, s, \phi_2-\phi_1) = 0\,. \tag{14.157}$$

The left-hand side of (14.157) is a rational periodic function in s. Therefore, assuming that $e^{sT} = z$ or $e^{-sT} = \zeta$, from (14.157) we obtain an algebraic characteristic equation.

Example 14.11 Construct the characteristic equation of the system shown in Fig. 12.6 without exogenous disturbance. This system is a special case of the system shown in Fig. 14.11 with $W_{d1}(s) = W_{d2}(s) = 1$, $F_1(s) = s^{-1}$, $F_2(s) = (s+1)^{-1}$, $\phi_1 = 0.5T$ and $\phi_2 = 0$. Equation (14.157) takes the form

$$1 + \mathcal{D}_{F_1M_1}(T, s, 0.5T)\mathcal{D}_{F_2M_2}(T, s, -0.5T) = 0\,. \tag{14.158}$$

Since

$$\mathcal{D}_{F_2M_2}(T, s, -0.5T) = \mathcal{D}_{F_2M_2}(T, s, 0.5T)e^{-sT}$$

using (9.102), we have

$$\begin{aligned}
\mathcal{D}_{F_1M_1}(T, s, 0.5T) &= \frac{M_1(0)}{e^{sT}-1} + \int_0^{0.5T} \mu_1(\tau)\,d\tau \\
\mathcal{D}_{F_2M_2}(T, s, -0.5T) &= \left[\frac{M_2(-1)}{e^{sT}e^{T}-1} + \int_0^{0.5T} e^{\tau}\mu_2(\tau)\,d\tau\right] e^{-0.5T}e^{-sT}\,.
\end{aligned} \tag{14.159}$$

Substituting (14.159) into (14.158) and changing the variable e^{sT} for z, we obtain the equation

$$a_0z^3 + a_1z^2 + a_2z + a_3 = 0 \tag{14.160}$$

where

$$\begin{aligned}
a_0 &= e^{T} \\
a_1 &= -1 - e^{T} + e^{0.5T}\int_0^{0.5T} \mu_1(\tau)\,d\tau \int_0^{0.5T} e^{\tau}\mu_2(\tau)\,d\tau \\
a_2 &= 1 + e^{0.5T}\int_{0.5T}^{T} \mu_1(\tau)\,d\tau \int_0^{0.5T} e^{\tau}\mu_2(\tau)\,d\tau + e^{-0.5T}\int_0^{0.5T} \mu_1(\tau)\,d\tau \int_{0.5T}^{T} e^{\tau}\mu_2(\tau)\,d\tau \\
a_3 &= e^{-0.5T}\int_{0.5T}^{T} \mu_1(\tau)\,d\tau \int_{0.5T}^{T} e^{\tau}\mu_2(\tau)\,d\tau\,.
\end{aligned}$$

The system is stable if all roots of Eq. (14.160) are located inside the unit circle. □

14.12 Characteristic equations for feedback multirate systems

Without exogenous disturbances the ascending and descending systems shown in Fig. 12.8 and 12.10, respectively, coincide. Therefore, we construct only the characteristic equation for the descending system shown in Fig. 12.10.

Using the equivalent block-diagram of Fig. 12.11 for $x(t) = 0$, we take

$$y(t) = \mathrm{e}^{\lambda t} f(t)\,, \quad f(t) = f(t + NT) \tag{14.161}$$

where the function $f(t)$ is continuous. Then, using the stroboscopic effect, we have

$$z_k(t) = -\mathrm{e}^{\lambda t} \breve{\varphi}_{F_1 M_1 W_{d1}}(NT, \lambda, t - kT) f(kT)\,.$$

Using the stroboscopic effect once again, from Fig. 12.11 we find

$$y(t) = -\mathrm{e}^{\lambda t} W_{d2}(\lambda) \varphi_{F_2 M_2}(NT, \lambda, t) \sum_{k=0}^{N-1} \breve{\varphi}_{F_1 M_1 W_{d1}}(NT, \lambda, -kT) f(kT)\,. \tag{14.162}$$

Comparing Eqs. (14.161) and (14.162) for $y(t)$ and substituting s for λ, we find

$$f(t) = -W_{d2}(s) \varphi_{F_2 M_2}(NT, s, t) \sum_{k=0}^{N-1} \breve{\varphi}_{F_1 M_1 W_{d1}}(NT, s, -kT) f(kT)\,. \tag{14.163}$$

A solution of Eq. (14.163) must have the form

$$f(t) = W_{d2}(s) \varphi_{F_2 M_2}(NT, s, t) A(s)$$

where $A(s)$ is an unknown function. Substituting the latter equation into (14.163), after cancellation we find

$$A(s) = -W_{d2}(s) \sum_{k=0}^{N-1} \breve{\varphi}_{F_1 M_1 W_{d1}}(NT, s, t - kT) \varphi_{F_2 M_2}(NT, s, kT) A(s)\,.$$

The last relation is compatible if

$$1 + W_{d2}(s) \sum_{k=0}^{N-1} \breve{\varphi}_{F_1 M_1 W_{d1}}(NT, s, -kT) \varphi_{F_2 M_2}(NT, s, kT) = 0 \tag{14.164}$$

which is the required equation determining the characteristic equation. Equation (14.164) can be represented in the form

$$1 + W_{d2}(s) \sum_{k=0}^{N-1} \breve{\mathcal{D}}_{F_1 M_1 W_{d1}}(NT, s, -kT) \mathcal{D}_{F_2 M_2}(NT, s, kT) = 0 \tag{14.165}$$

where the left-hand side is a rational periodic function in s with period $\mathrm{j}\omega/N$. Taking into account that for $0 \le k \le N - 1$ we have

$$\breve{\mathcal{D}}_{F_1 M_1 W_{d1}}(NT, s, -kT) = \breve{\mathcal{D}}_{F_1 M_1 W_{d1}}(NT, s, NT - kT) \mathrm{e}^{-sNT}\,.$$

Equation (14.165) can be written in the form

$$1 + W_{d2}(s)e^{-sNT} \sum_{k=0}^{N-1} \breve{\mathcal{D}}_{F_1 M_1 W_{d1}}(NT, s, NT - kT)\mathcal{D}_{F_2 M_2}(NT, s, kT) = 0 . \tag{14.166}$$

With the substitution $e^{sNT} = z$ or $e^{-sNT} = \zeta$ in (14.165), we obtain an algebraic characteristic equation.

Example 14.12 Construct the characteristic equation for the system with $N = 2$, $W_{d1}(s) = W_{d2}(s) = 1$,

$$F_1(s) = \frac{\gamma_1}{s - a}, \quad F_2(s) = \frac{\gamma_2}{s - b}$$

with an arbitrary pulse form. The corresponding block-diagram is shown in Fig. 14.12. In this case, Eq. (14.165) has the form

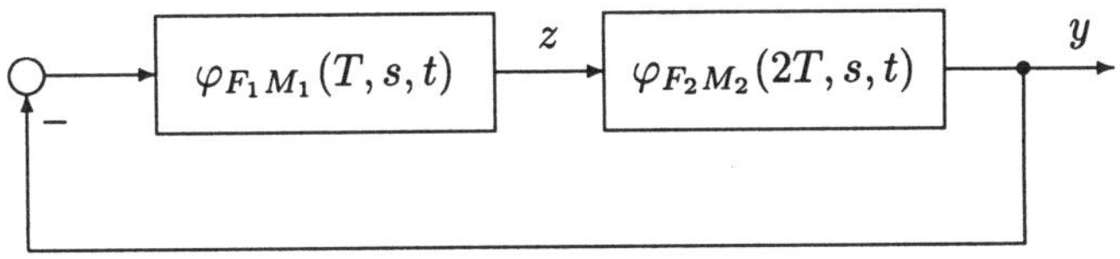

Figure 14.12: Feedback multirate system from Example 14.12

$$1 + \breve{\mathcal{D}}_{F_1 M_1}(2T, s, 0)\mathcal{D}_{F_2 M_2}(2T, s, 0) + \breve{\mathcal{D}}_{F_1 M_1}(2T, s, -T)\mathcal{D}_{F_2 M_2}(2T, s, T) = 0 .$$

From (9.102) and (9.103) for $0 \le t \le 2T$ we find the following equivalent expressions

$$\mathcal{D}_{F_2 M_2}(2T, s, t) = \begin{cases} \dfrac{\gamma_2 M_2(b)e^{bt}}{1 - e^{2(b-s)T}} - \gamma_2 \displaystyle\int_t^{2T} e^{b(t-\tau)}\mu_2(\tau)\,d\tau \\ \dfrac{\gamma_2 M_2(b)e^{bt}}{e^{2(s-b)T} - 1} + \gamma_2 \displaystyle\int_0^t e^{b(t-\tau)}\mu_2(\tau)\,d\tau \end{cases}$$

where

$$M_2(s) = \int_0^{2T} e^{-s\tau}\mu_2(\tau)\,d\tau .$$

Therefore,

$$\begin{aligned} \mathcal{D}_{F_2 M_2}(2T, s, 0) &= \gamma_2 \frac{e^{2(b-s)T} M_2(b)}{1 - e^{2(b-s)T}} \\ \mathcal{D}_{F_2 M_2}(2T, s, T) &= \gamma_2 \frac{M_2(b)e^{bT}}{1 - e^{2(b-s)T}} - \gamma_2 e^{bT} \int_T^{2T} e^{-b\tau}\mu_2(\tau)\,d\tau . \end{aligned} \tag{14.167}$$

Using (12.47), we also have

$$\breve{D}_{p1}^{2T}(s,t) = \\ = \breve{\mathcal{D}}_{F_1M_1}(2T,s,t) = \begin{cases} \dfrac{\gamma_1 M_1(a)\mathrm{e}^{at}}{1-\mathrm{e}^{2(a-s)T}} - \gamma_1 \int\limits_t^T \mathrm{e}^{a(t-\tau)}\mu_1(\tau)\,\mathrm{d}\tau, & 0 \le t \le T \\ \dfrac{\gamma_1 M_1(a)\mathrm{e}^{at}}{1-\mathrm{e}^{2(a-s)T}}, & T \le t \le 2T \end{cases}$$

where

$$M_1(s) = \int_0^T \mathrm{e}^{-s\tau}\mu_1(\tau)\,\mathrm{d}\tau\,.$$

Therefore,

$$\begin{aligned} \breve{\mathcal{D}}_{F_1M_1}(2T,s,0) &= \gamma_1 \frac{\mathrm{e}^{2(a-s)T}M_1(a)}{1-\mathrm{e}^{2(a-s)T}} \\ \breve{\mathcal{D}}_{F_1M_1}(2T,s,T) &= \gamma_1 \frac{M_1(a)\mathrm{e}^{aT}}{1-\mathrm{e}^{2(a-s)T}}\,. \end{aligned} \tag{14.168}$$

Since $\breve{\mathcal{D}}_{F_1M_1}(2T,s,-T) = \breve{\mathcal{D}}_{F_1M_1}(2T,s,T)\mathrm{e}^{-2sT}$, we find

$$\breve{\mathcal{D}}_{F_1M_1}(2T,s,-T) = \gamma_1 \frac{M_1(a)\mathrm{e}^{aT}\mathrm{e}^{-2sT}}{1-\mathrm{e}^{2(a-s)T}}\,. \tag{14.169}$$

As a result, taking account of (14.167)–(14.169), Eq. (14.165) can be written in the form

$$1 + \gamma_1\gamma_2 \frac{M_1(a)M_2(b)\mathrm{e}^{2(a+b)T}\mathrm{e}^{-4sT}}{\left[1-\mathrm{e}^{2(a-s)T}\right]\left[1-\mathrm{e}^{2(b-s)T}\right]} + \\ + \gamma_1\gamma_2 \frac{M_1(a)\mathrm{e}^{(a+b)T}\mathrm{e}^{-2sT}}{1-\mathrm{e}^{2(a-s)T}} \left[\frac{M_2(b)}{1-\mathrm{e}^{2(b-s)T}} - \int_T^{2T} \mathrm{e}^{-b\tau}\mu_2(\tau)\,\mathrm{d}\tau\right] = 0$$

which is reduced, after the substitution $\mathrm{e}^{-2sT} = \zeta$, to the quadratic equation

$$a_0\zeta^2 + a_1\zeta + 1 = 0$$

where

$$a_0 = \mathrm{e}^{2(a+b)T} + \gamma_1\gamma_2 M_1(a)M_2(b)\mathrm{e}^{2(a+b)T} + \gamma_1\gamma_2\mathrm{e}^{(a+3b)T}M_1(a) \int\limits_T^{2T} \mathrm{e}^{-b\tau}\mu_2(\tau)\,\mathrm{d}\tau$$

$$a_1 = -\mathrm{e}^{2aT} - \mathrm{e}^{2bT} + \gamma_1\gamma_2 M_1(a)\mathrm{e}^{(a+b)T} \int_0^T \mathrm{e}^{-b\tau}\mu_2(\tau)\,\mathrm{d}\tau\,.$$

The system is stable, if the characteristic equation has no roots inside or on the unit circle. □

Chapter 15

Analysis of SD systems under stochastic disturbances

15.1 Analysis of open-loop sampled-data systems

In this section we consider the system shown in Fig. 15.1, where $G(s)$ and

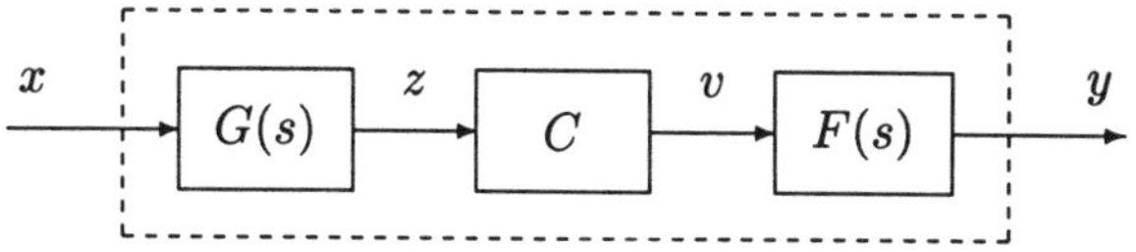

Figure 15.1: Open-loop SD system with prefilter

$F(s)$ are transfer functions of LTI elements, and C is a computer described by Eqs. (10.1)–(10.3). For concreteness, we shall assume that all the characteristic indices $\tilde{\mu}_1, \ldots, \tilde{\mu}_r$ of the system are negative, though, in principle, all the following results are valid for an e-dichotomic system, which has non-zero characteristic indices $\tilde{\mu}_i$. The PTF of the system from the input $x(t)$ to the output $y(t)$ is given by

$$W(s,t) = G(s)W_d(s)\varphi_{FM}(T,s,t)\,. \tag{15.1}$$

If the function $G(s)$ is strictly proper, then, according to Section 6.8, the integral

$$h(t,\tau) = \frac{1}{2\pi \mathrm{j}} \int_{c-\mathrm{j}\infty}^{c+\mathrm{j}\infty} W(s,t)\mathrm{e}^{s(t-\tau)}\,\mathrm{d}s\,, \quad c > \tilde{\mu}_r \tag{15.2}$$

defines the Green function of the following periodic causal e-dichotomic integral operator

$$y(t) = \int_{-\infty}^{t} h(t,\tau)x(\tau)\,\mathrm{d}\tau\,. \tag{15.3}$$

Let $x(t)$ be a centered stationary stochastic process. Then, the associated response of the operator (15.3) will be called the ***steady-state response*** of the system to the input signal $x(t)$. If the spectral density of the input is denoted by $R_x(s)$, by virtue of (8.41) and (15.1) the correlation function of the output is given by the integral

$$K_y(t_1,t_2) = \frac{1}{2\pi\mathrm{j}}\int_{-\mathrm{j}\infty}^{\mathrm{j}\infty} R_x(s)G(s)G(-s)W_d(s)W_d(-s)\varphi_{FM}(T,-s,t_1)\varphi_{FM}(T,s,t_2)\mathrm{e}^{s(t_2-t_1)}\,\mathrm{d}s \tag{15.4}$$

which can be represented in the form

$$K_y(t_1,t_2) = \frac{1}{2\pi\mathrm{j}}\int_{-\mathrm{j}\infty}^{\mathrm{j}\infty} R_x(s)G(s)G(-s)W_d(s)W_d(-s)\mathcal{D}_{FM}(T,-s,t_1)\mathcal{D}_{FM}(T,s,t_2)\,\mathrm{d}s\,. \tag{15.5}$$

Moreover, if the spectral density $R_x(s)$ decreases sufficiently fast for $|s| \to \infty$, Eqs. (15.4) and (15.5) hold when $G(s)$ is proper. Discretizing the integral (15.5) gives

$$K_y(t_1,t_2) = \frac{T}{2\pi\mathrm{j}}\int_{-\mathrm{j}\omega/2}^{\mathrm{j}\omega/2} \mathcal{D}_{FM}(T,-s,t_1)\mathcal{D}_{FM}(T,s,t_2)W_d(s)W_d(-s)\mathcal{D}_{R_z}(T,s,0)\,\mathrm{d}s \tag{15.6}$$

where
$$\mathcal{D}_{R_z}(T,s,0) = \frac{1}{T}\sum_{k=-\infty}^{\infty} R_z(s+k\mathrm{j}\omega) \tag{15.7}$$

with the notation

$$R_z(s) \triangleq G(s)G(-s)R_x(s)\,. \tag{15.8}$$

As follows from Section 8.4, the function $\mathcal{D}_{R_z}(s)$ can be considered as the discrete spectral density of a stochastic sequence $\{z_m\} = \{z(mT)\}$ acting on the input of the discrete-time element.

For $t_1 = t_2 = t$ Eq. (15.6) yields a formula for the variance of the steady-state output

$$d_y(t) = \frac{T}{2\pi\mathrm{j}}\int_{-\mathrm{j}\omega/2}^{\mathrm{j}\omega/2} \mathcal{D}_{FM}(T,s,t)\mathcal{D}_{FM}(T,-s,t)W_d(s)W_d(-s)\mathcal{D}_{R_z}(T,s,0)\,\mathrm{d}s\,. \tag{15.9}$$

Since $d_y(t) = d_y(t+T)$, the mean output variance over the period

$$\bar{d}_y = \frac{1}{T}\int_0^T d_y(t)\,\mathrm{d}t \tag{15.10}$$

can be found from

$$\bar{d}_y = \frac{T}{2\pi\mathrm{j}}\int_{-\mathrm{j}\omega/2}^{\mathrm{j}\omega/2} \Psi_{FM}(T,s)W_d(s)W_d(-s)\mathcal{D}_{R_z}(T,s,0)\,\mathrm{d}s \tag{15.11}$$

where

$$\Psi_{FM}(T,s) \triangleq \frac{1}{T}\int_0^T \mathcal{D}_{FM}(T,-s,t)\mathcal{D}_{FM}(T,s,t)\,\mathrm{d}t\,. \tag{15.12}$$

Using (4.106), we can write

$$\Psi_{FM}(T,s) = \frac{1}{T^2}\sum_{k=-\infty}^{\infty} F(s+k\mathrm{j}\omega)F(-s-k\mathrm{j}\omega)M(s+k\mathrm{j}\omega)M(-s-k\mathrm{j}\omega) \tag{15.13}$$

which can, with reference to (4.107), be written in a shorter form as

$$\Psi_{FM}(T,s) = \frac{1}{T}\mathcal{D}_{F\underline{F}M\underline{M}}(T,s,0) \tag{15.14}$$

where the underbar denotes the change of variable $-s$ for s. The integrands in Eqs. (15.9) and (15.11) are rational periodic functions in s, therefore Theorems A.3 and A.4 given in Appendix A can be used for their calculation.
First, we consider the integral (15.9) assuming that the function $W_d(s)$ is limited, i.e., there are finite limits

$$\ell^+[W_d(s)] = \ell_d^+\,, \quad \ell^-[W_d(s)] = \ell_d^-\,. \tag{15.15}$$

Let also the real rational function $R_z(s)$ (15.8) decrease as $|s|^{-2}$ for $|s| \to \infty$. Then, using (15.7) and the equality

$$\mathcal{D}_{R_z}(T,s,0) = \mathcal{D}_{R_z}(T,-s,0)$$

we obtain

$$\ell^-\left[\mathcal{D}_{R_z}(T,s,0)\right] = \ell^+\left[\mathcal{D}_{R_z}(T,s,0)\right] = 0\,.$$

Moreover, assuming that the function $F(s)$ is strictly proper, we may use Eqs. (9.139) and (9.140). As a result, we find that, under the given assumptions, the integrand in (15.9) tends to zero as $\operatorname{Re} s \to \pm\infty$. Then, using (A.83), we obtain

$$d_y(t) = T\sum_i \operatorname*{Res}_{\tilde{s}_i}\left[\mathcal{D}_{FM}(T,s,t)\mathcal{D}_{FM}(T,-s,t)W_d(s)W_d(-s)\mathcal{D}_{R_z}(T,s,0)\right] \tag{15.16}$$

where the sum extends over all primary poles of the expression in the square brackets, which are in the left half-plane.

Remark 1 Sometimes it is advantageous to use Eq. (A.91) for calculating the integral (15.9).

Remark 2 If the real rational function $R_z(s)$ is not strictly proper, with account for (8.129) Eq.(15.9) can be changed for

$$d_y(t) = \frac{T}{2\pi \mathrm{j}} \int\limits_{-\mathrm{j}\omega/2}^{\mathrm{j}\omega/2} \mathcal{D}_{FM}(T,-s,t)\mathcal{D}_{FM}(T,s,t)W_d(s)W_d(-s)R_z^*(s)\,\mathrm{d}s \quad (15.17)$$

where $R_z^*(s)$ is the discrete spectral density of the stochastic sequence acting on the input of the discrete-time element. In this case the integrand in (15.17), does not, generally speaking, vanish for $\mathrm{Re}\, s \to \pm\infty$, and therefore the general relations given in Sections A.6–A.8 must be used for its calculation. This reasoning pertains also to the calculation of the integral (15.11). The corresponding derivation is demonstrated below in examples.

Example 15.1 Let us apply the received relations for analysing the stochastic processes in the elementary open-loop SD system shown in Fig. 15.2, where

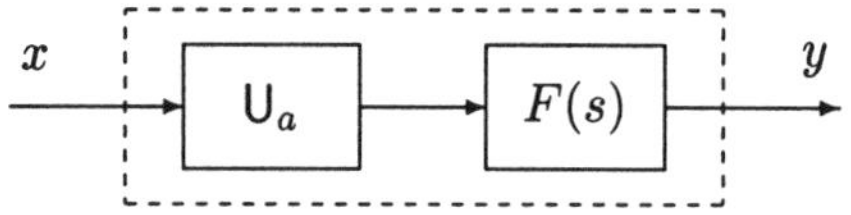

Figure 15.2: Simplest open-loop SD system for Example 15.1

$$F(s) = \frac{\gamma}{s-a}, \quad M(s) = \frac{1-\mathrm{e}^{-sT}}{s}, \quad W_d(s) = 1, \quad G(s) = 1, \quad a < 0.$$

The spectral density of the input has the form

$$R_x(s) = \beta\left(\frac{1}{s-\eta} - \frac{1}{s+\eta}\right), \quad \eta < 0, \quad \eta \neq a. \quad (15.18)$$

In this case, the integral (15.9) takes the form

$$d_y(t) = \frac{T}{2\pi \mathrm{j}} \int_{-\mathrm{j}\omega/2}^{\mathrm{j}\omega/2} \mathcal{D}_{FM}(T,s,t)\mathcal{D}_{FM}(T,-s,t)\mathcal{D}_{R_x}(T,s,0)\,\mathrm{d}s\,. \quad (15.19)$$

To calculate the integral (15.19) we use formula (A.91). Denote

$$d(s) \triangleq \mathcal{D}_{FM}(T,s,t)\mathcal{D}_{FM}(T,-s,t)$$

$$r(s) \triangleq \mathcal{D}_{R_x}(T,s,0)\,.$$

Taking (9.115) into account, for $0 \le t \le T$ we obtain

$$\ell^-[\mathcal{D}_{FM}(T,s,t)] = \gamma\,\frac{\mathrm{e}^{at}-1}{a}\,, \quad \ell^+[\mathcal{D}_{FM}(T,s,t)] = \gamma\,\frac{\mathrm{e}^{a(t-T)}-1}{a}\,.$$

Therefore

$$\ell^+[d(s)] = \ell^-[d(s)] = \gamma^2\,\frac{\mathrm{e}^{2at}\mathrm{e}^{-aT} - \mathrm{e}^{at} - \mathrm{e}^{at}\mathrm{e}^{-aT} + 1}{a^2} \triangleq \Gamma(t)\,. \qquad (15.20)$$

Moreover, in the given case we have

$$\mathcal{D}_{R_x}(T,s,0) = \beta\left[\frac{1}{1-\mathrm{e}^{(\eta-s)T}} - \frac{1}{1-\mathrm{e}^{-(\eta+s)T}}\right]$$

and, in accordance with the aforesaid,

$$\ell^-\left[\mathcal{D}_{R_x}(T,s,0)\right] = \ell^+\left[\mathcal{D}_{R_x}(T,s,0)\right] = 0\,.$$

The integrand in (15.19) has two simple primary poles in the left half-plane at $s = a$ and $s = \eta$. Since

$$\mathcal{D}_{FM}(T,s,t) = \frac{\gamma}{T}\sum_{k=-\infty}^{\infty}\frac{M(s+k\mathrm{j}\omega)}{s+k\mathrm{j}\omega-a}\,\mathrm{e}^{(s+k\mathrm{j}\omega)t}$$

we find

$$\operatorname*{Res}_{a}\mathcal{D}_{FM}(T,s,t) = \frac{\gamma}{T}M(a)\mathrm{e}^{at}\,.$$

Hence,

$$\operatorname*{Res}_{a} d(s) = \frac{\gamma}{T}M(a)\mathrm{e}^{at}\mathcal{D}_{FM}(T,-a,t) = \frac{\gamma}{T}M(a)\varphi_{FM}(T,-a,t) \triangleq a_1\,.$$

Moreover, we have

$$\operatorname*{Res}_{\eta} r(s) = \frac{\beta}{T} \triangleq b_1\,.$$

In this case Eq. (A.91) yields

$$d_y(t) = T\ell_d b_1 - T^2 a_1 b_1 \coth\frac{(a+\eta)T}{2}$$

which, taking account of the preceding relations, gives

$$d_y(t) = \beta\Gamma(t) - \gamma\beta M(a)\varphi_{FM}(T,-a,t)\coth\frac{(a+\eta)T}{2}\,, \quad 0 \le t \le T\,. \qquad (15.21)$$

Using (9.118), in the interval $0 \le t \le T$ we have

$$\varphi_{FM}(T,-a,t) = -\gamma\left[\frac{\mathrm{e}^{2at}}{a\,(1+\mathrm{e}^{-aT})} - \frac{\mathrm{e}^{2at}-\mathrm{e}^{at}}{a}\right]\,.$$

Substituting (15.21) into (15.10) and taking into account that

$$\int_0^T \varphi_{FM}(T, -a, t)\,\mathrm{d}t = F(-a)M(-a)$$

we find the mean variance

$$\bar{d}_y = \beta\bar{\Gamma} + \frac{\gamma^2}{T}\frac{M(a)M(-a)}{2a}\coth\frac{(a+\eta)T}{2} \tag{15.22}$$

where

$$\bar{\Gamma} \triangleq \frac{1}{T}\int_0^T \Gamma(t)\,\mathrm{d}t\,.$$

Figure 15.3 demonstrates the plots of the variance, computed by (15.21) for $a = -\tau^{-1}$, $\gamma = \tau^{-1}$, $\beta = 1$ and $\eta = -2\tau^{-1}$, versus the dimensionless parameter $\bar{t} = t/T$. In Fig. 15.4 the plots of the mean variance (15.22) versus the

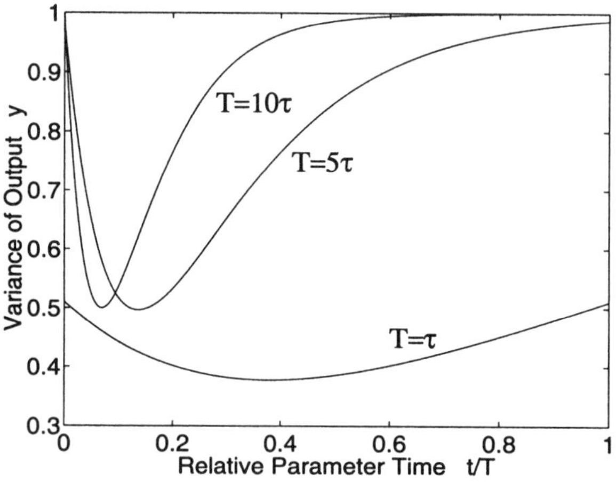

Figure 15.3: Variance of output signal for Example 15.1

parameter $\overline{T} = T/\tau$ are shown. ∎

Example 15.2 In the special case, where under the conditions of Example 15.1 is assumed $\mathcal{D}_{R_x}(T, s, 0) = 1$, the method for the computation of the output variance must be changed a little. In this case from (15.19) follows

$$d_y(t) = \frac{T}{2\pi\mathrm{j}}\int_{-\mathrm{j}\omega/2}^{\mathrm{j}\omega/2} \mathcal{D}_{FM}(T, s, t)\mathcal{D}_{FM}(T, -s, t)\,\mathrm{d}s\,. \tag{15.23}$$

To calculate the integral (15.23) we use formula (A.74). In this case the value ℓ^- is defined by (15.20). The integrand has a single primary pole $s = a$ in the left half-plane, the associated residue was obtained in Example 15.1. As a result, Eq. (A.74) yields

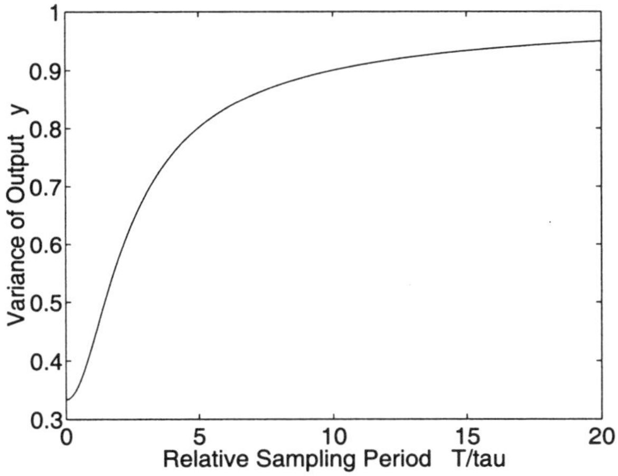

Figure 15.4: Average variance of output for Example 15.1

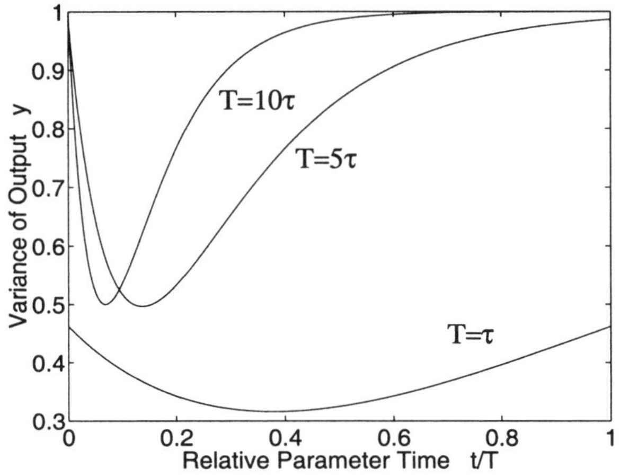

Figure 15.5: Variance of the output for Example 15.2

$$d_y(t) = \Gamma(t) + \gamma M(a)\varphi_{FM}(T, -a, t)\,, \quad 0 \le t \le T\,. \tag{15.24}$$

The mean variance $\overline{d}_y$ appears as

$$\overline{d}_y = \overline{\Gamma} - \frac{\gamma^2}{T}\,\frac{M(a)M(-a)}{2a}\,. \tag{15.25}$$

In Fig. 15.5 and 15.6 the curves for the variance and mean variance are shown, which are calculated for the same conditions as in Example 15.1. ∎

Next, we consider the calculation of the integral (15.9) for the case when the function $W_d(s)$ is not limited. Using (A.27), we obtain

$$W_d(s) = W_{d1}(s) + W_{d\tau}(s) \tag{15.26}$$

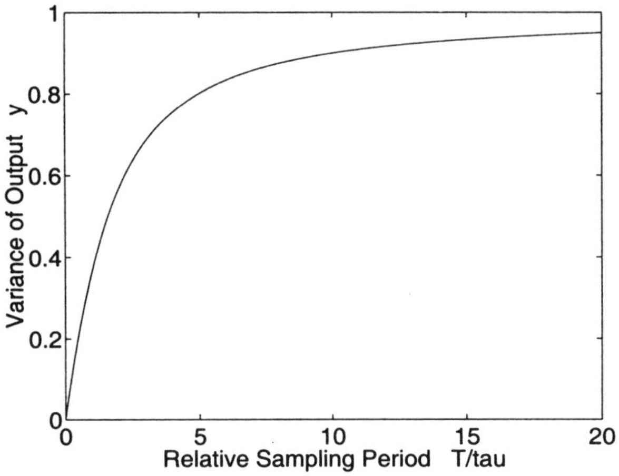

Figure 15.6: Mean variance of the output for Example 15.2

where the function $W_{d1}(s)$ is limited and

$$W_{d\mathcal{T}}(s) = l_\kappa e^{-\kappa sT} + \ldots + l_1 e^{-sT}. \tag{15.27}$$

Substituting (15.26) into (15.9), we obtain

$$d_y(t) = d_1(t) + d_2(t) + d_3(t) + d_4(t) \tag{15.28}$$

where

$$d_1(t) = \frac{T}{2\pi j}\int_{-j\omega/2}^{j\omega/2} \mathcal{D}_{FM}(T,s,t)\mathcal{D}_{FM}(T,-s,t)W_{d1}(s)W_{d1}(-s)\mathcal{D}_{R_z}(T,s,0)\,ds \tag{15.29}$$

$$d_2(t) = \frac{T}{2\pi j}\int_{-j\omega/2}^{j\omega/2} \mathcal{D}_{FM}(T,s,t)\mathcal{D}_{FM}(T,-s,t)W_{d\mathcal{T}}(s)W_{d1}(-s)\mathcal{D}_{R_z}(T,s,0)\,ds \tag{15.30}$$

$$d_3(t) = \frac{T}{2\pi j}\int_{-j\omega/2}^{j\omega/2} \mathcal{D}_{FM}(T,s,t)\mathcal{D}_{FM}(T,-s,t)W_{d1}(s)W_{d\mathcal{T}}(-s)\mathcal{D}_{R_z}(T,s,0)\,ds \tag{15.31}$$

$$d_4(t) = \frac{T}{2\pi j}\int_{-j\omega/2}^{j\omega/2} \mathcal{D}_{FM}(T,s,t)\mathcal{D}_{FM}(T,-s,t)W_{d\mathcal{T}}(s)W_{d\mathcal{T}}(-s)\mathcal{D}_{R_z}(T,s,0)\,ds\,. \tag{15.32}$$

Since the function $W_{d1}(s)$ is limited, the integral (15.29) can be computed by means of the above relations. Moreover,

$$d_2(t) = d_3(t) \tag{15.33}$$

which can be proved by changing $-s$ for s. Thus, it remains to investigate the calculation of the integrals (15.31) and (15.32).
Substituting (15.27) into (15.32), we find that the calculation of $d_4(t)$ reduces to an integral of the form

$$\tilde{d}_m(t) \triangleq \frac{T}{2\pi \mathrm{j}} \int_{-\mathrm{j}\omega/2}^{\mathrm{j}\omega/2} \mathcal{D}_{FM}(T,s,t)\mathcal{D}_{FM}(T,-s,t)\mathcal{D}_{R_z}(T,s,0)\mathrm{e}^{msT}\,\mathrm{d}s \quad (15.34)$$

where m is an integer. Moreover, without loss of generality, we can assume that $m > 0$, because the integral (15.34) does not change with the substitution of $-m$ for m. Similarly, the calculation of the function $d_3(t)$ reduces to integrals of the form

$$\hat{d}_m(t) \triangleq \frac{T}{2\pi \mathrm{j}} \int_{-\mathrm{j}\omega/2}^{\mathrm{j}\omega/2} \mathcal{D}_{FM}(T,s,t)\mathcal{D}_{FM}(T,-s,t)W_{d1}(s)\mathcal{D}_{R_z}(T,s,0)\mathrm{e}^{msT}\,\mathrm{d}s \quad (15.35)$$

where $m > 0$, and $W_{d1}(s)$ is a limited function. The integrals (15.34) and (15.35) can be computed by Theorem (A.3).

Example 15.3 As an example for the computation of the output variance of the system in which the transfer function of the discrete-time controller is not limited, let us consider the open-loop system shown in Fig. 15.1, where

$$G(s) = 1\,, \quad \mathcal{D}_{R_z}(T,s,0) = R_z^*(s) = 1\,, \quad W_d(s) = \mathrm{e}^{-sT} + \frac{\mathrm{e}^{-sT}}{1-0.5\mathrm{e}^{-sT}} \quad (15.36)$$

and

$$F(s) = \frac{\gamma}{s-a}\,, \quad M(s) = \frac{1-\mathrm{e}^{-sT}}{s}\,, \quad a < 0\,, \quad \mathrm{e}^{aT} \neq 0.5\,. \quad (15.37)$$

The output variance can be computed by (15.28), where

$$d_1(t) = \frac{T}{2\pi \mathrm{j}} \int_{-\mathrm{j}\omega/2}^{\mathrm{j}\omega/2} \mathcal{D}_{FM}(T,s,t)\mathcal{D}_{FM}(T,-s,t)\frac{1}{1-0.5\mathrm{e}^{-sT}}\,\frac{1}{1-0.5\mathrm{e}^{sT}}\,\mathrm{d}s \quad (15.38)$$

$$d_2(t) = \frac{T}{2\pi \mathrm{j}} \int_{-\mathrm{j}\omega/2}^{\mathrm{j}\omega/2} \mathcal{D}_{FM}(T,s,t)\mathcal{D}_{FM}(T,-s,t)\frac{1}{1-0.5\mathrm{e}^{sT}}\,\mathrm{d}s \quad (15.39)$$

$$d_3(t) = \frac{T}{2\pi \mathrm{j}} \int_{-\mathrm{j}\omega/2}^{\mathrm{j}\omega/2} \mathcal{D}_{FM}(T,s,t)\mathcal{D}_{FM}(T,-s,t)\frac{1}{1-0.5\mathrm{e}^{-sT}}\,\mathrm{d}s \quad (15.40)$$

$$d_4(t) = \frac{T}{2\pi \mathrm{j}} \int_{-\mathrm{j}\omega/2}^{\mathrm{j}\omega/2} \mathcal{D}_{FM}(T,s,t)\mathcal{D}_{FM}(T,-s,t)\,\mathrm{d}s\,. \quad (15.41)$$

The integral (15.41) has been computed in Example 15.2, while the integral (15.39) can be obtained by Eq. (A.74). The integrand has a single simple pole $s = a$ in the left half-plane. Therefore, using (A.74) and (15.20), we obtain

$$d_2(t) = \Gamma(t) + \gamma M(a)\varphi_{FM}(T,-a,t)\,\frac{1}{1-0.5\mathrm{e}^{aT}}\,. \quad (15.42)$$

To calculate the integral (15.38) it is convenient to employ formula (A.91). A reasoning, similar to that of Example 15.1, leads to the result

$$d_1(t) = \frac{4}{3}\Gamma(t) - \frac{4}{3}\gamma M(a)\varphi_{FM}(T,-a,t)\,\frac{0.5\mathrm{e}^{aT}+1}{0.5\mathrm{e}^{aT}-1}\,. \quad (15.43)$$

□

15.2 Parallel connection of continuous-time and SD systems

Consider the system shown in Fig. 15.7, under all the assumptions made in

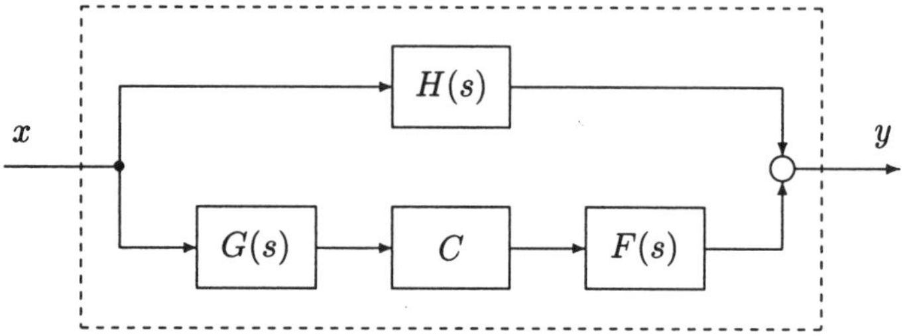

Figure 15.7: Parallel connection of continuous-time and SD systems

Sections 15.1, where $H(s)$ is a strictly proper stable real rational function, with all its poles in the left half-plane. The problem is rather interesting, because it can hardly be solved by a standard continuous-time or discrete-time approach. This is caused by the fact that the system is hybrid-time, continuous-discrete. But the methods based on the PTF preserve the full strength of its meaning. The PTF of the system at hand is given by

$$W(s,t) = H(s) + G(s)W_d(s)\varphi_{FM}(T,s,t)\,. \tag{15.44}$$

If the input signal has spectral density $R_x(s)$, the output variance is equal to

$$d_y(t) = \frac{1}{2\pi \mathrm{j}} \int_{-\mathrm{j}\infty}^{\mathrm{j}\infty} W(s,t)W(-s,t)R_x(s)\,\mathrm{d}s\,. \tag{15.45}$$

Substituting (15.44) into (15.45), we find

$$d_y(t) = d_1 + d_2(t) + d_3(t) + d_4(t) \tag{15.46}$$

where

$$d_1 = \frac{1}{2\pi \mathrm{j}} \int_{-\mathrm{j}\infty}^{\mathrm{j}\infty} H(s)H(-s)R_x(s)\,\mathrm{d}s \tag{15.47}$$

$$d_2(t) = \frac{1}{2\pi \mathrm{j}} \int_{-\mathrm{j}\infty}^{\mathrm{j}\infty} H(-s)G(s)W_d(s)\varphi_{FM}(T,s,t)R_x(s)\,\mathrm{d}s \tag{15.48}$$

$$d_3(t) = \frac{1}{2\pi \mathrm{j}} \int_{-\mathrm{j}\infty}^{\mathrm{j}\infty} H(s)G(-s)W_d(-s)\varphi_{FM}(T,-s,t)R_x(s)\,\mathrm{d}s \tag{15.49}$$

$$d_4(t) = \frac{1}{2\pi \mathrm{j}} \int_{-\mathrm{j}\infty}^{\mathrm{j}\infty} G(s)G(-s)W_d(s)W_d(-s)\varphi_{FM}(T,s,t)\varphi_{FM}(T,-s,t)R_x(s)\,\mathrm{d}s\,. \tag{15.50}$$

The integral (15.47) gives the steady-state variance of the continuous-time element. The technique of its computation can be found in Åström (1970). Since the integral (15.50) coincides with (15.9) after discretization, it remains to consider the integrals (15.48) and (15.49). First of all, we note that

$$d_2(t) = d_3(t) \tag{15.51}$$

which can be proved by changing $-s$ for s in any of the integrals (15.48) or (15.49). Since

$$\begin{aligned} d_3(t) &= \frac{T}{2\pi \mathrm{j}} \int_{-\mathrm{j}\omega/2}^{\mathrm{j}\omega/2} \mathcal{D}_{FM}(T,-s,t) W_d(-s) \mathcal{D}_{H\underline{G}R_x}(T,s,t))\,\mathrm{d}s \\ &= \frac{T}{2\pi \mathrm{j}} \int_{-\mathrm{j}\omega/2}^{\mathrm{j}\omega/2} \mathcal{D}_{FM}(T,s,t) W_d(s) \mathcal{D}_{H\underline{G}R_x}(T,-s,t))\,\mathrm{d}s \end{aligned} \tag{15.52}$$

where the underbar denotes the substitution of $-s$ for s. The integrand in (15.52) is a rational periodic function in the variable s, therefore the function (15.52) can be found by Theorem A.3.

Example 15.4 Let us consider the obtained relations for the system of Fig. 15.7 in the special case with

$$G(s) = W_d(s) = 1\,, \quad F(s) = \frac{\gamma}{s-a}\,, \quad H(s) = \frac{\delta}{s-b}\,, \quad M(s) = \frac{1-\mathrm{e}^{-sT}}{s}$$

where $a < 0$, $b < 0$, and $a \neq b$. The spectral density of the input signal has the form (15.18), where $\eta \neq a$ and $\eta \neq b$. By the residue theorem, it immediately follows that

$$d_1 = \frac{\delta^2 \beta}{b(\eta + b)}$$

while $d_4(t)$ was determined in Example 15.1 by Eq. (15.24).
Since the real rational function $H(s)R_x(s)$ is strictly proper, the integrand in the first integral in (15.52) vanishes as $\operatorname{Re} s \to -\infty$. In this case the left-hand side has primary poles in the left half-plane at the points $s = b$ and $s = \eta$. Since

$$\operatorname*{Res}_{b} \mathcal{D}_{HR_x}(T,s,t) = \frac{\delta}{T} R_x(b) \mathrm{e}^{bt}\,, \quad \operatorname*{Res}_{\eta} \mathcal{D}_{HR_x}(T,s,t) = \frac{\beta\delta}{T} \frac{\mathrm{e}^{\eta t}}{\eta - b}$$

using (A.83) yields

$$d_3(t) = \delta R_x(b) \varphi_{FM}(T,-b,t) + \frac{\beta\delta}{\eta - b} \varphi_{FM}(T,-\eta,t)\,.$$

The mean value of the function $d_3(t)$ over the period is given as

$$\overline{d}_3(t) = \frac{1}{T}\int_0^T d_3(t)\,\mathrm{d}t = -\frac{\delta\gamma}{T} \frac{R_x(b)M(-b)}{b+a} - \frac{\beta\delta\gamma}{T} \frac{M(-\eta)}{(\eta - b)(\eta + a)}\,.$$

The desired output variance appears as

$$d_y(t) = d_1 + 2d_3(t) + d_4(t)\,.$$

Figure 15.8 presents the mean variance $\overline{d}_y$ versus the parameter $\overline{T} = T/\tau$ for

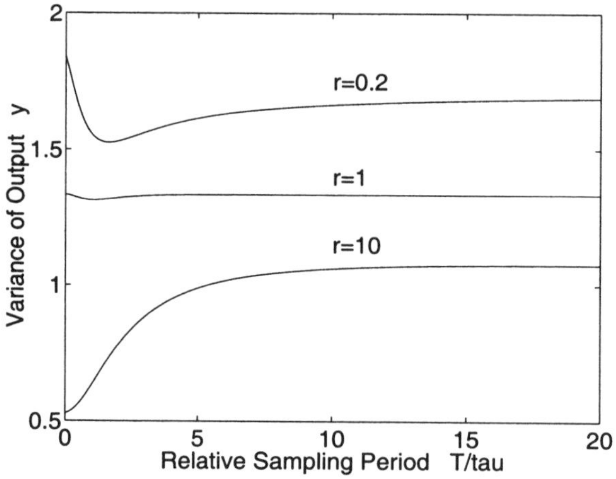

Figure 15.8: Mean variance of output for Example 15.4

$\beta = 1, \eta = -2\tau^{-1}, a = -\tau^{-1}, \gamma = \tau^{-1}, \delta = \gamma/r$ and $b = a/r$ with various r. □

15.3 Stochastic processes in feedback SD systems

In this section we consider the steady-state response of the single-loop system shown in Fig. 11.1 to a centered stochastic input signal. For concreteness, we shall consider the processes for the input $x(t)$ and the output $y(t)$. In this case, the PTF of the system is given by (14.2)

$$W(s) = R(s) - \frac{R(s)W_d(s)\varphi_{FM}(T,s,t)}{1 - W_d(s)\varphi_{FM}(T,s,0)} \tag{15.53}$$

where

$$W_d(s) = \frac{b(s)}{a(s)}, \quad \varphi_{FM}(T,s,0) = \mathcal{D}_{FM}(t,s,0) = \frac{e^{-sT}\chi(s)}{\alpha(s)} \tag{15.54}$$

and the rational periodic functions in (15.54) are assumed to be irreducible. The considered feedback system is assumed to be stable, i.e., the roots s_i of the characteristic equation

$$\Delta(s) = \alpha(s)a(s) + e^{-sT}\chi(s)b(s) = 0 \tag{15.55}$$

lie in the left half-plane.

Let the transfer function $R(s)$ be strictly proper, so that the integral

$$h_\gamma(t,\tau) = \frac{1}{2\pi \mathrm{j}} \int_{c-\mathrm{j}\infty}^{c+\mathrm{j}\infty} W(s,t) \mathrm{e}^{s(t-\tau)} \, \mathrm{d}s \tag{15.56}$$

converges, where c is a constant such that all roots of Eq. (15.55) are in the half-plane $\operatorname{Re} s < c$. Then, the system under consideration is associated with a linear T−periodic operator

$$y(t) = \mathsf{U}_0[x(t)] = \int_{-\infty}^{t} h_\gamma(t,\tau) x(\tau) \, \mathrm{d}\tau \tag{15.57}$$

which is causal and exponentially dichotomic. Let $x(t)$ be a centered stationary stochastic signal. The response $y(t)$ of the operator (15.57) to this signal is called the *steady-state response* of the system to the signal $x(t)$. As follows from Section 8.2, $y(t)$ is a periodically non-stationary stochastic process with the variance

$$d_y(t) = \frac{1}{2\pi \mathrm{j}} \int_{-\mathrm{j}\infty}^{\mathrm{j}\infty} W(s,t) W(-s,t) P_x(s) \, \mathrm{d}s \tag{15.58}$$

where $P_x(s)$ is the spectral density of the input signal and $W(s,t)$ is the PTF (15.53). Substituting (15.53) into (15.58), after discretization we find

$$\begin{aligned} d_y(t) = \frac{T}{2\pi \mathrm{j}} \int_{-\mathrm{j}\omega/2}^{\mathrm{j}\omega/2} \Bigg[& \mathcal{D}_{R\underline{R}P_x}(T,s,0) - \frac{W_d(s)\mathcal{D}_{FM}(T,s,t)\mathcal{D}_{R\underline{R}P_x}(T,-s,t)}{\lambda(s)} - \\ & - \frac{W_d(-s)\mathcal{D}_{FM}(T,-s,t)\mathcal{D}_{R\underline{R}P_x}(T,s,t)}{\lambda(-s)} + \\ & + \frac{W_d(s)W_d(-s)\mathcal{D}_{FM}(T,s,t)\mathcal{D}_{FM}(T,-s,t)\mathcal{D}_{R\underline{R}P_x}(T,s,0)}{\lambda(s)\lambda(-s)} \Bigg] \mathrm{d}s \end{aligned} \tag{15.59}$$

where

$$\lambda(s) = 1 + \frac{b(s)}{a(s)} \frac{\mathrm{e}^{-sT}\chi(s)}{\alpha(s)} \tag{15.60}$$

$$\mathcal{D}_{R\underline{R}P_x}(T,s,t) = \frac{1}{T} \sum_{k=-\infty}^{\infty} R(s+k\mathrm{j}\omega) R(-s-k\mathrm{j}\omega) P_x(s+k\mathrm{j}\omega) \mathrm{e}^{(s+k\mathrm{j}\omega)t} . \tag{15.61}$$

Since the variance is periodic, it suffices to consider only the interval $0 \le t \le T$. Denote the integrand of (15.59) by $f(s,t)$. It can be easily checked that $f(s,t) = f(s,t+T)$.

Theorem 15.1 *Let the real rational function $R(s)R(-s)P_x(s)$ be strictly proper and the conditions of the Theorems 14.1 and 14.2 hold, i.e., if $W_d(s)$ is limited, Eq. (14.15) holds, otherwise Eq. (14.18) holds. Then, the function $f(s,t)$ is limited and there exist finite limits*

$$\ell^{\pm}[f(s,t)] = \lim_{\operatorname{Re} s \to \pm\infty} f(s,t) \stackrel{\triangle}{=} \ell_f(t) < \infty . \tag{15.62}$$

Proof Since the function $f(s,t)$ is periodic and continuous with respect to t, we take $0 \le t \le T$. Then, using the propositions of Sections 4.8 and 9.7 we find that the functions $\mathcal{D}_{FM}(T,s,t)$ and $\mathcal{D}_{R\underline{R}P_x}(T,s,t)$ are both limited. Moreover, under the conditions of Theorems 14.1 and 14.2, the rational periodic function

$$\tilde{W}_d(s) = \frac{W_d(s)}{1 + W_d(s)\mathcal{D}_{FM}(T,s,0)} \tag{15.63}$$

is limited. Then, taking the limit in (15.59) as $\mathrm{Re}\, s \to \pm\infty$, we prove the proposition. ∎

Corollary Under the conditions of Theorem 15.1,

$$d_y(t) = T \sum_i \operatorname*{Res}_{\tilde{s}_{i0}} f(s,t) + \ell_f(t) \tag{15.64}$$

where $\tilde{s}_{i0}$ are the primary poles of $f(s,t)$ which are in the left half-plane.

Remark The function $\ell_f(t)$ in (15.64) can be computed as

$$\ell_f(t) = -\tilde{\ell}_d^- h_p^*(t) h_{R\underline{R}P_x}(t)$$

where

$$h_{R\underline{R}P_x}(t) = \frac{1}{2\pi\mathrm{j}} \int_{c-\mathrm{j}\infty}^{c+\mathrm{j}\infty} R(s)R(-s)P_x(s)\mathrm{e}^{st}\,\mathrm{d}s$$

and the abscissa c is chosen so that all poles of the integrand are left from the integration line. Then,

$$\tilde{\ell}_d^- = \lim_{\mathrm{Re}\, s \to -\infty} \tilde{W}_d(s)$$

and the function $h_p^*(t)$ is given by (9.98).

For practical computations of the integral (15.59) it is useful to represent it in a different form. Let the conditions of Theorem 11.1 hold, so that

$$W(s,t) = \frac{n(s,t)}{\Delta(s)} \tag{15.65}$$

is an entire function in s

$$n(s,t) = R(s)\Delta(s) - R(s)b(s)\alpha(s)\varphi_{FM}(T,s,t)\,. \tag{15.66}$$

Substituting (15.65) and (15.66) into (15.58), after discretization we obtain

$$d_y(t) = \frac{T}{2\pi\mathrm{j}} \int_{-\mathrm{j}\omega/2}^{\mathrm{j}\omega/2} f(s,t)\,\mathrm{d}s \tag{15.67}$$

where

$$f(s,t) \triangleq \frac{1}{T} \sum_{k=-\infty}^{\infty} \frac{n(s+k\mathrm{j}\omega,t)n(-s-k\mathrm{j}\omega,t)P_x(s+k\mathrm{j}\omega)}{\Delta(s)\Delta(-s)}\,. \tag{15.68}$$

Under the given assumptions the series (15.68) converges absolutely and uniformly in any range free of the roots of $\Delta(s)$ and $\Delta(-s)$ and the poles of $P_x(s+kj\omega)$. Therefore, under the given assumptions the function $f(s,t)$ has singular points (poles) at $\pm s_i$ and at $\eta_i + kj\omega$, where η_i are the poles of $P_x(s)$ and k is an integer. Substituting (15.66) into (15.68), we find

$$f(s,t) = \frac{m(s,t)}{\Delta(s)\Delta(-s)} \tag{15.69}$$

where

$$\begin{aligned} m(s,t) &= \mathcal{D}_{R\underline{R}P_x}(T,s,0)\Delta(s)\Delta(-s) - \mathcal{D}_{R\underline{R}P_x}(T,s,t)\Delta(s)b(-s)\alpha(-s) * \\ &\quad *\mathcal{D}_{FM}(T,-s,t) - \mathcal{D}_{R\underline{R}P_x}(T,-s,t)\Delta(-s)b(s)\alpha(s)\mathcal{D}_{FM}(T,s,t) + \\ &\quad + \mathcal{D}_{R\underline{R}P_x}(T,s,0)\alpha(s)\alpha(-s)b(s)b(-s)\mathcal{D}_{FM}(T,s,t)\mathcal{D}_{FM}(T,-s,t)\,. \end{aligned} \tag{15.70}$$

By construction, the function $m(s,t)$ is analytic for all s except for the points $s = \eta_i + kj\omega$. Henceforth, we assume that all poles of the function $f(s,t)$ are simple. Let $\tilde{s}_{i0}$ be the primary poles of Eq. (15.55), $\tilde{\eta}_m$ be the primary poles of $\mathcal{D}_{P_x}(T,s,t)$, and there are no equal numbers among $\tilde{s}_{i0}$ and $\tilde{\eta}_m$. Then,

$$\operatorname*{Res}_{\tilde{s}_{i0}} f(s,t) = \frac{q_i(t)}{\Delta'(\tilde{s}_{i0})\Delta(-\tilde{s}_{i0})} \tag{15.71}$$

where

$$\begin{aligned} q_i(t) &= -\mathcal{D}_{R\underline{R}P_x}(T,\tilde{s}_{i0},t)\Delta(-\tilde{s}_{i0})b(\tilde{s}_{i0})\alpha(\tilde{s}_{i0})\mathcal{D}_{FM}(T,\tilde{s}_{i0},t) + \\ &\quad + \mathcal{D}_{R\underline{R}P_x}(T,\tilde{s}_{i0},0)\alpha(-\tilde{s}_{i0})\alpha(-\tilde{s}_{i0})b(\tilde{s}_{i0})b(-\tilde{s}_{i0})\mathcal{D}_{FM}(T,\tilde{s}_{i0},t)\mathcal{D}_{FM}(T,-\tilde{s}_{i0},t). \end{aligned} \tag{15.72}$$

Hence,

$$\operatorname*{Res}_{\eta_m} f(s,t) = \frac{1}{T}\,\frac{r_m(t)}{\Delta(\eta_m)\Delta(-\eta_m)} \tag{15.73}$$

where

$$\begin{aligned} r_m(t) &= \beta_m R(\eta_m)R(-\eta_m)\Delta(\eta_m)\Delta(-\eta_m) \; - \\ &\quad -\beta_m R(\eta_m)R(-\eta_m)\Delta(\eta_m)b(-\eta_m)\alpha(-\eta_m)\varphi_{FM}(T,-\eta_m,t) \; - \\ &\quad -\beta_m R(\eta_m)R(-\eta_m)\Delta(-\eta_m)b(\eta_m)\alpha(\eta_m)\varphi_{FM}(T,\eta_m,t) \; + \\ &\quad +\beta_m R(\eta_m)R(-\eta_m)\alpha(\eta_m)\alpha(-\eta_m)b(\eta_m)b(-\eta_m)\varphi_{FM}(T,\eta_m,t)\mathcal{D}_{FM}(T,-\eta_m,t) \end{aligned} \tag{15.74}$$

with the notation

$$P_x(s) = \sum_m \beta_m \left(\frac{1}{s-\eta_m} - \frac{1}{s+\eta_m} \right).$$

Substituting (15.71)–(15.75) into (15.64), we obtain the required output variance. The mean output variance $\overline{d}_y$ can be obtained by (15.10).

For $P_x(s) = 1$, in accordance with (8.61) we have

$$\overline{d}_y = \overline{r}_y = \|\mathsf{U}_0\|_2^2$$

where $\|\mathsf{U}_0\|_2$ is the $\mathcal{H}_2$–norm of the operator (15.57), which will be called the $\mathcal{H}_2$–norm of the system under consideration.

Example 15.5 Find the $\mathcal{H}_2$–norm of the SD system shown in Fig. 15.9, where

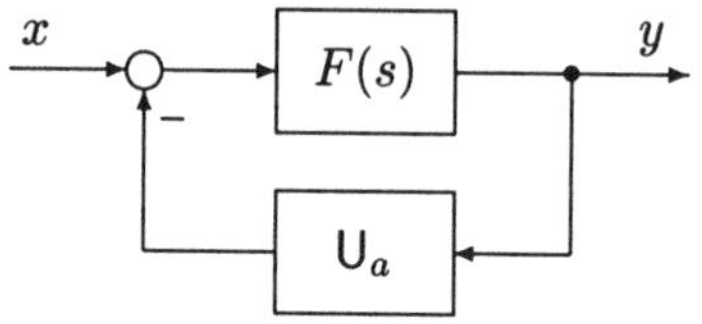

Figure 15.9: Simple SD system for Example 15.5

$$F(s) = \frac{\gamma}{s-a}, \quad W_d(s) = 1$$

and the form of the pulse is arbitrary. As follows from the above relations,

$$\|\mathsf{U}\|_2^2 = \frac{T}{2\pi \mathrm{j}} \int_{-\mathrm{j}\omega/2}^{\mathrm{j}\omega/2} \overline{f}(s)\, \mathrm{d}s$$

where

$$\overline{f}(s) = \mathcal{D}_{F\underline{F}}(T,s,0) - \frac{1}{T}\frac{\mathcal{D}_{F^2\underline{F}M}(T,-s,0)}{\lambda(-s)} - \frac{1}{T}\frac{\mathcal{D}_{F^2\underline{F}M}(T,s,0)}{\lambda(s)} + \\ + \frac{1}{T}\frac{\mathcal{D}_{F\underline{F}}(T,s,0)\mathcal{D}_{F\underline{F}M\underline{M}}(T,s,0)}{\lambda(s)\lambda(-s)}$$

and

$$\lambda(s) = \frac{\mathrm{e}^{sT} - \mathrm{e}^{aT} + \gamma M(a)\mathrm{e}^{aT}}{\mathrm{e}^{sT} - \mathrm{e}^{aT}}\,.$$

Under the given assumptions,

$$\ell^-\left[\overline{f}(s)\right] = \frac{1}{T}\frac{\ell^-[\mathcal{D}_{F^2\underline{F}M}(T,s,0)]}{1-\gamma M(a)}\,. \tag{15.75}$$

The function $\overline{f}(s)$ has in the left half-plane a single primary pole $s = \tilde{s}$, for which

$$\mathrm{e}^{\tilde{s}T} = [1 - \gamma M(a)]\,\mathrm{e}^{aT}$$

and

$$\operatorname*{Res}_{\tilde{s}} \overline{f}(s) = \frac{\gamma}{T^2}\frac{\mathcal{D}_{F^2\underline{F}M}(T,\tilde{s},0)M(a)}{1-\gamma M(a)} - \frac{\gamma}{T^2}\frac{\mathcal{D}_{F\underline{F}}(T,\tilde{s},0)\mathcal{D}_{F\underline{F}M\underline{M}}(T,\tilde{s},0)M(a)}{[1-\gamma M(a)]\lambda(-\tilde{s})}\,. \tag{15.76}$$

Then, using (A.74) we arrive at the desired result. □

15.4 Deadbeat systems

In this section we consider stochastic processes in the system considered in Section 14.4. This case demands a special investigation, because the calculation techniques presented in Section 15.3 are not applicable here. Assuming that all the assumptions of Section 14.4 hold, we obtain the DTF from the input $x_d(t)$ to the output $y(t)$ for $0 \le t \le T$

$$D(s,t) = \sum_{k=0}^{2n-1} d_k(t)\mathrm{e}^{-ksT} = \beta(s,t)b(s) \tag{15.77}$$

where $d_k(t)$ are known functions. From (15.77) it follows that

$$D(-s,t) = \sum_{k=0}^{2n-1} d_k(t)\mathrm{e}^{ksT}.$$

Therefore,

$$D(s,t)D(-s,t) = \sum_{m=0}^{2n-1}\sum_{k=0}^{2n-1} d_k(t)d_m(t)\mathrm{e}^{(m-k)sT}. \tag{15.78}$$

Since the feedback system is stable, the variance of the steady-state output can be obtained by means of Eq. (8.129), which in this case takes the form

$$d_y(t) = \frac{T}{2\pi \mathrm{j}}\int_{-\mathrm{j}\omega/2}^{\mathrm{j}\omega/2} D(s,t)D(-s,t)R_z^*(s)\,\mathrm{d}s \tag{15.79}$$

where $R_z^*(s)$ is the discrete spectral density of the stochastic sequence acting on the system input. Substituting (15.78) into (15.79), we find

$$d_y(t) = \sum_{m=0}^{2n-1}\sum_{k=0}^{2n-1} d_k(t)d_m(t)\frac{T}{2\pi \mathrm{j}}\int_{-\mathrm{j}\omega/2}^{\mathrm{j}\omega/2} R_z^*(s)\mathrm{e}^{(m-k)sT}\,\mathrm{d}s. \tag{15.80}$$

But, due to (8.20),

$$R_z^*(s) = \sum_{i=-\infty}^{\infty} K_z^*(iT)\mathrm{e}^{-isT} \tag{15.81}$$

where $K_z^*(nT)$ denotes the associated correlation function. Therefore,

$$\frac{T}{2\pi \mathrm{j}}\int_{-\mathrm{j}\omega/2}^{\mathrm{j}\omega/2} R_z^*(s)\mathrm{e}^{(m-k)sT}\,\mathrm{d}s = K_z^*(kT - mT).$$

Using this equation in (15.80) gives

$$d_y(t) = \sum_{m=0}^{2n-1}\sum_{k=0}^{2n-1} d_k(t)d_m(t)K_z^*(kT - mT). \tag{15.82}$$

If, as a special case, the input signal is a discrete white noise with unit intensity, i.e.,

$$K_z^*(0) = 1\,, \quad K_z^*(iT) = 0\,, \text{for}\, i \neq 0$$

from (15.82) we find

$$d_y(t) = \sum_{k=0}^{2n-1} d_k^2(t)\,. \tag{15.83}$$

The mean variance over a period obtained by (15.82) takes the form

$$\bar{d}_y = \sum_{m=0}^{2n-1} \sum_{k=0}^{2n-1} \lambda_{km} K_z^*(kT - mT) \tag{15.84}$$

where

$$\lambda_{km} \triangleq \frac{1}{T} \int_0^T d_k(t) d_m(t)\,\mathrm{d}t\,. \tag{15.85}$$

Example 15.6 Let us consider the procedure for calculating the output variance of the deadbeat system considered in Example 14.2. The system is assumed to be excited by an unit discrete white noise. Then, taking account of (15.83) and the results of Example 14.2,

$$d_y(t) = d_0^2(t) + d_1^2(t)$$

$$\text{with} \quad d_0(t) = \frac{1}{M(a)} \int_0^t \mathrm{e}^{a(t-\tau)} \mu(\tau)\,\mathrm{d}\tau, \quad d_1(t) = \frac{\mathrm{e}^{aT}}{M(a)} \int_t^T \mathrm{e}^{a(t-\tau)} \mu(\tau)\,\mathrm{d}\tau.$$

Since

$$d_0(0) = 0\,, \quad d_0(T) = \mathrm{e}^{aT}\,, \quad d_1(0) = \mathrm{e}^{aT}\,, \quad d_1(T) = 0$$

we have, in particular,

$$d_y(0) = d_y(T) = \mathrm{e}^{2aT}\,.$$

□

15.5 $\mathcal{H}_2$−norm for standard multivariable systems

As an example of the use of the present approach, we consider the calculation of the $\mathcal{H}_2$−norm for the multivariable system shown in Fig. 11.10. The PTM from the input x to the output y is given by (11.132)

$$\boldsymbol{W}_c(s,t) = \boldsymbol{K}(s) + \varphi_{\boldsymbol{F}M}(T,s,t)\tilde{\boldsymbol{W}}_{\boldsymbol{N}}(s)\boldsymbol{R}(s)\,. \tag{15.86}$$

Assume that the real rational matrices $\boldsymbol{K}(s)$ and $\boldsymbol{R}(s)$ are strictly proper and all poles of $\boldsymbol{W}_c(s,t)$ are in the left half-plane. In this case the following integral converges

$$\boldsymbol{H}(t,\tau) = \frac{1}{2\pi \mathrm{j}} \int_{c-\mathrm{j}\infty}^{c+\mathrm{j}\infty} \boldsymbol{W}_c(s,t) \mathrm{e}^{s(t-\tau)}\,\mathrm{d}s \tag{15.87}$$

where c is a constant such that all poles of $\boldsymbol{W}_c(s,t)$ are located in the half-plane $\operatorname{Re} s < c$. In this case,

$$\boldsymbol{H}(t,\tau) = \boldsymbol{0} \quad \text{for } t < \tau\,. \tag{15.88}$$

The matrix $\boldsymbol{H}(t,\tau)$ is the Green matrix of the causal T−periodic e-dichotomic integral operator

$$y(t) \triangleq \mathsf{U}_0\left[\boldsymbol{x}(t)\right] = \int_{-\infty}^{t} \boldsymbol{H}(t,\tau)\boldsymbol{x}(\tau)\,\mathrm{d}\tau\,. \tag{15.89}$$

Let $\boldsymbol{x}(t)$ be a centered stationary stochastic vector. Then, the corresponding response of the operator (15.89) will be called the steady-state response of the standard system to the input signal $\boldsymbol{x}(t)$. The mean variance $\bar{d}_{\boldsymbol{y}}$ of the steady-state output $\boldsymbol{y}(t)$ is given by (8.100)

$$\bar{d}_{\boldsymbol{y}} = \frac{1}{2\pi\mathrm{j}}\int_{-\mathrm{j}\infty}^{\mathrm{j}\infty} \operatorname{tr}\left[\boldsymbol{B}_0(s)\boldsymbol{R}_{\boldsymbol{x}\boldsymbol{x}}(s)\right]\,\mathrm{d}s = \frac{1}{2\pi\mathrm{j}}\int_{-\mathrm{j}\infty}^{\mathrm{j}\infty} \operatorname{tr}\left[\boldsymbol{R}_{\boldsymbol{x}\boldsymbol{x}}(s)\boldsymbol{B}_0(s)\right]\,\mathrm{d}s \tag{15.90}$$

where $\boldsymbol{R}_{\boldsymbol{x}\boldsymbol{x}}(s)$ is the spectral density matrix of the vector $\boldsymbol{x}(t)$, and the matrix $\boldsymbol{B}_0(s)$ is expressed in terms of the PTM (15.86) as

$$\boldsymbol{B}_0(s) = \frac{1}{T}\int_0^T \boldsymbol{W}'_c(s,t)\boldsymbol{W}_c(-s,t)\,\mathrm{d}t \tag{15.91}$$

where the prime denotes the transpose operator. Under the given assumptions, the $\mathcal{H}_2$−norm of the operator (15.89) will be called the $\mathcal{H}_2$−norm of the standard multivariable system. As follows from Section 8.3,

$$\|\mathsf{U}_0\|_2^2 = \bar{r}_{\boldsymbol{y}} \tag{15.92}$$

where $\bar{r}_{\boldsymbol{y}}$ is the mean variance of the steady-state output (15.89) for the case when $\boldsymbol{R}_{\boldsymbol{x}\boldsymbol{x}}(s) = \boldsymbol{I}_m$, and $\boldsymbol{I}_m$ is the identity matrix of the appropriate dimension. Hence,

$$\|\mathsf{U}_0\|_2^2 = \frac{1}{2\pi\mathrm{j}}\int_{-\mathrm{j}\infty}^{\mathrm{j}\infty} \operatorname{tr}\left[\boldsymbol{B}_0(s)\right]\,\mathrm{d}s\,. \tag{15.93}$$

Since $\boldsymbol{K}(s)$ and $\boldsymbol{L}(s)$ are strictly proper, all elements of the matrix $\boldsymbol{B}_0(s)$ decrease as $|s|^{-2}$ for $s \to \infty$. Therefore, the integral can be discretized, which leads to Eq. (8.108)

$$\|\mathsf{U}_0\|_2^2 = \frac{T}{2\pi\mathrm{j}}\int_{-\mathrm{j}\omega/2}^{\mathrm{j}\omega/2} \operatorname{tr}\left[\mathcal{D}_{\boldsymbol{B}0}(T,s,0)\right]\,\mathrm{d}s \tag{15.94}$$

where

$$\mathcal{D}_{\boldsymbol{B}0}(T,s,0) = \frac{1}{T}\sum_{k=-\infty}^{\infty} \boldsymbol{B}_0(s+k\mathrm{j}\omega)\,. \tag{15.95}$$

Next, we calculate the matrix $\mathcal{D}_{\boldsymbol{B}0}(T,s,0)$ for the PTM (15.86). We note that

$$\begin{aligned} \boldsymbol{W}_c(-s,t) &= \boldsymbol{K}(-s) + \varphi_{\boldsymbol{L}M}(T,-s,t)\tilde{\boldsymbol{W}}_{\boldsymbol{N}}(-s)\boldsymbol{R}(-s) \\ \boldsymbol{W}'_c(s,t) &= \boldsymbol{K}'(s) + \boldsymbol{R}'(s)\tilde{\boldsymbol{W}}'_{\boldsymbol{N}}(s)\varphi'_{\boldsymbol{L}M}(T,s,t)\,. \end{aligned} \tag{15.96}$$

Hence,

$$\begin{aligned} \boldsymbol{W}'_c(s,t)\boldsymbol{W}_c(-s,t) = \boldsymbol{K}'(s)\boldsymbol{K}(-s) + \boldsymbol{K}'(s)\varphi_{\boldsymbol{L}M}(T,-s,t)\tilde{\boldsymbol{W}}_{\boldsymbol{N}}(-s)\boldsymbol{R}(-s) + \\ + \boldsymbol{R}'(s)\tilde{\boldsymbol{W}}'_{\boldsymbol{N}}(s)\varphi'_{\boldsymbol{L}M}(T,s,t)\boldsymbol{K}(-s) + \\ + \boldsymbol{R}'(s)\tilde{\boldsymbol{W}}'_{\boldsymbol{N}}(s)\varphi'_{\boldsymbol{L}M}(T,s,t)\varphi_{\boldsymbol{L}M}(T,-s,t)\tilde{\boldsymbol{W}}_{\boldsymbol{N}}(-s)\boldsymbol{R}(-s)\,. \end{aligned} \tag{15.97}$$

It can be easily verified, that

$$\begin{aligned} \frac{1}{T}\int_0^T \varphi'_{\boldsymbol{L}M}(T,s,t)\,\mathrm{d}t &= \frac{1}{T}\boldsymbol{L}'(s)M(s) \\ \frac{1}{T}\int_0^T \varphi_{\boldsymbol{L}M}(T,-s,t)\,\mathrm{d}t &= \frac{1}{T}\boldsymbol{L}(-s)M(-s) \end{aligned} \tag{15.98}$$

and

$$\frac{1}{T}\int_0^T \varphi'_{\boldsymbol{L}M}(T,s,t)\varphi_{\boldsymbol{L}M}(T,-s,t)\,\mathrm{d}t = \frac{1}{T}\mathcal{D}_{\boldsymbol{L}'\underline{\boldsymbol{L}}M\underline{M}}(T,s,0)\,. \tag{15.99}$$

Substituting (15.97)–(15.99) into (15.91), we find

$$\begin{aligned} \boldsymbol{B}_0(s) = \;& \boldsymbol{K}'(s)\boldsymbol{K}(-s) + \frac{1}{T}\boldsymbol{K}'(s)\boldsymbol{L}(-s)M(-s)\tilde{\boldsymbol{W}}_{\boldsymbol{N}}(-s)\boldsymbol{R}(-s) + \\ & + \frac{1}{T}\boldsymbol{R}'(s)\tilde{\boldsymbol{W}}'_{\boldsymbol{N}}(s)\boldsymbol{L}(s)M(s)\boldsymbol{K}(-s) + \\ & + \frac{1}{T}\boldsymbol{R}'(s)\tilde{\boldsymbol{W}}'_{\boldsymbol{N}}(s)\mathcal{D}_{\boldsymbol{L}'\underline{\boldsymbol{L}}M\underline{M}}(T,s,0)\tilde{\boldsymbol{W}}_{\boldsymbol{N}}(-s)\boldsymbol{R}(-s)\,. \end{aligned} \tag{15.100}$$

From (15.100) it follows that

$$\begin{aligned} \operatorname{tr}\boldsymbol{B}_0(s) = \frac{1}{T}\operatorname{tr}\Big[&\mathcal{D}_{\boldsymbol{L}'\underline{\boldsymbol{L}}M\underline{M}}(t,s,0)\tilde{\boldsymbol{W}}_{\boldsymbol{N}}(-s)\boldsymbol{R}(-s)\boldsymbol{R}'(s)\tilde{\boldsymbol{W}}'_{\boldsymbol{N}}(s) + \\ &+\tilde{\boldsymbol{W}}_{\boldsymbol{N}}(-s)\boldsymbol{R}(-s)\boldsymbol{K}'(s)\boldsymbol{L}(-s)M(s) + \boldsymbol{L}'(s)\boldsymbol{K}(-s)\boldsymbol{R}'(s)M(s)\tilde{\boldsymbol{W}}'_{\boldsymbol{N}}(s)\Big] + \\ &+ \operatorname{tr}\left[\boldsymbol{K}'(s)\boldsymbol{K}(-s)\right]\,. \end{aligned} \tag{15.101}$$

Then, Eq. (15.95) yields

$$\begin{aligned} \operatorname{tr}\mathcal{D}_{\boldsymbol{B}0}(T,s,0) = \;& \frac{1}{T}\operatorname{tr}\Big[\mathcal{D}_{\boldsymbol{L}'\underline{\boldsymbol{L}}M\underline{M}}(T,s,0)\tilde{\boldsymbol{W}}_{\boldsymbol{N}}(-s)\mathcal{D}_{\underline{\boldsymbol{R}}\boldsymbol{R}'}(T,s,0)\tilde{\boldsymbol{W}}'_{\boldsymbol{N}}(s) + \\ & +\tilde{\boldsymbol{W}}_{\boldsymbol{N}}(-s)\mathcal{D}_{\underline{\boldsymbol{R}}\boldsymbol{K}'\underline{\boldsymbol{L}}M}(T,s,0) + \mathcal{D}_{\boldsymbol{L}'\underline{\boldsymbol{K}}\boldsymbol{R}'M}(T,s,0)\Big] + \\ & + \operatorname{tr}\mathcal{D}_{\boldsymbol{K}'\underline{\boldsymbol{K}}}(T,s,0)\,. \end{aligned} \tag{15.102}$$

Substituting (15.101) into (15.93), we find the desired value of $\|\mathsf{U}_0\|_2^2$.

Part V

Direct synthesis of SD systems

Introduction The problem of finding a discrete-time control for a continuous-time processes is solved mostly by one of the following two methods.

In the first approach, a continuous-time control is to be found for the system at hand, which is further replaced by a discrete approximation according with a numerical integration formula. The second approach involves the determination of a discrete-time control for a specially constructed discrete-time model of the plant. Both approaches have important shortcomings.

The first postpones the question of a discrete-time approximation of the continuous-time control law, which leads to problems no less involved than the investigation of the initial continuous-time systems. Moreover, it is not yet known, whether it is possible to obtain optimal discrete control as an approximation of a known continuous control. The second approach ignores intersample behaviour, though this is inadmissible in many cases. Moreover, it is not always possible to construct an adequate discrete model of a continuous plant.

Therefore, the problem of constructing an optimal discrete-time control based directly on a known continuous-time model of the process to be controlled is crucial for many applications. Following Chen and Francis (1995), this problem will be termed *direct design* of sampled-data systems.

This part of the book presents frequency methods for the direct sampled-data system design. We consider three direct design problems. The first one is the quadratic optimization of sampled-data systems in the interval $0 \leq t < \infty$ (LQ–problem). The second type of problem ($\mathcal{H}_2-$problem) involves the quadratic optimization of sampled-data systems on the whole $t-$axis $-\infty < t < \infty$. The third type is connected with a robust optimization of sampled-data systems on the basis of an auxiliary $\mathcal{H}_\infty-$problem.

In all these cases, the solution is based on some quadratic functionals defined on the set of real rational functions in the indeterminate $\zeta = \mathrm{e}^{-sT}$. For the LQ–problem such a functional is constructed by the Laplace transform in continuous time. For the $\mathcal{H}_2-$problem, the quadratic functional is constructed by means of the PTF technique.

Chapter 16

Quadratic optimization of open-loop SD systems

16.1 Statement of the optimization problem in the interval $0 < t < \infty$ (LQ–problem)

The LQ–optimization problem can be formulated, in a fairly general form, as follows.
Let us consider two systems shown in Fig. 16.1.

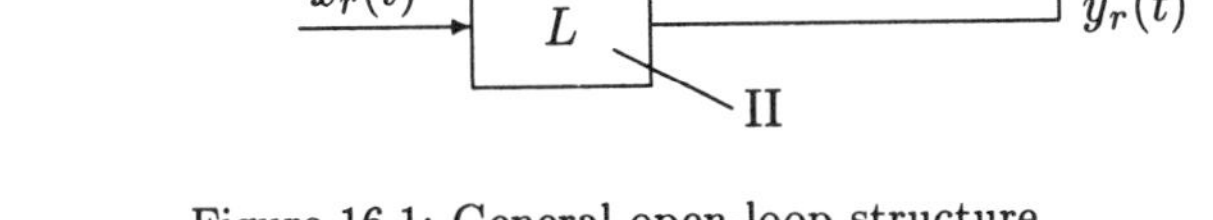

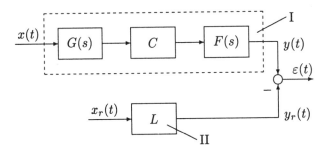

Figure 16.1: General open-loop structure

The number I in Fig. 16.1 denotes the designed open-loop system, and the number II denotes a linear T–periodic system with an arbitrary form of mathematical description. The latter system will be called the *reference system*. As a special case, it can be time invariant. In Fig. 16.1 $G(s)$ and $F(s)$ are the transfer functions of LTI elements, the properties of which will depend on the problem under consideration. It will also be assumed that the input signals $x(t)$ and $x_r(t)$ are zero for $t < 0$ and have for $\operatorname{Re} s > 0$ Laplace transforms $X(s)$ and $X_r(s)$, respectively, which are real rational functions in s. Besides,

C in Fig. 16.1 denotes a computer described by Eqs. (10.1)–(10.3) with sampling period T. Both systems are assumed to be stable, so that for any initial conditions without exogenous disturbances we have

$$\lim_{t\to\infty} y(t) = 0, \quad \lim_{t\to\infty} y_r(t) = 0.$$

The value

$$\varepsilon(t) \triangleq y(t) - y_r(t) \tag{16.1}$$

will be called the *control error*. The proximity of the systems I and II can be evaluated by the criterion

$$I_c \triangleq \int_0^\infty \varepsilon^2(t)\,dt \tag{16.2}$$

or, in some cases,

$$I_d \triangleq \sum_{k=0}^{\infty} \varepsilon^2(kT+0). \tag{16.3}$$

The following combined criterion may also be used

$$I_k \triangleq I_c + \lambda^2 I_d \tag{16.4}$$

where λ^2 is a positive constant.

We will demonstrate that the value I_k in (16.4) can be represented as a quadratic functional in the transfer function of the control program $W_d(s)$. Assume the initial conditions of the elements $F(s)$ and $G(s)$, and the control program to be zero. Then, for $t > 0$ the output $y(t)$ has a Laplace transform that, due to (10.104), has the form

$$Y(s) = U(s)W_d(s) \tag{16.5}$$

where

$$U(s) = F(s)M(s)\mathcal{D}_{GX}(T,s,+0) \tag{16.6}$$

taking account of the fact that the function $G(s)X(s)$ is proper and the function $\mathcal{D}_{GX}(T,s,t)$ is piece-wise continuous. Let the function $F(s)$ be strictly proper, then the DLT of the output is given by the relations following from (10.105)

$$\mathcal{D}_y(T,s,t) = \mathcal{D}_Y(T,s,t) = \mathcal{D}_{FM}(T,s,t)\mathcal{D}_{GX}(T,s,+0). \tag{16.7}$$

We also assume that the reference output $y_r(t) \in \Lambda_+(0,\infty)$, and its DLT $\mathcal{D}_{y_r}(T,s,t)$ is a rational periodic function in s. Then, for $\operatorname{Re} s > 0$ the error $\varepsilon(t)$ has the Laplace transform

$$E(s) = U(s)W_d(s) - Y_r(s) \tag{16.8}$$

where $Y_r(s)$ is the image of the reference output. The DLT of the output takes the form

$$\mathcal{D}_\varepsilon(T,s,t) = \mathcal{D}_y(T,s,t) - \mathcal{D}_{y_r}(T,s,t) \tag{16.9}$$

and is a rational periodic function in s. Additionally, we assume that the image $E(s)$ is analytic on the imaginary axis. Then, from (16.2) and (4.102) it follows that

$$I_c = \frac{T}{2\pi \mathrm{j}} \int_{-\mathrm{j}\omega/2}^{\mathrm{j}\omega/2} \left[\int_0^T \mathcal{D}_\varepsilon(T,s,t)\mathcal{D}_\varepsilon(T,-s,t)\,\mathrm{d}t \right] \mathrm{d}s\,. \tag{16.10}$$

To calculate the internal integral in (16.10)

$$J \triangleq \int_0^T \mathcal{D}_\varepsilon(T,s,t)\mathcal{D}_\varepsilon(T,-s,t)\,\mathrm{d}t \tag{16.11}$$

we substitute Eqs. (16.7) and (16.9) into (16.11). Then, taking into account that, under the given assumptions,

$$\begin{gathered}\int_0^T \mathcal{D}_y(T,s,t)\mathcal{D}_{y_r}(T,-s,t)\,\mathrm{d}t = \int_0^T \mathcal{D}_Y(T,s,t)\mathcal{D}_{Y_r}(T,-s,t)\,\mathrm{d}t = \mathcal{D}_{Y\underline{Y_r}}(T,s,0)\\ \int_0^T \mathcal{D}_{y_r}(T,s,t)\mathcal{D}_{y_r}(T,-s,t)\,\mathrm{d}t = \mathcal{D}_{Y_r\underline{Y_r}}(T,s,0)\end{gathered} \tag{16.12}$$

where the underbar denotes the substitution of $-s$ for s, we find

$$J = \tilde{A}(s)W_d(s)W_d(-s) - \tilde{B}(s)W_d(s) - \tilde{B}(s)W_d(-s) + \tilde{C}(s) \tag{16.13}$$

where

$$\begin{aligned}\tilde{A}(s) &= \mathcal{D}_{F\underline{F}M\underline{M}}(T,s,0)\mathcal{D}_{GX}(T,s,+0)\mathcal{D}_{GX}(T,-s,+0) \triangleq \tilde{A}_c(s)\\ \tilde{B}(s) &= \mathcal{D}_{F\underline{Y_r}M}(T,s,0)\mathcal{D}_{GX}(T,s,+0) \triangleq \tilde{B}_c(s)\\ \tilde{C}(s) &= \mathcal{D}_{Y_r\underline{Y_r}}(T,s,0) \triangleq \tilde{C}_c(s)\,.\end{aligned} \tag{16.14}$$

Henceforth it is assumed that Eq. (16.14) defines rational periodic functions in s. Substituting (16.13) into (16.10), we obtain

$$I_c = \frac{T}{2\pi \mathrm{j}} \int_{-\mathrm{j}\omega/2}^{\mathrm{j}\omega/2} \left[\tilde{A}(s)W_d(s)W_d(-s) - \tilde{B}(s)W_d(s) - \tilde{B}(-s)W_d(-s) + \tilde{C}(s)\right] \mathrm{d}s\,. \tag{16.15}$$

To construct a similar relation for the functional (16.3), we use Eq. (4.87)

$$I_d = \frac{T}{2\pi \mathrm{j}} \int_{-\mathrm{j}\omega/2}^{\mathrm{j}\omega/2} \mathcal{D}_\varepsilon(T,s,+0)\mathcal{D}_\varepsilon(T,-s,+0)\,\mathrm{d}s\,. \tag{16.16}$$

Using (16.9) gives the latter integral in the form (16.15), where

$$\begin{aligned}
\tilde{A}(s) &= \mathcal{D}_{FM}(T,s,0)\mathcal{D}_{FM}(T,-s,0)\mathcal{D}_{GX}(T,s,+0)\mathcal{D}_{GX}(T,-s,+0) \triangleq \tilde{A}_d(s) \\
\tilde{B}(s) &= \mathcal{D}_{FM}(T,s,0)\mathcal{D}_{GX}(T,s,+0)\mathcal{D}_{y_r}(T,-s,+0) \triangleq \tilde{B}_d(s) \\
\tilde{C}(s) &= \mathcal{D}_{y_r}(T,s,+0)\mathcal{D}_{y_r}(T,-s,+0) \triangleq \tilde{C}_d(s)\,.
\end{aligned} \tag{16.17}$$

Taking these relations in the aggregate, the functional I_k (16.4) can be written in the form (16.15) with known rational periodic functions

$$\tilde{A}(s) = \tilde{A}_c(s)+\lambda^2\tilde{A}_d(s), \quad \tilde{B}(s) = \tilde{B}_c(s)+\lambda^2\tilde{B}_d(s), \quad \tilde{C}(s) = \tilde{C}_c(s)+\lambda^2\tilde{C}_d(s)\,. \tag{16.18}$$

By construction,

$$\tilde{A}(s) = \tilde{A}(-s)\,, \quad \tilde{C}(s) = \tilde{C}(-s)\,. \tag{16.19}$$

Hence, assuming that

$$e^{-sT} = \zeta \tag{16.20}$$

in (16.15), the functional (16.4) can be represented in the form

$$I_k = \frac{T}{2\pi j}\oint \left[A(\zeta)W_d(\zeta)W_d(\zeta^{-1}) - B(\zeta)W_d(\zeta) - B(\zeta^{-1})W_d(\zeta^{-1}) + C(\zeta)\right]\frac{d\zeta}{\zeta} \tag{16.21}$$

where

$$A(\zeta) \triangleq \tilde{A}(s)\,\big|_{e^{-sT}=\zeta}\,, \quad B(\zeta) \triangleq \tilde{B}(s)\,\big|_{e^{-sT}=\zeta}\,, \quad C(\zeta) \triangleq \tilde{C}(s)\,\big|_{e^{-sT}=\zeta} \tag{16.22}$$

are known real rational functions. The integration in (16.21) is performed along the unit circle anti-clockwise. Moreover, in (16.21)

$$W_d(\zeta) \triangleq W_d(s)\,\big|_{e^{-sT}=\zeta} \tag{16.23}$$

is a real rational function. By construction,

$$A(\zeta) = A(\zeta^{-1})\,, \quad C(\zeta) = C(\zeta^{-1})\,. \tag{16.24}$$

Thus, the value I_k is represented in form of a quadratic functional (16.21) with known real rational functions (16.22) and unknown real rational function $W_d(\zeta)$.

According to the given assumptions, the discrete controller must be stable, i.e., the poles of the rational periodic function $W_d(s)$ must be located in the left half-plane. Therefore, from (16.23) it follows that the poles of the real rational function $W_d(\zeta)$ must be outside the unit circle. Henceforth, such real rational functions will be called *stable*.

Assume that the SD system is to be designed so that the functional (16.21) is minimal. Then, we obtain the following optimal control problem.

LQ–problem ($0 \le t < \infty$). Given all characteristics of the systems I and II which define uniquely the coefficients (16.22), find the stable real rational function $W_d(\zeta)$ minimizing the functional (16.21).

Remark 1. All stable real rational functions $W_d(\zeta)$ are causal, as follows from Section A.2. Therefore, in the optimal control problem formulated above the additional assumption on causality of the optimal controller is dropped.

Remark 2. The special case of the problem under consideration, when

$$x(t) = x_r(t) \tag{16.25}$$

in Fig. 16.1, will be called the *LQ–redesign* problem.

16.2 General technique of the construction of optimal control programs

Dropping the terms independent of $W_d(\zeta)$ in (16.21), we obtain a minimization problem for the integral

$$\tilde{I}_k = \frac{T}{2\pi \mathrm{j}} \oint \left[A(\zeta)W_d(\zeta)W_d(\zeta^{-1}) - B(\zeta)W_d(\zeta) - B(\zeta^{-1})W_d(\zeta^{-1})\right] \frac{\mathrm{d}\zeta}{\zeta}. \tag{16.26}$$

The following theorem gives a method for the construction of the stable control program $W_{d0}(\zeta)$ minimizing the functional (16.26).

Theorem 16.1 (Chang (1961)) *Let $A(\zeta)$ and $B(\zeta)$ in (16.26) be known real rational functions, such that $A(\zeta)$ satisfies the condition (16.24). Let also $A(\zeta)$ have no poles and zeros on the unit circle, and $B(\zeta)$ have no poles there. Then, the stable real rational function $W_{d0}(\zeta)$ minimizing the functional (16.26) is defined as follows.*

1. *Perform the factorization*

$$A(\zeta) = K(\zeta)K(\zeta^{-1}) \tag{16.27}$$

where all roots and poles of the function $K(\zeta)$ are located outside the unit circle.

2. *For the real rational function*

$$u(\zeta) \triangleq \frac{B(\zeta^{-1})}{K(\zeta^{-1})} \tag{16.28}$$

perform the separation

$$u(\zeta) = u_+(\zeta) + u_-(\zeta) \tag{16.29}$$

where $u_+(\zeta)$ has poles outside the unit circle, and $u_-(\zeta)$ has poles inside the unit circle with

$$\lim_{\zeta \to \infty} u_-(\zeta) = 0\,. \tag{16.30}$$

3. *The optimal function $W_{d0}(\zeta)$ is given as*

$$W_{d0}(\zeta) = \frac{u_+(\zeta)}{K(\zeta)}\,. \tag{16.31}$$

Proof If (16.27) holds, taking account of (16.28), we have

$$f(\zeta) \triangleq A(\zeta)W_d(\zeta)W_d(\zeta^{-1}) - B(\zeta)W_d(\zeta) - B(\zeta^{-1})W_d(\zeta^{-1}) =$$

$$= [K(\zeta)W_d(\zeta) - u(\zeta)]\left[K(\zeta^{-1})W_d(\zeta^{-1}) - u(\zeta^{-1})\right] - \frac{B(\zeta)B(\zeta^{-1})}{A(\zeta)}.$$

Using (16.29), the latter relation can be transformed as

$$\begin{aligned} f(\zeta) = {} & [K(\zeta)W_d(\zeta) - u_+(\zeta)]\left[K(\zeta^{-1})W_d(\zeta^{-1}) - u_+(\zeta^{-1})\right] - \\ & - [K(\zeta)W_d(\zeta) - u_+(\zeta)]\,u_-(\zeta^{-1}) - \left[K(\zeta^{-1})W_d(\zeta^{-1}) - u_+(\zeta^{-1})\right]u_-(\zeta) \\ & + u_-(\zeta)u_-(\zeta^{-1}) - \frac{B(\zeta)B(\zeta^{-1})}{A(\zeta)}. \end{aligned} \tag{16.32}$$

Substituting (16.32) into (16.26), we find

$$\tilde{I}_k = I_1 + I_2 + I_3 + I_4 \tag{16.33}$$

where

$$I_1 = \frac{T}{2\pi \mathrm{j}} \oint [K(\zeta)W_d(\zeta) - u_+(\zeta)]\left[K(\zeta^{-1})W_d(\zeta^{-1}) - u_+(\zeta^{-1})\right] \frac{\mathrm{d}\zeta}{\zeta} \tag{16.34}$$

$$I_2 = -\frac{T}{2\pi \mathrm{j}} \oint [K(\zeta)W_d(\zeta) - u_+(\zeta)]\,u_-(\zeta^{-1}) \frac{\mathrm{d}\zeta}{\zeta} \tag{16.35}$$

$$I_3 = -\frac{T}{2\pi \mathrm{j}} \oint \left[K(\zeta^{-1})W_d(\zeta^{-1}) - u_+(\zeta^{-1})\right]u_-(\zeta) \frac{\mathrm{d}\zeta}{\zeta} \tag{16.36}$$

$$I_4 = \frac{T}{2\pi \mathrm{j}} \oint \left[u_-(\zeta)u_-(\zeta^{-1}) - \frac{B(\zeta)B(\zeta^{-1})}{A(\zeta)}\right] \frac{\mathrm{d}\zeta}{\zeta}. \tag{16.37}$$

Next, we consider the integrals (16.34)–(16.37). First of all, we find the relation

$$I_2 = I_3$$

because each of these integrals is obtained from the other by substitution of ζ^{-1} for ζ. Then, we demonstrate that

$$I_2 = I_3 = 0. \tag{16.38}$$

Indeed, the integrand in (16.35) is, by construction, analytic inside and on the unit circle except for, possibly, the point $\zeta = 0$. Nevertheless, from (16.30) it follows that

$$\lim_{\zeta \to 0} u_-(\zeta^{-1}) = \lim_{\zeta \to \infty} u_-(\zeta) = 0.$$

Hence,

$$u_-(\zeta^{-1}) = \zeta \tilde{u}_-(\zeta) \tag{16.39}$$

where $\tilde{u}_-(\zeta)$ is a real rational function analytic for $\zeta = 0$. Substituting (16.39) into (16.35), we find that the integrand in (16.35) is analytic inside and on the unit circle. Hence, because of the residue theorem, (16.38) follows. Therefore, (16.33) takes the form

$$\tilde{I}_k = I_1 + I_4 \,.$$

As follows from (16.37), the integral I_4 is independent of $W_d(\zeta)$. As is evident from (16.34), the integral I_1 is non-negative and assumes the minimal value zero if $W_d(\zeta)$ is selected according to (16.31). ■

Remark. We note that the minimal value of the integral (16.26) is equal to I_4, therefore the minimal value $I_{k\,\min}$ of the initial functional (16.21) equals

$$I_{k\,\min} = \frac{1}{2\pi \mathrm{j}} \oint \left[u_-(\zeta)u_-(\zeta^{-1}) - \frac{B(\zeta)B(\zeta^{-1})}{A(\zeta)} + C(\zeta) \right] \frac{\mathrm{d}\zeta}{\zeta} \,. \tag{16.40}$$

Example 16.1 Let us consider in detailed the technique for solving the LQ–problem for the open-loop system shown in Fig. 16.2, which is a special

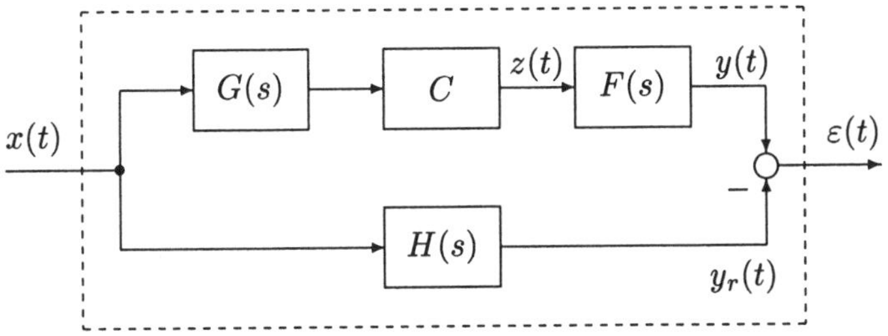

Figure 16.2: Open-loop redesign structure

case of the system shown in Fig. 16.1, where

$$F(s) = \frac{1}{s-a}\,, \quad G(s) = \frac{1}{s-b}\,, \quad H(s) = \frac{1}{s-d}\,, \quad M(s) = \frac{1-\mathrm{e}^{-sT}}{s}$$

with real negative numbers a, b and d. The initial conditions are assumed to be zero, and the input of the compared systems is acted upon by the unit impulse $x(t) = \delta(t)$, i.e., $X(s) = 1$. Then, from (16.6) we have

$$U(s) = F(s)M(s)\mathcal{D}_G(T, s, +0)$$

where, by (4.49),

$$\mathcal{D}_G(T, s, +0) = \frac{1}{1 - \mathrm{e}^{bT}\mathrm{e}^{-sT}}$$

Moreover, it is evident that

$$Y_r(s) = \frac{1}{s-d}\,.$$

The integral error I_c (16.2) is taken as an optimality criterion. Then, using (16.14), we obain a functional of the form (16.15), where

$$\begin{aligned}\tilde{A}(s) &= \mathcal{D}_{F\underline{F}M\underline{M}}(T,s,0)\,\frac{1}{1-e^{bT}e^{-sT}}\,\frac{1}{1-e^{bT}e^{sT}}\\ \tilde{B}(s) &= \mathcal{D}_{F\underline{H}M}(T,s,0)\,\frac{1}{1-e^{bT}e^{-sT}}\\ \tilde{C}(s) &= \mathcal{D}_{H\underline{H}}(T,s,0)\,.\end{aligned} \tag{16.41}$$

In this case, by (9.115), for $0 \le t \le T$ we obtain

$$\begin{aligned}\mathcal{D}_{FM}(T,s,t) &= \frac{1-e^{-aT}}{a}\,\frac{e^{at}}{1-e^{aT}e^{-sT}} + \frac{e^{a(t-T)}-1}{a}\\ \mathcal{D}_{FM}(T,-s,t) &= \frac{1-e^{-aT}}{a}\,\frac{e^{at}}{1-e^{aT}e^{sT}} + \frac{e^{a(t-T)}-1}{a}\,.\end{aligned} \tag{16.42}$$

Similarly to (16.12), we find

$$\mathcal{D}_{F\underline{F}M\underline{M}}(T,s,0) = \int_0^T \mathcal{D}_{FM}(T,s,t)\mathcal{D}_{FM}(T,-s,t)\,dt\,.$$

Using (16.42), we obtain

$$\mathcal{D}_{F\underline{F}M\underline{M}}(T,s,0) = \frac{\alpha}{\left[1-e^{(a-s)T}\right]\left[1-e^{(a+s)T}\right]} + \frac{\beta}{1-e^{(a-s)T}} + \frac{\beta}{1-e^{(a+s)T}} + \gamma$$

where $\alpha > 0$, β, and $\gamma > 0$ are known constants. Moreover, since

$$F(s)H(-s) = \frac{1}{a+d}\left(\frac{1}{s+d} - \frac{1}{s-a}\right)$$

using (9.102) for $t = 0$ gives

$$\mathcal{D}_{F\underline{H}M}(T,s,0) = \frac{q_1}{e^{sT}-e^{aT}} + \frac{q_2}{e^{sT}-e^{-dT}}$$

where q_1 and q_2 are known constants. Therefore, taking account of (16.41)

$$\tilde{B}(s) = \left(\frac{q_1}{e^{sT}-e^{aT}} + \frac{q_2}{e^{sT}-e^{-dT}}\right)\frac{e^{sT}}{e^{sT}-e^{bT}}\,.$$

With the change of variable $\zeta = e^{-sT}$, we obtain a functional of the form (16.26), where

$$A(\zeta) = \frac{\alpha + \beta\left(1-e^{aT}\zeta\right) + \beta\left(1-e^{aT}\zeta^{-1}\right) + \gamma\left(1-e^{aT}\zeta\right)\left(1-e^{aT}\zeta^{-1}\right)}{\left(1-e^{aT}\zeta\right)\left(1-e^{aT}\zeta^{-1}\right)\left(1-e^{bT}\zeta\right)\left(1-e^{bT}\zeta^{-1}\right)} \tag{16.43}$$

$$B(\zeta) = \frac{\zeta(r_1\zeta + r_2)}{\left(1-e^{aT}\zeta\right)\left(1-e^{bT}\zeta\right)\left(1-e^{-dT}\zeta\right)} \tag{16.44}$$

where r_1 and r_2 are new known constants. It can be shown that under the conditions of this example the function $A(\zeta)$ is not zero for $|\zeta| = 1$. Then, from (16.43) follows the representation

$$A(\zeta) = N^2 \frac{(1-\mu\zeta)(1-\mu\zeta^{-1})}{(1-e^{aT}\zeta)(1-e^{bT}\zeta)(1-e^{aT}\zeta^{-1})(1-e^{bT}\zeta^{-1})} \tag{16.45}$$

where N^2 is a positive constant, and μ is the real root of the equation

$$1 - \frac{\alpha + 2\beta + \gamma(1+e^{2aT})}{e^{aT}(\beta+\gamma)}\zeta + \zeta^2 = 0$$

with $|\mu| < 1$. Using (16.45), we obtain the factorization (16.27), where

$$\begin{aligned} K(\zeta) &= N \frac{1-\mu\zeta}{(1-e^{aT}\zeta)(1-e^{bT}\zeta)} \\ K(\zeta^{-1}) &= N \frac{\zeta(\zeta-\mu)}{(\zeta-e^{aT})(\zeta-e^{bT})} . \end{aligned} \tag{16.46}$$

Equation (16.44) yields

$$B(\zeta^{-1}) = \frac{\zeta(r_1 + r_2\zeta)}{(\zeta-e^{aT})(\zeta-e^{bT})(\zeta-e^{-dT})} .$$

Therefore, the function $u(\zeta)$ given by (16.28) appears as

$$u(\zeta) = \frac{1}{N} \frac{r_1 + r_2\zeta}{(\zeta-\mu)(\zeta-e^{-dT})} . \tag{16.47}$$

The real rational function (16.47) is strictly proper and has a single pole $\zeta = e^{-dT}$ outside the unit circle. Therefore, the separation (16.29) gives

$$u_+(\zeta) = \frac{m}{\zeta - e^{-dT}} \tag{16.48}$$

where m is a known constant. Using (16.46) and (16.48), we obtain the optimal control program

$$W_{d0}(\zeta) = \frac{u_+(\zeta)}{K(\zeta)} = L \frac{(1-e^{aT}\zeta)(1-e^{bT}\zeta)}{(1-e^{dT}\zeta)(1-\mu\zeta)}$$

where L is a known constant. In the special case, when $a = -1$, $b = -2$, $d = -0.5$, $T = 0.2$, the optimal control has the form

$$W_{d0}(\zeta) = 6.4718 \frac{(1-0.8187\zeta)(1-0.6703\zeta)}{(1-0.9048\zeta)(1+0.2673\zeta)} .$$

Figure 16.3 demonstrates the outputs of the reference and the designed systems, while Fig. 16.4 and 16.5 show the corresponding control signal $z(t)$ and the control error $\varepsilon(t)$. □

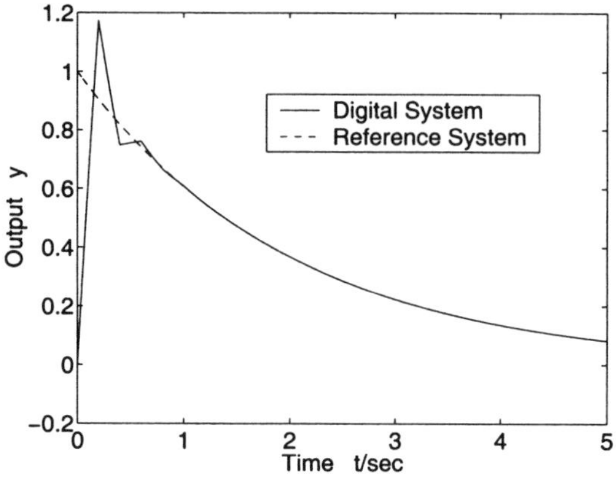

Figure 16.3: Transients for LQ–redesign

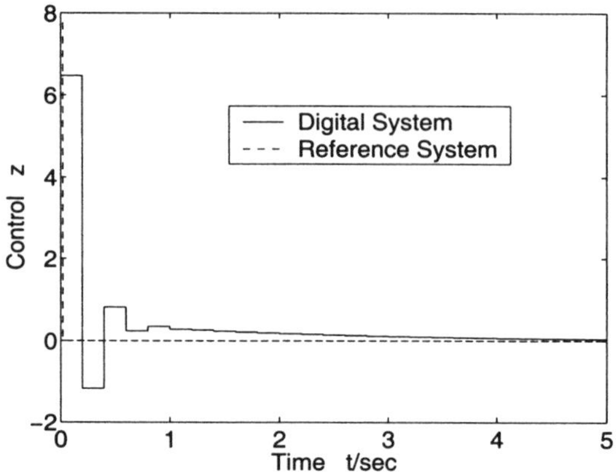

Figure 16.4: Control signal for LQ–redesign

16.3 Optimization in the interval $-\infty < t < \infty$ ($\mathcal{H}_2$–problem)

Assume that in Fig. 16.1 $x(t)$ and $x_r(t)$ are centered, stationary and stationary-connected processes forming the stochastic vector

$$\boldsymbol{x}(t) \stackrel{\triangle}{=} \begin{bmatrix} x(t) \\ x_r(t) \end{bmatrix} \tag{16.49}$$

with the autocorrelation matrix

$$\boldsymbol{K}_{\boldsymbol{xx}}(\tau) = \begin{bmatrix} K_{xx}(\tau) & K_{xx_r}(\tau) \\ K_{x_rx}(\tau) & K_{x_rx_r}(\tau) \end{bmatrix} \tag{16.50}$$

where

$$K_{xx}(\tau) = K_{xx}(-\tau)\,, \quad K_{x_rx_r}(\tau) = K_{x_rx_r}(-\tau)\,, \quad K_{xx_r}(\tau) = K_{x_rx}(-\tau)\,.$$

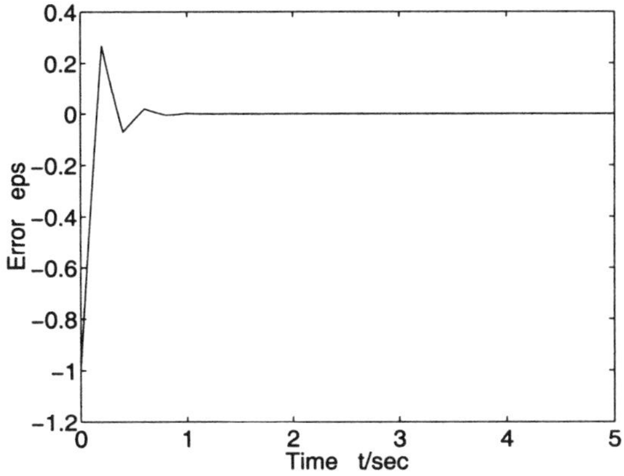

Figure 16.5: Control error for LQ–redesign

The correlation matrix (16.50) is associated with a spectral density matrix

$$\boldsymbol{R_{xx}}(s) = \begin{bmatrix} R_{xx}(s) & R_{xx_r}(s) \\ R_{x_rx}(s) & R_{x_rx_r}(s) \end{bmatrix}$$

that satisfies the first relation in (8.81).
Consider the aggregate of the systems I and II in Fig. 16.1 as a multivariable system with two inputs and a single output $\varepsilon(t)$. Assume that the reference system has the PTF $W_r(s,t)$. Then, the PTM $\boldsymbol{W}(s,t)$ of the system, with regard to the input vector (16.49) is a row matrix of the form

$$\boldsymbol{W}(s,t) = \begin{bmatrix} W(s,t) & -W_r(s,t) \end{bmatrix} \tag{16.51}$$

where

$$W(s,t) = W_1(s,t)W_d(s)\,, \quad W_1(s,t) = G(s)\varphi_{FM}(T,s,t)\,. \tag{16.52}$$

Henceforth we assume that the conditions of the applicability of Eq. (8.97) hold for the system at hand, so that

$$d_\varepsilon(t) = \frac{1}{2\pi \mathrm{j}} \int_{-\mathrm{j}\infty}^{\mathrm{j}\infty} \boldsymbol{W}(-s,t)\boldsymbol{R_{xx}}(s)\boldsymbol{W}'(s,t)\,\mathrm{d}s = \frac{1}{2\pi \mathrm{j}} \int_{-\mathrm{j}\infty}^{\mathrm{j}\infty} N(s,t)\,\mathrm{d}s \tag{16.53}$$

where the integrand is a scalar function given by

$$\begin{aligned} N(s,t) \stackrel{\Delta}{=}\; & \boldsymbol{W}(-s,t)\boldsymbol{R_{xx}}(s)\boldsymbol{W}'(s,t) = \\ =\; & W(s,t)W(-s,t)R_{xx}(s) - W_r(s,t)W(-s,t)R_{xx_r}(s) - \\ & -W(s,t)W_r(-s,t)R_{x_rx}(s) + W_r(s,t)W_r(-s,t)R_{x_rx_r}(s)\,. \end{aligned} \tag{16.54}$$

To calculate the mean variance

$$\bar{d}_\varepsilon = \frac{1}{T}\int_0^T d_\varepsilon(t)\,\mathrm{d}t$$

we take the mean value of the function $N(s,t)$ over the time. As follows from (16.54) and (16.51),

$$\overline{N}(s) \triangleq \frac{1}{T}\int_0^T N(s,t)\,\mathrm{d}t =$$

$$= \hat{A}(s)W_d(s)W_d(-s) - \hat{B}(s)W_d(s) - \hat{B}(-s)W_d(-s) + \hat{C}(s)$$

where

$$\hat{A}(s) \triangleq R_{xx}(s)\frac{1}{T}\int_0^T W_1(s,t)W_1(-s,t)\,\mathrm{d}t$$

$$\hat{B}(s) \triangleq R_{x_r x}(s)\frac{1}{T}\int_0^T W_1(s,t)W_r(-s,t)\,\mathrm{d}t$$

$$\hat{C}(s) \triangleq R_{x_r x_r}(s)\frac{1}{T}\int_0^T W_r(s,t)W_r(-s,t)\,\mathrm{d}t\,.$$

Hence, using (16.52)

$$\begin{aligned}
\hat{A}(s) &= \frac{1}{T}G(s)G(-s)R_{xx}(s)\mathcal{D}_{F\underline{F}M\underline{M}}(T,s,0)\\
\hat{B}(s) &= \frac{1}{T}G(s)R_{x_r x}(s)\int_0^T \varphi_{FM}(T,s,t)W_r(-s,t)\,\mathrm{d}t \qquad (16.55)\\
\hat{C}(s) &= \frac{1}{T}R_{x_r x_r}(s)\int_0^T W_r(s,t)W_r(-s,t)\,\mathrm{d}t\,,
\end{aligned}$$

so that the mean output variance is given as

$$\bar{d}_\varepsilon = \frac{1}{2\pi \mathrm{j}}\int_{-\mathrm{j}\infty}^{\mathrm{j}\infty}\left[\hat{A}(s)W_d(s)W_d(-s) - \hat{B}(s)W_d(s) - \hat{B}(-s)W_d(-s) + \hat{C}(s)\right]\mathrm{d}s\,.$$

Bearing in mind the possible discretization of this integral, we obtain

$$\bar{d}_\varepsilon = \frac{T}{2\pi \mathrm{j}}\int_{-\mathrm{j}\omega/2}^{\mathrm{j}\omega/2}\left[\tilde{A}(s)W_d(s)W_d(-s) - \tilde{B}(s)W_d(s) - \tilde{B}(-s)W_d(-s) + \tilde{C}(s)\right]\mathrm{d}s \qquad (16.56)$$

where

$$\begin{aligned}
\tilde{A}(s) &= \frac{1}{T}\mathcal{D}_{F\underline{F}M\underline{M}}(T,s,0)\mathcal{D}_{G\underline{G}R_{xx}}(T,s,0)\\
\tilde{B}(s) &= \mathcal{D}_{\hat{B}}(T,s,0) = \frac{1}{T}\sum_{k=-\infty}^{\infty}\hat{B}(s+k\mathrm{j}\omega) \qquad (16.57)\\
\tilde{C}(s) &= \mathcal{D}_{\hat{C}}(T,s,0) = \frac{1}{T}\sum_{k=-\infty}^{\infty}\hat{C}(s+k\mathrm{j}\omega)\,.
\end{aligned}$$

Henceforth, the reference system is assumed to be such that the functions $\tilde{A}(s)$, $\tilde{B}(s)$, and $\tilde{C}(s)$ are rational periodic. Then, the change of variable $\zeta = e^{-sT}$ gives

$$\overline{d}_\varepsilon = \frac{1}{2\pi j} \oint \left[A(\zeta)W_d(\zeta)W_d(\zeta^{-1}) - B(\zeta)W_d(\zeta) - B(\zeta^{-1})W_d(\zeta^{-1}) + C(\zeta)\right] \frac{d\zeta}{\zeta} \tag{16.58}$$

where $A(\zeta)$, $B(\zeta)$ and $C(\zeta)$ are known real rational function in ζ

$$A(\zeta) = \tilde{A}(s)\,\big|_{e^{-sT}=\zeta}\,, \quad B(\zeta) = \tilde{B}(s)\,\big|_{e^{-sT}=\zeta}\,, \quad C(\zeta) = \tilde{C}(s)\,\big|_{e^{-sT}=\zeta}$$

and $W_d(\zeta)$ is an unknown real rational function, which is related to the transfer function $W_d(s)$ by (16.23). In this case Eq. (16.24) holds by construction. Using the above relations, we can formulate a general optimization problem.

CMV-problem. Given all characteristics of the systems I and II except for the control program $W_d(s)$, and the spectral density matrix $R_{xx}(s)$, find the stable real rational function $W_{d0}(\zeta)$ (16.23) minimizing the value $\overline{d}_\varepsilon$ (16.58). Henceforth this problem of minimization of the mean variance will be called the *CMV-problem* (continuous minimum variance).

Together with the CMV-problem, which is connected with the minimization of the mean output variance, we can consider a different statement of the problem. Assume that the PTM $\boldsymbol{W}(s,t)$ (16.51) is continuous with respect to t. Then, the variance $d_\varepsilon(t)$ is continuous with respect to t and Eq. (16.53) yields

$$d_\varepsilon(kT) = d_\varepsilon(0) = \frac{1}{2\pi j} \int_{-j\infty}^{j\infty} \boldsymbol{W}(-s,0)\boldsymbol{R}_{xx}(s)\boldsymbol{W}'(s,0)\, ds \tag{16.59}$$

which holds for all integers k. The value $d_\varepsilon(0)$ can also be employed as a criterion of the proximity of the systems I and II. Therefore, under the above conditions it is possible to state the problem of determining the stable control program $W_{d0}(\zeta)$ which minimizes the functional (16.59). This problem will be referred to as the *MV-problem* [(discrete) minimum variance]. Similarly to the aforesaid, it can be shown that the MV-problem reduces to the minimization of an integral of the form (16.58), where

$$A(\zeta) = \left[\mathcal{D}_{G\underline{G}R_{xx}}(T,s,0)\mathcal{D}_{FM}(T,s,0)\mathcal{D}_{FM}(T,-s,0)\right]\big|_{e^{-sT}=\zeta}$$

$$B(\zeta) = \Big[\mathcal{D}_{FM}(T,s,0) \,*$$
$$*\frac{1}{T}\sum_{k=-\infty}^{\infty} G(s+kj\omega)R_{x_r x}(s+kj\omega)W_r(-s-kj\omega,0)\Big]\big|_{e^{-sT}=\zeta}$$

$$C(\zeta) = \Big[\frac{1}{T}\sum_{k=-\infty}^{\infty} W_r(s+kj\omega,0)W_r(-s-kj\omega,0)\Big]\big|_{e^{-sT}=\zeta}\,.$$

The statement of the problem given above is fairly general and involves many important cases for application. In the special case, when (16.25) holds, the associated design problem will be called the $\mathcal{H}_2$−redesign problem.

16.4 Optimal discrete filtering of continuous processes

In this section we consider the system shown in Fig. 16.6, where the reference

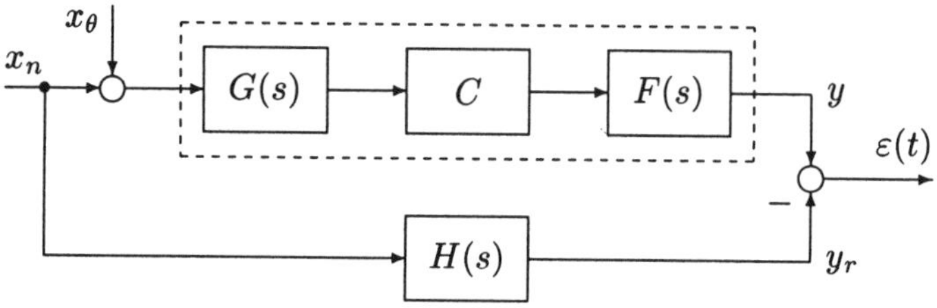

Figure 16.6: Hybrid filter structure

system II is assumed to be linear and time-invariant, its transfer function being $H(s)$. The input signal of the system I is the sum of the reference signal $x_n(t)$ and the noise $x_\theta(t)$, which are both centered, stationary and stationary connected. The output $y(t)$ of the system I is perturbed by the noise and amplitude modulation in the system. Moreover, the output becomes corrupted due to the inertial properties of the elements $G(s)$ and $F(s)$. At the same time, it is often required (in communications engineering) to obtain, for known output $y(t)$, information as complete as possible pertaining to the properties of the input $x_n(t)$ or of its linear transformation $y_r(t)$. The difference between $y(t)$ and the desired signal $y_r(t)$ is defined as the error

$$\varepsilon(t) = y(t) - y_r(t)\,. \tag{16.60}$$

Figure 16.6 corresponds to the situation, when the ideal (desired) transformation of the reference signal is stationary and has a transfer function $H(s)$. In this case the stochastic input vector (16.49) can be defined as

$$\boldsymbol{x}(t) = \begin{bmatrix} x_n(t) + x_\theta(t) \\ x_n(t) \end{bmatrix}. \tag{16.61}$$

Then, we have

$$\boldsymbol{x}(t)\boldsymbol{x}'(t+\tau) = \tag{16.62}$$
$$\begin{bmatrix} [x_n(t) + x_\theta(t)][x_n(t+\tau) + x_\theta(t+\tau)] & [x_n(t) + x_\theta(t)]x_n(t+\tau) \\ x_n(t)[x_n(t+\tau) + x_\theta(t+\tau)] & x_n(t)x_n(t+\tau) \end{bmatrix}.$$

Further, for simplicity of derivation, we assume that the reference signal and the noise are statistically independent. Then, taking the mathematical expectation of both sides of Eq. (16.62), we find the correlation matrix of the vector (16.61) as

$$\boldsymbol{K_{xx}}(\tau) = \begin{bmatrix} K_n(\tau) + K_\theta(\tau) & K_n(\tau) \\ K_n(\tau) & K_n(\tau) \end{bmatrix} \tag{16.63}$$

where $K_n(\tau)$ and $K_\theta(\tau)$ are the correlation matrices of the reference signal and the noise, respectively. Using the bilateral Laplace transformation in (16.63), the spectral density matrix of the input can be found as

$$\boldsymbol{R_{xx}}(s) = \begin{bmatrix} R_n(s) + R_\theta(s) & R_n(s) \\ R_n(s) & R_n(s) \end{bmatrix}$$

where $R_n(s)$ and $R_\theta(s)$ are the spectral densities of the reference signal and the noise, respectively. Taking into account that in the given case the PTM $\boldsymbol{W}(s,t)$ (16.51) has the form

$$\boldsymbol{W}(s,t) = \begin{bmatrix} G(s)\varphi_{FM}(T,s,t)W_d(s) & -H(s) \end{bmatrix} \tag{16.64}$$

from (16.55) we obtain

$$\begin{aligned} \hat{A}(s) &= \frac{1}{T}\mathcal{D}_{F\underline{F}M\underline{M}}(T,s,0)G(s)G(-s)[R_n(s) + R_\theta(s)] \\ \hat{B}(s) &= \frac{1}{T}F(s)G(s)H(-s)M(s)R_n(s) \\ \hat{C}(s) &= H(s)H(-s)R_n(s) . \end{aligned}$$

Hence, the functions (16.57) appear as

$$\begin{aligned} \tilde{A}(s) &= \frac{1}{T}\mathcal{D}_{F\underline{F}M\underline{M}}(T,s,0)\mathcal{D}_{G\underline{G}(R_n+R_\theta)}(T,s,0) \\ \tilde{B}(s) &= \frac{1}{T}\mathcal{D}_{FG\underline{H}R_nM}(T,s,0) \\ \tilde{C}(s) &= \mathcal{D}_{H\underline{H}R_n}(T,s,0) \end{aligned} \tag{16.65}$$

where

$$\mathcal{D}_{G\underline{G}(R_n+R_\theta)}(T,s,0) = \frac{1}{T}\sum_{k=-\infty}^{\infty} G(s+kj\omega)G(-s-kj\omega)[R_n(s+kj\omega)+R_\theta(s+kj\omega)] \tag{16.66}$$

$$\begin{aligned} &\mathcal{D}_{FG\underline{H}R_nM}(T,s,0) = \\ &= \frac{1}{T}\sum_{k=-\infty}^{\infty} F(s+kj\omega)G(s+kj\omega)H(-s-kj\omega)R_n(s+kj\omega)M(s+kj\omega) . \end{aligned} \tag{16.67}$$

Consider some special cases of the above relations. If

$$H(s) = 1 \tag{16.68}$$

we obtain the problem of optimal reconstruction of a reference signal in continuous time by discrete measurements, which is a continuous-discrete analogue of the classical filtering problem. If (16.68) holds, Eq. (16.65) gives

$$\tilde{B}(s) = \frac{1}{T}\mathcal{D}_{FGR_nM}(T, s, 0)\,.$$

If

$$H(s) = \mathrm{e}^{qsT} \tag{16.69}$$

where $q > 0$ is an integer, we obtain the problem of optimal prediction by q periods. In this case $H(-s) = \mathrm{e}^{-qsT}$, and Eqs. (16.65) and (16.67) yield

$$\tilde{B}(s) = \frac{1}{T}\mathcal{D}_{FGR_nM}(T, s, 0)\mathrm{e}^{-qsT}\,.$$

In the case when

$$H(s) = \mathrm{e}^{-qsT} \tag{16.70}$$

where $q > 0$ is an integer, we obtain the optimal smoothing problem, so that

$$\tilde{B}(s) = \frac{1}{T}\mathcal{D}_{FGR_nM}(T, s, 0)\mathrm{e}^{qsT}\,.$$

As a result, the CMV-optimization filtering problem reduces to the minimization of a functional of the form (16.58), where

$$\begin{aligned} A(\zeta) &= \Big[\frac{1}{T}\mathcal{D}_{G\underline{G}(R_n+R_\theta)}(T, s, 0)\mathcal{D}_{F\underline{F}M\underline{M}}(T, s, 0)\Big]\,\Big|_{\mathrm{e}^{-sT}=\zeta} \\ B(\zeta) &= \Big[\frac{1}{T}\mathcal{D}_{FG\underline{H}R_nM}(T, s, 0)\Big]\,\Big|_{\mathrm{e}^{-sT}=\zeta} \\ C(\zeta) &= \Big[\mathcal{D}_{H\underline{H}R_n}(T, s, 0)\Big]\,\Big|_{\mathrm{e}^{-sT}=\zeta}\,. \end{aligned} \tag{16.71}$$

Example 16.2 Construct the optimal filter program for the system shown in Fig. 16.6 with $G(s) = H(s) = 1$ and an arbitrary form of the pulse. Assume that

$$F(s) = \frac{1}{s-a}\,, \qquad R_n(s) = \frac{2\beta\eta}{s^2-\eta^2}$$

where $a < 0$, $\eta < 0$, and $\beta > 0$ are constants, and

$$\mathcal{D}_{R_\theta}(T, s, 0) = d_\theta > 0\,. \tag{16.72}$$

Condition (16.72) means that the noise is a discrete white noise with variance d_θ. Using the above relation, we obtain the problem of minimizing the integral (16.58), where, taking due account of (16.71),

$$\begin{aligned} A(\zeta) &= \Big[\frac{1}{T}(\mathcal{D}_{R_n}(T, s, 0) + d_\theta)\mathcal{D}_{F\underline{F}M\underline{M}}(T, s, 0)\Big]\,\Big|_{\mathrm{e}^{-sT}=\zeta} \\ B(\zeta) &= \Big[\frac{1}{T}\mathcal{D}_{FR_nM}(T, s, 0)\Big]\,\Big|_{\mathrm{e}^{-sT}=\zeta} \end{aligned}$$

To construct the optimal control program the general procedure given in Section 16.2 is used. In this case, taking account of (8.27),

$$\left[\mathcal{D}_{R_n}(T,s,0)+d_\theta\right]\big|_{e^{-sT}=\zeta} = \frac{d_\theta + 2(\beta\sinh\eta T - d_\theta\cosh\eta T)\zeta + d_\theta\zeta^2}{1-2\zeta\cosh\eta T+\zeta^2}. \tag{16.73}$$

Let $\lambda_1 = \lambda$ and $\lambda_2 = \lambda^{-1}$ be the roots of the equation

$$d_\theta + 2(\beta\sinh\eta T - d_\theta\cosh\eta T)\zeta + d_\theta\zeta^2 = 0$$

such that $|\lambda| < 1$. Then, the right-hand side of (16.73) can be represented in the form

$$\left[\mathcal{D}_{R_n}(T,s,0)+d_\theta\right]\big|_{e^{-sT}=\zeta} = Q^2\,\frac{(1-\lambda\zeta)(1-\lambda\zeta^{-1})}{(1-e^{\eta T}\zeta)(1-e^{\eta T}\zeta^{-1})}$$

where Q^2 is a positive constant. Moreover, as follows from the results of Example 16.1,

$$\frac{1}{T}\mathcal{D}_{F\underline{E}M\underline{M}}(T,s,0)\big|_{e^{-sT}=\zeta} = N_1^2\,\frac{(1-\mu\zeta)(1-\mu\zeta^{-1})}{(1-e^{aT}\zeta)(1-e^{aT}\zeta^{-1})}$$

where $|\mu| < 1$. Therefore, there exists the factorization (16.27), where

$$\begin{aligned} K(\zeta) &= QN_1\,\frac{(1-\lambda\zeta)(1-\mu\zeta)}{(1-e^{aT}\zeta)(1-e^{\eta T}\zeta)} \\ K(\zeta^{-1}) &= QN_1\,\frac{(\zeta-\lambda)(\zeta-\mu)}{(\zeta-e^{aT})(\zeta-e^{\eta T})}. \end{aligned} \tag{16.74}$$

For further calculations, we consider the partial fraction expansion

$$F(s)R_n(s) = \frac{\xi_1}{s-a} + \frac{\xi_2}{s-\eta} + \frac{\xi_3}{s+\eta}$$

where ξ_i are known constants and $a \neq \eta$. Hence, using (9.102) for $t=0$, we find

$$\mathcal{D}_{FR_nM}(T,s,0) = \frac{\xi_1 M(a)}{e^{sT}e^{-aT}-1} + \frac{\xi_2 M(\eta)}{e^{sT}e^{-\eta T}-1} + \frac{\xi_3 M(-\eta)}{e^{sT}e^{\eta T}-1}$$

and with regard to (16.71)

$$B(\zeta^{-1}) = \frac{\alpha_0+\alpha_1\zeta+\alpha_2\zeta^2}{(\zeta-e^{aT})(\zeta-e^{\eta T})(\zeta-e^{-\eta T})} \tag{16.75}$$

where α_i are known constants. Using (16.74) and (16.75) gives

$$u(\zeta) = \frac{1}{QN_1}\,\frac{\alpha_0+\alpha_1\zeta+\alpha_2\zeta^2}{(\zeta-e^{-\eta T})(\zeta-\lambda)(\zeta-\mu)}.$$

The real rational function $u(\zeta)$ is strictly proper, therefore performing the separation (16.29) and (16.30) yields

$$u_+(\zeta) = \frac{l}{1-e^{\eta T}\zeta} \tag{16.76}$$

where l is a known constant. From (16.76) and (16.74) the optimal control program can be found as

$$W_{d0}(\zeta) = \frac{u_+(\zeta)}{K(\zeta)} = \frac{l}{QN_1} \frac{1 - e^{aT}\zeta}{(1-\lambda\zeta)(1-\mu\zeta)}.$$

Returning back to the variable s, the optimal control can be written in the form

$$W_{d0}(s) = \frac{l}{QN_1} \frac{1 - e^{aT}e^{-sT}}{(1-\lambda e^{-sT})(1-\mu e^{-sT})}.$$

□

Example 16.3 Consider a special case of the problem of Example 16.2 for $T = 0.1$ and

$$F(s) = \frac{1}{s+1}, \quad \mathcal{D}_{R_\theta}(T,s,0) = 1\,, R_n(s) = \frac{4}{4-s^2}, \quad M(s) = \frac{1-e^{-sT}}{s}.$$

Then, calculation by the above formulae yields

$$\left[\mathcal{D}_{R_n}(T,s,0) + \mathcal{D}_{R_\theta}(T,s,0)\right]\Big|_{e^{-sT}=\zeta} = \frac{1.574(1+0.520\zeta)(1+0.520\zeta^{-1})}{(1-0.819\zeta)(1-0.819\zeta^{-1})}$$

$$\left[\frac{1}{T}\mathcal{D}_{F\underline{F}M\underline{M}}(T,s,0)\right]\Big|_{e^{-sT}=\zeta} = \frac{0.00563(1+0.268\zeta)(1+0.268\zeta^{-1})}{(1-0.905\zeta)(1-0.905\zeta^{-1})}.$$

The factorization gives

$$K(\zeta) = \frac{0.0942(1+0.520\zeta)(1+0.268\zeta)}{(1-0.819\zeta)(1-0.905\zeta)}$$

$$K(\zeta^{-1}) = \frac{0.0942(\zeta+0.520)(\zeta+0.268)}{(\zeta-0.819)(\zeta-0.905)}.$$

To find $B(\zeta^{-1})$ we use Eq.(16.75), which leads to

$$B(\zeta^{-1}) = \frac{-0.00652\zeta^2 - 0.0255\zeta - 0.00620}{(\zeta-0.819)(\zeta-1.221)(\zeta-0.905)}.$$

Then,

$$u(\zeta) = \frac{-0.0692\zeta^2 - 0.271\zeta - 0.0658}{(\zeta-0.520)(\zeta+0.268)(\zeta-1.221)}$$

and the separation (16.29) and (16.30) gives

$$u_+(\zeta) = \frac{0.392}{1-0.819\zeta}.$$

The transfer function of the CMV-optimal control program $W_{CMV}(\zeta)$ appears as

$$W_{CMV}(\zeta) = \frac{4.158 - 3.762\zeta}{1 - 0.252\zeta - 0.139\zeta^2}. \tag{16.77}$$

The mean error variance is equal to

$$\overline{d}_\varepsilon = 0.535\,.$$

The error variance $d_\varepsilon(0)$ at the point $t = 0$ using the controller (16.77) equals 0.590. Similar computation gives the following transfer function of the MV-optimal control program

$$W_{MV}(\zeta) = \frac{3.138 - 2.839\zeta}{1 - 0.520\zeta}\,. \qquad (16.78)$$

Using the controller (16.78) we have the minimal value of $d_\varepsilon(0) = 0.574$ and the mean variance $\overline{d}_\varepsilon = 0.544$. In this example the advantage of the CMV-optimal control is not great, but the following examples demonstrate that this is not always the case. □

Example 16.4 As an example in which the optimal CMV and MV regulators generate very distinct behaviour, let us consider, under the conditions of Example 16.3,

$$F(s) = \frac{1}{s^2 + s + 1} \qquad (16.79)$$

and the period T should be a parameter. Consider the influence of the sampling period on the performance of the MV- and CMV-optimal systems. Figure 16.7 presents the mean error variance for the two types of optimal systems. The curves show that for small sampling periods the MV-control gives far greater

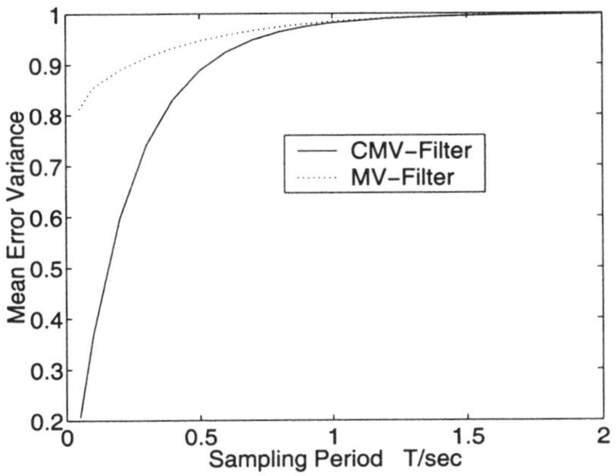

Figure 16.7: Mean error variance with MV- and CMV-control

mean variances than the CMV-control. Moreover, this difference increases as the sampling period decreases. This is caused by the fact that the transfer function (16.79) has an even pole excess (of 2). As was shown in Åström et al. (1984), in this case the function

$$\mathcal{D}^o_{FM}(T,\zeta,0) \triangleq \mathcal{D}_{FM}(T,s,0)\,\big|_{e^{-sT}=\zeta}$$

has for $T \to 0$ a (stable or unstable) zero that tends to $\zeta = -1$. This zero (or, if it is unstable, the corresponding stable zero, 'reflected' from the unit circle) becomes a pole of the MV-optimal controller, so that the system tends to the stability bound and the mean error variance increases. □

Example 16.5 Consider the CMV-redesign problem for the system shown in Fig. 16.8, where

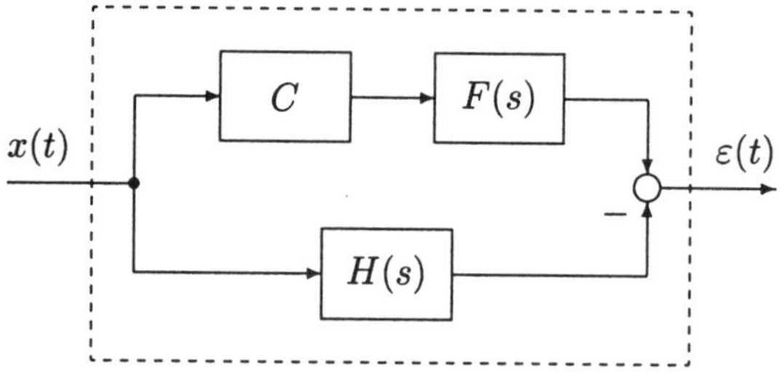

Figure 16.8: Hybrid structure for the redesign problem

$$F(s) = H(s) = \frac{1}{s+1}, \quad M(s) = \frac{1-e^{-sT}}{s}, \quad R_{xx}(s) = \frac{4}{4-s^2}.$$

Under these conditions the CMV-optimal controller takes the form

$$W_{CMV}(\zeta) = \frac{1.518 - 0.261\zeta}{1 - 0.268\zeta}.$$

Figure 16.9 shows the curves of the mean error variance versus the sampling

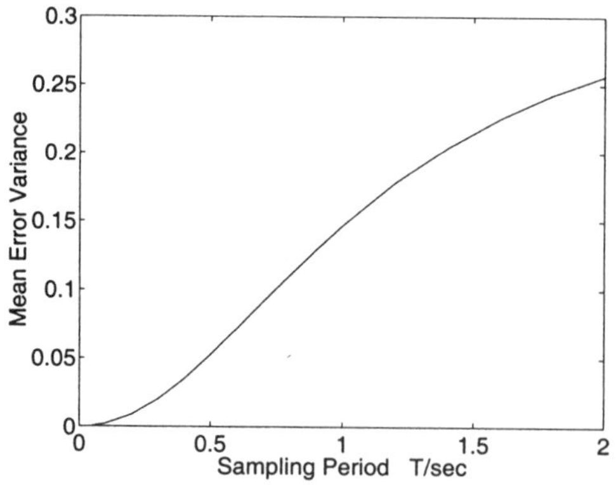

Figure 16.9: Mean variance with CMV-control

period T. □

16.5 Optimal filtering in time-delayed systems

In this section we consider in detail the influence of the computational delay on the quality of the optimal filtering. Consider the system shown in Fig. 16.10, with the notations and assumptions of Sections 16.4. Moreover, the value τ

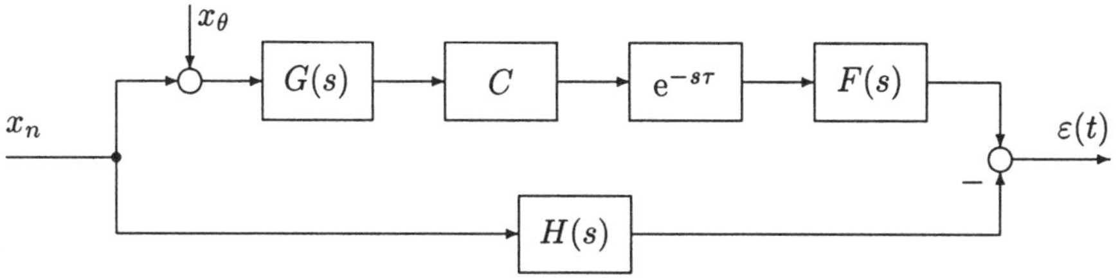

Figure 16.10: Hybrid structure for filter with time-delay

in Fig. 16.10 denotes a pure delay reduced to the continuous-time part of the system, and describes approximately the computational delay and possible delay in the continuous-time elements after the computer. In this case, if all the assumptions of Section 16.4 remain valid, the transfer function $F(s)$ is replaced by $F_1(s) = F(s)e^{-s\tau}$. Therefore, taking into account that

$$F_1(s)F_1(-s) = F(s)F(-s)$$

it is possible to obtain a functional of the form (16.56), in which the function $\tilde{A}(s)$ is defined by Eq. (16.65) as before, but the coefficient $\tilde{B}(s)$ is given by

$$\tilde{B}(s) = \frac{1}{T}\mathcal{D}_{F_1 G\underline{H} R_n M}(T, s, 0) = \frac{1}{T}\mathcal{D}_{FG\underline{H} R_n M}(T, s, \delta T)e^{-msT} \tag{16.80}$$

and, as before,

$$\tau = mT - \delta T$$

where $0 < \delta < 1$ and $m \geq 0$ is an integer. The following transition to the variable $\zeta = e^{-sT}$ leads to a functional of the form (16.58).
Henceforth, for simplicity of formulae, the continuous-time elements of the pulse part may have a transfer function with simple poles, so that

$$F(s) = \sum_{i=1}^{n} \frac{\phi_i}{s - f_i}, \quad G(s) = g_0 + \sum_{j=1}^{\ell} \frac{\gamma_j}{s - g_j} \tag{16.81}$$

where $\operatorname{Re} f_i < 0$ and $\operatorname{Re} g_j < 0$ for all i and j.
The ideal operator $H(s)$ is assumed to be

$$H(s) = H_0(s)e^{qsT} \tag{16.82}$$

where q is an integer and

$$H_0(s) = h_0 + \sum_{k=1}^{p} \frac{\chi_k}{s - h_k} \tag{16.83}$$

with $\operatorname{Re} h_k < 0$ for all k. The spectral densities of the reference signal and noise are assumed to be real rational functions with simple poles

$$\begin{aligned} R_n(s) &= \sum_{\rho=1}^{\nu} \frac{\alpha_\rho}{s - r_\rho} - \sum_{\rho=1}^{\nu} \frac{\alpha_\rho}{s + r_\rho} \\ R_\theta(s) &= \sum_{\eta=1}^{\sigma} \frac{\beta_\eta}{s - s_\eta} - \sum_{\eta=1}^{\sigma} \frac{\beta_\eta}{s + s_\eta} \end{aligned} \tag{16.84}$$

with $\operatorname{Re} r_\rho < 0$ and $\operatorname{Re} s_\eta < 0$ for all ρ and η . If $g_0 = 0$ in (16.81), we can consider the problem when $x_\theta(t)$ is a white noise with the intensity $\tilde{d}_\theta$, i.e., $R_\theta(s) = \tilde{d}_\theta$. In the case of $G(s) = g_0 =$const we may take $\mathcal{D}_{R_\theta}(T, s, 0) = d_\theta =$ const as in Example 16.2. The aggregate of the poles of the functions (16.81), (16.83), and (16.84) will be denoted by $\mathcal{K}$. It will be assumed that the elements of the set $\mathcal{K}$ satisfy the non-pathological conditions (9.161) and (9.162).
Henceforth, for any real rational function $G(s)$ in the form (16.81) we denote

$$P_G(\zeta) \triangleq \prod_{j=1}^{\ell} \left(1 - \mathrm{e}^{q_j T}\zeta\right), \qquad Q_G(\zeta) \triangleq \prod_{j=1}^{\ell} \left(\zeta - \mathrm{e}^{q_j T}\right). \tag{16.85}$$

For two real rational functions $G_1(s)$ and $G_2(s)$ with an irreducible product, introduce the notation

$$P_{G_1G_2}(\zeta) \triangleq P_{G_1}(\zeta)P_{G_2}(\zeta), \qquad Q_{G_1G_2}(\zeta) \triangleq Q_{G_1}(\zeta)Q_{G_2}(\zeta). \tag{16.86}$$

A similar notation will be used for the spectral densities, though only stable poles are included in the polynomials (16.85). Then,

$$\begin{aligned} P_{R_n}(\zeta) &\triangleq \prod_{\rho=1}^{\nu} \left(1 - \mathrm{e}^{r_\rho T}\zeta\right), \qquad & Q_{R_n}(\zeta) &\triangleq \prod_{\rho=1}^{\nu} \left(\zeta - \mathrm{e}^{r_\rho T}\right) \\ P_{R_\theta}(\zeta) &\triangleq \prod_{\eta=1}^{\sigma} \left(1 - \mathrm{e}^{s_\eta T}\zeta\right), \qquad & Q_{R_\theta}(\zeta) &\triangleq \prod_{\eta=1}^{\sigma} \left(\zeta - \mathrm{e}^{s_\eta T}\right). \end{aligned}$$

First, we consider the minimization of the mean variance, assuming the pulse form $\mu(t)$ to be arbitrary.
Using the formula

$$\mathcal{D}_{F\underline{F}M\underline{M}}(T, s, 0) = \int_0^T \mathcal{D}_{FM}(T, s, t)\mathcal{D}_{FM}(T, -s, t)\,\mathrm{d}t$$

and (9.105), it is possible to obtain the representation

$$\frac{1}{T}\mathcal{D}_{F_1\underline{F_1}M\underline{M}}(T,s,0)\,\big|_{e^{-sT}=\zeta} = \frac{1}{T}\mathcal{D}_{F\underline{F}M\underline{M}}(T,s,0)\,\big|_{e^{-sT}=\zeta} = \frac{A_1(\zeta)A_1(\zeta^{-1})}{P_F(\zeta)P_F(\zeta^{-1})}$$

where $A_1(\zeta)$ is a stable polynomial of degree n. Moreover, using (4.44) yields

$$\mathcal{D}_{G\underline{G}(R_n+R_\theta)}(T,s,0)\,\big|_{e^{-sT}=\zeta} = \frac{A_2(\zeta)A_2(\zeta^{-1})}{P_{GR_nR_\theta}(\zeta)P_{GR_nR_\theta}(\zeta^{-1})}$$

where $A_2(\zeta)$ is also a stable polynomial. Then, if the function

$$L(s) \triangleq G(s)G(-s)[R_n(s)+R_\theta(s)]$$

is strictly proper, the degree of the polynomial $A_2(\zeta)$ is equal to

$$\deg A_2(\zeta) = \ell+\nu+\sigma-1\,.$$

If $G(s)=g_0$ and the noise $x_\theta(t)$ is the (discrete) white noise, then

$$\deg A_2(\zeta) = \nu\,.$$

Hence, factoring the function $A(\zeta)$ in (16.71) gives

$$K(\zeta) = \frac{A_1(\zeta)A_2(\zeta)}{P_{FGR_nR_\theta}(\zeta)} \qquad K(\zeta^{-1}) = \frac{A_3(\zeta)}{Q_{FGR_nR_\theta}(\zeta)} \tag{16.87}$$

where the polynomial $A_3(\zeta)$ is given by

$$A_3(\zeta) = \zeta^{n+\ell+\nu+\sigma}A_1(\zeta^{-1})A_2(\zeta^{-1})\,.$$

By construction, the polynomial $A_3(\zeta)$ has no zeros outside the unit circle. Consider the calculation of the function

$$B(\zeta^{-1}) = \left[\frac{1}{T}\mathcal{D}_{F_1G\underline{H}R_nM}(T,s,0)\right]\big|_{e^{sT}=\zeta^{-1}}$$

assuming that the product

$$U(s) \triangleq F(s)G(s)H_0(-s)R_n(s)$$

is irreducible and has simple poles satisfying the non-pathological conditions (9.161) and (9.162). Then, the use of (16.80) and (9.99) yields

$$\frac{1}{T}\mathcal{D}_{F_1G\underline{H}R_nM}(T,s,0) = \frac{1}{T}\mathcal{D}_{FG\underline{H_0}R_nM}(T,s,\delta T)e^{-(m+q)sT} =$$

$$= \left[\sum_{i=1}^{n}\frac{a_i}{e^{sT}-e^{f_iT}} + \sum_{j=1}^{\ell}\frac{b_j}{e^{sT}-e^{g_jT}} + \sum_{k=1}^{p}\frac{c_k}{e^{sT}-e^{-h_kT}} + \right. \tag{16.88}$$

$$\left. + \sum_{\rho=1}^{\nu}\frac{d_\rho}{e^{sT}-e^{r_\rho T}} + \sum_{\rho=1}^{\nu}\frac{l_\rho}{e^{sT}-e^{-r_\rho T}} + \gamma\right] e^{-(m+q)sT}$$

where

$$\gamma = \int_0^{\delta T} h_U(\delta T - \tau)\mu(\tau)\,\mathrm{d}\tau$$

and $h_U(t)$ is the original for the image $U(s)$ in the right half-plane, and a_i, b_j, c_k, d_ρ, and l_ρ are known constants. Equations (16.88) and (16.82) give

$$B(\zeta^{-1}) = \frac{A_4(\zeta)\zeta^{-(m+q)}}{Q_{FGR_n}(\zeta)P_{H_0R_n}(\zeta)} \tag{16.89}$$

where $A_4(\zeta)$ is a polynomial of degree $n+\ell+2\nu+p$. From (16.87) and (16.89) we obtain

$$u(\zeta) = \frac{A_4(\zeta)Q_{R_\theta}(\zeta)\zeta^{-(m+q)}}{A_3(\zeta)P_{H_0R_n}(\zeta)}\,.$$

The poles of $u(\zeta)$ outside the unit circle are the roots of the polynomial $P_{H_0R_n}(\zeta)$. Therefore, the separation (16.29)–(16.30) gives

$$u_+(\zeta) = \frac{u_1(\zeta)}{P_{H_0R_n}(\zeta)} \tag{16.90}$$

where $u_1(\zeta)$ is a polynomial, whose degree depends on the value of the delay (i.e., on m) and on q appearing in (16.82). From (16.90) and (16.87) the optimal CMV-control can be found as

$$W_{CMV}(\zeta) = \frac{u_+(\zeta)}{K(\zeta)} = \frac{u_1(\zeta)P_{FGR_\theta}(\zeta)}{P_{H_0}(\zeta)A_1(\zeta)A_2(\zeta)}\,. \tag{16.91}$$

Relation (16.91) makes it possible to formulate the following conclusions.

1. If the factorization (16.27) is valid, the above control program always gives a stable CMV-optimal control (16.91).

2. The poles of the transfer function of the CMV-optimal controller are the numbers $\zeta_k = \mathrm{e}^{-h_k T}$, where h_k are the poles of the transfer function of the ideal operator and the zeros of the function $A(\zeta)$ outside the unit circle.

3. The set of the zeros of the transfer function of the CMV-optimal controller contains the numbers $\mathrm{e}^{-f_i T}$, $\mathrm{e}^{-g_j T}$, and $\mathrm{e}^{-r_\rho T}$, defined by the poles of the transfer functions $F(s)$ and $G(s)$ and by those of the spectral density of the reference signal $R_n(s)$.

4. The degree of the numerator of the transfer function (16.91) can be higher than that of the denominator, for instance, in the optimal smoothing problem, i.e., the function $W_{CMV}(\zeta)$ may not be limited.

The MV-optimization problem can be considered in a similar way . The above formulae can be employed if we take

$$\begin{aligned} A(\zeta) &= \left[\mathcal{D}_{FM}(T,s,\delta T)\mathcal{D}_{FM}(T,-s,\delta T)\mathcal{D}_{G\underline{G}(R_n+R_\theta)}(T,-s,0)\right]\big|_{\mathrm{e}^{-sT}=\zeta} \\ B(\zeta) &= \left[\mathcal{D}_{FM}(T,s,\delta T)\mathcal{D}_{G\underline{H_0}R_n}(T,s,0)\mathrm{e}^{-(m+q)sT}\right]\big|_{\mathrm{e}^{-sT}=\zeta}\,. \end{aligned} \tag{16.92}$$

Example 16.6 Investigate the influence of the delay τ on the performance of the optimal systems. Consider the system shown in Fig. 16.11, where $T = 0.1$,

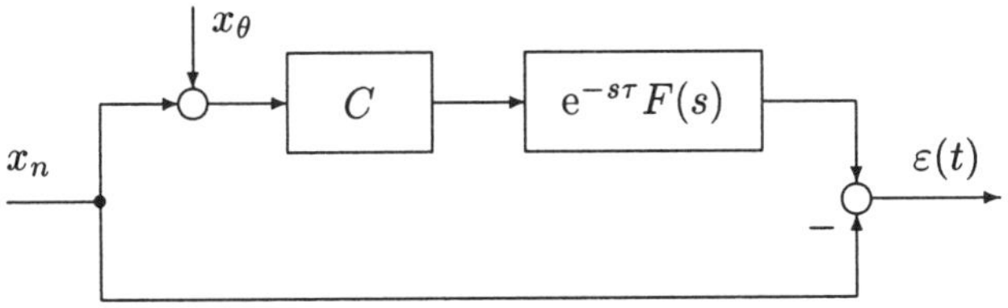

Figure 16.11: Simple hybrid filter with time-delay

$0 < \tau < T$ and

$$F(s) = \frac{1}{s+1}, \quad M(s) = \frac{1-e^{-sT}}{s}, \quad R_n(s) = \frac{4}{4-s^2}$$
$$G(s) = H(s) = 1, \quad \mathcal{D}_{R_\theta}(T,s,0) = 1. \tag{16.93}$$

Figure 16.12 presents the graph of the mean variance $\overline{d}_\varepsilon$ versus the value of

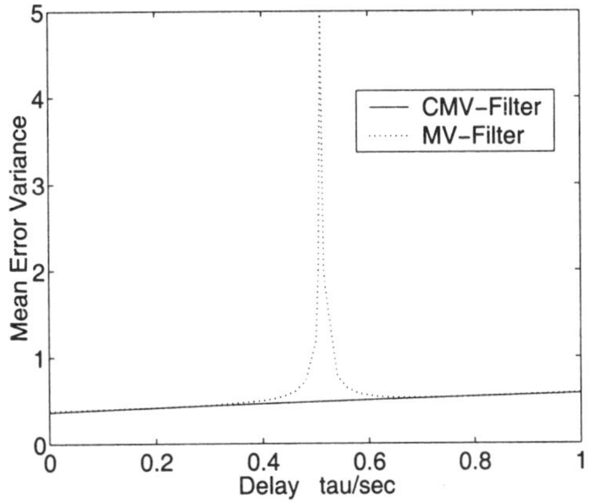

Figure 16.12: Optimal filter with time-delay

the time delay for CMV- and MV-optimal control. As is evident, for some τ the MV-optimal control gives an impermissibly large mean variance. This is caused by the fact that the function

$$\mathcal{D}^o_{FM}(T,\zeta,\delta T) \triangleq \mathcal{D}_{FM}(T,s,\delta T)\,\big|_{e^{-sT}=\zeta}$$

has, for some δ, a zero near to the stability boundary, Hara et al. (1989). Let us demonstrate this effect for the given example. As follows from (16.93),

$$\mathcal{D}^o_{FM}(T,\zeta,\delta T) = \frac{1-e^{-aT}}{a}\,\frac{e^{\delta T}}{1-e^{aT}\zeta} + \frac{e^{a(\delta T - T)}-1}{a}.$$

It can be immediately verified that the equation

$$\mathcal{D}^o_{FM}(T,\zeta,\delta T) = 0$$

has the root

$$\zeta_0 = \frac{1 - e^{a\delta T}}{e^{aT} - e^{a\delta T}}$$

which can lie either inside or outside the unit circle. But if $|\zeta_0| > 1$, then ζ_0 is a pole of the MV-optimal controller transfer function, and if $|\zeta_0| < 1$, then the controller has a pole at the point $\zeta = \zeta_0^{-1}$. Therefore, if $\zeta_0 \to -1$, the MV-optimal controller has a pole near the stability boundary, and the mean variance of the error increases, though $d_\varepsilon(0)$ decreases. If

$$2e^{a\delta T} = 1 + e^{aT} \tag{16.94}$$

then $\zeta_0 = -1$ and the MV-optimization problem has no solution, because the factorization is impossible. Consider Eq. (16.94) as an equation in the parameter δ. This equation always has a single root $\delta = \delta_0$, where $0 < \delta_0 < 1$, i.e., this effect appears for any $a < 0$ and any T. For $T \to 0$ we have

$$\lim_{T\to 0} \delta_0 = \frac{1}{2}$$

i.e., for small sampling periods the 'dangerous' zero appears if the delay is approximately equal to half of the sampling period. Moreover, there exists the limit

$$\lim_{T\to\infty} \delta_0 T = \frac{\ln 0.5}{a} = -\frac{0.693}{a}\,.$$

Figure 16.13 shows the graphs of the variance $d_\varepsilon(t)$ for the case of (16.93) with $T = 0.1$ and $\tau = 0.051$. The chosen delay is near to the critical value, for which $\zeta_0 = -1$. With the MV-control the value $d_\varepsilon(0)$ is minimal, though a great increase in the variance $d_\varepsilon(t)$ inside the sampling interval is evident, which causes the sharp increase in the mean variance. □

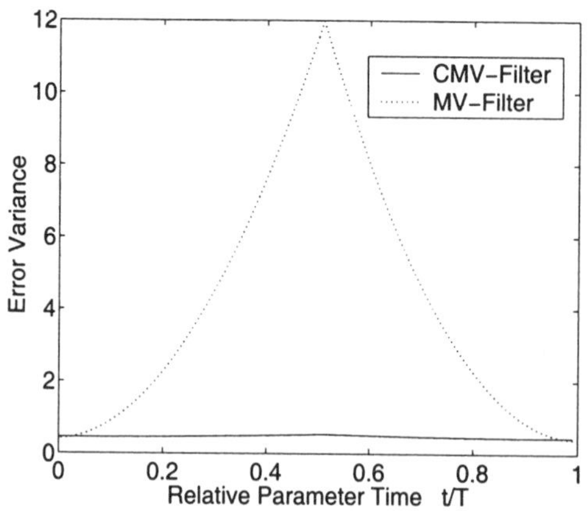

Figure 16.13: Variance of the optimal filters

Chapter 17

Direct design of feedback SD systems

17.1 Problem statement of LQ–design for feedback systems

Let us have the feedback sampled-data system shown in Fig. 17.1, with the

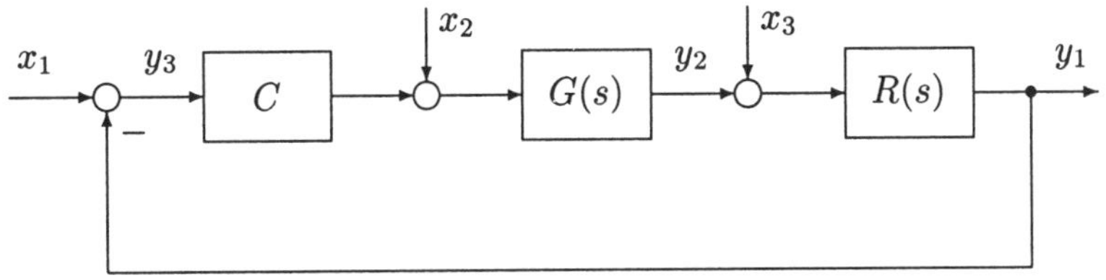

Figure 17.1: Feedback SD system

notation and assumptions of Section 11.1–11.5, so that

$$W_d(s) = \frac{b(s)}{a(s)}, \quad \mathcal{D}_{FM}(T, s, 0) = \frac{\mathrm{e}^{-sT}\chi(s)}{\alpha(s)} \tag{17.1}$$

where $F(s) = G(s)R(s)$. Consider the rational periodic functions

$$\tilde{W}_d(s) \triangleq \frac{W_d(s)}{1 + W_d(s)\mathcal{D}_{FM}(T, s, 0)} \tag{17.2}$$

$$\hat{W}_d(s) \triangleq \frac{1}{1 + W_d(s)\mathcal{D}_{FM}(T, s, 0)}. \tag{17.3}$$

Denote by

$$x(t) \triangleq \begin{bmatrix} x_1(t) \\ x_2(t) \\ x_3(t) \end{bmatrix}, \quad y(t) \triangleq \begin{bmatrix} y_1(t) \\ y_2(t) \\ y_3(t) \end{bmatrix} \tag{17.4}$$

the input and output vectors of the multivariable system under consideration, which are zero for $t < 0$. Denote the Laplace transforms of these vectors by

$$x(s) \triangleq \begin{bmatrix} x_1(s) \\ x_2(s) \\ x_3(s) \end{bmatrix}, \quad y(s) \triangleq \begin{bmatrix} y_1(s) \\ y_2(s) \\ y_3(s) \end{bmatrix}.$$

Henceforth, the initial conditions of the control algorithm are assumed to be zero. For appropriate conditions on the continuous-time elements, using the reasoning of Chapter 11 we get

$$y(s) = u_1(s)\tilde{W}_d(s) + u_2(s)\hat{W}_d(s) + u_3(s) \tag{17.5}$$

where the 3×1 vectors $u_i(s)$ $(i = 1, 2, 3)$ depend on the vector of exogenous disturbances and initial conditions of the continuous elements. Taking into account that

$$\hat{W}_d(s) = 1 - \tilde{W}_d(s)\mathcal{D}_{FM}(T, s, 0)$$

Eq. (17.5) gives

$$y(s) = u(s)\tilde{W}_d(s) + v(s) \tag{17.6}$$

where $u(s)$ and $v(s)$ are vectors dependent on the vector $x(s)$ and the initial conditions of the continuous elements. Assume that there is a vector of images $y_r(s)$ associated with the desired system outputs. Then, the vector

$$e(s) \triangleq y(s) - y_r(s) \tag{17.7}$$

defines the transform of the distance between the actual and the desired outputs, $y(t)$ and $y_r(t)$, respectively. Assume that the norm of the error vector $e(t)$ satisfies the estimate

$$\|e(t)\| < Ce^{-\lambda t} \tag{17.8}$$

where C and λ are positive constants. Then, the quality of reconstruction of the desired output by the designed system can be evaluated by the functional

$$I_c \triangleq \int_0^\infty \left[k_1^2 e_1^2(t) + k_2^2 e_2^2(t) + k_3^2(t) e_3^2(t)\right] \mathrm{d}t \tag{17.9}$$

where $e_i(t)$ are components of the error vector $e(t)$, and k_i^2 are non-negative constants. Using Parseval's formula (1.68), the functional (17.9) can be written in the form

$$I_c = \frac{1}{2\pi \mathrm{j}} \int_{-\mathrm{j}\infty}^{\mathrm{j}\infty} \left[k_1^2 e_1(s)e_1(-s) + k_2^2 e_2(s)e_2(-s) + k_3^2(t)e_3(s)e_3(-s)\right] \mathrm{d}s \tag{17.10}$$

where

$$e_i(s) = \int_0^\infty e_i(t)\mathrm{e}^{-st}\,\mathrm{d}t\,, \quad \mathrm{Re}\, s \geq 0\,.$$

Introduce the diagonal matrix

$$\boldsymbol{K} \triangleq \begin{bmatrix} k_1^2 & 0 & 0 \\ 0 & k_2^2 & 0 \\ 0 & 0 & k_3^2 \end{bmatrix} = \mathrm{diag}\{k_1^2 \quad k_2^2 \quad k_3^2\}\,. \tag{17.11}$$

Then, Eq. (17.10) can be represented as the trace of the matrix

$$I_c = \frac{1}{2\pi \mathrm{j}} \int_{-\mathrm{j}\infty}^{\mathrm{j}\infty} \mathrm{tr}\left[\boldsymbol{K}\boldsymbol{e}(s)\boldsymbol{e}'(-s)\right]\,\mathrm{d}s\,.$$

Substituting (17.11) and (17.7), we find

$$I_c = \frac{1}{2\pi \mathrm{j}} \int\limits_{-\mathrm{j}\infty}^{\mathrm{j}\infty} \left[\hat{A}(s)\tilde{W}_d(s)\tilde{W}_d(-s) - \hat{B}(s)\tilde{W}_d(s) - \hat{B}(-s)\tilde{W}_d(-s) + \hat{C}(s)\right]\,\mathrm{d}s \tag{17.12}$$

where

$$\hat{A}(s) = \mathrm{tr}\left[\boldsymbol{K}\boldsymbol{u}(s)\boldsymbol{u}'(-s)\right]\,, \qquad \hat{B}(s) = \mathrm{tr}\left[\boldsymbol{K}\boldsymbol{u}(s)\tilde{\boldsymbol{v}}'(-s)\right]$$

$$\hat{C}(s) = \mathrm{tr}\left[\boldsymbol{K}\tilde{\boldsymbol{v}}(s)\tilde{\boldsymbol{v}}'(-s)\right]\,, \qquad \tilde{\boldsymbol{v}}(s) = \boldsymbol{v}(s) - \boldsymbol{y}_r(s)\,.$$

Assume that the integral (17.12) may be discretized as

$$I_c = \frac{T}{2\pi \mathrm{j}} \int\limits_{c-\mathrm{j}\omega/2}^{c+\mathrm{j}\omega/2} \left[\tilde{A}(s)\tilde{W}_d(s)\tilde{W}_d(-s) - \tilde{B}(s)\tilde{W}_d(s) - \tilde{B}(-s)\tilde{W}_d(-s) + \tilde{C}(s)\right]\,\mathrm{d}s \tag{17.13}$$

where

$$\begin{aligned} \tilde{A}(s) &= \mathcal{D}_{\hat{A}}(T,s,0) = \frac{1}{T}\sum_{k=-\infty}^{\infty} \hat{A}(s+k\mathrm{j}\omega) \\ \tilde{B}(s) &= \mathcal{D}_{\hat{B}}(T,s,0) = \frac{1}{T}\sum_{k=-\infty}^{\infty} \hat{B}(s+k\mathrm{j}\omega) \\ \tilde{C}(s) &= \mathcal{D}_{\hat{C}}(T,s,0) = \frac{1}{T}\sum_{k=-\infty}^{\infty} \hat{C}(s+k\mathrm{j}\omega) \end{aligned} \tag{17.14}$$

are known functions dependent on the form of the input signals and the initial conditions of the continuous elements. Henceforth, the functions (17.14) are assumed to be rational periodic. Then, performing the substitution of $\zeta = \mathrm{e}^{-sT}$ for s in (17.13), we obtain

$$I_c = \frac{1}{2\pi j} \oint \left[A(\zeta)\tilde{W}_d(\zeta)\tilde{W}_d(\zeta^{-1}) - B(\zeta)\tilde{W}_d(\zeta) - B(\zeta^{-1})\tilde{W}_d(\zeta^{-1}) + C(\zeta) \right] \frac{d\zeta}{\zeta} \tag{17.15}$$

where the integration is performed along the unit circle anti-clockwise. Moreover, in (17.15)

$$A(\zeta) \triangleq \tilde{A}(s)\,\big|_{e^{-sT}=\zeta} \quad B(\zeta) \triangleq \tilde{B}(s)\,\big|_{e^{-sT}=\zeta}\,, \quad C(\zeta) \triangleq \tilde{C}(s)\,\big|_{e^{-sT}=\zeta} \tag{17.16}$$

are known real rational functions, and

$$\tilde{W}_d(\zeta) \triangleq \frac{W_d(\zeta)}{1 + W_d(\zeta)\mathcal{D}^o_{FM}(T,\zeta,0)} \tag{17.17}$$

with the notation

$$W_d(\zeta) \triangleq W_d(s)\,\big|_{e^{-sT}=\zeta}\,, \quad \mathcal{D}^o_{FM}(T,\zeta,0) \triangleq \mathcal{D}_{FM}(T,s,0)\,\big|_{e^{-sT}=\zeta}\,. \tag{17.18}$$

From (17.17) and (17.18) it follows that the integral I_c can be regarded as a functional on the set of real rational functions $W_d(\zeta)$.
Using the above relations, the optimization problem in the interval $0 < t < \infty$ (LQ−problem) can be formulated as follows:

LQ−problem ($0 \leq t < \infty$). Given all the data determining the coefficients (17.16), find the real rational function $W_{d0}(\zeta)$ which provides the stability of the system and the minimal value of the functional (17.15).

Remark 1 Recall that this statement of the problem assumes that the initial conditions of the control algorithm are zero.

Remark 2 It is not necessary specially to require the causality of the function $W_{d0}(\zeta)$, because, as was proved in Section 14.5, under the given assumptions all stabilizing transfer functions are causal.

Remark 3 Using the reasoning of Section 16.1, the more general functional (16.4) can be constructed for the feedback system and will have the form (17.15).

Remark 4 In principle, the dimensions of the chosen input and output vectors can be reduced for special optimization problems. The procedure of constructing the functional in this case remains the same.

17.2 Quadratic optimization in the interval $-\infty < t < \infty$ ($\mathcal{H}_2$−problem)

As regards the chosen input and output vectors $x(t)$ and $y(t)$, the system under consideration can be described by the vector operator equation

$$y = W(s,t)x \tag{17.19}$$

where $W(s,t)$ is the PTM (11.77). Assume that there is a reference system of the same dimensions

$$y_r = W_r(s,t)x_r \tag{17.20}$$

where $W_r(s,t)$ is the PTM of the reference system, and the problem is to find the best approximation of the output $y(t)$ to the desired output $y_r(t)$ in the interval $-\infty < t < \infty$ (Fig. 17.2). This is equivalent to the minimization of a

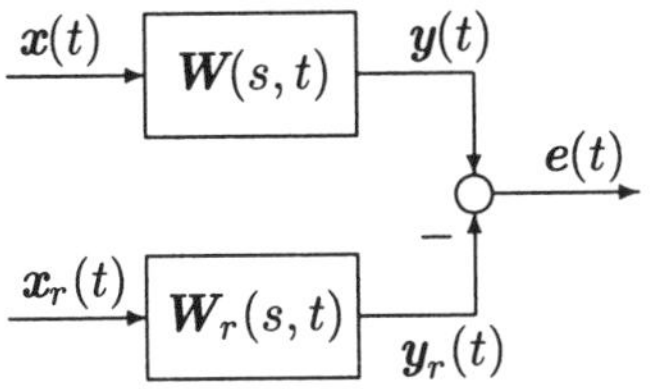

Figure 17.2: General block-diagram for error definition

norm of the error

$$e(t) = y(t) - y_r(t) .$$

Introducing the PTM of the system shown in Fig. 17.2

$$W_a(s,t) \triangleq \begin{bmatrix} W(s,t) & -W_r(s,t) \end{bmatrix} \tag{17.21}$$

and the augmented input vector

$$\tilde{x}(t) \triangleq \begin{bmatrix} x(t) \\ x_r(t) \end{bmatrix}$$

the operator equation of the error can be written as

$$e = W_a(s,t)\tilde{x} . \tag{17.22}$$

In what follows, both systems to be compared will be assumed stable with elements (17.21) satisfying the conditions of Theorem 6.1. Then, the integral

$$H_e(t,\tau) = \frac{1}{2\pi \mathrm{j}} \int_{-\mathrm{j}\infty}^{\mathrm{j}\infty} W_a(s,t)\mathrm{e}^{s(t-\tau)}\,\mathrm{d}s$$

converges, and the system (17.22) is associated with a causal e-dichotomic operator

$$e(t) \triangleq \mathsf{U}_e[\tilde{x}(t)] = \int_{-\infty}^{t} H_e(t,\tau)\tilde{x}(\tau)\,\mathrm{d}\tau . \tag{17.23}$$

Let $\tilde{x}(t)$ be a centered stationary stochastic vector. Then, the response of the operator (17.23) to the input $\tilde{x}(t)$ will be called the *steady-state error* of the system. The variance of the steady-state error $d_e(t)$ by (8.97) is given as

$$d_e(t) = \frac{1}{2\pi \mathrm{j}} \int_{-\mathrm{j}\infty}^{\mathrm{j}\infty} \mathrm{tr}\left[\boldsymbol{W}_a(-s,t)\boldsymbol{R}_{\tilde{x}\tilde{x}}(s)\boldsymbol{W}'_a(s,t)\right] \mathrm{d}s$$

where $\boldsymbol{R}_{\tilde{x}\tilde{x}}$ is the spectral density matrix of the vector $\tilde{\boldsymbol{x}}$. In this case, the average error variance is given by (8.100) as

$$\overline{d}_e = \frac{1}{2\pi \mathrm{j}} \int_{-\mathrm{j}\infty}^{\mathrm{j}\infty} \mathrm{tr}\left[\boldsymbol{B}_0(s)\boldsymbol{R}_{\tilde{x}\tilde{x}}(s)\right] \mathrm{d}s = \frac{1}{2\pi \mathrm{j}} \int_{-\mathrm{j}\infty}^{\mathrm{j}\infty} \mathrm{tr}\left[\boldsymbol{R}_{\tilde{x}\tilde{x}}(s)\boldsymbol{B}_0(s)\right] \mathrm{d}s \quad (17.24)$$

where

$$\boldsymbol{B}_0(s) = \frac{1}{T} \int_0^T \boldsymbol{W}'_a(s,t)\boldsymbol{W}_a(-s,t)\,\mathrm{d}t\,. \quad (17.25)$$

If $\boldsymbol{R}_{\tilde{x}\tilde{x}}(s)$ is the identity matrix,

$$\overline{d}_e = \overline{r}_e = \|\mathsf{U}_e\|_2^2 = \frac{1}{2\pi \mathrm{j}} \int_{-\mathrm{j}\infty}^{\mathrm{j}\infty} \mathrm{tr}\,\boldsymbol{B}_0(s)\,\mathrm{d}s\,. \quad (17.26)$$

Taking the value $\overline{r}_e$ as a cost function of the designed system, the following optimization problem can be formulated:

$\mathcal{H}_2$−problem $(-\infty < t < \infty)$**.** Given all characteristics of the two systems, except for the discrete controller $W_d(s)$, find the optimal control algorithm $W_{d0}(s)$ for which the system at hand is stable and the value $\overline{r}_e$ is minimal.

Remark 1 Using appropriate filters, the general problem of minimizing the functional (17.24) can be reduced to the above $\mathcal{H}_2$−problem.

Remark 2 In principle, not all of the components of the vectors $\tilde{\boldsymbol{x}}(t)$ and $\boldsymbol{e}(t)$ must be employed in the statement of an $\mathcal{H}_2$−problem. In this case the dimensions of the matrices in (17.25) becomes reduced.

Remark 3 In some cases the operator corresponding to the system shown in Fig. 17.2 is not an integral operator. Nevertheless, all the above results remain valid, if the integral in (17.26) converges absolutely.

Next, we demonstrate that the above optimization problem can be reduced to a quadratic functional of the form (17.15). With this aim in view, we note that the equations of Sections 11.2 and 11.3 establish that the PTM (11.77) can be represented in the form

$$\boldsymbol{W}(s,t) = \boldsymbol{U}(s,t)\tilde{W}_d(s,t) + \boldsymbol{V}(s,t) \quad (17.27)$$

where $\boldsymbol{U}(s,t)$ and $\boldsymbol{V}(s,t)$ are known 3×3−matrices. From the preceding relations,

$$\begin{aligned}\tilde{B}(s,t) &\triangleq \operatorname{tr}\left[\boldsymbol{W}'_a(s,t)\boldsymbol{W}_a(-s,t)\right] = \\ &= \operatorname{tr}\left[\boldsymbol{W}'(s,t)\boldsymbol{W}(-s,t)\right] + \operatorname{tr}\left[\boldsymbol{W}'_r(s,t)\boldsymbol{W}_r(-s,t)\right] .\end{aligned}$$

Using (17.27) yields

$$\tilde{B}(s,t) = A(s,t)\tilde{W}_d(s)\tilde{W}_d(-s) - B(s,t)\tilde{W}_d(s) - B(-s,t)\tilde{W}_d(-s) + C(s,t) \tag{17.28}$$

where $A(s,t)$, $B(s,t)$, and $C(s,t)$ are known scalar functions. Substituting (17.25) and (17.28) into (17.26), after averaging we obtain

$$\bar{r}_e = \frac{1}{2\pi \mathrm{j}} \int\limits_{-\mathrm{j}\infty}^{\mathrm{j}\infty} \left[\hat{A}(s)\tilde{W}_d(s)\tilde{W}_d(-s) - \hat{B}(s)\tilde{W}_d(s) - \hat{B}(-s)\tilde{W}_d(-s) + \hat{C}(s)\right] \mathrm{d}s \tag{17.29}$$

where

$$\begin{aligned}\hat{A}(s) &\triangleq \frac{1}{T}\int_0^T A(s,t)\,\mathrm{d}t\,, \quad \hat{B}(s) \triangleq \frac{1}{T}\int_0^T B(s,t)\,\mathrm{d}t \\ \hat{C}(s) &\triangleq \frac{1}{T}\int_0^T C(s,t)\,\mathrm{d}t\end{aligned}$$

are known functions. Discretizing the integral (17.29) with the substitution $\zeta = \mathrm{e}^{-sT}$, we obtain the equation similar to (17.15)

$$\bar{r}_e = \frac{1}{2\pi \mathrm{j}} \oint \left[A(\zeta)\tilde{W}_d(\zeta)\tilde{W}_d(\zeta^{-1}) - B(\zeta)\tilde{W}_d(\zeta) - B(\zeta^{-1})\tilde{W}_d(\zeta^{-1}) + C(\zeta)\right] \frac{\mathrm{d}\zeta}{\zeta} \tag{17.30}$$

where

$$\begin{aligned}&A(\zeta) \triangleq \mathcal{D}_{\hat{A}}(T,s,0)\,\big|_{\mathrm{e}^{-sT}=\zeta}\,, \quad B(\zeta) \triangleq \mathcal{D}_{\hat{B}}(T,s,0)\,\big|_{\mathrm{e}^{-sT}=\zeta} \\ &C(\zeta) \triangleq \mathcal{D}_{\hat{C}}(T,s,0)\,\big|_{\mathrm{e}^{-sT}=\zeta}\,.\end{aligned} \tag{17.31}$$

Henceforth, the PTM $\boldsymbol{W}_r(s,t)$ is assumed to have real rational coefficients (17.31).

Remark 4 The problem formulated above is an analog of the CMV-problem for open-loop systems, described in Section 16.3. Taking the value

$$d_e(0) = \frac{1}{2\pi \mathrm{j}} \int_{-\mathrm{j}\infty}^{\mathrm{j}\infty} \operatorname{tr}\left[\boldsymbol{W}'_a(s,0)\boldsymbol{W}_a(-s,0)\right] \mathrm{d}s \tag{17.32}$$

as an optimality criterion, we obtain the corresponding MV-problem for feedback systems, which can be reduced to a functional of the form (17.30) by means of the above technique.

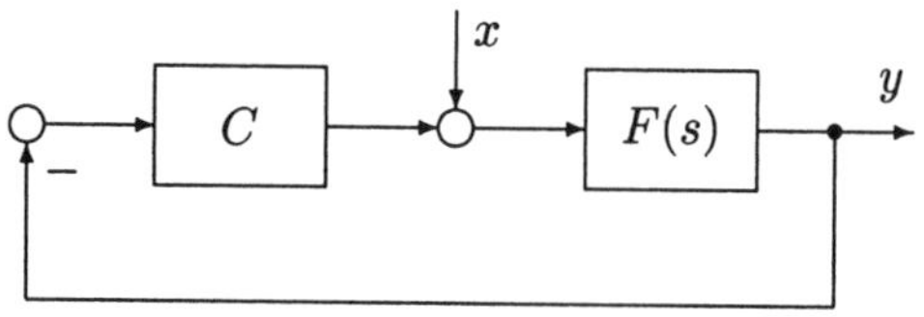

Figure 17.3: Simple digital feedback system

Example 17.1 Let us build the cost function (17.30) for the system shown in Fig. 17.3. The PTF $W_{yx}(s,t)$ from the input $x(t)$ to the output $y(t)$ is equal to

$$W_{yx}(s,t) = F(s) - F(s)\varphi_{FM}(T,s,t)\tilde{W}_d(s) \tag{17.33}$$

where

$$\tilde{W}_d(s) = \frac{W_d(s)}{1 + W_d(s)\mathcal{D}_{FM}(T,s,0)}\,. \tag{17.34}$$

With the chosen input and output, the $\mathcal{H}_2$−norm is defined as

$$\|\mathsf{U}_y\|_2^2 = \overline{r}_y = \frac{1}{2\pi \mathrm{j}}\int_{-\mathrm{j}\infty}^{\mathrm{j}\infty} B_0(s)\,\mathrm{d}s$$

where

$$B_0(s) = \frac{1}{T}\int_0^T W_{yx}(s,t)W_{yx}(-s,t)\,\mathrm{d}t\,.$$

Substituting (17.33) in the last equation, we find

$$\begin{aligned} B_0(s) = F(s)F(-s) - \frac{1}{T}F^2(s)F(-s)M(s)\tilde{W}_d(s) - \\ -\frac{1}{T}F^2(-s)F(s)M(-s)\tilde{W}_d(-s) + \\ +\frac{1}{T}\mathcal{D}_{F\underline{F}M\underline{M}}(T,s,0)F(s)F(-s)\tilde{W}_d(s)\tilde{W}_d(-s)\,. \end{aligned}$$

After routine transformations the value $\overline{r}_y$ can be defined in the form (17.30), where

$$\begin{aligned} A(\zeta) &= \frac{1}{T}\left[\mathcal{D}_{F\underline{F}M\underline{M}}(T,s,0)\mathcal{D}_{F\underline{F}}(T,s,0)\right]\big|_{\mathrm{e}^{-sT}=\zeta} \\ B(\zeta) &= \frac{1}{T}\mathcal{D}_{F^2\underline{F}M}(T,s,0)\,\big|_{\mathrm{e}^{-sT}=\zeta} \\ C(\zeta) &= \frac{1}{T}\mathcal{D}_{F\underline{F}}(T,s,0)\,\big|_{\mathrm{e}^{-sT}=\zeta} \end{aligned}$$

□

17.3 Peculiarities of solutions to feedback system design problems. Robust optimization

As follows from the aforesaid, the optimization of the controller in LQ− and $\mathcal{H}_2$−problems reduces to the minimization of an integral of the form

$$\tilde{I} \triangleq \frac{1}{2\pi \mathrm{j}} \oint \left[A(\zeta)\tilde{W}_d(\zeta)\tilde{W}_d(\zeta^{-1}) - B(\zeta)\tilde{W}_d(\zeta) - B(\zeta^{-1})\tilde{W}_d(\zeta^{-1}) + C(\zeta) \right] \frac{\mathrm{d}\zeta}{\zeta} \tag{17.35}$$

where

$$\tilde{W}_d(\zeta) = \frac{W_d(\zeta)}{1 + W_d(\zeta)\mathcal{D}^o_{FM}(T,\zeta,0)} \tag{17.36}$$

with the notation

$$W_d(\zeta) = \frac{b(\zeta)}{a\zeta)}, \quad \mathcal{D}^o_{FM}(T,\zeta,0) = \mathcal{D}_{FM}(T,s,0)\,\big|_{\mathrm{e}^{-sT}=\zeta} = \frac{\zeta\chi^o(\zeta)}{\alpha^o(\zeta)}. \tag{17.37}$$

In (17.35)–(17.37) the $b(\zeta)$, $a(\zeta)$, $\chi^0(\zeta)$, and $\alpha^o(\zeta)$ are polynomials, where the last two are known, $A(\zeta)$, $B(\zeta)$, and $C(\zeta)$ are known real rational functions, and $W_d(\zeta)$ is a real rational function to be found in such a way that the integral (17.35) becomes minimal. Moreover, the characteristic polynomial of the feedback system

$$\Delta^o(\zeta) = \alpha^o(\zeta)a(\zeta) + \zeta\chi^o(\zeta)b(\zeta)$$

has to be stable.

Lemma 17.1 *If the polynomial $\Delta^o(\zeta)$ is stable, the function $\tilde{W}_d(\zeta)$ is also stable.*

Proof From (17.36) and (17.37) we have

$$\tilde{W}_d(\zeta) = \frac{b(\zeta)\alpha^o(\zeta)}{\Delta^o(\zeta)}.$$

Therefore, $\tilde{W}_d(\zeta)$ is free of poles inside and on the unit circle. ∎

Using Lemma 17.1, we may try to solve the optimization problem in the following way. Assume that we managed, by means of the algorithm described in Section 16.2, to find a stable real rational function $\tilde{W}_{d0}(\zeta)$ minimizing the integral (17.35). Then, the corresponding transfer function of the optimal discrete controller $W_{d0}(\zeta)$ can allegedly be found from the equality

$$\tilde{W}_{d0}(\zeta) = \frac{W_{d0}(\zeta)}{1 + \zeta W_{d0}(\zeta)\frac{\chi^o(\zeta)}{\alpha^o(\zeta)}}$$

that leads to

$$W_{d0}(\zeta) = \frac{\tilde{W}_{d0}(\zeta)}{1 - \zeta\tilde{W}_{d0}(\zeta)\frac{\chi^o(\zeta)}{\alpha^o(\zeta)}}. \tag{17.38}$$

We will demonstrate, that in some cases Eq. (17.38) yields an incorrect result. Indeed, let

$$W_{d0}(\zeta) = \frac{q(\zeta)}{r(\zeta)}$$

where $q(\zeta)$ and $r(\zeta)$ are co-prime polynomials. Then, Eq. (17.38) gives

$$W_{d0}(\zeta) = \frac{q(\zeta)\alpha^o(\zeta)}{r(\zeta)\alpha^o(\zeta) - \zeta q(\zeta)\chi^o(\zeta)}\,. \tag{17.39}$$

Denote

$$\tilde{a}(\zeta) \stackrel{\triangle}{=} r(\zeta)\alpha^o(\zeta) - \zeta q(\zeta)\chi^o(\zeta)\,, \quad \tilde{b}(\zeta) \stackrel{\triangle}{=} q(\zeta)\alpha^o(\zeta)\,.$$

Then, the characteristic polynomial of the feedback system with the controller (17.38) takes the form

$$\Delta^o(\zeta) = \alpha^o(\zeta)\tilde{a}(\zeta) + \zeta\chi^o(\zeta)\tilde{b}(\zeta) = [\alpha^o(\zeta)]^2 r(\zeta)\,.$$

Hence, if $\alpha^o(\zeta)$ is not stable, the polynomial $\Delta^o(\zeta)$ is also unstable. But, as follows from (9.151), the polynomial $\alpha^o(\zeta)$ is stable iff all the continuous networks in the designed system are stable. Thus, if all the continuous elements are stable, Eq. (17.38) yields the desired optimal controller. If at least one of the transfer functions $G(s)$ or $R(s)$ has poles in the closed right half-plane, Eq. (17.38) does not give a solution of the problem, because the corresponding transfer function $W_{d0}(\zeta)$ is not stabilizing. Therefore, to solve the problem in the general case, we shall employ a different approach.

The idea of this approach consists in the use of the parametrization of the set of stabilizing controllers, considered in Section 14.5. Let $\{\tilde{a}^o(\zeta), \tilde{b}^o(\zeta)\}$ be a basic stabilizing polynomial pair. Then, the set of all stabilizing controllers is given by Eqs. (14.91) and (14.92)

$$W_d(\zeta) = \frac{\tilde{b}^o(\zeta) + \phi(\zeta)\alpha^o(\zeta)}{\tilde{a}^o(\zeta) - \zeta\phi(\zeta)\chi^o(\zeta)} \tag{17.40}$$

where $\phi(\zeta)$ is an arbitrary stable irreducible real rational function. Substituting (17.40) into (17.36), we find

$$\tilde{W}_d(\zeta) = [\alpha^o(\zeta)]^2\phi(\zeta) + \alpha^o(\zeta)\tilde{b}^o(\zeta)\,. \tag{17.41}$$

Substituting (17.41) into (17.35), we obtain the problem of minimizing the functional

$$\tilde{I}_1 \stackrel{\triangle}{=} \frac{1}{2\pi \mathrm{j}} \oint \left[A_1(\zeta)\phi(\zeta)\phi(\zeta^{-1}) - B_1(\zeta)\phi(\zeta) - B_1(\zeta^{-1})\phi(\zeta^{-1}) + C_1(\zeta)\right] \frac{\mathrm{d}\zeta}{\zeta} \tag{17.42}$$

over the set of all stable real rational functions $\phi(\zeta)$, where

$$\begin{aligned} A_1(\zeta) &= A(\zeta)\left[\alpha^o(\zeta)\alpha^o(\zeta^{-1})\right]^2 \\ B_1(\zeta) &= A(\zeta)\left[\alpha^o(\zeta)\right]^2 - A(\zeta)\left[\alpha^o(\zeta)\right]^2\alpha^o(\zeta^{-1})\tilde{b}^o(\zeta^{-1}) \\ C_1(\zeta) &= A(\zeta)\alpha^o(\zeta)\alpha^o(\zeta^{-1}) - B(\zeta)\alpha^o(\zeta)\tilde{b}^o(\zeta) - B(\zeta^{-1})\alpha^o(\zeta^{-1})\tilde{b}^o(\zeta^{-1})\,. \end{aligned} \tag{17.43}$$

Let

$$\phi_0(\zeta) = \frac{c_0(\zeta)}{\Delta_0(\zeta)}$$

be the stable real rational function minimizing the functional (17.42). Then, the desired transfer function of the optimal controller is given by

$$W_{d0}(\zeta) = \frac{\tilde{b}^o(\zeta)\Delta_0(\zeta) + c_0(\zeta)\alpha^o(\zeta)}{\tilde{a}^o(\zeta)\Delta_0(\zeta) - \zeta c_0(\zeta)\chi^o(\zeta)}. \tag{17.44}$$

By construction, the function (17.44) is stabilizing. In this case the characteristic polynomial of the feedback system is given by

$$\Delta^o(\zeta) = \Delta_0(\zeta).$$

Remark We note that, using the parametrization (14.103), all the aforesaid can be extended to time-delayed systems.

As follows from Section 16.2, the optimal function $\phi_0(\zeta)$ does exist and is unique, if the function $A_1(\zeta)$ is free of poles and zeros on the unit circle and the function $B_1(\zeta)$ is free of poles there. To check these conditions, it is useful to take some extra ideas into account. Assuming that $\zeta = e^{-sT}$ in (17.41), we obtain

$$\tilde{W}_d(s) = \alpha^2(s)\phi(s) + \alpha(s)\tilde{b}(s) \tag{17.45}$$

where $\alpha(s)$ and $\tilde{b}(s)$ are quasi-polynomials in $\zeta = e^{-sT}$, and $\phi(s)$ is a stable rational periodic function. Consider the system shown in Fig. 17.1 without exogenous disturbance. Substituting (17.45) into (17.6), we find

$$\boldsymbol{y}(s) = \boldsymbol{u}_1(s)\phi(s) + \boldsymbol{v}_1(s) \tag{17.46}$$

where the known vector functions

$$\boldsymbol{u}_1(s) \stackrel{\triangle}{=} \boldsymbol{u}(s)\alpha^2(s), \quad \boldsymbol{v}_1(s) \stackrel{\triangle}{=} \boldsymbol{v}(s) + \alpha(s)\tilde{b}(s)\boldsymbol{u}(s)$$

depend on the initial conditions of the continuous elements.

Theorem 17.1 *The vector functions $\boldsymbol{u}_1(s)$ and $\boldsymbol{v}_1(s)$ are entire, i.e. have no poles in any finite part of the complex plane.*

Proof Let the function $\phi(s)$ be chosen as an arbitrary quasi-polynomial in $\zeta = e^{-sT}$. Then, the characteristic quasi-polynomial (11.73) degenerates into a constant, so that the natural oscillations in the system die down within a finite time interval. This may happen only if the image $\boldsymbol{y}(s)$ (17.46) has no poles for any quasi-polynomials $\phi(s)$, i.e., the vector functions $\boldsymbol{u}_1(s)$ and $\boldsymbol{v}_1(s)$ are entire. ■

Now let $\boldsymbol{W}(s,t)$ be the PTM of the designed system given by (11.77). As follows from Sections 11.2 and 11.3, the matrix $\boldsymbol{W}(s,t)$ admits the representation

$$\boldsymbol{W}(s,t) = \boldsymbol{Q}(s,t)\tilde{W}_d(s) + \boldsymbol{P}(s,t)$$

where $\boldsymbol{Q}(s,t)$ and $\boldsymbol{P}(s,t)$ are known 3×3 –matrices. Using (17.45), we find

$$\boldsymbol{W}(s,t) = \boldsymbol{Q}_1(s,t)\phi(s) + \boldsymbol{P}_1(s,t) \tag{17.47}$$

where

$$\boldsymbol{Q}_1(s,t) \stackrel{\triangle}{=} \boldsymbol{Q}(s,t)\alpha^2(s)\,, \quad \boldsymbol{P}_1(s,t) \stackrel{\triangle}{=} \boldsymbol{Q}(s,t)\alpha(s)\tilde{b}(s) + \boldsymbol{P}(s,t)\,. \tag{17.48}$$

Theorem 17.2 *The matrices $\boldsymbol{Q}_1(s,t)$ and $\boldsymbol{P}_1(s,t)$ are entire functions in the variable s for any t.*

Proof If $\phi(s)$ is an arbitrary quasi-polynomial, the characteristic quasi-polynomial (11.73) appears as

$$\Delta(s) = \Delta^o(\zeta)\left|_{\zeta=\mathrm{e}^{-sT}}\right. = \text{const.}$$

and, by (11.77), the matrix $\boldsymbol{W}(s,t)$ is an entire function in s. Thus, the matrix (17.47) is an entire function in s for any quasi-polynomial $\phi(s)$, i.e., the matrices (17.48) are entire functions in s. ■

Example 17.2 Let us directly examine the statement of Theorem 17.2 in an example. Consider for instance the system shown in Fig. 17.3 with the PTF (17.33). Then

$$Q(s,t) = -F(s)\varphi_{FM}(T,s,t)\,, \quad P(s,t) = F(s)\,.$$

Therefore, taking account of (17.47),

$$Q_1(s,t) = -[\varphi_{FM}(T,s,t)F(s)]\alpha^2(s)$$
$$P_1(s,t) = F(s) - \varphi_{FM}(T,s,t)\alpha(s)\tilde{b}(s)F(s)\,.$$

The function $Q_1(s,t)$ is entire, because the poles of the expression in the square brackets are cancelled by the zeros of the quasi-polynomial $\alpha^2(s) = \left(1 - \mathrm{e}^{aT}\mathrm{e}^{-sT}\right)^2$.
Let us demonstrate that the function $P_1(s,t)$ is also entire. Taking account of

$$\alpha(s)\tilde{a}(s) + \mathrm{e}^{-sT}\chi(s)\tilde{b}(s) = 1$$

we obtain

$$P_1(s,t) = F(s)\alpha(s)\tilde{a}(s) + \alpha(s)\tilde{b}(s)F(s)\left[\varphi_{FM}(T,s,0) - \varphi_{FM}(T,s,t)\right]\,.$$

The right-hand side of the last equation is an entire function, as follows from the proof of Theorem 11.1. □

Remark If the continuous part of the designed system is stable, to solve the optimization problem we may use the parametrization (14.99), which gives

$$\tilde{W}_d(\zeta) = \alpha^o(\zeta)\tilde{\phi}(\zeta) \tag{17.49}$$

where $\tilde{\phi}(\zeta)$ is an arbitrary stable real rational function. It can easily be shown, Rosenwasser (1994c), that the use of (17.49) leads to the solution (17.38).

17.4 Associated $\mathcal{H}_\infty$−problem

Lastly, we show that, based on the functional (17.42), another general optimization problem can be formulated. With this aim in view, we rewrite (17.42) in the form

$$\tilde{I}_1 = \frac{1}{2\pi \mathrm{j}} \oint X(\zeta)\,\frac{d\zeta}{\zeta} \tag{17.50}$$

where

$$X(\zeta) = A_1(\zeta)\phi(\zeta)\phi(\zeta^{-1}) - B_1(\zeta)\phi(\zeta) - B_1(\zeta^{-1})\phi(\zeta^{-1}) + C_1(\zeta)$$

is a quadratic polynomial in ϕ. Equation (17.50) can formally be considered as the $\mathcal{H}_2$−norm of an equivalent discrete-time system. Then, as easily follows from Grimble (1995), the value

$$I_\infty \triangleq \sup_{|\zeta|=1} X(\zeta) \tag{17.51}$$

can be used as a criterion for the robust optimization of this equivalent discrete-time system.Thus, we can formulate the following robust optimization problem.

> **Associated $\mathcal{H}_\infty$−problem.** Given all parameters of the initial SD system determining the coefficients (17.43), find the stable real rational function $\phi_{r0}(\zeta)$, such that $I_\infty \to \min$.

If such a function ϕ_{r0} is defined, using (17.44) we can obtain the corresponding robust controller. A polynomial technique for robust optimization is presented in Chapter 18.

17.5 LQ−redesign for time-delayed systems

Let us consider the stable continuous-time system shown in Fig. 17.4, where $W(s)$, $K(s)$, $L(s)$, and $N(s)$ are real rational functions, Rosenwasser et al. (1999a). The element with transfer function $W(s)$ is to be replaced by a controlling computer, thus obtaining the SD system shown in Fig. 17.5, where $K(s)$, $L(s)$, and $N(s)$ are the same as in Fig. 17.4. Moreover, from Fig. 17.5 it follows that

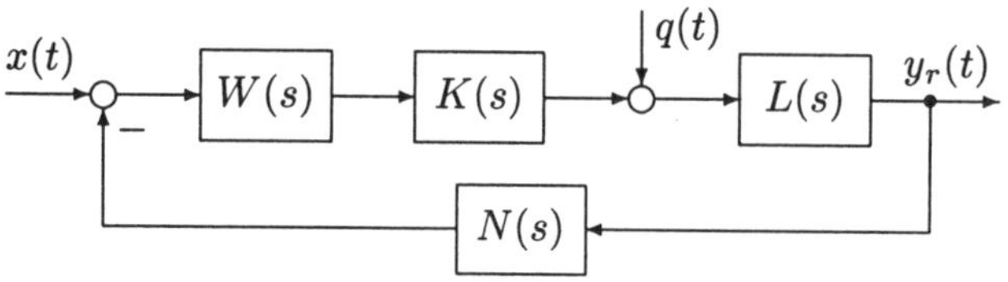

Figure 17.4: Continuous reference system

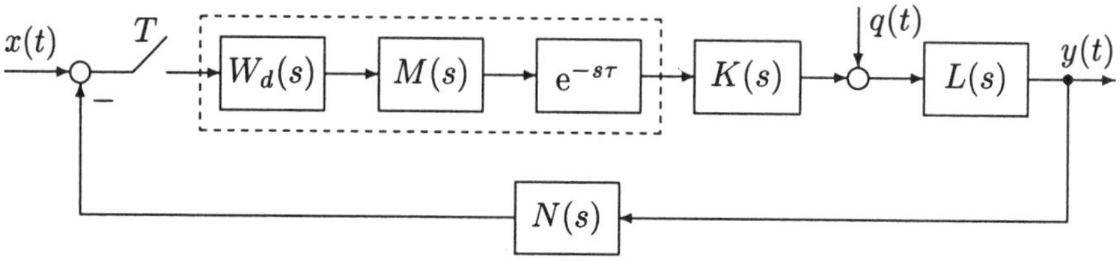

Figure 17.5: Redesigned SD system

$$M(s) = \frac{1 - e^{-sT}}{s}$$

and the value τ represents the computational delay and the pure delay in the continuous-time elements. The input signals in both systems are the same, and have for $\operatorname{Re} s > 0$ the image

$$X(s) = \frac{1}{s}, \quad Q(s) = \frac{1}{s}$$

i.e., $x(t)$ and $q(t)$ are unit steps (1.42). Assuming that the initial conditions of all elements of both systems are zero, take the integral quadratic error between the transients of the systems to be compared

$$I_c = \int_0^\infty [y(t) - y_r(t)]^2 \, dt \tag{17.52}$$

as a cost function. Under the given assumptions, the LQ–redesign problem can be formulated as follows.

LQ–redesign problem ($0 \le t < \infty$). Given the real rational functions $L(s)$, $K(s)$, $N(s)$, $W(s)$, the hold $M(s)$ and the constants T and τ, find the transfer function of the controller $W_d(\zeta)$, which ensures the stability of the feedback system, as well as the convergence and the minimal value of the integral (17.52).

Assume that the designed system is stable and the error

$$\varepsilon(t) = y(t) - y_r(t)$$

satisfies an estimate of the form (17.18). Then, the integral (17.52) converges, and using Parseval's formula, we have

$$I_c = \frac{1}{2\pi \mathrm{j}} \int_{-\mathrm{j}\infty}^{\mathrm{j}\infty} [Y(s) - Y_r(s)][Y(-s) - Y_r(-s)] \,\mathrm{d}s \tag{17.53}$$

where $Y(s)$ and $Y_r(s)$ are the Laplace transforms of the outputs $y(t)$ and $y_r(t)$, respectively. Calculate the Laplace transforms appearing in (17.53) for two cases:

Case 1 $x(t) = \mathbb{1}(t)\,, \quad q(t) = 0$

Case 2 $x(t) = 0\,, \quad q(t) = \mathbb{1}(t)\,.$

In **Case 1** the transfer function of the continuous-time system from the input $x(t)$ to the output $y(t)$ is given as

$$W_{yx}(s) = \frac{W(s)K(s)L(s)}{1 + W(s)K(s)L(s)N(s)} \tag{17.54}$$

and the image of the output $Y_{r1}(s)$ takes the form

$$Y_{r1}(s) = \frac{W(s)K(s)L(s)}{1 + W(s)K(s)L(s)N(s)} \, \frac{1}{s} \,. \tag{17.55}$$

To calculate the Laplace image of the output of the SD system $Y_1(s)$, let

$$G(s) \triangleq K(s)L(s)\mathrm{e}^{-s\tau}\,, \quad R(s) \triangleq N(s)\,.$$

Then, the use of the formulae given in Section 11.7 leads to

$$Y_1(s) = K(s)L(s)M(s)\mathrm{e}^{-s\tau}\tilde{W}_d(s)\mathcal{D}_x(T, s, +0) \tag{17.56}$$

where

$$\tilde{W}_d(s) = \frac{W_d(s)}{1 + W_d(s)\mathcal{D}_{KLNM}(T, s, -\tau)}$$

and, by virtue of (4.49),

$$\mathcal{D}_x(T, s, t) = \frac{1}{1 - \mathrm{e}^{-sT}} \,.$$

In **Case 2** the image of the output of the continuous system $Y_{r2}(s)$ has the form

$$Y_{r2}(s) = W_{yq}(s) \, \frac{1}{s} \tag{17.57}$$

where

$$W_{yq}(s) = \frac{L(s)}{1 + W(s)K(s)L(s)N(s)} \tag{17.58}$$

while the image of the output of the SD system is equal to

$$Y_2(s) = L(s)Q(s) - L(s)K(s)M(s)\mathrm{e}^{-s\tau}\tilde{W}_d(s)\mathcal{D}_{LNQ}(T, s, 0)\,. \tag{17.59}$$

Substituting (17.55) and (17.56) into (17.53), after transformation we obtain an integral of the form (17.35), where

$$A(\zeta) = \left[\mathcal{D}_{K\underline{K}L\underline{L}M\underline{M}}(T,s,0)\mathcal{D}_x(T,s,+0)\mathcal{D}_x(T,-s,+0)\right]\big|_{e^{-sT}=\zeta}$$

$$B(\zeta) = \left[\mathcal{D}_{KL\underline{Y}_{r1}M}(T,s,-\tau)\mathcal{D}_x(T,s,+0)\right]\big|_{e^{-sT}=\zeta}\,. \tag{17.60}$$

Similarly, in **Case 2**, using (17.57) and (17.59), we obtain the integral (17.35), where

$$A(\zeta) = \left[\mathcal{D}_{K\underline{K}L\underline{L}M\underline{M}}(T,s,0)\mathcal{D}_{LNQ}(T,s,0)\mathcal{D}_{LNQ}(T,-s,0)\right]\big|_{e^{-sT}=\zeta}$$

$$B(\zeta) = \left[\mathcal{D}_{KL\underline{Y}_{r2}M}(T,s,-\tau)\mathcal{D}_{LNQ}(T,s,0)\right]\big|_{e^{-sT}=\zeta}\,. \tag{17.61}$$

To solve the optimization problem, we construct the set of admissible controllers. Let the real rational function

$$F(s) \triangleq K(s)L(s)N(s)$$

be irreducible and denote its poles by $s_1,\ldots,s_\ell$. Assume that the function $F(s)$ satisfies the non-pathological conditions, which, taking account of (17.51), have the form

$$e^{s_m T} \neq e^{s_k T}, \quad (k \neq m, \quad k,m = 1,\ldots,\ell)\,. \tag{17.62}$$

If (17.62) holds, using (9.197) we have

$$\mathcal{D}_{KLNM}(T,s,-\tau)\big|_{e^{-sT}=\zeta} = \mathcal{D}_{FM}(T,s,-\tau)\big|_{e^{-sT}=\zeta} = \frac{\zeta^m\beta^o(\zeta,\delta T)}{\alpha^o(\zeta)}$$

where the ratio at the right-hand side is irreducible. Then, the set of all stabilizing controllers can be written in the form (14.103)

$$W_d(\zeta) = \frac{\tilde{b}^o(\zeta) + \phi(\zeta)\alpha^o(\zeta)}{\tilde{a}^o(\zeta) - \zeta^m\phi(\zeta)\beta^o(\zeta,\delta T)} \tag{17.63}$$

where $\{\tilde{a}^o(\zeta),\tilde{b}^o(\zeta)\}$ is an arbitrary basic pair, satisfying (14.100). Nevertheless, not all controllers (17.63) satisfy the conditions of the problem. Indeed, for the integral (17.52) to converge it is necessary that the following condition hold

$$\lim_{t\to\infty}[y(t) - y_r(t)] = 0$$

which is equivalent to the estimate (17.8), or, in the space of images,

$$\lim_{s\to 0} s[Y(s) - Y_r(s)] = 0\,. \tag{17.64}$$

Condition (17.64) means that the static gains of both systems are equal. Thus, in the given case the minimization of the functional (17.52) should be performed over a subset of the set of stabilizing controllers (17.63), satisfying the

additional condition (17.64). Denote by $\mathcal{W}_d$ the set of all stabilizing controllers, and by $\mathcal{W}_d^0$ its subset satisfying the condition (17.64).
To formulate the theorem defining the subset $\mathcal{W}_d^0$, we introduce an additional notation. Any rational function $A(s)$ will be written as

$$A(s) = \frac{m_A(s)}{d_A(s)} \tag{17.65}$$

where $m_A(s)$ and $d_A(s)$ are co-prime polynomials.

Theorem 17.3 (Rosenwasser et al. (1999a)) *Consider Case 1. Then, if*

$$d_N(0)d_K(0)d_L(0)m_K(0)m_L(0) = 0 \tag{17.66}$$

the sets $\mathcal{W}_d$ and $\mathcal{W}_d^0$ coincide. If the product (17.66) is not zero, to satisfy (17.64) it is necessary and sufficient that the parameter $\phi(\zeta)$ in (17.63) satisfy the relation

$$\phi(\zeta) = k_\phi + (1-\zeta)\phi_1(\zeta) \tag{17.67}$$

where $\phi(\zeta)$ is an arbitrary stable real rational function, and

$$k_\phi = \frac{m_W(0)\tilde{a}^o(1) - d_W(0)\tilde{b}^o(1)}{m_W(0)\beta^o(1,\delta T) + d_W(0)\alpha^o(1)} \tag{17.68}$$

is a finite value.
Similarly for Case 2 the sets $\mathcal{W}_d$ and $\mathcal{W}_d^0$ coincide, if

$$d_N(0)d_K(0)d_L(0)m_K(0)m_L(0) = 0\,. \tag{17.69}$$

If the last expression is not zero, to satisfy (17.64) it is necessary and sufficient to provide (17.67)–(17.68).

Proof First, consider Case 1. Under the given assumptions there exists the finite limit

$$y_{r\infty} \triangleq \lim_{s\to 0} sY_{r1}(s) = \frac{W(0)K(0)L(0)}{1+W(0)K(0)L(0)N(0)}$$

or, taking account of the notation (17.65),

$$y_{r\infty} = \frac{d_N(0)m_W(0)m_K(0)m_L(0)}{d_W(0)d_K(0)d_L(0)d_N(0)+m_W(0)m_K(0)m_L(0)m_N(0)}\,.$$

On the other hand, there exists the limit

$$y_\infty \triangleq \lim_{s\to 0} sY_1(s) = \frac{W_d(0)K(0)L(0)}{1+W_d(0)K(0)L(0)N(0)}$$

which can also be obtained by means of (14.39). If we have

$$W_d(s) = \frac{b(s)}{a(s)} \tag{17.70}$$

with irreducible polynomials $b(s)$ and $a(s)$, we may write

$$y_\infty = \frac{d_N(0)b(0)m_K(0)m_L(0)}{a(0)d_K(0)d_L(0)d_N(0) + b(0)m_K(0)m_L(0)m_N(0)}\,.$$

Hence, the condition (17.64) reduces to the equation

$$\frac{d_N(0)m_W(0)m_K(0)m_L(0)}{d_W(0)d_K(0)d_L(0)d_N(0) + m_W(0)m_K(0)m_L(0)m_N(0)} =$$

$$= \frac{d_N(0)b(0)m_K(0)m_L(0)}{a(0)d_K(0)d_L(0)d_N(0) + b(0)m_K(0)m_L(0)m_N(0)}\,. \tag{17.71}$$

If

$$\gamma \triangleq d_N(0)m_K(0)m_L(0) = 0$$

then (17.71) is satisfied for all $W(s)$ and $W_d(s)$. In this case $y_\infty = y_{r\infty} = 0$ and the sets $\mathcal{W}_d$ and $\mathcal{W}_d^0$ are equal. Then, let

$$\gamma = d_N(0)m_K(0)m_L(0) \neq 0\,.$$

In this case Eq. (17.71) gives

$$\frac{m_W(0)}{d_W(0)d_K(0)d_L(0)d_N(0) + m_W(0)m_K(0)m_L(0)m_N(0)} =$$

$$= \frac{b(0)}{a(0)d_K(0)d_L(0)d_N(0) + b(0)m_K(0)m_L(0)m_N(0)}\,.$$

The denominators of both sides of the last equality are not zero. Therefore,

$$d_K(0)d_L(0)d_N(0)[m_W(0)a(0) - d_W(0)b(0)] = 0\,. \tag{17.72}$$

If

$$\rho \triangleq d_K(0)d_L(0) = 0$$

the equality (17.72) is satisfied for all $W(s)$ and $W_d(s)$, and

$$y_\infty = y_{r\infty} = \frac{d_N(0)}{m_N(0)}\,.$$

In the case of $\rho \neq 0$ from (17.72) it follows that

$$m_W(0)a(0) - d_W(0)b(0) = 0\,. \tag{17.73}$$

Consider all the arising possibilities. If

$$m_W(0) \neq 0\,, \quad d_W(0) \neq 0$$

then

$$a(0) \neq 0\,, \quad b(0) \neq 0$$

because the ratio (17.70) is irreducible. Hence,

$$W_d(0) = W(0) \triangleq k_d\,. \tag{17.74}$$

If

$$m_W(0) = 0\,, \quad d_W(0) \neq 0$$

from (17.73) we find

$$b(0) = 0\,, \quad a(0) \neq 0\,. \tag{17.75}$$

Lastly, if

$$m_W(0) \neq 0\,, \quad d_W(0) = 0$$

Eq. (17.73) gives

$$b(0) \neq 0\,, \quad a(0) = 0\,. \tag{17.76}$$

Using (17.73), we obtain

$$W_d(0) = W_d(\zeta)\,|_{\zeta=1} = \frac{\tilde{b}^o(1) + \phi(1)\alpha^o(1)}{\tilde{a}^o(1) - \phi(1)\beta^o(1,\delta T)}$$

where the numerator and denominator are not simultaneously zero. In the case of (17.74)

$$\frac{\tilde{b}^o(1) + \phi(1)\alpha^o(1)}{\tilde{a}^o(1) - \phi(1)\beta^o(1,\delta T)} = \frac{m_W(0)}{d_W(0)} = k_d \tag{17.77}$$

whence follows

$$\phi(1) = \frac{k_d\tilde{a}^o(1) - \tilde{b}(1)}{k_d\beta^o(1,\delta T) + \alpha^o(1)} = k_\phi\,.$$

Similarly to the above analysis of Eq. (14.124), it can be shown that the value k_ϕ is finite, because the denominator of the last equation is not zero. In the case of (17.75), from (17.77) it follows that

$$\tilde{b}^o(1) + \phi(1)\alpha^o(1) = 0\,.$$

We note that in the given case $\alpha^o(1) \neq 0$, because otherwise the characteristic equation of the system would have a root $\zeta = 1$, which contradicts the stability assumption. Then,

$$\phi(1) = -\frac{\tilde{b}^o(1)}{\alpha^o(1)} = k_\phi$$

where k_ϕ is finite
In the case of (17.76) we obtain the equation

$$\tilde{a}^o(1) - \phi(1)\beta^o(1,\delta T) = 0\,.$$

In a similar way it can be shown that $\beta^o(1,\delta T) \neq 0$, therefore

$$\phi(1) = \frac{\tilde{a}^o(1)}{\beta^o(1, \delta T)} = k_\phi\,.$$

It can be immediately checked that all these cases are combined in (17.68). Thus, the claim for Case 1 is proved. Analogously, the claim for Case 2 can be substantiated. ■

If (17.67) is valid, we obtain from (17.41)

$$\tilde{W}_d(s) = \left(1 - e^{-sT}\right)\alpha^2(s)\phi_1(s) + \alpha^2(s)k_\phi + \alpha(s)\tilde{b}(s)\,. \tag{17.78}$$

Using the parametrization (17.78) and changing the variable for $\zeta = e^{-sT}$, we obtain a functional of the form (17.42) in the stable parameter $\phi_1(\zeta)$. Then, if the optimal value $\phi_{10}(\zeta)$ is found, the optimal function $\phi_0(\zeta)$ is given as

$$\phi_0(\zeta) = k_\phi + (1 - \zeta)\phi_{10}(\zeta)$$

and the optimal controller is computed by (17.40).

Example 17.3 Let us demonstrate the application of the described optimization method in Case 1 assuming that we have in the continuous-time reference system

$$L(s) = \frac{1}{s-1}\,, \quad K(s) = N(s) = 1\,, \quad W(s) = \frac{s+2}{s+1}\,, \quad Q(s) = 0.$$

It is required to redesign this system on the basis of a computer with sampling period $T = 0.5$ and delay $\tau = 0.2$, taking the integral (17.52) as a cost function. In this case the transfer function (17.54) appears as

$$W_{yx}(s) = \frac{s+2}{s^2+s+1}\,.$$

Find the real rational functions appearing in the functional (17.35). Using the formulas of Sections 4.7 and 4.9, we have

$$\mathcal{D}_{LM}(T, s, -\tau)\,\big|_{e^{-sT}=\zeta} = \frac{-\zeta(-0.181\zeta + 0.212)}{\zeta - 0.607}$$

$$\mathcal{D}_x(T, s, 0)\,\big|_{e^{-sT}=\zeta} = \frac{1}{1-\zeta}$$

$$\mathcal{D}_{L\underline{L}M\underline{M}}(T, s, 0)\,\big|_{e^{-sT}=\zeta} = \frac{0.00919(\zeta + 3.786)(\zeta^{-1} + 3.786)}{(\zeta - 1.649)(\zeta^{-1} - 1.649)}$$

$$\mathcal{D}_{L\underline{Y_{r1}}M}(T, s, -\tau)\,\big|_{e^{-sT}=\zeta} =$$

$$= \frac{10^{-2}(-0.133\zeta^4 - 4.948\zeta^3 - 4.579\zeta^2 + 1.915\zeta + 0.164)}{(\zeta - 1)(\zeta - 0.607)(\zeta^2 - 1.414\zeta + 0.607)}\,.$$

Minimizing the functional (17.35) with these data by the polynomial equation technique given in Appendix B, we find the optimal controller

$$W_{d0}(\zeta) = \frac{2.893 - 5.230\zeta + 3.599\zeta^2 - 0.974\zeta^3}{1 - 1.120\zeta + 0.0467\zeta^2 + 0.398\zeta^3 - 0.177\zeta^4}.$$

We note that the replacement of the continuous controller by a discrete one by means of the Tustin transform

$$s = \frac{2}{T}\frac{1-\zeta}{1+\zeta}$$

leads to the discrete-time controller

$$W_T(\zeta) = \frac{1.2 - 0.4\zeta}{1 - 0.6\zeta} \tag{17.79}$$

or, equivalently,

$$W_T(s) = \frac{1.2 - 0.4e^{-sT}}{1 - 0.6e^{-sT}}.$$

The transients of the initial continuous-time and the redesigned system are shown in Fig. 17.6. As follows from the figures the proposed method ensures stability of the feedback system and close approximation of the reference transient response, while the SD system with the controller (17.79) is unstable. □

Example 17.4 Let us demonstrate the declared methods for Case 2. Consider for instance the continuous-time system with

$$L(s) = \frac{1}{s(s-1)}, \quad K(s) = N(s) = 1, \quad W(s) = \frac{5s+1}{s+3}$$

which is to be replaced by an equivalent digital system with a zero-order hold, sampling period $T = 0.5$ and pure delay $\tau = 0.2$. Using Eq. (17.61) and similar calculations gives the optimal controller

$$W_{d0}(\zeta) = \frac{14.321 - 31.566\zeta + 34.318\zeta^2 - 23.697\zeta^3 + 8.560\zeta^4 - 1.293\zeta^5}{1 + 1.231\zeta - 1.083\zeta^2 + 1.373\zeta^3 - 0.688\zeta^4 + 0.070\zeta^5 + 0.024\zeta^6}. \tag{17.80}$$

The Tustin transform of the continuous controller has the form

$$W_T(\zeta) = \frac{3 - 2.714\zeta}{1 - 0.143\zeta}. \tag{17.81}$$

The transients of the digital system with the controllers (17.80) and (17.81) as compared with that of the reference system are shown in Fig. 17.7. As follows from Fig. 17.7, the feedback system with the Tustin-redesigned controller (17.81) is unstable, while the optimal controller (17.80) ensures stability and practically exact matching of the transients of the initial and the redesigned system. □

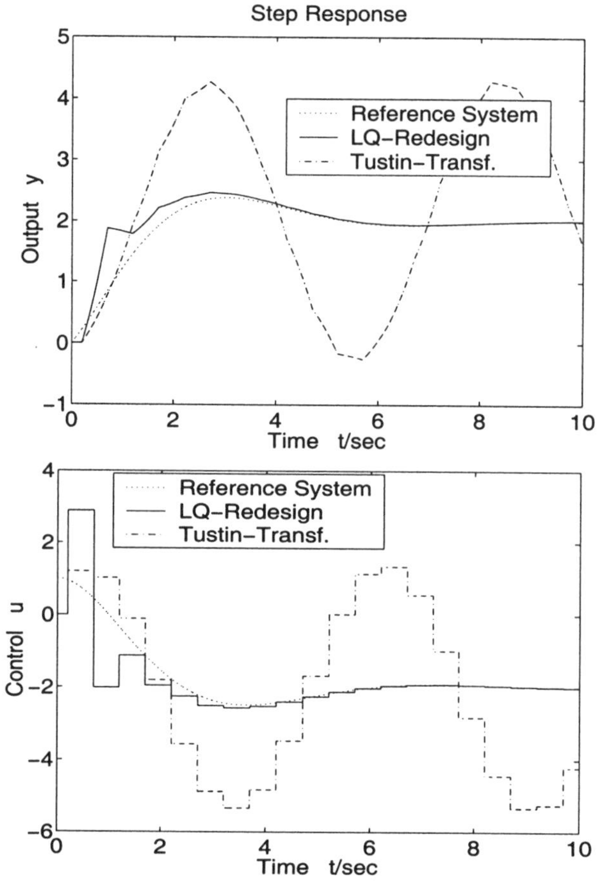

Figure 17.6: Transients for Example 17.3

17.6 $\mathcal{H}_2$–optimization of feedback servosystems with time-delay

Consider the feedback SD servosystem shown in Fig.17.8, Rosenwasser et al. (1997), where all the exogenous signals, viz.: reference signal $r(t)$, measurement noise $\theta(t)$ and disturbance $q(t)$ are centered stationary stochastic processes. The continuous part of the system is formed by the plant $L(s)$, actuator $K(s)$, prefilter $L_0(s)$, dynamic feedback $N(s)$, and pure delay elements. Henceforth, it is assumed that $L_0(s)$, $K(s)$, and $N(s)$ are proper functions, and $L(s)$ is strictly proper. The designed system has to follow the reference signal $r(t)$ with additional restriction on the control action $u(t)$. The quality of the system can be assessed by the properties of the error

$$\varepsilon(t) = r(t) - y(t)\,.$$

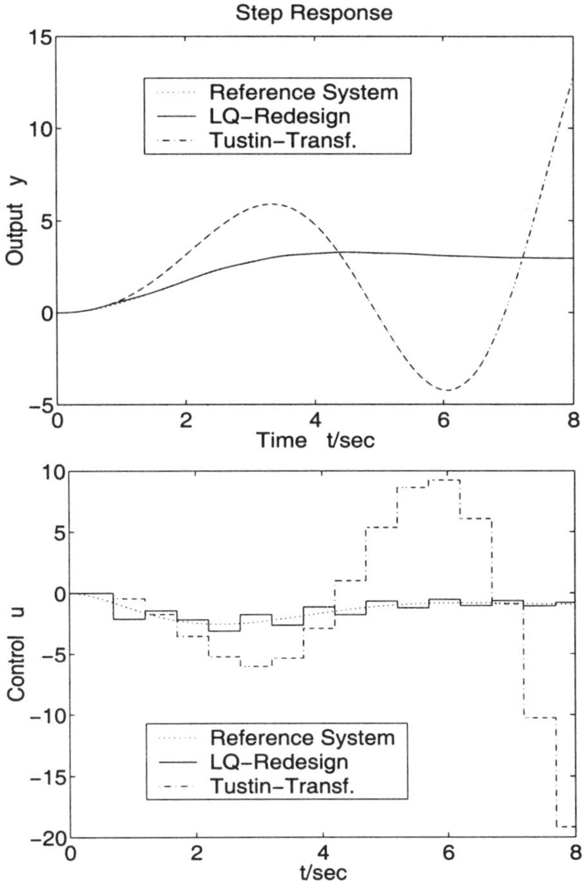

Figure 17.7: Transients for Example 17.4

If the closed-loop system is stable, the steady-state outputs $\varepsilon(t)$ and $u(t)$ are periodically non-stationary stochastic processes, so that the corresponding variances are periodic

$$d_\varepsilon(t) = d_\varepsilon(t+T)\,, \quad d_u(t) = d_u(t+T)$$

where T is the sampling period. The weighted sum of these variances, which characterizes the performance of the system, has the form

$$d(t) \triangleq d_\varepsilon(t) + \lambda^2 d_u(t) \tag{17.82}$$

where λ^2 is a non-negative constant and $d(t) = d(t+T)$. The function (17.82) can be, in principle, minimized for any t, for instance for $t = 0$. Nevertheless, in this case we ignore the intersample values of $d(t)$. Therefore, it is reasonable to employ the following averaged value as a cost function

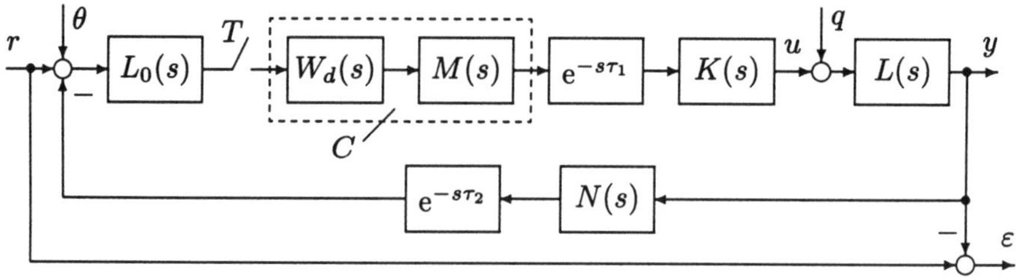

Figure 17.8: Stochastic feedback servosystem

$$\overline{d} \triangleq \overline{d}_\varepsilon + \lambda^2 \overline{d}_u \tag{17.83}$$

where

$$\overline{d}_\varepsilon = \frac{1}{T}\int_0^T d_\varepsilon(t)\,\mathrm{d}t\,, \quad \overline{d}_u = \frac{1}{T}\int_0^T d_u(t)\,\mathrm{d}t$$

are the corresponding mean variances. Taking the aforesaid into account, the following optimization problem can be formulated.

$\mathcal{H}_2$–servoproblem ($-\infty < t < \infty$). Given the sampling period T, the transfer functions of the continuous-time elements of the system and that of the hold, and the spectral densities of the input signals, find the transfer function of the digital controller, which stabilizes the system and minimizes the cost function (17.83).

For comparison, we consider the problem of minimizing the value $d(t)$ for $t = 0$, i.e., $d(0)$. Using the above terminology, the controller minimizing $\overline{d}$ will be called the CMV-regulator, while its counterpart minimizing $d(0)$ under the same conditions will be referred to as the MV-regulator.

Next, using the reasoning of Section 17.2, we construct an integral representation of the cost function (17.83). Calculating the transfer functions of the system at hand from the inputs $x_1(t) = r(t)$, $x_2(t) = \theta(t)$, and $x_3(t) = q(t)$ to the outputs $y_1(t) = \varepsilon(t)$ and $y_2(t) = u(t)$ by the formulae of Sections 11.2–11.4, we obtain

$$\begin{aligned}
W_{11}(s,t) &= 1 - \tilde{W}_d(s)L_0(s)\varphi_{LK_1M}(T,s,t)\\
W_{12}(s,t) &= -\tilde{W}_d(s)L_0(s)\varphi_{LK_1M}(T,s,t)\\
W_{13}(s,t) &= \tilde{W}_d(s)L_0(s)L(s)N_1(s)\varphi_{LK_1M}(T,s,t) - L(s)\\
W_{21}(s,t) &= W_{22}(s,t) = \tilde{W}_d(s)L_0(s)\varphi_{K_1M}(T,s,t)\\
W_{23}(s,t) &= -\tilde{W}_d(s)L_0(s)L(s)N_1(s)\varphi_{K_1M}(T,s,t)
\end{aligned} \tag{17.84}$$

with the notation

$$\begin{aligned}
K_1(s) &= K(s)e^{-s\tau_1}\,, \quad N_1(s) = N(s)e^{-s\tau_2}\\
\tilde{W}_d(s) &= \frac{W_d(s)}{1 + W_d(s)\mathcal{D}_{L_0LK_1N_1M}(T,s,0)}
\end{aligned}$$

where $W_{ik}(s,t)$ denotes the PTF from the k–th input to the i–th output. The PTFs (17.84) taken in the aggregate form the PTM

$$\boldsymbol{W}(s,t) = \begin{bmatrix} W_{11}(s,t) & W_{12}(s,t) & W_{13}(s,t) \\ W_{21}(s,t) & W_{22}(s,t) & W_{23}(s,t) \end{bmatrix}.$$

Henceforth, all the input signals are assumed to be statistically independent, and the spectral density matrix of the input vector $\boldsymbol{x} = [\, x_1 \; x_2 \; x_3 \,]'$ should possess the form

$$\boldsymbol{R}_{\boldsymbol{x}\boldsymbol{x}}(s) = \operatorname{diag}\left\{R_r(s) \;\; R_\theta(s) \;\; R_q(s)\right\}.$$

Then, the diagonal elements of the matrix

$$\boldsymbol{K}_{\boldsymbol{y}\boldsymbol{y}}(t) = \frac{1}{2\pi \mathrm{j}} \int_{-\mathrm{j}\infty}^{\mathrm{j}\infty} \boldsymbol{W}(-s,t)\boldsymbol{R}_{\boldsymbol{x}\boldsymbol{x}}(s)\boldsymbol{W}'(s,t)\,\mathrm{d}s$$

give the variances of the outputs $\varepsilon(t)$ and $u(t)$

$$\begin{aligned} d_\varepsilon(t) &= \frac{1}{2\pi \mathrm{j}} \int_{-\mathrm{j}\infty}^{\mathrm{j}\infty} \big[W_{11}(s,t)W_{11}(-s,t)R_r(s) + W_{12}(s,t)W_{12}(-s,t)R_\theta(s) + \\ &\qquad + W_{13}(s,t)W_{13}(-s,t)R_q(s)\big]\,\mathrm{d}s \\ d_u(t) &= \frac{1}{2\pi \mathrm{j}} \int_{-\mathrm{j}\infty}^{\mathrm{j}\infty} \big[W_{21}(s,t)W_{21}(-s,t)\big(R_r(s) + R_\theta(s)\big) + \\ &\qquad + W_{23}(s,t)W_{23}(-s,t)R_q(s)\big]\,\mathrm{d}s\,. \end{aligned} \tag{17.85}$$

Substituting (17.85) into (17.82) and passing to finite integration limits, we can represent the function $d(t)$ in the form

$$d(t) = d_0 + d_1(t)$$

where

$$\begin{aligned} d_0 &= \frac{1}{2\pi \mathrm{j}} \int_{-\mathrm{j}\infty}^{\mathrm{j}\infty} \big[R_r(s) + L(s)L(-s)R_q(s)\big]\,\mathrm{d}s \\ d_1(t) &= \frac{T}{2\pi \mathrm{j}} \int_{-\mathrm{j}\omega/2}^{\mathrm{j}\omega/2} \big[A(s,t)\tilde{W}_d(s)\tilde{W}_d(-s) - B(s,t)\tilde{W}_d(s) - B(-s,t)\tilde{W}_d(-s)\big]\,\mathrm{d}s \end{aligned} \tag{17.86}$$

with the notation

$$\begin{aligned} A(s,t) &= \big[\mathcal{D}_{LK_1M}(T,s,t)\mathcal{D}_{LK_1M}(T,-s,t) + \\ &\qquad + \lambda^2\mathcal{D}_{K_1M}(T,s,t)\mathcal{D}_{K_1M}(T,-s,t)\big]\mathcal{D}_U(T,s,0) \\ B(s,t) &= \mathcal{D}_{\underline{V}}(T,-s,t)\mathcal{D}_{LK_1M}(T,s,t) \\ U(s) &= L_0(s)L_0(-s)\big[R_r(s) + R_\theta(s) + L(s)L(-s)N(s)N(-s)R_q(s)\big] \\ V(s) &= L_0(s)\big[R_r(s) + L(s)L(-s)N_1(s)R_q(s)\big]\,. \end{aligned} \tag{17.87}$$

Substituting (17.87) into (17.86) and averaging over time, we find

$$\begin{aligned} \bar{d}_1 &\triangleq \frac{1}{T}\int_0^T d_1(t)\,\mathrm{d}t \\ &= \frac{T}{2\pi \mathrm{j}}\int_{-\mathrm{j}\omega/2}^{\mathrm{j}\omega/2} \left[\tilde{A}(s)\tilde{W}_d(s)\tilde{W}_d(-s) - \tilde{B}(s)\tilde{W}_d(s) - \tilde{B}(-s)\tilde{W}_d(-s)\right] \mathrm{d}s \end{aligned} \tag{17.88}$$

where

$$\begin{aligned} \tilde{A}(s) &= \frac{1}{T}\left[\mathcal{D}_{L\underline{L}K\underline{K}M\underline{M}}(T,s,0) + \lambda^2 \mathcal{D}_{K\underline{K}M\underline{M}}(T,s,0)\right]\mathcal{D}_U(T,s,0) \\ \tilde{B}(s) &= \frac{1}{T}\mathcal{D}_{LK_1VM}(T,s,0)\,. \end{aligned}$$

Using the change of variables $\zeta = \mathrm{e}^{-sT}$ in (17.88), we obtain an integral of the form (17.35), where

$$A(\zeta) = \tilde{A}(s)\,\big|_{\mathrm{e}^{sT}=\zeta}\,, \quad B(\zeta) = \tilde{B}(s)\,\big|_{\mathrm{e}^{sT}=\zeta}\,.$$

The minimization of the corresponding functional (17.35) is performed by means of the polynomial equations method developed in Appendix B using the notation

$$\mathcal{D}_{L_0LK_1N_1M}(T,s,0)\,\big|_{\mathrm{e}^{sT}=\zeta} = \frac{\zeta^m \beta^o(\zeta,\delta T)}{\alpha^o(\zeta)} \tag{17.89}$$

where the numbers m and δ are determined by (11.59).

Example 17.5 Consider the $\mathcal{H}_2$–optimal design problem for the system shown in Fig. 17.9, which is a special case of the system shown in Fig. 17.8

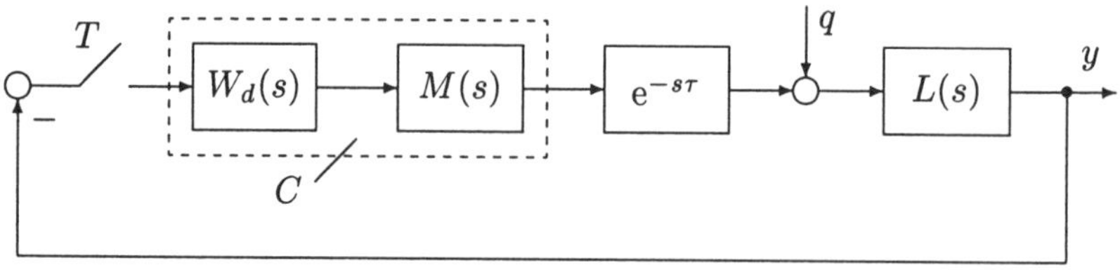

Figure 17.9: Control system with time delay

with

$$L_0(s) = K(s) = N(s) = 1\,, \quad r(t) = \theta(t) = 0\,, \quad \tau_2 = 0\,.$$

Let $T = 0.2$, $\tau = 0.093$, and

$$L(s) = \frac{s+0.5}{s^2-s}\,, \quad M(s) = \frac{1-\mathrm{e}^{-sT}}{s}\,, \quad R_q(s) = \frac{4}{4-s^2}\,.$$

In this case the plant transfer function has an unstable and a neutral pole, and the zero-order hold is used. The aim of the control is to minimize the mean output variance $\overline{d}_y$ without restrictions imposed on the magnitude of the control signal, i.e., for $\lambda^2 = 0$ in the cost function (17.83). For these conditions, the calculations give

$$\begin{aligned}
\zeta^m \beta^o(\zeta, \delta T) &= -0.0867\zeta^3 + 0.00994\zeta^2 + 0.0949\zeta \\
\alpha^o(\zeta) &= \zeta^2 - 1.819\zeta + 0.819 \\
\frac{1}{T}\mathcal{D}_{\underline{L}L M\underline{M}}(T, s, 0)\,\big|_{e^{sT}=\zeta} &= \\
&= \frac{-0.00667(\zeta + 0.267)(\zeta - 0.905)(\zeta + 3.739)(\zeta - 1.105)}{(\zeta - 1)^2(\zeta - 0.819)(\zeta - 1.221)} \\
\mathcal{D}_U(T, s, 0)\,\big|_{e^{sT}=\zeta} &= \frac{-0.0538\zeta(\zeta + 0.265)(\zeta - 0.905)(\zeta + 3.773)(\zeta - 1.105)}{(\zeta - 1)^2(\zeta - 0.819)(\zeta - 0.670)(\zeta - 1.221)(\zeta - 1.492)}\,.
\end{aligned}$$

Then,

$$\mathcal{D}_{FK_1VM}(T, s, 0)\,\big|_{e^{sT}=\zeta} = \frac{\beta_1(\zeta)}{\alpha_1(\zeta)}$$

where

$$\begin{aligned}
\beta_1(\zeta) &= 10^{-3}\Big(0.0240\zeta^8 + 0.473\zeta^7 + 1.945\zeta^6 - 5.090\zeta^5 - 0.171\zeta^4 + \\
&\qquad + 4.957\zeta^3 - 1.723\zeta^2 - 0.400\zeta - 0.00138\Big) \\
\alpha_1(\zeta) &= (\zeta - 1)^3(\zeta - 1.221)^2(\zeta - 0.819)(\zeta - 1.492)(\zeta - 0.670)\,.
\end{aligned}$$

Performing the factorization (16.27), we obtain

$$K(\zeta) = \frac{0.00144(\zeta - 1.105)^2(\zeta + 3.739)(\zeta + 3.773)}{\zeta - 1.492}\,.$$

Using Theorem 17.2, it can be shown that the poles $\zeta = 1$ and $\zeta = 0.819$ do not appear in the separation. Using this fact, the transfer function of the CMV-regulator takes the form

$$W_{CMV}(\zeta) = \frac{14.198 - 19.551\zeta + 6.008\zeta^2}{1 - 0.701\zeta - 0.760\zeta^2 + 0.521\zeta^3} \tag{17.90}$$

which provides the average output variance $\overline{d}_\varepsilon = \overline{d}_y = 0.0348$ and the variance at the sampling instants $d_\varepsilon(0) = d_y(0) = 0.0256$. The corresponding MV-regulator designed for the same conditions has the form

$$W_{MV}(\zeta) = \frac{19.294 - 25.304\zeta + 7.038\zeta^2}{1 - 0.569\zeta - 0.978\zeta^2 + 0.610\zeta^3}\,. \tag{17.91}$$

The system with the MV-regulator gives $\overline{d}_y = 0.8551$ and $d_y(0) = 0.0181$. Thus, a decrease in the variance at the sampling instants by 27 percent leads to an increase in the average output variance by 24.6 times. Figures 17.10 and 17.11 demonstrate the transients of the systems with the controllers (17.90)

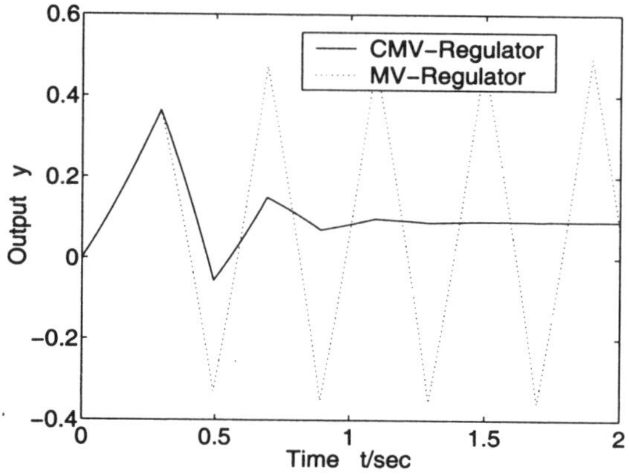

Figure 17.10: Comparison of the outputs for Example 17.5

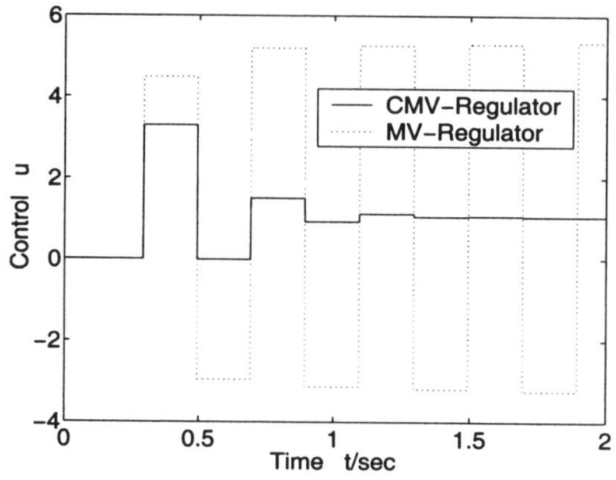

Figure 17.11: Comparison of the control for Example 17.5

and (17.91), respectively, under the unit step acting at the input $q(t)$. As follows from Fig. 17.10 and 17.11, with the MV-regulator the transient is essentially oscillatory and dies down very slowly, while the transient with the CMV-regulator damps much faster and has a large stability margin.

Consider the influence of the pure delay $0 < \tau_1 < T$ on the performance of the optimal systems in the given case. The curves of the average variance $\overline{d}_y$ versus the value of pure delay τ_1 is shown in Fig. 17.12 for the systems with CMV- and MV-regulators. For some τ_1 the system with the MV-regulator exhibits unacceptably large mean variance and is, in practice, inoperable. This is caused by the zero of the function (17.89) near the oscillatory stability boundary ($\zeta = -1$), the effect of which is similar to that considered in Example 16.6. □

An example of the design of the optimal digital controller maintaining a ship's course under irregular waves is given in Rosenwasser et al. (1996).

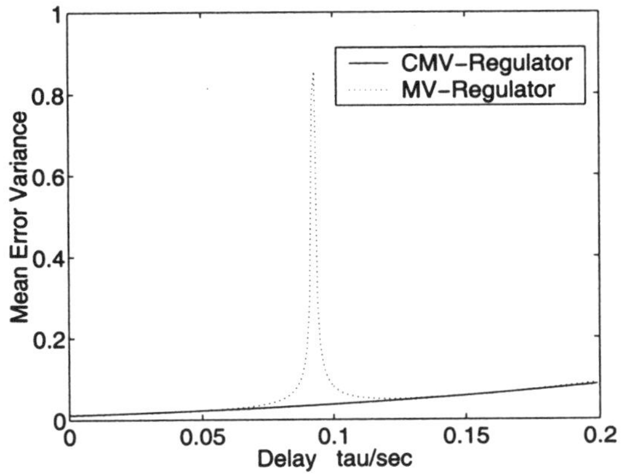

Figure 17.12: Influence of pure delay on performance of optimal systems

17.7 $\mathcal{H}_2$–optimization of multivariable systems

The above optimization technique can be extended to multivariable systems. In this section, following Lampe and Rosenwasser (1996), an example of the statement of an $\mathcal{H}_2$–optimization problem is given for a wide class of SD systems, Anderson (1993). The main results are given without proofs. The block-diagram of the system under consideration is shown in Fig. 17.13, where

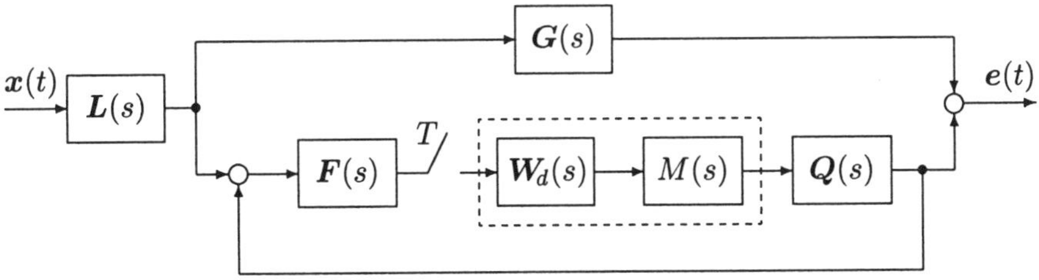

Figure 17.13: Hybrid multivariable system with digital controller

$\boldsymbol{L}(s)$, $\boldsymbol{G}(s)$, $\boldsymbol{F}(s)$, and $\boldsymbol{Q}(s)$ are real rational matrices of the corresponding dimensions. The dashed line in Fig. 17.13 denotes a computer described by the multivariable linear model given in Section 10.9 with the transfer matrix of the discrete controller $\boldsymbol{W}_d(s)$ given by (10.111) and the transfer function $M(s)$ of the scalar pulse-forming element. The system is a special case of the standard system shown in Fig. 11.10. The PTM $\boldsymbol{W}(s,t)$ of the system under consideration from the input $\boldsymbol{x}(t)$ to the output $\boldsymbol{e}(t)$ takes the form

$$\boldsymbol{W}(s,t) = \boldsymbol{G}(s)\boldsymbol{L}(s) + \varphi_{\boldsymbol{Q}M}(T,s,t)\tilde{\boldsymbol{W}}_d(s)\boldsymbol{F}(s)\boldsymbol{L}(s) \tag{17.92}$$

where

$$\tilde{\boldsymbol{W}}_d(s) \triangleq \boldsymbol{W}_d(s)\left[\boldsymbol{I} - \varphi_{\boldsymbol{FQ}M}(T,s,0)\boldsymbol{W}_d(s)\right]^{-1} . \tag{17.93}$$

Comparing the above relation with Eqs. (11.132) and (11.133), we find that the corresponding standard system has the transfer matrix $\boldsymbol{P}(s)$ of the form

$$\boldsymbol{P}(s)=\begin{bmatrix} \boldsymbol{G}(s)\boldsymbol{L}(s) & \boldsymbol{Q}(s) \\ \boldsymbol{F}(s)\boldsymbol{L}(s) & \boldsymbol{F}(s)\boldsymbol{Q}(s) \end{bmatrix}.$$

Henceforth, the systems $\boldsymbol{L}(s)$ and $\boldsymbol{G}(s)$ as well as the closed loop in Fig. 17.13 are assumed to be stable. Then, the $\mathcal{H}_2$−norm of the system with the input $\boldsymbol{x}(t)$ and the output $\boldsymbol{e}(t)$ is given by

$$\|\mathsf{U}\|_2^2=\overline{r_e}=\frac{T}{2\pi \mathrm{j}}\int_{-\mathrm{j}\omega/2}^{\mathrm{j}\omega/2}\operatorname{tr}\left[\mathcal{D}_{\boldsymbol{B}_0}(T,s,0)\right]\mathrm{d}s \tag{17.94}$$

where

$$\boldsymbol{B}_0(s)=\frac{1}{T}\int_0^T \boldsymbol{W}'(s,t)\boldsymbol{W}(-s,t)\,\mathrm{d}t\,.$$

In this case, using (17.92) and (15.101) we get

$$\begin{aligned}\operatorname{tr}\mathcal{D}_{\boldsymbol{B}_0}(T,s,0)&=\frac{1}{T}\operatorname{tr}\Big[\mathcal{D}_{\boldsymbol{Q}'\boldsymbol{Q}M\underline{M}}(T,s,0)\tilde{\boldsymbol{W}}_d(-s)\mathcal{D}_{\underline{\boldsymbol{F}}\underline{\boldsymbol{L}}\boldsymbol{L}'\boldsymbol{F}'}(T,s,0)\tilde{\boldsymbol{W}}'_d(s)+\\&+\mathcal{D}_{\boldsymbol{Q}'\underline{\boldsymbol{G}}\underline{\boldsymbol{L}}\boldsymbol{L}'\boldsymbol{F}'M}(T,s,0)\tilde{\boldsymbol{W}}'_d(s)+\tilde{\boldsymbol{W}}_d(-s)\mathcal{D}_{\underline{\boldsymbol{F}}\underline{\boldsymbol{L}}\boldsymbol{L}'\boldsymbol{G}'\underline{\boldsymbol{Q}}M}(T,s,0)\Big]+\\&+\operatorname{tr}\mathcal{D}_{\boldsymbol{L}'\boldsymbol{G}'\underline{\boldsymbol{G}}\underline{\boldsymbol{L}}}(T,s,0)\,.\end{aligned} \tag{17.95}$$

Substituting (17.95) into (17.94) and changing the variable for $\zeta=\mathrm{e}^{-sT}$, we obtain the problem of minimizing the integral

$$I_2\triangleq \frac{1}{2\pi\mathrm{j}}\oint\operatorname{tr}\left[\boldsymbol{A}(\zeta)\tilde{\boldsymbol{W}}_d(\zeta^{-1})\boldsymbol{C}(\zeta)\tilde{\boldsymbol{W}}'_d(\zeta)+\boldsymbol{B}(\zeta)\tilde{\boldsymbol{W}}'_d(\zeta)+\tilde{\boldsymbol{W}}_d(\zeta^{-1})\boldsymbol{B}'(\zeta^{-1})\right]\frac{\mathrm{d}\zeta}{\zeta} \tag{17.96}$$

where

$$\begin{aligned}\boldsymbol{A}(\zeta)&=\frac{1}{T}\mathcal{D}_{\boldsymbol{Q}'\boldsymbol{Q}M\underline{M}}(T,s,0)\,|_{\mathrm{e}^{-sT}=\zeta}\\ \boldsymbol{C}(\zeta)&=\mathcal{D}_{\underline{\boldsymbol{F}}\underline{\boldsymbol{L}}\boldsymbol{L}'\boldsymbol{F}'}(T,s,0)\,|_{\mathrm{e}^{-sT}=\zeta}\\ \boldsymbol{B}(\zeta)&=\mathcal{D}_{\boldsymbol{Q}'\underline{\boldsymbol{G}}\underline{\boldsymbol{L}}\boldsymbol{L}'\boldsymbol{F}'M}(T,s,0)\,|_{\mathrm{e}^{-sT}=\zeta}\end{aligned}$$

and the term independent of the controller is dropped. Moreover, in (17.96)

$$\tilde{\boldsymbol{W}}_d(\zeta)\triangleq\left\{\boldsymbol{W}_d(s)\left[\boldsymbol{I}-\mathcal{D}_{\boldsymbol{F}\boldsymbol{Q}M}(T,s,0)\boldsymbol{W}_d(s)\right]^{-1}\right\}|_{\mathrm{e}^{-sT}=\zeta}$$

is introduced. The minimization of the integral (17.96) is to be performed over the set of controllers $\boldsymbol{W}_d(s)$ that stabilize the closed loop in Fig. 17.13.

For an effective solution of the problem a parametrization of the set of stabilizing controllers similar to (14.91) can be used. The procedure of constructing such a parametrization is greatly simplified if, following Lampe and Rosenwasser (1996), we make some assumptions of little importance for applications. Let $\mathbf{\Gamma}(\lambda)$ be a $n \times n$ polynomial matrix with the determinant

$$\det \mathbf{\Gamma}(\lambda) \stackrel{\triangle}{=} d_{\mathbf{\Gamma}}(\lambda) = K(\lambda - \lambda_1)^{\nu_1} \cdots (\lambda - \lambda_q)^{\nu_q}, \quad K = \text{const.}$$

The matrix $\mathbf{\Gamma}(\lambda)$ will be called *simple*, if

$$\operatorname{rank} \mathbf{\Gamma}(\lambda_i) = n - 1, \quad (i = 1, \ldots, q).$$

Any real rational matrix $\boldsymbol{P}(s)$ can be written as an irreducible ratio

$$\boldsymbol{P}(s) = \frac{\boldsymbol{M_P}(s)}{d_{\boldsymbol{P}}(s)}$$

where $\boldsymbol{M_P}(s)$ is a polynomial matrix and $d_{\boldsymbol{P}}(s)$ is a scalar polynomial. Then, as a special case,

$$\boldsymbol{F}(s) = \frac{\boldsymbol{M_F}(s)}{d_{\boldsymbol{F}}(s)}, \quad \boldsymbol{Q}(s) = \frac{\boldsymbol{M_Q}(s)}{d_{\boldsymbol{Q}}(s)}.$$

The product

$$\boldsymbol{R}(s) \stackrel{\triangle}{=} \boldsymbol{F}(s)\boldsymbol{Q}(s) = \frac{\boldsymbol{M_F}(s)\boldsymbol{M_Q}(s)}{d_{\boldsymbol{F}}(s)d_{\boldsymbol{Q}}(s)} = \frac{\boldsymbol{M_R}(s)}{d_{\boldsymbol{R}}(s)}$$

will be assumed to be an irreducible strictly proper real rational matrix in what follows. Introduce the left co-prime MFD (matrix fraction description), Kailath (1980)

$$\boldsymbol{F}(s) = \boldsymbol{a_F}^{-1}(s)\boldsymbol{b_F}(s), \quad \boldsymbol{Q}(s) = \boldsymbol{a_Q}^{-1}(s)\boldsymbol{b_Q}(s)$$

where

$$\begin{gathered} \boldsymbol{F}(s) = \boldsymbol{a_F}^{-1}(s)\boldsymbol{b_F}(s), \quad \boldsymbol{Q}(s) = \boldsymbol{a_Q}^{-1}(s)\boldsymbol{b_Q}(s) \\ \text{with} \quad \det \boldsymbol{a_F}(s) = d_{\boldsymbol{F}}(s), \quad \det \boldsymbol{a_Q}(s) = d_{\boldsymbol{Q}}(s). \end{gathered} \tag{17.97}$$

As was shown in Rosenwasser (1994b), under the condition (17.97) the matrices $\boldsymbol{a_F}(s)$ and $\boldsymbol{a_Q}(s)$ are simple. Let

$$d_{\boldsymbol{R}}(s) = (s - \tilde{s}_1)^{\mu_1} \cdots (s - \tilde{s}_\rho)^{\mu_\rho}, \quad \sum_{i=1}^{\rho} \mu_i = \sigma$$

assuming that the numbers $\tilde{s}_i$ $(i = 1, \ldots, \rho)$ satisfy the non-pathological conditions (9.161)–(9.162). Then, it can be proved, Rosenwasser (1994b) that, similarly to the scalar case,

$$\mathcal{D}_{\boldsymbol{FQ}M}(T, s, 0)\,\big|_{\mathrm{e}^{-sT} = \zeta} = \frac{\zeta \chi^o(\zeta)}{\alpha^o(\zeta)} \tag{17.98}$$

where $\chi^o(\zeta)$ is a polynomial matrix with $\deg \chi^o(\zeta) \leq \sigma - 1$ and $\alpha^o(\zeta)$ is a scalar polynomial

$$\alpha^o(\zeta) = \left(1 - e^{\tilde{s}_1 T}\zeta\right)^{\mu_1} \cdots \left(1 - e^{\tilde{s}_\rho T}\zeta\right)^{\mu_\rho} .$$

In this case the ratio at the right-hand side of (17.98) is irreducible, i.e.,

$$\chi^o(e^{-\tilde{s}_i T}) \neq \mathbf{0}$$

where the matrix at the right-hand side is the zero matrix of the corresponding type. Moreover, the following left co-prime MFD holds

$$\mathcal{D}_{FQM}(T, s, 0)\,|_{e^{-sT}=\zeta} = \zeta \boldsymbol{\alpha}^{-1}(\zeta)\boldsymbol{\chi}(\zeta) \tag{17.99}$$

where $\boldsymbol{\alpha}(\zeta)$ is a simple matrix such that

$$\det \boldsymbol{\alpha}(\zeta) = \alpha^o(\zeta) .$$

Let a co-prime left MFD for the discrete controller $\boldsymbol{W}_d(\zeta)$ have the form

$$\boldsymbol{W}_d(\zeta) \triangleq \boldsymbol{W}_d(s)\,|_{e^{-sT}=\zeta} = \boldsymbol{a}^{-1}(\zeta)\boldsymbol{b}(\zeta) \tag{17.100}$$

obtained from (10.111) with $e^{-sT} = \zeta$.

Theorem 17.4 *Under the above assumptions the closed loop shown in Fig. 17.13 is stable as defined in Section 14.5 if and only if the matrix*

$$\boldsymbol{V}(\zeta, \boldsymbol{a}, \boldsymbol{b}) \triangleq \begin{bmatrix} \boldsymbol{\alpha}(\zeta) & -\zeta\boldsymbol{\chi}(\zeta) \\ -\boldsymbol{b}(\zeta) & \boldsymbol{a}(\zeta) \end{bmatrix}$$

is free of eigenvalues inside or on the unit circle. ■

Theorem 17.5 *Let the assumptions of Theorem 17.4 hold. Then, there exist matrix polynomials $\boldsymbol{a}_*(\zeta)$ and $\boldsymbol{b}_*(\zeta)$ such that*

$$\det \boldsymbol{V}(\zeta, \boldsymbol{a}_*, \boldsymbol{b}_*) = 1 . \tag{17.101}$$

In this case the set of all stabilizing controllers is given by

$$\boldsymbol{W}_d(\zeta) = \left[\boldsymbol{a}_*(\zeta) - \zeta\boldsymbol{\phi}(\zeta)\boldsymbol{\chi}(\zeta)\right]^{-1}\left[\boldsymbol{b}_*(\zeta) - \boldsymbol{\phi}(\zeta)\boldsymbol{\alpha}(\zeta)\right] \tag{17.102}$$

where $\boldsymbol{\phi}(\zeta)$ is an arbitrary real rational matrix free of poles inside or on the unit circle. As a special case, $\boldsymbol{\phi}(\zeta)$ can be a polynomial matrix of the corresponding dimensions. If its left co-prime MFD has the form

$$\boldsymbol{\phi}(\zeta) = \boldsymbol{a_\phi}^{-1}(\zeta)\boldsymbol{b_\phi}(\zeta)$$

then the characteristic polynomial of the feedback system is equal to $\det \boldsymbol{a_\phi}(\zeta)$. ■

Theorem 17.6 *Let Eqs. (17.99) and (17.100) hold and $\boldsymbol{a}_*(\zeta)$ and $\boldsymbol{b}_*(\zeta)$ be a matrix pair satisfying the condition (17.101). Denote*

$$\boldsymbol{V}^{-1}(\zeta, \boldsymbol{a}_*, \boldsymbol{b}_*) = \begin{bmatrix} \boldsymbol{v}_{11}(\zeta) & \boldsymbol{v}_{12}(\zeta) \\ \boldsymbol{v}_{21}(\zeta) & \boldsymbol{v}_{22}(\zeta) \end{bmatrix}$$

where the matrix $\boldsymbol{v}_{22}(\zeta)$ is the same type as $\boldsymbol{a}(\zeta)$. Then, the matrices

$$\hat{\boldsymbol{W}}_d(\zeta) = \left[\boldsymbol{I} - \zeta\boldsymbol{\alpha}^{-1}(\zeta)\boldsymbol{\chi}(\zeta)\boldsymbol{W}_d(\zeta)\right]^{-1}$$

$$\tilde{\boldsymbol{W}}_d(\zeta) = \boldsymbol{W}_d(\zeta)\left[\boldsymbol{I} - \zeta\boldsymbol{\alpha}^{-1}(\zeta)\boldsymbol{\chi}(\zeta)\boldsymbol{W}_d(\zeta)\right]^{-1}$$

can be represented in the form

$$\hat{\boldsymbol{W}}_d(\zeta) = \left[\boldsymbol{v}_{11}(\zeta) - \boldsymbol{v}_{12}(\zeta)\boldsymbol{\phi}(\zeta)\right]\boldsymbol{a}(\zeta) \tag{17.103}$$

$$\tilde{\boldsymbol{W}}_d(\zeta) = \left[\boldsymbol{v}_{21}(\zeta) - \boldsymbol{v}_{22}(\zeta)\boldsymbol{\phi}(\zeta)\right]\boldsymbol{a}(\zeta)\,. \tag{17.104}$$

■

Substituting (17.104) into (17.96) and dropping the terms independent of the controller, we obtain a functional of the form

$$\tilde{I}_2 = \frac{1}{2\pi \mathrm{j}}\oint \mathrm{tr}\left[\boldsymbol{A}_1(\zeta)\boldsymbol{\phi}(\zeta^{-1})\boldsymbol{C}_1(\zeta)\boldsymbol{\phi}'(\zeta) + \boldsymbol{B}_1(\zeta)\boldsymbol{\phi}'(\zeta) + \boldsymbol{\phi}(\zeta^{-1})\boldsymbol{B}'_1(\zeta^{-1})\right]\frac{\mathrm{d}\zeta}{\zeta} \tag{17.105}$$

where $\boldsymbol{A}_1(\zeta)$, $\boldsymbol{B}_1(\zeta)$ and $\boldsymbol{C}_1(\zeta)$ are known real rational matrices such that $\boldsymbol{A}_1(\zeta^{-1}) = \boldsymbol{A}'_1(\zeta)$ and $\boldsymbol{C}_1(\zeta^{-1}) = \boldsymbol{C}'_1(\zeta)$. The functional (17.105) is to be minimized over the set of all stable real rational matrices $\boldsymbol{\phi}$. The methods of solution of such a problem are given in Kučera (1979); Grimble (1995); Grimble and Kučera (1996); Weinmann (1991). If the optimal matrix $\boldsymbol{\phi}_0$ is found, from (17.103) and (17.104) we can find the optimal functions

$$\begin{aligned} \hat{\boldsymbol{W}}_{d0}(\zeta) &= \left[\boldsymbol{v}_{11}(\zeta) - \boldsymbol{v}_{12}(\zeta)\boldsymbol{\phi}_0(\zeta)\right]\boldsymbol{a}(\zeta) \\ \tilde{\boldsymbol{W}}_{d0}(\zeta) &= \left[\boldsymbol{v}_{21}(\zeta) - \boldsymbol{v}_{22}(\zeta)\boldsymbol{\phi}_0(\zeta)\right]\boldsymbol{a}(\zeta) \end{aligned} \tag{17.106}$$

and the optimal controller takes the form

$$\boldsymbol{W}_{d0}(\zeta) = \tilde{\boldsymbol{W}}_{d0}\hat{\boldsymbol{W}}_{d0}^{-1}(\zeta)\,. \tag{17.107}$$

17.8 $\mathcal{L}_2$–optimization of SD systems

In this section we consider the multivariable sampled-data system shown in Fig. 17.14, where $\boldsymbol{x}(t)$ and $\boldsymbol{y}(t)$ are vectors of dimensions $\ell \times 1$ and $n \times 1$, respectively; $\boldsymbol{F}(s)$ and $\boldsymbol{G}(s)$, are strictly proper real rational matrices, and $\boldsymbol{N}(s)$ is a proper one. Moreover, $\boldsymbol{C}$ in Fig. 17.14 denotes a computer as in Fig. 17.13.
The PTM of the system from the input $\boldsymbol{x}(t)$ to the output $\boldsymbol{y}(t)$ has the form

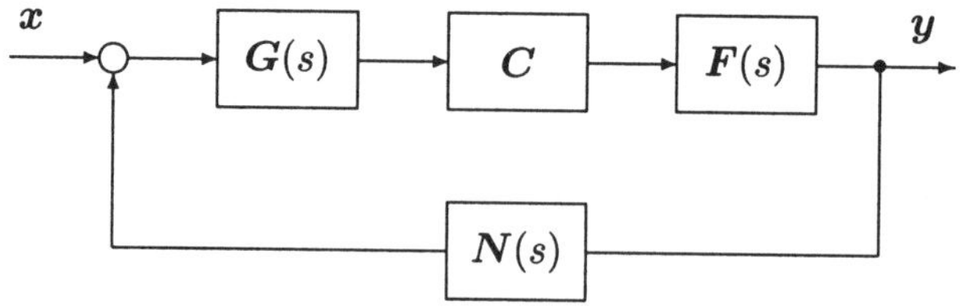

Figure 17.14: Multivariable system for Section 17.8

$$\boldsymbol{W}_{\boldsymbol{yx}}(s,t) = \varphi_{\boldsymbol{F}M}(T,s,t)\tilde{\boldsymbol{W}}_d(s)\boldsymbol{G}(s) \tag{17.108}$$

where

$$\tilde{\boldsymbol{W}}_d(s) = \boldsymbol{W}_d(s)\left[\boldsymbol{I} - \varphi_{\boldsymbol{GNF}M}(T,s,0)\boldsymbol{W}_d(s)\right]^{-1}. \tag{17.109}$$

Let the system in Fig. 17.14 be stable in the sense of Section 14.5. Then, under the given assumptions, the integral

$$\boldsymbol{H}_0(t,\tau) = \frac{1}{2\pi \mathrm{j}}\int_{-\mathrm{j}\infty}^{\mathrm{j}\infty} \boldsymbol{W}_{\boldsymbol{yx}}(s,t)\mathrm{e}^{s(t-\tau)}\,\mathrm{d}s$$

defines the Green function of the causal e-dichotomic operator

$$\boldsymbol{y} = \mathsf{U}_0[\boldsymbol{x}] = \int_{-\infty}^{t} \boldsymbol{H}_0(t,\tau)\boldsymbol{x}(\tau)\,\mathrm{d}\tau. \tag{17.110}$$

Before formulating the main theorem, let us introduce some general notions. Let $\boldsymbol{f}(t)$ be a complex vector defined for $-\infty < t < \infty$ and

$$\boldsymbol{f}^*(t) \triangleq \overline{\boldsymbol{f}}'(t)$$

be the conjugate hermitian vector. Denote by $\mathcal{L}_{2m}(-\infty,\infty)$ the space of $m \times 1$ – vectors $\boldsymbol{f}(t)$ with the norm

$$\|\boldsymbol{f}\|_{\mathcal{L}_2} \triangleq \sqrt{\int_{-\infty}^{\infty} \boldsymbol{f}^*(t)\boldsymbol{f}(t)\,\mathrm{d}t}\,.$$

In this section it is demonstrated that the linear operator (17.110) maps from the space $\mathcal{L}_{2\ell}(-\infty,\infty)$ to $\mathcal{L}_{2n}(-\infty,\infty)$ and is bounded. In this case we can calculate the corresponding norm of the operator U_0, which is equal to the $\mathcal{H}_\infty$–norm of a finite-dimensional matrix, for which closed expressions are given.
Consider some auxiliary relations. Denote

$$\boldsymbol{Q}_{\boldsymbol{G}}(s) \triangleq \mathcal{D}_{\boldsymbol{G}\underline{\boldsymbol{G}}'}(T,s,0) = \frac{1}{T}\sum_{k=-\infty}^{\infty} \boldsymbol{G}(s+k\mathrm{j}\omega)\boldsymbol{G}'(-s-k\mathrm{j}\omega)\,. \tag{17.111}$$

As follows from the relations of Section 4.7, the matrix $\boldsymbol{Q}_{\boldsymbol{G}}(s)$ is rational periodic, therefore the matrix

$$Q_G^o(\zeta) = Q_G(s)\,\big|_{e^{-sT}=\zeta} \tag{17.112}$$

is real rational. From (17.111) it immediately follows that

$$Q_G(s) = Q'_G(-s)\,.$$

Then, from (17.112) we have

$$Q_G^o(\zeta) = Q_G^o(\zeta^{-1})\,.$$

For $s = j\nu$, where ν is a real parameter, the matrix $Q_G(s)$ is non-negative, as follows directly from (17.111). Therefore, the matrix $Q_G^o(\zeta)$ is non-negative on the unit circle $|\zeta| = 1$. Hence, there exists the factorization

$$Q_G^o(\zeta) = B(\zeta)B'(\zeta^{-1}) \tag{17.113}$$

where $B(\zeta)$ is a real rational matrix, taking real values for real argument. Denote

$$R_{FM}(s) \triangleq \int_0^T \varphi_{F'M}(T,-s,t)\varphi_{FM}(T,s,t)\,\mathrm{d}t = \mathcal{D}_{\underline{F}'FM\underline{M}}(T,s,0)$$

and

$$R^o{}_{FM}(\zeta) = R_{FM}(s)\,\big|_{e^{-st}=\zeta} = \mathcal{D}^o{}_{\underline{F}F'M\underline{M}}(T,\zeta,0)\,.$$

As follows from the aforesaid, the matrix $R^o{}_{FM}(\zeta)$ is real rational, and

$$R^o{}_{FM}(\zeta) = R^{o\prime}{}_{FM}(\zeta^{-1})\,.$$

Moreover, the matrix $R^o{}_{FM}(\zeta)$ is non-negative for $|\zeta| = 1$. Therefore, similarly to (17.113), there exists the factorization

$$R^o{}_{FM}(\zeta) = A'(\zeta^{-1})A(\zeta) \tag{17.114}$$

where $A(\zeta)$ is a real rational matrix, taking real values for real argument.

Theorem 17.7 *Let the matrices $F(s)$ and $G(s)$ be strictly proper, the matrix $N(s)$ be proper, and the matrix $G(s)$ have no poles on the imaginary axis. Then, the operator $\mathsf{U}_0[x]$ maps from the space $\mathcal{L}_{2\ell}(-\infty,\infty)$ into $\mathcal{L}_{2n}(-\infty,\infty)$ and is bounded. The corresponding norm $||\mathsf{U}_0||_{\mathcal{L}_2}$ of the operator U_0 is given by*

$$||\mathsf{U}_0||_{\mathcal{L}_2} = \sup_{|\zeta|=1} \sigma_{\max}[\Gamma(\zeta)] \tag{17.115}$$

where $\Gamma(\zeta)$ is a real rational matrix defined as

$$\Gamma(\zeta) = A(\zeta)\tilde{W}_d^o(\zeta)B(\zeta)$$

and

$$\tilde{W}_d^o(\zeta) = \tilde{W}_d(s)\,\big|_{e^{-sT}=\zeta}\,.$$

At this, $\sigma_{\max}$ in (17.115) denotes the maximal singular value of the matrix.

The singular values of an arbitrary matrix A are the square roots from the eigenvalues of the smaller one of the matrices AA^* or A^*A, Stengel (1986). The proof of Theorem 17.7 is given in Rosenwasser (1996b). ■

Remark 1 Equation (17.115) can be represented in the form

$$\|\mathsf{U}_0\|_{\mathcal{L}_2} = \|\boldsymbol{\Gamma}(\zeta)\|_\infty \tag{17.116}$$

where the right hand side is the $\mathcal{H}_\infty$−norm of the real rational matrix $\boldsymbol{\Gamma}(\zeta)$ with respect to the unit circle.

Remark 2 A generalization of Theorem 17.7 is given in Lampe and Rosenwasser (1997b).

Taking the value $\|\mathsf{U}_0\|_{\mathcal{L}_2}$ as a cost function of the system to be designed, we can formulate the following optimization problem.

> **$\mathcal{L}_2$−problem** $(-\infty < t < \infty)$**.** Given all the characteristics of the system at hand, defining the matrices $\boldsymbol{A}(\zeta)$, $\boldsymbol{B}(\zeta)$, and $\mathcal{D}^o{}_{\boldsymbol{GNFM}}(T,\zeta,0)$, find the transfer function of the discrete-time controller $\boldsymbol{W}_d(\zeta)$, which stabilizes the closed-loop system and provides the minimal value of $\|\boldsymbol{\Gamma}(\zeta)\|_\infty$.

As follows from Remark 1 to Theorem 17.7, the solution of the $\mathcal{L}_2$−problem in this case can be reduced to the $\mathcal{H}_\infty$−optimization problem for an equivalent discrete-time system. Consider this question in detail. Let us have a left coprime MFD

$$\mathcal{D}^o{}_{\boldsymbol{GNFM}}(T,\zeta,0) = \mathcal{D}_{\boldsymbol{GNFM}}(T,s,0)\,\big|_{e^{-sT}=\zeta} = \zeta\boldsymbol{a}^{-1}(\zeta)\boldsymbol{b}(\zeta)\,.$$

In this case there hold relations similar to (17.103) and (17.104), where the corresponding matrices $\boldsymbol{v}_{ik}(\zeta)$ are calculated by the formulas of Section 17.7. From (17.115) and (17.104) we have

$$\|\mathsf{U}_0\|_{\mathcal{L}_2} = \|\boldsymbol{A}(\zeta)\boldsymbol{v}_{21}(\zeta)\boldsymbol{a}(\zeta)\boldsymbol{B}(\zeta) - \boldsymbol{A}(\zeta)\boldsymbol{v}_{22}(\zeta)\boldsymbol{\phi}(\zeta)\boldsymbol{a}(\zeta)\boldsymbol{B}(\zeta)\|_\infty\,. \tag{17.117}$$

Assume that we have managed to find a stable real rational matrix $\boldsymbol{\phi}_0(\zeta)$, for which the value $\|\mathsf{U}_0\|_{\mathcal{L}_2}$ is minimal. Then, computing the optimal functions (17.106) by (17.107), we can obtain the desired transfer matrix $\boldsymbol{W}_{d0}(\zeta)$ of the $\mathcal{L}_2$−optimal discrete-time controller.

Remark If all continuous-time parts of the system are stable, the $\mathcal{L}_2$−optimization problem has the trivial solution $\boldsymbol{W}_{d0}(\zeta) = 0$, because this controller is stabilizing. If there are unstable continuous-time networks, the minimization of the criterion (17.117) gives a non-trivial controller.

Example 17.6 Consider the scalar system shown in Fig. (17.14), where

$$G(s) = \frac{1}{s-a}\,, \quad F(s) = \frac{1}{s-c}\,, \quad N(s) = -1\,.$$

It is assumed that $a \neq 0$, $a \neq c$ and $c \neq 0$, though the final result is independent of the latter assumption. The form of the controlling pulse is assumed to be arbitrary. Since

$$G(s)G(-s) = \frac{1}{2a}\left(\frac{1}{s+a} - \frac{1}{s-a}\right)$$

from (4.44) and (17.111) we have

$$Q_G(s) = \frac{e^{aT}\sinh aT}{a}\,\frac{1}{(1-e^{aT}e^{-sT})\,(1-e^{aT}e^{sT})}\,. \tag{17.118}$$

Comparing (17.118) with (17.113), we obtain

$$B(\zeta) = \frac{K_1}{1-e^{aT}\zeta}$$

where K_1 is an arbitrary constant. Moreover, using the results of Example 16.1, we have

$$\mathcal{D}^o_{F\underline{F}M\underline{M}}(T,\zeta,0) = N^2\,\frac{(1-\mu\zeta)(1-\mu\zeta^{-1})}{(1-e^{cT}\zeta)(1-e^{cT}\zeta^{-1})}\,.$$

Hence, we may take

$$A(\zeta) = \frac{N(1-\mu\zeta)}{1-e^{cT}\zeta}\,.$$

Using (9.160) and (9.151) yields

$$\mathcal{D}^o_{FGM}(T,\zeta,0) = \frac{\zeta\chi^o(\zeta)}{\alpha^o(\zeta)}$$

where

$$\alpha^o(\zeta) = (1-e^{aT}\zeta)(1-e^{cT}\zeta)$$

and $\chi^o(\zeta)$ is a polynomial such that $\deg\chi^o(\zeta) \le 1$. Let the poles $s = a$ and $s = c$ satisfy the non-pathological conditions (9.161) and (9.162). Let also $\tilde{a}^o(\zeta)$, $\tilde{b}^o(\zeta)$ be an arbitrary solution of the Diophantine equation (14.86). Then, from (17.41) we find that, on the set of stabilizing controllers,

$$\tilde{W}^o_d(\zeta) = [\alpha^o(\zeta)]^2\phi(\zeta) + \alpha^o(\zeta)\tilde{b}^o(\zeta)$$

where $\phi(\zeta)$ is an arbitrary stable real rational function. Using the above relations, we find the matrix

$$\Gamma(\zeta) = K_1N(1-\mu\zeta)[\alpha^o(\zeta)\phi(\zeta) + \tilde{b}^o(\zeta)]\,.$$

The $\mathcal{L}_2$–norm of the system under investigation is equal to

$$||\mathsf{U}_0||_{\mathcal{L}_2} = K_1N\sup_{|\zeta|=1}\left|(1-\mu\zeta)[\alpha^o(\zeta)\phi(\zeta) + \tilde{b}^o(\zeta)]\right| \tag{17.119}$$

In this case the above $\mathcal{L}_2$–optimization problem reduces to the search for the stable real rational function $\phi(\zeta)$ which assures the minimal value of the norm (17.119), which is defined also for $c = 0$. This is linked with the fact that for $a \neq 0$ the operator U_0 in (17.110) is causal independently of c. □

Chapter 18

Polynomial methods for direct SD system design

18.1 Introduction

Immediate utilization of the formulae given in Chapter 17 for the direct design of feedback sampled-data systems involves some technical difficulties. This results from the fact that the coefficients of the functionals to be minimized depend on the choice of the basis stabilizing pair $a^0(\zeta), b^0(\zeta)$. Even though the final result does not depend, from the theoretical viewpoint, on the choice of this pair, this procedure is very important for practical computations, because in many cases the solution is not robust with respect to computational errors in the coefficients of the polynomials a^0 and b^0. Therefore, for practical design it is desirable to use a minimization technique that does not use the polynomials a^0 and b^0 and is numerically stable. For this purpose we have to eliminate a^0, b^0 and the parameter-function $\phi(\zeta)$ from the equations of Chapter 17, which determine the optimal controller. Various conducting of such procedures are considered in Rosenwasser et al. (1996); Rosenwasser (1995b).
A different approach called the *polynomial (equation) method*, that attains the same goal and does not use a^0, b^0 and ϕ, was introduced in Kučera (1979); Volgin (1986). In this appendix we present a novel direct optimal SD design method based on the use of Diophantine polynomial equation technique. This approach is fairly general and can be the basis for numerically robust optimization of SD systems. The polynomial equation method makes it possible to reduce the quadratic optimization problems for scalar SD systems to the solution of a pair of Diophantine equations which give the numerator and denominator of the optimal controller transfer function immediately. Moreover, the design procedure does not use a basic controller at all; therefore it is possible to determine the *order* of the optimal controller directly by initial data. Previously, the polynomial equation method was employed by different authors for solving $\mathcal{H}_2$-and $\mathcal{H}_\infty$-optimization problems for LTI continuous-

time or discrete-time systems, Kučera (1979, 1991); Kwakernaak (1985, 1990); Grimble (1995). In this appendix we generalize the polynomial technique to the direct design of optimal SD systems.

Henceforth, for notational simplicity we use the following notation.
Denote the degree of the denominator of any irreducible real rational function $F(s)$ by δ_F. If the function F has no poles on the imaginary axis, then

$$\delta_F = \delta_F^+ + \delta_F^-$$

where δ_F^+ and δ_F^- denote the number of its stable and unstable poles (taking account of miltiplicities), respectively. Denote by $d_F(\zeta)$ the denominator of the real rational function

$$\mathcal{D}_F(T,\zeta,t) \triangleq \mathcal{D}_F(T,s,t)|_{e^{-sT}=\zeta} = \frac{m_F(\zeta)}{d_F(\zeta)}\,.$$

For functions of the variable s that are, in fact, functions in $\zeta = e^{-sT}$, we denote

$$F(\zeta) \triangleq F(s)|_{e^{-sT}=\zeta}\,.$$

The asterisk * for functions in ζ denotes the substitution of ζ^{-1} for ζ, i.e., $F^*(\zeta) = F(\zeta^{-1})$.
The stable and anti-stable factors of any polynomial $f(\zeta)$ having no roots on the unit circle will be denoted by $f^+(\zeta)$ and $f^-(\zeta)$, respectively. Thus, $f(\zeta) = f^+ f^-$. [1]
The tilde denotes the *conjugate polynomial* which has the reverse order of coefficients

$$\tilde{f} = f^* \zeta^{\deg f}\,.$$

In what follows, we use the evident equalities

$$ff^* = \tilde{f}\tilde{f}^*, \qquad \frac{f^*}{\tilde{f}^*} = \frac{\tilde{f}}{f}\,. \tag{18.1}$$

18.2 Solution of the $\mathcal{H}_2$–optimization problem

Consider the SD system shown in Fig. 18.1. The control loop consists of a plant to be controlled that is formed by two networks with tranfer functions $F_1(s)$ and $F_2(s)$, actuator $H(s)$, dynamic feedback $G(s)$, and a digital controller with transfer function $W_d(\zeta)$. The exogenous signals, viz.: the disturbance $w(t)$ and measurement noise $m(t)$, pass through stable forming filters $F_w(s)$ and $F_m(s)$, respectively. The pure delay networks $e^{-s\tau_1}$, $e^{-s\tau_2}$ modeling time delays in continuous-time elements and computational delays.
Introduce the following assumptions:

[1] In what follows, function arguments are omitted for the sake of brevity if no confusion is possible.

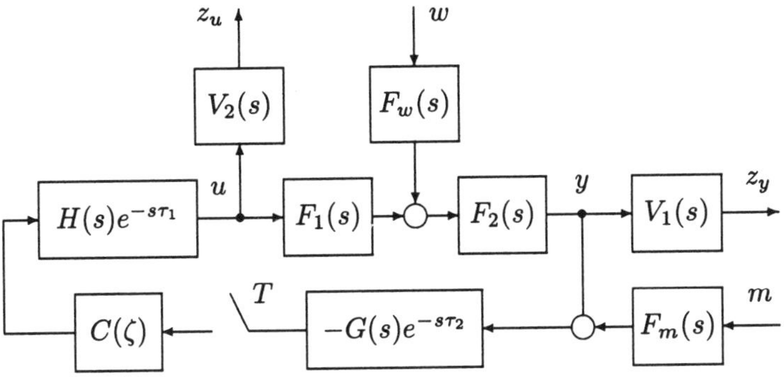

Figure 18.1: Block-diagram of SD system

i) The function F_1F_2HG is irreducible and strictly proper.
ii) The system in Fig. 18.1 is non-pathological, i.e., the poles of the function F_1F_2HG satisfy the conditions (9.161) and (9.162).
iii) The function F_1HG does not have any poles on the imaginary axis.
iv) The functions

$$\begin{aligned}\tilde{A}_1(s) &= (F_w\underline{F_w}F_2\underline{F_2} + F_m\underline{F_m})G\underline{G}\\ \tilde{B}_2(s) &= F_w\underline{F_w}F_2\underline{F_2}G\underline{V_1}F_1F_2HV_1\\ \tilde{C}_2(s) &= F_wF_2V_1\end{aligned}$$

are strictly proper.
v) The transfer functions of the forming filters $F_w(s)$ and $F_m(s)$ are asymptotically stable.
vi) The signals $w(t)$ and $m(t)$ are described by independent white noises with unit intensities.

For this system we consider the linear quadratic optimization problem under stochastic disturbances ($\mathcal{H}_2$–problem), the aim of which is to find a discrete controller $W_d(\zeta)$ that stabilizes the feedback system and ensures the minimal value of the cost function

$$I = \bar{d}_y + \bar{d}_u \quad \to \min \tag{18.2}$$

where $\bar{d}_y$ and $\bar{d}_u$ are the mean variances of the signals $z_y(t)$ and $z_u(t)$, respectively

$$\bar{d}_y = \int_0^T d_y(t)\,\mathrm{d}t\,, \qquad \bar{d}_u = \int_0^T d_u(t)\,\mathrm{d}t\,.$$

Thus, to solve the optimization problem by the criterion (18.2) it is required to find a controller $W_d(\zeta)$ for the system shown in Fig. 18.1 such that the feedback system is stable and $I \to \min$. As was shown above, such a problem reduces to the $\mathcal{H}_2$–optimization problem for an equivalent discrete-time system. Using the technique developed in Chapter 17, we may represent the cost function (18.2) in the form (17.35) as

$$I = \frac{1}{2\pi j}\oint_\Gamma X(\zeta)\,\frac{d\zeta}{\zeta} = \frac{1}{2\pi j}\oint_\Gamma \left[A\tilde{W}_d\tilde{W}_d^* - B\tilde{W}_d - B^*\tilde{W}_d^* + C\right]\frac{d\zeta}{\zeta} \tag{18.3}$$

where

$$A(\zeta) = \mathcal{D}_{\tilde{A}_1}(T,s,0)\,\mathcal{D}_{\tilde{A}_2}(T,s,0)\,\big|_{e^{-sT}=\zeta} \tag{18.4}$$
$$B(\zeta) = \mathcal{D}_{\tilde{B}_1}(T,s,-(\tau_1+\tau_2))\,\big|_{e^{-sT}=\zeta} \tag{18.5}$$
$$C(\zeta) = \mathcal{D}_{\tilde{C}_1}(T,s,0)\,\big|_{e^{-sT}=\zeta} \tag{18.6}$$

are the discrete Laplace transforms of the functions

$$\tilde{A}_1(s) = (F_w\underline{F_w}F_2\underline{F_2} + F_m\underline{F_m})G\underline{G} \tag{18.7}$$
$$\tilde{A}_2(s) = \frac{1}{T}\left(F_1\underline{F_1}F_2\underline{F_2}V_1\underline{V_1} + V_2\underline{V_2}\right) H\underline{H}M\underline{M} \tag{18.8}$$
$$\tilde{B}_1(s) = \frac{1}{T}F_w\underline{F_w}F_2\underline{F_2}G\underline{V_1}F_1F_2HV_1M \tag{18.9}$$
$$\tilde{C}_1(s) = F_w\underline{F_w}F_2\underline{F_2}V_1\underline{V_1} \tag{18.10}$$

and

$$\tilde{W}_d(\zeta) = \frac{W_d}{1 + W_d\mathcal{D}_{F_1F_2HGGh}(T,\zeta,-(\tau_1+\tau_2))}\,.$$

To construct the set of stabilizing controllers, we find the function

$$\mathcal{D}_{F_1F_2HGM}(T,\zeta,-(\tau_1+\tau_2)) = \frac{\zeta\chi_\tau}{\alpha} \tag{18.11}$$

where $\chi_\tau(\zeta)$ and $\alpha(\zeta)$ are polynomials, and $\alpha(\zeta) = d_{F_1F_2HG}(\zeta)$. In this case, as was shown in Section 17.3,

$$\tilde{W}_d(\zeta) = \alpha(\alpha\phi + \tilde{b}^0) \tag{18.12}$$

where ϕ is a stable real rational function, and the polynomial $\tilde{b}^0$ satisfies the polynomial equation

$$\tilde{b}^0\zeta\chi_\tau + \tilde{a}^0\alpha = 1 \tag{18.13}$$

with unknown $\tilde{a}^0$ and $\tilde{b}^0$. Being substituted into (18.12), the integrand in (18.3) takes the form

$$X = A_1\phi\phi^* - B_1\phi - B_1^*\phi^* + C_1 \tag{18.14}$$

with

$$A_1(\zeta) = (\alpha\alpha^*)^2 A \tag{18.15}$$
$$B_1(\zeta) = \alpha^2(B - (\tilde{b}^0)^*\alpha^* A) \tag{18.16}$$
$$C_1(\zeta) = C + \tilde{b}^0(\tilde{b}^0)^*\alpha\alpha^* A - \tilde{b}^0\alpha B - (\tilde{b}^0)^*\alpha^* B^*\,. \tag{18.17}$$

Theorem 18.1 (**$\mathcal{H}_2$–problem**) *Under the assumptions i) – vi) the optimal stabilizing controller, which minimizes the cost function (18.3), can be obtained as follows:*

1. *Find the polynomials* $\chi_\tau(\zeta)$ *and* $\alpha(\zeta)$ *in (18.11).*
2. *Find the discrete Laplace transforms*

$$\mathcal{D}_{\tilde{A}_1}(T,\zeta,0) = \frac{\alpha_1}{d_{F_w F_m F_2 G} d^*_{F_w F_m F_2 G}} \tag{18.18}$$

$$\mathcal{D}_{\tilde{A}_2}(T,\zeta,0) = \frac{\alpha_2}{d_{F_1 F_2 H V_1 V_2} d^*_{F_1 F_2 H V_1 V_2}} \tag{18.19}$$

$$\mathcal{D}_{\tilde{B}_1}(T,\zeta,-(\tau_1+\tau_2)) = \frac{\beta}{d_{F_w F_1 F_2 F_2 H G V_1} d^*_{F_w F_2 V_1}} \tag{18.20}$$

for the functions (18.7), (18.8), (18.9).

3. *Find a stable polynomial* $g(\zeta)$ *as a result of the factorization with respect to the unit circle*

$$gg^* = \alpha_1\alpha_2\,. \tag{18.21}$$

4. *Construct the polynomial*

$$\omega(\zeta) = \frac{(gg^* - \beta^*\zeta\chi_\tau d_{F_m V_2} d^*_{F_m V_2})\zeta^\rho}{d_{F_2}} \tag{18.22}$$

where ρ *is the minimal non-negative integer such that*

$$\eta_1(\zeta) = \beta^* d^*_{F_m V_2}\zeta^\rho\,, \qquad \eta_2(\zeta) = g^*\zeta^\rho \tag{18.23}$$

become polynomials in the indeterminate ζ*.*

5. *Find the polynomials* $N(\zeta)$*,* $D(\zeta)$ *and* $\pi(\zeta)$ *as the solution of the coupled Diophantine equations*

$$g^*\zeta^\rho N + \pi d_{F_w F_2 V_1} = \beta^* \tilde{d}_1^- d^*_{F_m V_2}\zeta^\rho \tag{18.24}$$

$$g^*\zeta^\rho d_1^- D - \pi\zeta\chi_\tau d_{F_w F_m V_1 V_2} = \tilde{d}_1^-\omega \tag{18.25}$$

where

$$d_1(\zeta) = \frac{\alpha}{d_{F_2}} = d_{F_1 HG}$$

and the polynomial $\pi(\zeta)$ *has the minimal possible degree (such a solution will henceforth be called* minimal *with respect to* π*).*

6. *The transfer function of the optimal controller takes the form*

$$W_d(\zeta) = \frac{d_1^+ d_{F_m V_2} N}{D}\,. \tag{18.26}$$

7. *The characteristic polynomial of the closed-loop system is* $\Delta = g d_1^+ \tilde{d}_1^-$*.*

Proof To minimize the functional (18.3) with the quadratic form (18.14) over the set of stable real rational functions ϕ we shall use the algorithm described in Section 16.2.

Using the notation introduced in Section 18.1 the functions $\mathcal{D}_{\tilde{A}_1}(T,\zeta,0)$, $\mathcal{D}_{\tilde{A}_2}(T,\zeta,0)$ and $\mathcal{D}_{\tilde{B}_1}(T,\zeta,0)$ may be written in the form (18.18), (18.19) and

(18.20), where the functions $\alpha_1(\zeta)$, $\alpha_2(\zeta)$ and $\beta(\zeta)$ are quasipolynomials of the form

$$a_n\zeta^n + a_{n-1}\zeta^{n-1} + \cdots + a_1\zeta + a_0 + a_{-1}\zeta^{-1} + \cdots + a_{-m+1}\zeta^{-m+1} + a_{-m}\zeta^{-m}.$$

Moreover, since the functions $\tilde{A}_1(s)$ and $\tilde{A}_2(s)$ are para-conjugate hermitian, we have

$$\alpha_1(\zeta) = \alpha_1(\zeta^{-1}), \qquad \alpha_2(\zeta) = \alpha_2(\zeta^{-1}).$$

Factoring Eq. (18.15), and taking account of (18.4), (18.18) and (18.19) gives

$$K(\zeta) = \frac{g d_1^+ \tilde{d}_1^-}{d_{F_w F_m V_1 V_2}} \tag{18.27}$$

where the stable polynomial $g(\zeta)$ is obtained as a result of the factorization (18.21).

Substituting (18.5) and (18.4) into Eq. (18.16) to calculate $\tilde{B}$ and using (18.27) for $K(\zeta)$, according to (16.28) after cancellations with respect to (18.1) we obtain the function $u(\zeta)$ (16.28) in the form

$$u(\zeta) = \frac{\tilde{d}_1^- (\beta^* d_{F_m V_2} d^*_{F_m V_2} - \tilde{b}^0 g g^*)\zeta^\rho}{d_{F_w F_m F_2 V_1 V_2} d_1^- g^* \zeta^\rho} \tag{18.28}$$

where the non-negative number ρ is chosen so that the functions (18.23) become polynomials. Moreover, in this case the numerator and denominator of the function $u(\zeta)$ are also polynomials.

As follows from Theorem 17.2, the expression inside the brackets in the numerator of (18.28) is divisible by d_{F_2}, and the roots of d_{F_2} are not poles of the function $u(\zeta)$. Taking account of this fact, the separation of the function $u(\zeta)$ gives

$$u(\zeta) = \{u\}_+ + \{u\}_- = \frac{\vartheta}{d_{F_w F_m V_1 V_2}} + \frac{\pi}{d_1^- g^* \zeta^\rho}. \tag{18.29}$$

The optimal function ϕ_0 has the form

$$\phi_0(\zeta) = \frac{\{u\}_+}{K} = \frac{\vartheta}{g d_1^+ \tilde{d}_1^-}.$$

Its denominator $\Delta(\zeta) = g d_1^+ \tilde{d}_1^-$ is the characteristic polynomial of the closed-loop system.

According to (17.40) the transfer function of the optimal controller is given as the ratio of the functions $W_1(\zeta)$ and $W_2(\zeta)$

$$W_1(\zeta) = \tilde{b}^0 + \alpha\phi = \frac{\tilde{b}^0 g d_1^+ \tilde{d}_1^- + \alpha\vartheta}{\Delta} = \frac{d_1^+ \left(\tilde{b}^0 g \tilde{d}_1^- + d_1^- d_{F_2}\vartheta\right)}{\Delta} \tag{18.30}$$

$$W_2(\zeta) = \tilde{a}^0 - \zeta\chi_\tau\phi = \frac{\tilde{a}^0 g d_1^+ \tilde{d}_1^- - \zeta\chi_\tau\vartheta}{\Delta}. \tag{18.31}$$

As follows from (18.30) and (18.31), the controller can be written as the ratio of two polynomials $W_d(\zeta) = b/a$, where

$$b(\zeta) = d_1^+ \left(\tilde{b}^0 g \tilde{d}_1^- + d_1^- d_{F_2} \vartheta\right) \tag{18.32}$$

$$a(\zeta) = \tilde{a}^0 g d_1^+ \tilde{d}_1^- - \zeta \chi_\tau \vartheta \,. \tag{18.33}$$

We note that the numerator of the optimal controller transfer function contains the factor d_1^+, so that $b(\zeta) = d_1^+ N_0$, where

$$N_0(\zeta) = \tilde{b}^0 g \tilde{d}_1^- + d_1^- d_{F_2} \vartheta \,.$$

Then, we get the polynomial equations, which make it possible to obtain immediately the polynomials b and a without calculating the function ϕ. Combining (18.28) and (18.29), we write the polynomial equation defining the separation

$$\tilde{d}_1^- \left(\beta^* d_{F_m V_2} d_{F_m V_2}^* - \tilde{b}^0 g g^*\right) \zeta^\rho = \vartheta d_{F_2} d_1^- g^* \zeta^\rho + \pi d_{F_w F_m F_2 V_1 V_2} \,. \tag{18.34}$$

By the conditions of the separation, the function $\{u\}_-$ must be strictly proper, therefore we have to find a solution of this equation with respect to the unknown polynomials ϑ and π which is minimal with respect to π. Regrouping the terms on the right and left-hand sides and using Eqs. (18.32) and (18.33), we obtain the equation determining N_0

$$g^* \zeta^\rho N_0 + \pi d_{F_w F_m F_2 V_1 V_2} = \beta^* \tilde{d}_1^- d_{F_m V_2} d_{F_m V_2}^* \zeta^\rho \,. \tag{18.35}$$

Note that the second term in the left part and the right part of Eq. (18.35) has the factor $d_{F_m V_2}$, therefore the first term on the left side must also have this factor. Therefore, the polynomial N_0 is divisible by $d_{F_m V_2}$, i.e., $b = d_1^+ d_{F_m V_2} N$, where $N(\zeta)$ is a polynomial. From (18.35) we can obtain the equation for the unknown polynomial N

$$g^* \zeta^\rho N + \pi d_{F_w F_2 V_1} = \beta^* \tilde{d}_1^- d_{F_m V_2}^* \zeta^\rho \,.$$

Since the characteristic polynomial of the closed-loop system can be written in two forms

$$\Delta = \zeta \chi_\tau b + \alpha a = g d_1^+ \tilde{d}_1^-$$

where $b(\zeta) = d_1^+ N_0$, then, dividing the latter equality by d_1^+, we easily find the equation linking the polynomials N_0 and $D = a$

$$\zeta \chi_\tau N_0 + d_1^- d_{F_2} D = g \tilde{d}_1^- \,. \tag{18.36}$$

Multiplying (18.35) by $\zeta \chi_\tau$ and substituting $g \tilde{d}_1^- - d_1^- d_{F_2} D$ for $\zeta \chi_\tau N_0$ in accordance with (18.36), after regrouping the terms and division by d_{F_2} we obtain

$$g^* \zeta^\rho d_1^- D - \pi \zeta \chi_\tau d_{F_w F_m V_1 V_2} = \tilde{d}_1^- \omega$$

where $\omega(\zeta)$ is given by (18.22). Using Theorem 17.2, it can be easily shown that the function $\omega(\zeta)$ (18.22) is a polynomial. ∎

Equations (18.24) and (18.25) allow us to determine the degrees of the polynomials b and a.

Theorem 18.2 *Under the assumptions i) – vi) the degrees of the numerator and denominator of the optimal controller transfer function are given by the formulae*

$$\deg b = \delta_{F_w F_m F_1 F_2 HGV_1 V_2} - 1 \tag{18.37}$$

$$\deg a = \delta_{F_w F_m F_1 F_2 HGV_1 V_2} + N_\tau - 1 \tag{18.38}$$

where the integer number N_τ is defined by the equality

$$\tau_1 + \tau_2 = N_\tau T - \delta T, \qquad 0 \le \delta < 1 .$$

The proof is given in Rosenwasser and Polyakov (1996). ∎

Example 18.1 For the system shown in Fig. 18.1 with

$$F_1(s) = H(s) = G(s) = 1, \qquad F_2(s) = \frac{1}{s-1}$$
$$F_w(s) = 1, \quad F_m(s) = 0, \qquad V_1(s) = 1, \quad V_2(s) = 0$$
$$T = 1, \qquad \tau_1 = \tau_2 = 0, \qquad M(s) = \frac{1 - e^{-sT}}{s}$$

which corresponds to the system shown in Fig. 17.3, design the optimal digital controller minimizing the cost function (18.2).
With the above data we get

$$\mathcal{D}_{F_2 M}(T,\zeta,0) = \frac{-0.6321\zeta}{\zeta - 0.3679}$$

$$\mathcal{D}_{\tilde{A}_1}(T,\zeta,0) = \mathcal{D}_{F_2 \underline{F_2}}(T,\zeta,0) = \frac{0.4323}{(\zeta - 0.3679)(\zeta^{-1} - 0.3679)}$$

$$\mathcal{D}_{\tilde{A}_2}(T,\zeta,0) = \frac{1}{T}\mathcal{D}_{F_2 \underline{F_2} M \underline{M}}(T,\zeta,0) = \frac{0.0163(\zeta + 3.9461)(\zeta^{-1} + 3.9461)}{(\zeta - 0.3679)(\zeta^{-1} - 0.3679)}$$

$$\mathcal{D}_{\tilde{B}_1}(T,\zeta,-(\tau_1+\tau_2)) = \frac{1}{T}\mathcal{D}_{F_2 F_2 \underline{F_2} M}(T,\zeta,0)$$
$$= \frac{-0.0513\zeta^2 - 0.1704\zeta - 0.0309}{(\zeta - 0.3679)^2(\zeta^{-1} - 0.3679)} .$$

Comparing these equations with (18.18)–(18.20) gives

$$\begin{aligned} &\alpha_1 = 0.4323, \qquad \alpha_2 = 0.0163(\zeta + 3.9461)(\zeta^{-1} + 3.9461) \\ &\beta = -0.0513\zeta^2 - 0.1704\zeta - 0.0309 . \end{aligned} \tag{18.39}$$

Hence, owing to (18.21),

$$g = 0.0840\zeta + 0.3316\,. \tag{18.40}$$

In order that the functions (18.23) become polynomials, we have to take $\rho = 2$. By (18.22) we find

$$\omega(\zeta) = 0.0083\zeta^2 + 0.0124\zeta\,.$$

Therefore, Eqs. (18.24)–(18.25) take the form

$$(0.3316\zeta^2 + 0.0840\zeta)N + (\zeta - 0.3679)\pi = -0.0309\zeta^2 - 0.1704\zeta - 0.0513 \tag{18.41}$$

$$(0.3316\zeta^2 + 0.0840\zeta)D - 0.6321\zeta\pi = 0.0083\zeta^2 + 0.0124\zeta\,. \tag{18.42}$$

Comparing the degrees of all the polynomials and taking account of the cancellation in (18.42) by ζ yields that there exists the minimal solution with respect to π of this system of Diophantine equations, such that $\deg \pi = 1$, $\deg N = 0$, and $\deg D = 0$. Solving the equations gives $\pi = 0.4860\zeta + 0.1394$, $N = -1.5589$, and $D = -0.9014$. Substituting these polynomials into (18.26) gives $W_d(\zeta) = 1.7295$. This controller ensures the minimal cost (18.2) equal to 2.3105. In the given case the optimal controller is a static gain, which corresponds to (18.37)–(18.38). □

18.3 Design of optimal tracking systems

Consider the tracking system with a single controller shown in Fig. 18.2. This system should follow, as closely as possible, the input signal defined as a result

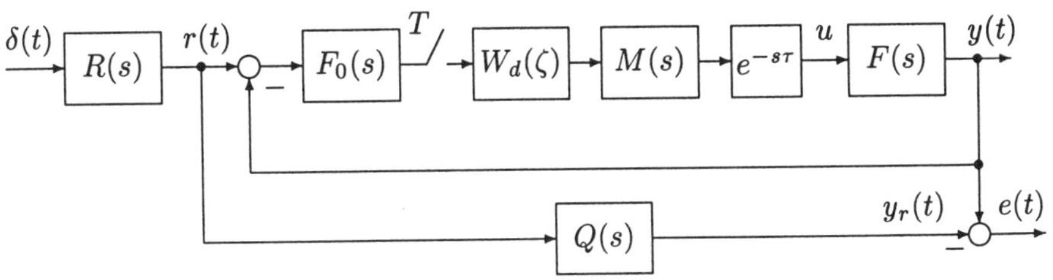

Figure 18.2: Block-diagram of tracking system

of the passage of a delta function $\delta(t)$ via a forming filter $R(s)$. A desired linear transformation of the input signal is given by a reference system with transfer function $Q(s)$. The control loop consists of a plant to be controlled with transfer function $F(s)$ and prefilter at the input of the computer $F_0(s)$. The computational delay and pure delay in continuous-time networks is modelled by the element $e^{-s\tau}$.

Introduce the following assumptions:

i) The functions F, F_0F, $FQ\underline{R}$ and F_0R are irreducible and strictly proper.
ii) The system shown in Fig. 18.2 is non-pathological, i.e., the poles of the function F_0F satisfy the non-pathological conditions (9.161) and (9.162).
iii) The function F_0 has no poles on the imaginary axis, and the function F has no more than one neutral pole.
iv) The transform $R(s)$ of the reference signal is a proper function of the form

$$R(s) = \frac{1}{s}R_1(s) \tag{18.43}$$

where $R_1(s)$ is an asymptotically stable function.
v) The transfer function of the reference system $Q(s)$ is asymptotically stable.
v) A zero-order hold is used, i.e.

$$M(s) = \frac{1-\mathrm{e}^{-sT}}{s}.$$

As a cost function we employ the integral quadratic error between the outputs of the SD and the reference system, denoted by $y(t)$ and $y_r(t)$, respectively

$$I_c = \int_0^\infty e^2(t)\,\mathrm{d}t = \int_0^\infty [y(t) - y_r(t)]^2\,\mathrm{d}t. \tag{18.44}$$

The aim of the optimization problem for tracking SD system is to find a stabilizing controller such that $I_c \to \min$.

As was shown in Section 17.4, the cost function (18.44) can be written as

$$I_c = \frac{1}{2\pi\mathrm{j}}\int_{-\mathrm{j}\infty}^{\mathrm{j}\infty} (Y - Y_r)(\underline{Y} - \underline{Y_r})\,\mathrm{d}s$$

where $Y(s)$ and $Y_r(s)$ denote the Laplace images of the signals $y(t)$ and $y_r(t)$, respectively. Using the technique described in Section 17.4, the above criterion can be reduced to the form

$$I_c = \frac{1}{2\pi\mathrm{j}}\oint_\Gamma \left[A\tilde{W}_d\tilde{W}_d^* - B\tilde{W}_d - B^*\tilde{W}_d^* + C\right]\frac{\mathrm{d}\zeta}{\zeta} \tag{18.45}$$

where

$$A(\zeta) = \mathcal{D}_{\tilde{A}_1}(T,\zeta,0)\mathcal{D}_{\tilde{A}_2}(T,\zeta,+0)\mathcal{D}_{\tilde{A}_2}(T,\zeta^{-1},+0) \tag{18.46}$$
$$B(\zeta) = \mathcal{D}_{\tilde{B}_1}(T,\zeta,-\tau)\mathcal{D}_{\tilde{A}_2}(T,\zeta,+0) \tag{18.47}$$
$$C(\zeta) = \mathcal{D}_{\tilde{C}_1}(T,\zeta,0) \tag{18.48}$$
$$\tilde{A}_1(s) = F\underline{F}M\underline{M} \tag{18.49}$$
$$\tilde{A}_2(s) = F_0R \tag{18.50}$$
$$\tilde{B}_1(s) = F\underline{Q}\underline{R}M \tag{18.51}$$
$$\tilde{C}_1(s) = Q\underline{Q}R\underline{R}. \tag{18.52}$$

To ensure the stability of the closed loop, we use the set of stabilizing controllers (17.40), which leads to the substitution of (18.12) for $\tilde{W}_d$.
As was shown in Section 17.4, the cost function (18.44) is finite iff the steady-state outputs of both systems are equal, so that

$$\lim_{s\to 0} s\,[Y(s) - Y_r(s)] = 0\,. \tag{18.53}$$

To ensure (18.53), the optimization should be performed not over the whole set of stabilizing controllers (17.40), but over some of its subset. In this case we have

$$\phi = k_\phi + \theta\phi_1 \tag{18.54}$$

where k_ϕ is a known constant, $\theta = \zeta - 1$ and $\phi_1(\zeta)$ is an arbitrary stable real rational function. Substituting (18.12) into (18.45) and taking account of (18.54) yields

$$I_c = \frac{1}{2\pi \mathrm{j}} \oint_\Gamma [A_1\phi_1\phi_1^* - B_1\phi_1 - B_1^*\phi_1^* + C_1]\,\frac{\mathrm{d}\zeta}{\zeta}$$

where

$$\begin{aligned}
A_1(\zeta) &= (\alpha\alpha^*)^2\theta\theta^* A && (18.55)\\
B_1(\zeta) &= \alpha^2\theta(B - a_1^*\alpha^* A) && (18.56)\\
C_1(\zeta) &= C + b_1 a_1^*\alpha\alpha^* A - b_1\alpha B - a_1^*\alpha^* B^* &&\\
b_1(\zeta) &= \tilde{b}^0 + k_\phi\alpha\,. &&
\end{aligned}$$

The main result of this section is formulated below as a theorem.

Theorem 18.3 *Under the assumptions i) – vi) the optimal stabilizing controller which provides the minimum of the cost function (18.45) for the system at hand, is designed as follows:*

1. *Find the polynomials $\chi_\tau(\zeta)$ and $\alpha(\zeta)$ appearing in the function*

$$\mathcal{D}_{FF_0M}(T, s, -\tau) = \frac{\zeta\chi_\tau}{\alpha}\,.$$

2. *Calculate the discrete Laplace transforms*

$$\mathcal{D}_{\tilde{A}_1}(T, \zeta, 0) = \frac{\alpha_1}{d_F d_F^*} \tag{18.57}$$

$$\mathcal{D}_{\tilde{A}_2}(T, \zeta, 0) = \frac{\alpha_2}{d_{F_0R}} \tag{18.58}$$

$$\mathcal{D}_{\tilde{B}_1}(T, \zeta, -\tau) = \frac{\beta}{d_F d_{QR}^*} \tag{18.59}$$

for the functions (18.49)–(18.51).

3. *Find the stable polynomial* $g(\zeta)$ *as a result of the factorization*

$$gg^* = \alpha_1\alpha_2\alpha_2^* . \tag{18.60}$$

4. *Compute the polynomial*

$$\omega(\zeta) = (gg^* d_Q - \beta^* \alpha_2^* \zeta \chi_\tau)\zeta^\rho \tag{18.61}$$

where ρ *is the least non-negative number such that the functions*

$$\eta_1(\zeta) = \beta^* \alpha_2^* \zeta^\rho , \qquad \eta_2(\zeta) = g^* \zeta^\rho \tag{18.62}$$

become polynomials.

5. *Find the polynomials* $N(\zeta)$, $D(\zeta)$ *and* $\pi(\zeta)$ *which define the minimal solution with respect to* π *of the system of Diophantine equations*

$$g^* \zeta^\rho N + \pi d_{QR} = \beta^* \alpha_2^* \tilde{\alpha}^- \zeta^\rho \tag{18.63}$$

$$g^* \zeta^\rho \alpha^- D - \pi \zeta \chi_\tau d_{QR} = \tilde{\alpha}^- \omega . \tag{18.64}$$

6. *The transfer function of the optimal controller takes the form*

$$W_d(\zeta) = \frac{\alpha^+ N}{D} . \tag{18.65}$$

7. *The characteristic polynomial of the closed-loop system is given by*

$$\Delta = g\alpha^+ \tilde{\alpha}^- d_Q .$$

Proof Calculating the corresponding discrete Laplace transforms (18.57)–(18.59) the functions $A(\zeta)$ and $B(\zeta)$ may be written as

$$A(\zeta) = \frac{\alpha_1}{d_F d_F^*} \cdot \frac{\alpha_2}{d_{F_0 R}} \cdot \frac{\alpha_2^*}{d_{F_0 R}^*} \tag{18.66}$$

$$B(\zeta) = \frac{\beta}{d_F d_{QR}^*} \cdot \frac{\alpha_2}{d_{F_0 R}} \tag{18.67}$$

where the functions $\alpha_1(\zeta)$, $\alpha_2(\zeta)$ and $\beta(\zeta)$ are quasipolynomials. Moreover, since the function $\tilde{A}_1(s)$ is para-conjugate hermitian, we have $\alpha_1(\zeta) = \alpha_1(\zeta^{-1})$. Taking account of (18.66), Eq. (18.55) after cancellations is written in the form

$$A_1(\zeta) = \alpha\alpha^* gg^* \frac{\theta\theta^*}{d_R d_R^*} \tag{18.68}$$

where the stable polynomial $g(\zeta)$ is obtained as a result of the factorization (18.60). Since, by (18.43), the function $R(s)$ absorbs an integrator, we have

$$d_R = \theta d_{R_1} \tag{18.69}$$

(otherwise the factorization (18.68) would be impossible). Hence, the result of the factorization (18.55) has the form

$$K(\zeta) = \frac{g\alpha^+\tilde{\alpha}^-}{d_{R_1}}. \tag{18.70}$$

Substituting (18.67) and (18.66) in Eq. (18.56) for $\tilde{B}$ and using (18.70) for $K(\zeta)$, after cancellation taking account of (18.1) we obtain the function $u(\zeta)$ (16.28) as

$$u(\zeta) = \frac{\tilde{\alpha}^-(\beta^*\alpha_2^* - b_1 d_Q g g^*)\zeta^\rho}{d_{QR_1}\theta\alpha^- g^*\zeta^\rho} \tag{18.71}$$

where the non-negative number ρ is chosen so that the numerator and denominator of the function $u(\zeta)$ become polynomials.
Then, the function u should be separated into a stable and an anti-stable fractions. This is possible if and only if it has not poles on the unit circle. This means that the numerator of the function u is divisible by θ, i.e., the coefficient k_ϕ must be chosen in such a way that

$$\lim_{\zeta\to 1}(B^* - b_1\alpha A) = 0. \tag{18.72}$$

Using (18.46) and (18.47), it can be shown that the condition (18.72) may be written in the equivalent form

$$\lim_{\zeta\to 1}\zeta\chi_\tau(\tilde{b}^0 + k_\phi\alpha) = \lim_{s\to 0} Q(s). \tag{18.73}$$

Then, recalling that $\zeta = 1$ is not a pole of the function $u(\zeta)$, we obtain

$$u(\zeta) = \{u\}_+ + \{u\}_- = \frac{\vartheta}{d_{QR_1}} + \frac{\pi}{\alpha^- g^*\zeta^\rho}. \tag{18.74}$$

The optimal function ϕ_1 is given as

$$\phi_1(\zeta) = \frac{\{u\}_+}{K} = \frac{\vartheta}{g\alpha^+\tilde{\alpha}^- d_Q}.$$

Its denominator $\Delta(\zeta) = g\alpha^+\tilde{\alpha}^- d_Q$ is the characteristic polynomial of the optimal closed-loop system.
Using (18.54), in accordance with (17.40) the transfer function of the optimal controller can be represented in the form of the ratio of the functions $W_1(\zeta)$ and $W_2(\zeta)$

$$W_1(\zeta) = \tilde{b}^0 + \alpha\phi = \frac{b_1 g\alpha^+\tilde{\alpha}^- d_Q + \alpha\theta\vartheta}{\Delta} = \frac{\alpha^+(b_1 g\tilde{\alpha}^- d_Q + \alpha^-\theta\vartheta)}{\Delta} \tag{18.75}$$

$$W_2(\zeta) = \tilde{a}^0 - \zeta\chi_\tau\phi = \frac{(\tilde{a}^0 - \zeta\chi_\tau k_\phi)g\alpha^+\tilde{\alpha}^- d_Q - \zeta\chi_\tau\theta\vartheta}{\Delta}. \tag{18.76}$$

As follows from (18.75) and (18.76), the controller may also be written as the ratio of two polynomials $W_d(\zeta) = b/a$, where

$$b(\zeta) = \alpha^+\left(b_1 g\tilde{\alpha}^- d_Q + \alpha^-\theta\vartheta\right) \tag{18.77}$$

$$a(\zeta) = (\tilde{a}^0 - \zeta\chi_\tau k_\phi)g\alpha^+\tilde{\alpha}^- d_Q - \zeta\chi_\tau\theta\vartheta. \tag{18.78}$$

We note that the numerator of the optimal controller (18.77) contains the factor α^+, so that $b(\zeta) = \alpha^+ N$, where $N(\zeta) = b_1 g\tilde{\alpha}^- d_Q + \alpha^- \theta\vartheta$.

Now we can derive the polynomial equations defining the polynomials b and a directly, without determining ϕ_1 preliminarily. Combining (18.71) and (18.74), the polynomial equation associated with the separation can be written in the form

$$\tilde{\alpha}^- \left(\beta^* \alpha_2^* - b_1 d_Q g g^*\right) \zeta^\rho = \vartheta\theta\alpha^- g^* \zeta^\rho + \pi\theta d_{QR_1} \,. \tag{18.79}$$

By the conditions of the separation the function $\{u\}_-$ must be strictly proper, therefore we have to find the solution ϑ, π of this equation minimal with respect to π. Regrouping the terms on the left and right-hand sides and using (18.77), (18.78) and (18.69), we obtain the defining equation for N

$$g^* \zeta^\rho N + \pi d_{QR} = \beta^* \alpha_2^* \tilde{\alpha}^- \zeta^\rho \,. \tag{18.80}$$

Since the characteristic polynomial of the closed-loop system can be alternatively written as

$$\Delta = \zeta\chi_\tau b + \alpha a = g\alpha^+ \tilde{\alpha}^- d_Q$$

where $b(\zeta) = \alpha^+ N$, after division of the latter equality by α^+ it is easy to obtain the equation linking the polynomials N and $D = a$

$$\zeta\chi_\tau N + \alpha^- D = g\tilde{\alpha}^- d_Q \,. \tag{18.81}$$

Multiplying (18.80) by $\zeta\chi_\tau$ and replacing, according to (18.81), $\zeta\chi_\tau N$ by $g\tilde{\alpha}^- d_Q - \alpha^- D$, we obtain after regrouping of the terms

$$g^* \zeta^\rho \alpha^- D - \pi\zeta\chi_\tau d_{QR} = \tilde{\alpha}^- \omega$$

where $\omega(\zeta) = (g g^* d_Q - \beta^* \alpha_2^* \zeta\chi_\tau)\zeta^\rho$, which coincides with (18.64). ∎

Remark 1 It can be shown that under the condition (18.73), Eq. (18.53) is valid as well. Therefore, when using Theorem 18.3 we do not have to compute k_ϕ.

Remark 2 Equations (18.63) and (18.64) coincide with similar equations for the design of the optimal controller over the set of all stabilizing controllers (17.40) for an arbitrary stable $R(s)$.

Equations (18.63) and (18.64) allow us to find the degrees of the polynomials b and a.

Theorem 18.4 *Assume that the assumptions i) – vi) hold. Then*
a) *If the transfer function $F(s)$ has no poles on the imaginary axis, the degrees of the numerator and denominator of the optimal controller transfer function are given by*

$$\begin{aligned}\deg b &= \delta_{F_0 FQR} - 1\\ \deg a &= \delta_{F_0 FQR} + N_\tau - 1\end{aligned}$$

where the integer N_τ is defined by the equality

$$\tau = N_\tau T - \delta T, \qquad 0 \le \delta < 1\,.$$

b) *If $F(s)$ has a simple pole at $s = 0$, the polynomial D, which is a solution of (18.64), has a root at $\zeta = 1$. In this case*

$$\begin{aligned}\deg b &= \delta_{F_0 FQR} - 2\\ \deg a &= \delta_{F_0 FQR} + N_\tau - 2\,.\end{aligned}$$

The proof is given in Rosenwasser and Polyakov (1996). ■

Example 18.2 Let us design the optimal digital controller with the cost function (18.44) for the system shown in Fig. 18.2 for

$$\begin{aligned}&F(s) = \frac{1}{s}, \quad F_0(s) = 1, \quad R(s) = \frac{1}{s}, \quad Q(s) = \frac{1}{s+1}\\ &T = 1, \qquad \tau = 0\,.\end{aligned}$$

For these data

$$\begin{aligned}&\mathcal{D}_{FM}(T,\zeta,0) = \frac{-\zeta}{\zeta - 1}\\ &\mathcal{D}_{\tilde{A}_1}(T,\zeta,0) = \mathcal{D}_{F\underline{F}M\underline{M}}(T,\zeta,0) = \frac{0.1667\zeta + 0.6667 + 0.1667\zeta^{-1}}{(\zeta-1)(\zeta^{-1}-1)}\\ &\mathcal{D}_{\tilde{A}_2}(T,\zeta,0) = \mathcal{D}_R(T,\zeta,0) = \frac{-1}{\zeta-1}\\ &\mathcal{D}_{\tilde{B}_1}(T,\zeta,-\tau) = \mathcal{D}_{F\underline{QR}M}(T,\zeta,0) = \frac{-0.3591\zeta - 1.1408 - 0.2183\zeta^{-1}}{(\zeta-1)(\zeta^{-1}-2.7183)(\zeta^{-1}-1)}\,.\end{aligned}$$

Comparing the above relations with (18.57)–(18.59) gives

$$\begin{aligned}&\alpha_1 = 0.1667\zeta + 0.6667 + 0.1667\zeta^{-1}, \qquad \alpha_2 = -1\\ &\beta = -0.3591\zeta - 1.1408 - 0.2183\zeta^{-1}\end{aligned}$$

whence, due to (18.60), it follows that $g = 0.2113\zeta + 0.7887$. In order for the functions (18.62) to be polynomials, we take $\rho = 1$. Then, Eq. (18.61) yields

$$\omega(\zeta) = 0.3849\zeta^3 + 1.3545\zeta^2 - 1.2864\zeta - 0.4530$$

so that Eqs. (18.63-18.64) take the form

$$\begin{aligned}&(0.7887\zeta + 0.2113)N + (\zeta-1)(\zeta-2.7182)\pi = 0.2183\zeta^2 + 1.1408\zeta + 0.3591\\ &(0.7887\zeta + 0.2113)D - (\zeta-1)(\zeta-2.7182)\pi =\\ &\qquad\qquad = 0.3849\zeta^3 + 1.3545\zeta^2 - 1.2864\zeta - 0.4530\,.\end{aligned}$$

Comparing the degrees of the polynomials, we find the minimal solution with respect to π of this system of equation for $\deg \pi = 0$, $\deg N = 1$, and $\deg D = 2$. Solving the equations, we obtain $\pi = 0.01825$, $N = 0.2536\zeta + 1.4646$, and $D = 0.4649\zeta^2 + 1.6789\zeta - 2.1438$. Substituting these polynomials into (18.65), after cancellation of the common factor $\zeta - 1$ in the numerator and the denominator we obtain

$$W_d(\zeta) = \frac{0.6832 + 0.1183\zeta}{1 + 0.2169\zeta}.$$

In this case the cost function (18.44) is equal to 0.001856. The transient of the reference and the designed SD system are shown in Fig. 18.3. □

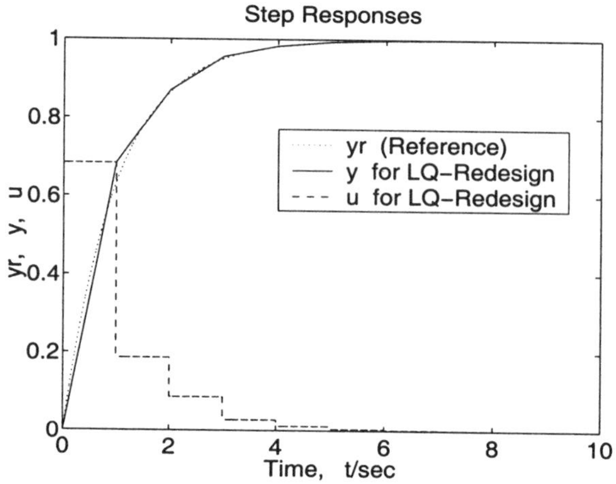

Figure 18.3: Transients in the optimal tracking system

18.4 Associated $\mathcal{H}_\infty$–problem

Consider the system shown in Fig. 18.1. In practical cases models of the plant and the exogenous disturbances are always approximations. To take account of uncertainties in the initial data *robust optimization* methods are used that make it possible to design 'the best system for the worst case'. As a cost function for robust optimization, a weighted sum of spectra of the sensitivity and complementary sensitivity functions is commonly employed, Kwakernaak (1985, 1990).

In this section for robust optimization of the system shown in Fig. 18.1 we use the cost function introduced in Section 17.3 - the $\mathcal{H}_\infty$–norm of the integrand in (18.3)

$$I_\infty = \sup_{|\zeta|=1} X(\zeta) = \sup_{|\zeta|=1} \left[A\tilde{W}_d\tilde{W}_d^* - B\tilde{W}_d - B^*\tilde{W}_d^* + C \right] \to \min \qquad (18.82)$$

where the functions A, B, and C are given by (18.4), (18.5) and (18.6), respectively. Such an approach is based on the method developed in Grimble (1995), where the close connection between the problems of $\mathcal{H}_\infty$−optimization and the linear quadratic Gaussian (LQG) problem is shown and the following lemma is proved.

Lemma 18.1 (Grimble (1995)) *Consider the auxiliary problem of minimization of the integral*

$$I_{aux} = \frac{1}{2\pi \mathrm{j}} \oint_{\Gamma} X(\zeta)\Sigma(\zeta) \frac{\mathrm{d}\zeta}{\zeta} \tag{18.83}$$

where

$$\Sigma(\zeta) = \Sigma(\zeta^{-1}) > 0 \,. \tag{18.84}$$

Assume that for a function $\Sigma(\zeta)$ the criterion (18.83) is minimized by an admissible controller, for which the quadratic form $X(\zeta) = X(\zeta^{-1})$ is such that

$$X(\zeta) = \lambda^2 = \text{const} \tag{18.85}$$

on the unit circle $|\zeta| = 1$. Then, such a controller minimizes also

$$\sup_{|\zeta|=1} X(\zeta) \,.$$

The proof is given in Grimble (1995). ∎

This lemma substantiates the following approach to the robust optimization problems: Find a real rational function Σ and λ such that the integral (18.83) is minimized by a controller for which the *equalizer principle* (18.85) holds.

Consider the auxiliary problem of the minization of the cost function

$$I_{aux} = \frac{1}{2\pi \mathrm{j}} \oint_{\Gamma} X(\zeta)\Sigma(\zeta) \frac{\mathrm{d}\zeta}{\zeta} \tag{18.86}$$

which, by Lemma 18.1, serves as a basis for robust optimal control synthesis. In this case the integrand in (18.86) takes the form

$$X\Sigma = A_2\phi\phi^* - B_2\phi - B_2^*\phi^* + C_2$$

where

$$\begin{aligned}
A_2(\zeta) &= (\alpha\alpha^*)^2 A\Sigma \\
B_2(\zeta) &= \alpha^2(B - (\tilde{b}^0)^*\alpha^* A)\Sigma \\
C_2(\zeta) &= (C + \tilde{b}^0(\tilde{b}^0)^*\alpha\alpha^* A - \tilde{b}^0\alpha B - (\tilde{b}^0)^*\alpha^* B^*)\Sigma \,.
\end{aligned}$$

Since the function $\Sigma(\zeta)$ possesses the property (18.84), it can be represented in the factorized form

$$\Sigma(\zeta) = \frac{n_\sigma n_\sigma^*}{d_\sigma d_\sigma^*}$$

where $n_\sigma(\zeta)$ and $d_\sigma(\zeta)$ are stable polynomials. The aim of this section is to find the optimal controller for known n_σ and d_σ.

Repeating the reasonings given in Section 18.2 and taking account of the weighting function Σ, we obtain the following minimization algorithm of the cost functional (18.86).

Theorem 18.5 *Under the assumptions i) – vi) of Section 18.2 the optimal stabilizing controller minimizing the criterion (18.83) is computed as follows:*

1. *Find the polynomials $\chi_\tau(\zeta)$ and $\alpha(\zeta)$ appearing in (18.11).*
2. *Compute the Laplace transforms (18.18), (18.19), and (18.20).*
3. *Find the stable polynomial $g(\zeta)$ as a result of the factorization (18.21).*
4. *Calculate the polynomial $\omega(\zeta)$ by Eq. (18.22).*
5. *Find the polynomials $N(\zeta)$, $D(\zeta)$, and $\pi(\zeta)$, which define the minimal solution with respect to π of the system of equations*

$$g^*\zeta^\rho N + \pi d_{\sigma F_w F_2 V_1} = \beta^* n_\sigma \tilde{d}_1^- d_{F_m V_2}^* \zeta^\rho \tag{18.87}$$

$$g^*\zeta^\rho d_1^- D - \pi\zeta\chi_\tau d_{\sigma F_w F_m V_1 V_2} = n_\sigma \tilde{d}_1^- \omega\,. \tag{18.88}$$

6. *The transfer function of the optimal controller takes the form*

$$W_d(\zeta) = \frac{d_1^+ d_{F_m V_2} N}{D}\,. \tag{18.89}$$

7. *The characteristic polynomial of the closed-loop system is given by*

$$\Delta = n_\sigma g d_1^+ \tilde{d}_1^-\,.$$

Now we are able to solve the initial robust optimization problem with the cost functional (18.82). The main result formulated below as a theorem allows us to find the *full-order solution* of the robust optimization problem, i.e., the solution which minimizes the cost function (18.83) for a given λ^2.

Theorem 18.6 (Full-order solution) *Under the assumptions i) – vi) of Section 18.2, for a given sufficiently large λ^2 the stabilizing controller which minimizes the cost function (18.83) and ensures the equalizer principle (18.85) is constructed as follows:*

1. *Find the polynomials $\chi_\tau(\zeta)$ and $\alpha(\zeta)$ according to (18.11).*
2. *Compute the discrete Laplace transforms*

$$\mathcal{D}_{\tilde{A}_1}(T,\zeta,0) = \frac{\alpha_1}{d_{F_w F_m F_2 G} d_{F_w F_m F_2 G}^*} \tag{18.90}$$

$$\mathcal{D}_{\tilde{A}_2}(T,\zeta,0) = \frac{\alpha_2}{d_{F_1 F_2 H V_1 V_2} d_{F_1 F_2 H V_1 V_2}^*} \tag{18.91}$$

$$\mathcal{D}_{\tilde{B}_1}(T,\zeta,0) = \frac{\beta}{d_{F_w F_1 F_2 F_2 H G V_1} d_{F_w F_2 V_1}^*} \tag{18.92}$$

$$\mathcal{D}_{\tilde{C}_1}(T,\zeta,0) = \frac{\gamma}{d_{F_w F_2 V_1} d_{F_w F_2 V_1}^*} \tag{18.93}$$

for the functions (18.7), (18.8), (18.9), and (18.10).

3. *Find the stable polynomial* $g(\zeta)$ *as a result of the factorization (18.21).*
4. *Calculate the polynomial* $\omega(\zeta)$ *(18.22).*
5. *Find the stable polynomial* $\sigma(\zeta)$ *as a result of the factorization*

$$\sigma\sigma^* = \lambda^2 gg^* d_{F_w V_1} d^*_{F_w V_1} - \psi \tag{18.94}$$

where the quasipolynomial $\psi(\zeta)$ *is given by*

$$\psi(\zeta) = \frac{\gamma gg^* - \beta\beta^* d_{F_m V_2} d^*_{F_m V_2}}{d_{F_2} d^*_{F_2}} . \tag{18.95}$$

6. *Find the polynomials* $N(\zeta)$, $D(\zeta)$, *and* $f(\zeta)$, *which define the minimal solution with respect to* $f(\zeta)$ *of the system of equations*

$$gg^*\zeta^\rho N + f\sigma d_{F_2} = \tilde{f}\beta^* d^*_{F_m V_2}\zeta^\rho \tag{18.96}$$

$$gg^*\zeta^\rho d_1^- D - f\sigma\zeta\chi_\tau d_{F_m V_2} = \tilde{f}\omega \tag{18.97}$$

where the polynomial $f(\zeta)$ *is strictly anti-stable (all its roots are located inside the unit circle in the* ζ−*plane). In order for the last condition to be valid, we must choose such a* λ, *that* $\lambda^2 > \lambda_0^2 = \|X(\zeta)\|_\infty$.

7. *The transfer function of the optimal robust controller takes the form*

$$W_d(\zeta) = \frac{d_1^+ d_{F_m V_2} N}{D} . \tag{18.98}$$

8. *The characteristic polynomial of the closed-loop system appears as*

$$\Delta = \tilde{f} d_1^+ .$$

Proof It can be shown that for the system with the optimal controller (18.89) obtained by Theorem 18.5 the integrand in (18.86) is equal to

$$X\Sigma = \{u\}_-\{u\}_-^* + \left(C - \frac{BB^*}{A}\right)\Sigma \tag{18.99}$$

where the strictly proper function $\{u\}_-$ is obtained as a result of the factorization (18.29). According to Lemma 18.1, the optimal controller minimizing the criterion (18.82) is *equalizing* i.e., $X(\zeta) = \lambda^2$. Hence, the function $\Sigma(\zeta)$ can be expressed from (18.99) as

$$\Sigma = \frac{\{u\}_-\{u\}_-^*}{\lambda^2 - \left(C - \frac{BB^*}{A}\right)} . \tag{18.100}$$

In this case

$$
\begin{aligned}
\{u\}_- &= \frac{\pi}{d_1^- g^* \zeta^\rho} \\
A(\zeta) &= \frac{gg^*}{d_{F_w F_m F_1 F_2 F_2 HGV_1 V_2} d^*_{F_w F_m F_1 F_2 F_2 HGV_1 V_2}} \\
B(\zeta) &= \frac{\beta}{d_{F_w F_1 F_2 F_2 HGV_1} d^*_{F_w F_2 V_1}} \\
C(\zeta) &= \frac{\gamma}{d_{F_w F_2 V_1} d^*_{F_w F_2 V_1}} \,.
\end{aligned}
\tag{18.101}
$$

Calculate the denominator of the function Σ (18.100) by

$$
\lambda^2 - \left(C - \frac{BB^*}{A}\right) = \lambda^2 - \frac{\gamma gg^* - \beta\beta^* d_{F_m V_2} d^*_{F_m V_2}}{gg^* d_{F_w F_2 V_1} d^*_{F_w F_2 V_1}} = \frac{\lambda^2 gg^* d_{F_w V_1} d^*_{F_w V_1} - \psi}{gg^* d_{F_w V_1} d^*_{F_w V_1}}
$$

where

$$
\psi(\zeta) = \frac{\gamma gg^* - \beta\beta^* d_{F_m V_2} d^*_{F_m V_2}}{d_{F_2} d^*_{F_2}} \,. \tag{18.102}
$$

Based on Theorem 17.2, it can be proved that the function $\psi(\zeta)$ (18.102) is a quasipolynomial, i.e., has no poles at the zeros of $d_{F_2} d^*_{F_2}$.
Thus, the function Σ has the form

$$
\Sigma = \frac{n_\sigma n_\sigma^*}{d_\sigma d_\sigma^*} = \frac{\pi\pi^* d_{F_w V_1} d^*_{F_w V_1}}{d_1^- (d_1^-)^* (\lambda^2 gg^* d_{F_w V_1} d^*_{F_w V_1} - \psi)} \,. \tag{18.103}
$$

Factorize the polynomial $\pi(\zeta)$ as $\pi(\zeta) = \pi^+\pi^-$, where $\pi^+(\zeta)$ and $\pi^-(\zeta)$ denotes the stable and anti-stable factors, respectively. Performing the factorization of (18.103) with the above notation gives

$$
n_\sigma = \pi^+ \tilde{f} d_{F_w V_1} \,, \qquad d_\sigma = \tilde{d}_1^- \sigma \tag{18.104}
$$

where $f(\zeta) = \pi^-(\zeta)$, and the stable polynomial $\sigma(\zeta)$ is a result of the factorization

$$
\sigma\sigma^* = \lambda^2 gg^* d_{F_w V_1} d^*_{F_w V_1} - \psi \,.
$$

Substitute (18.104) into Eqs. (18.87) and (18.88), and using the equality $\pi = \pi^+ f$ yields

$$
g^* \zeta^\rho N + \pi^+ f \sigma \tilde{d}_1^- d_{F_w F_2 V_1} = \pi^+ \tilde{f} \tilde{d}_1^- d_{F_w V_1} \beta^* d^*_{F_m V_2} \zeta^\rho \tag{18.105}
$$

$$
g^* \zeta^\rho d_1^- D - \pi^+ f \sigma \zeta \chi_\tau \tilde{d}_1^- d_{F_w F_m V_1 V_2} = \pi^+ \tilde{f} \tilde{d}_1^- d_{F_w V_1} \omega \,. \tag{18.106}
$$

As follows from Eqs. (18.105) and (18.106), the polynomials N and D are divisible by $\pi^+ \tilde{d}_1^- d_{F_w V_1}$, i.e.,

$$
N = \pi^+ \tilde{d}_1^- d_{F_w V_1} N_1 \,, \qquad D = \pi^+ \tilde{d}_1^- d_{F_w V_1} D_1 \,.
$$

The polynomials $N_1(\zeta)$ and $D_1(\zeta)$ are defined by the equations

$$g^*\zeta^\rho N_1 + f\sigma d_{F_2} = \tilde{f}\beta^* d^*_{F_m V_2}\zeta^\rho \tag{18.107}$$

$$g^*\zeta^\rho d_1^- D_1 - f\sigma\zeta\chi_\tau d_{F_m V_2} = \tilde{f}\omega\,. \tag{18.108}$$

As was shown in Rosenwasser and Polyakov (1996), the polynomials $N_1(\zeta)$ and $D_1(\zeta)$ are divisible by $g(\zeta)$, i.e.,

$$N_1(\zeta) = gN_2(\zeta), \qquad D_1(\zeta) = gD_2(\zeta)\,.$$

Using this fact in (18.107) and (18.108) and dropping the index '2' of the polynomials N_2 and D_2, we obtain (18.96) and (18.97). These equations are similar to those obtained in Grimble (1995) for discrete LTI systems. ■

For practical computations it is more convenient to use other forms of Eqs. (18.96-18.97) derived below.
Eliminating first $f(\zeta)$, and then $\tilde{f}(\zeta)$ from Eqs. (18.96)–(18.97), after regrouping the terms and taking account of (18.22) we obtain the following proposition.

Corollary 1 The polynomials $N(\zeta)$, $D(\zeta)$, and $f(\zeta)$, appearing in Theorem 18.6, can be found as a solution of the system of equations

$$\tilde{f} = \zeta\chi_\tau d_{F_m V_2} N + d_1^- d_{F_2} D \tag{18.109}$$

$$f\sigma = -\omega N + \beta^* d_1^- d^*_{F_m V_2}\zeta^\rho D\,. \tag{18.110}$$

Eliminating the polynomial σ from (18.109) and (18.110), we obtain the following equations, with a form similar to that of the robust optimization equations obtained for continuous-time LTI systems in Kwakernaak (1985), which contain λ in an explicit form.

Multiply each of the Eqs. (18.109))–((18.110) by the conjugate one

$$\begin{aligned} ff^* = {} & \chi_\tau\chi_\tau^* d_{F_m V_2} d^*_{F_m V_2} NN^* + d_1^-(d_1^-)^* d_{F_2} d^*_{F_2} DD^* + \\ & + \zeta\chi_\tau d_{F_m V_2}(d_1^-)^* d^*_{F_2} ND^* + \zeta^{-1}\chi_\tau^* d^*_{F_m V_2} d_1^- d_{F_2} N^* D \end{aligned} \tag{18.111}$$

$$\begin{aligned} ff^*\sigma\sigma^* = {} & \omega\omega^* NN^* + d_1^-(d_1^-)^*\beta\beta^* d_{F_m V_2} d^*_{F_m V_2} DD^* - \\ & - \omega\beta(d_1^-)^* d_{F_m V_2}\zeta^{-\rho} ND^* - \omega^*\beta^* d_1^- d^*_{F_m V_2}\zeta^\rho N^* D\,. \end{aligned} \tag{18.112}$$

Using (18.94), the right-hand side of the last equation can be written in the form

$$ff^*\sigma\sigma^* = ff^*(\lambda^2 gg^* d_{F_w V_1} d^*_{F_w V_1} - \psi)\,.$$

Hence,

$$ff^*\lambda^2 gg^* d_{F_w V_1} d^*_{F_w V_1} = ff^*\psi + ff^*\sigma\sigma^*\,.$$

Using (18.112) in the first summand of the right-hand side and (18.113) in the second one, after cancellation we obtain

$$\alpha_f f f^* \lambda^2 g g^* = g g^* \alpha_N N N^* + g g^* \alpha_D D D^* + g g^* \alpha_{ND} N D^* + g g^* \alpha_{ND}^* N^* D$$

where

$$\alpha_f = d_{F_w V_1} d_{F_w V_1}^* \tag{18.113}$$

$$\alpha_N = \frac{g g^* + d_{F_m V_2} d_{F_m V_2}^* (\gamma \chi_\tau \chi_\tau^* - \beta \zeta^{-1} \chi_\tau^* - \beta^* \zeta \chi_\tau)}{d_{F_2} d_{F_2}^*} \tag{18.114}$$

$$\alpha_D = d_1^- (d_1^-)^* \gamma \tag{18.115}$$

$$\alpha_{ND} = \frac{(d_1^-)^* d_{F_m V_2} (\gamma \zeta \chi_\tau - \beta)}{d_{F_2}} . \tag{18.116}$$

Cancelling both sides by gg^* gives the following result.

Corollary 2 The polynomials $N(\zeta)$, $D(\zeta)$, and $f(\zeta)$, appearing in Theorem 18.6, can be found as a solution of the system of equations

$$\tilde{f} = \zeta \chi_\tau d_{F_m V_2} N + d_1^- d_{F_2} D \tag{18.117}$$

$$\alpha_f \lambda^2 f f^* = \alpha_N N N^* + \alpha_D D D^* + \alpha_{ND} N D^* + \alpha_{ND}^* N^* D \tag{18.118}$$

where the quasipolynomials α_f, α_N, α_D, and α_{ND} are defined by (18.113)–(18.116).

Equations (18.117)–(18.118) allow us to obtain the order of the optimal controller. Denote

$$x = \delta_{F_w F_2 V_1} + \delta_{F_1 HG}^-$$
$$y = \delta_{F_w F_m F_1 F_2 HG V_1 V_2} + N_\tau$$
$$z = \delta_{F_w F_m F_1 F_2 F_2 HG V_1 V_2} + \delta_{F_1 HG}^- + N_\tau$$

where the integer N_τ is defined by the equality

$$\tau_1 + \tau_2 = N_\tau T - \delta T, \qquad 0 \le \delta < 1 .$$

Theorem 18.7 (Rosenwasser and Polyakov (1996)) *Under the assumptions i) – vi) of Section 18.2, for a sufficiently large λ Eqs. (18.96) and (18.97) have a solution $\{N, D, f\}$, such that*

$$\deg N = x , \qquad \deg D = y , \qquad \deg(f) = z \tag{18.119}$$

and the polynomial $f(\zeta)$ is strictly anti-stable. In this case, for a given λ, the degrees of the numerator and denominator of the full-order robust controller are defined by

$$\deg b = \delta_{F_w F_m F_1 F_2 HG V_1 V_2}$$
$$\deg a = \delta_{F_w F_m F_1 F_2 HG V_1 V_2} + N_\tau .$$

■

In order to find the *optimal* controller, i.e., to find the minimal λ, for which the polynomial f is strictly anti-stable (and, therefore, the closed-loop system is stable), numerical computations must be used.
Denote by λ_0 the first value λ, for which the right side of the equation

$$\sigma\sigma^* = \lambda^2 gg^* d_{F_w V_1} d^*_{F_w V_1} - \psi \tag{18.120}$$

possesses a root on the unit circle as λ decreases from ∞. If the factorization (18.21) is possible, then $gg^* d_{F_w V_1} d^*_{F_w V_1} > 0$ for $|\zeta| = 1$, hence $\lambda_0 < \infty$. Then, the optimal robust controller minimizing the cost functional (18.82) is given by the following theorem.

Theorem 18.8 (Rosenwasser and Polyakov (1996))
a) Generic case: *If the polynomial f_{λ_0}, which is the solution of Eqs. (18.96) and (18.97) (or, equivalently, (18.123)) for $\lambda = \lambda_0$, has roots outside the unit circle (stable), there is $\lambda > \lambda_0$, for which the solution of Eqs. (18.96) and (18.97) has the reduced degree*

$$\deg N = x - 1\,, \qquad \deg D = y - 1\,, \qquad \deg(f) = z - 1$$

and the corresponding polynomial $f(\zeta)$ is strictly anti-stable. The degrees of the numerator and denominator of the transfer function of the optimal robust controller are given by

$$\begin{aligned} \deg b &= \delta_{F_w F_m F_1 F_2 H G V_1 V_2} - 1 \\ \deg a &= \delta_{F_w F_m F_1 F_2 H G V_1 V_2} + \delta_\tau - 1\,. \end{aligned}$$

b) Non-generic case: *If the polynomial f_{λ_0} is strictly anti-stable, there are two optimal solutions (associated with two different factorizations (18.120) having the opposite signs). The degrees of the polynomials N, D, and f are defined by (18.119).* ■

Similarly to the continuous case, Kwakernaak (1985), in order to construct the optimal controller which ensures the minimal norm $\|X\|_\infty$, in the generic case we must find the *reduced-order solution*, i.e., find such λ, for which one of the roots of the polynomial $f(\zeta)$ crosses the stability boundary and $f(\zeta)$ becomes unstable. In this case the polynomials $f(\zeta)$, $N(\zeta)$, and $D(\zeta)$ have a common factor on the stability boundary, that can be cancelled. The robust optimization equations based on Theorem 18.6 and Corollaries 1 and 2 can be solved by various numerical methods, including those given in Brown et al. (1987); Saeki (1989).

For a numerical solution of the robust optimization problem it is often convenient to use a combined method, which, on the one hand, uses a basic controller a^0, b^0 and parameter-function $\phi(\zeta)$, and, on the other hand, is based on polynomial equations. In this case it is important that the degrees of the numerator and denominator of the optimal controller are known in advance from Theorem 18.8.

One advantage of this approach is, that for a given λ^2, the problem reduces to a single Diophantine equation. This makes it easier to find by the binary search method, such λ that the polynomial f becomes unstable as λ decreases from ∞. Having obtained the optimal function ϕ, we substitute it into Eq. (17.40), taking account of the fact that the degrees of the numerator and denominator of the optimal controller transfer function are known from Theorem 18.8.

Theorem 18.9 *Under the assumptions i) – vi) of Section 18.2, for a given sufficiently large λ^2 the robust stabilizing controller, which minimizes the cost function (18.83) and ensures (18.85), is computed as follows:*

1. *Find the polynomials $\chi_\tau(\zeta)$ and $\alpha(\zeta)$ according to (18.11).*
2. *Find the polynomials $\tilde{b}^0$ and $\tilde{a}^0$ as the minimal solution of the Diophantine equation (18.13).*
3. *Calculate the discrete Laplace transforms (18.90), (18.91), (18.92), and (18.93).*
4. *Find the stable polynomial $g(\zeta)$ as a result of the factorization (18.21).*
5. *Calculate the polynomial $\omega_1(\zeta)$ as*

$$\omega_1(\zeta) = \frac{\left(\beta^* d_{F_m V_2} d^*_{F_m V_2} - \tilde{b}^0 g g^*\right)\zeta^\rho}{d_{F_2}} \tag{18.121}$$

   where ρ is the minimal non-negative integer, such that the functions

$$\eta_1(\zeta) = \beta^* d^*_{F_m V_2}\zeta^\rho\,, \qquad \eta_2(\zeta) = g^*\zeta^\rho \tag{18.122}$$

   become polynomials in ζ.
6. *Find the stable polynomial $\sigma(\zeta)$ as a result of the factorization (18.103).*
7. *Find the polynomials $\vartheta(\zeta)$ and $f(\zeta)$, obtaining the minimal solution with respect to $f(\zeta)$ of the equation*

$$d_1^- g g^* \zeta^\rho \vartheta + f\sigma d_{F_m V_2} = \tilde{f}\omega_1 \tag{18.123}$$

   with a strictly anti-stable polynomial $f(\zeta)$ (having all roots inside the unit circle of the ζ-plane).
8. *The transfer function of the optimal controller has the form*

$$W_d(\zeta) = \frac{\tilde{b}^0 \tilde{f} d_1^+ + \alpha\vartheta}{\tilde{a}^0 \tilde{f} d_1^+ - \zeta\chi_\tau\vartheta}\,. \tag{18.124}$$

9. *The characteristic polynomial of the closed-loop system is given by*

$$\Delta = \tilde{f} d_1^+\,.$$

Proof As was shown above, the function ϕ, which is the solution of the weighted linear quadratic optimization problem, takes the form

$$\phi(\zeta) = \frac{\vartheta}{n_\sigma g d_1^+ \tilde{d}_1^-}.$$

The polynomial ϑ appearing in its numerator can be found as a result of the separation (18.29), or, equivalently, as the minimal solution with respect to π of the Diophantine equation (18.34), both sides of which can be divided by d_{F_2}, so that

$$d_1^- g^* \zeta^\rho \vartheta + \pi d_{\sigma F_w F_m V_1 V_2} = n_\sigma \tilde{d}_1^- \omega_1 \tag{18.125}$$

where $\omega_1(\zeta)$ is calculated by (18.121). Based on Theorem 17.2, we can easily prove that the function $\omega_1(\zeta)$ (18.121) is a polynomial.
To solve the robust design problem, we substitute (18.104)

$$n_\sigma = \pi^+ \tilde{f} d_{F_w V_1}\,, \qquad d_\sigma = \tilde{d}_1^- \sigma \tag{18.126}$$

into (18.125). Taking into account that $\pi = \pi^+ f$, we have

$$d_1^- g^* \zeta^\rho \vartheta + \pi^+ f \tilde{d}_1^- \sigma d_{F_w F_m V_1 V_2} = \pi^+ \tilde{f} d_{F_w V_1} \tilde{d}_1^- \omega_1\,.$$

We note that the first term on the lefthand and the right-hand side are divisible by $\pi^+ \tilde{d}_1^- d_{F_w V_1}$. Therefore, the first summand on the left-hand side or, more correctly, the polynomial ϑ, must also possess this factor. Thus,

$$\vartheta = \pi^+ \tilde{d}_1^- d_{F_w V_1} \vartheta_1 \tag{18.127}$$

where $\vartheta_1(\zeta)$ is a polynomial to be found as the minimal solution with respect to f of the polynomial equation

$$d_1^- g^* \zeta^\rho \vartheta_1 + f \sigma d_{F_m V_2} = \tilde{f} \omega_1\,.$$

Then, taking account of (18.127) and (18.126), the optimal function ϕ has the form

$$\phi = \frac{\vartheta_1}{\tilde{f} g d_1^+}.$$

Since the denominator of the function ϕ is the characteristic polynomial of the optimal closed-loop system, and the latter, as was shown above, is equal to $\Delta = \tilde{f} d_1^+$, the polynomial ϑ_1 is divisible by g. Therefore, $\vartheta_1 = g\vartheta_2$, where $\vartheta_2(\zeta)$ is a polynomial which can be found as the minimal solution with respect to $f(\zeta)$ of the equation

$$d_1^- g g^* \zeta^\rho \vartheta_2 + f \sigma d_{F_m V_2} = \tilde{f} \omega_1$$

which coincides with (18.123), if we drop the index '2' of ϑ_2. ∎

Example 18.3 Find the optimal robust controller for the system investigated in Example 18.1. The minimal solution of the Diophantine equation (18.13) appears as

$$\tilde{b}^0 = -4.3003\,, \qquad \tilde{a}^0 = -2.7183\,.$$

In this case Eq. (18.93) has the form

$$D_{C_1}(T,\zeta,0) = D_{F_2\underline{F_2}}(T,\zeta,0) = \frac{0.4323}{(\zeta - 0.3679)(\zeta^{-1} - 0.3679)}\,.$$

Hence, $\gamma = 0.4323$. Using (18.40) and (18.39), from (18.95) and (18.121) we find, respectively

$$\psi = 0.0043\zeta + 0.0186 + 0.043\zeta^{-1} \tag{18.128}$$

$$\omega_1 = 0.1198\zeta^2 + 0.5164\zeta + 0.1394\,. \tag{18.129}$$

With these data we obtain $\lambda_0 = 0.3972$. The corresponding polynomial $f = \zeta^2 + 4.4958\zeta + 1.0657$, which is a solution of (18.94) and (18.123), has a stable root $\zeta = -3.9832$. Therefore, we have the generic case. By numerical methods we find that the polynomial $f(\zeta)$ turns out to be strictly anti-stable for $\lambda = 1.5208$. Then, the corresponding full-order solution has the form

$$f = (\zeta + 0.2865)(\zeta + 1)\,, \qquad \vartheta = 1.2322(\zeta + 1)\,. \tag{18.130}$$

Both polynomials have a common root $(\zeta + 1)$ on the stability boundary, which can be cancelled. As a result, the reduced-order solution appears as

$$f = \zeta + 0.2865, \qquad \vartheta = 1.2322\,. \tag{18.131}$$

Substituting these polynomials into (18.124), we obtain the optimal robust controller $W_d(\zeta) = 1.7487$, which, as for the linear quadratic problem, is a static gain. □

Appendix A

Rational periodic functions

A.1 Basic definitions

Let us have an irreducible real rational function

$$f_o(z) \stackrel{\triangle}{=} \frac{q_o(z)}{r_o(z)} \tag{A.1}$$

with the polynomials

$$\begin{aligned} q_o(z) &= q_0 z^m + q_1 z^{m-1} + \ldots + q_m \\ r_o(z) &= r_0 z^m + r_1 z^{m-1} + \ldots + r_m \end{aligned} \tag{A.2}$$

where q_i and r_i are constants, some of which may be equal to zero, but $\max\{|q_0|, |r_0|\} > 0$. Using the change of variable

$$z = \mathrm{e}^{sT} \tag{A.3}$$

in (A.1), we obtain the function

$$f(s) \stackrel{\triangle}{=} f_o(z)\,|_{z=\mathrm{e}^{sT}} = \frac{q(s)}{r(s)}\,. \tag{A.4}$$

In (A.4) $q(s)$ and $r(s)$ are quasi-polynomials

$$\begin{aligned} q(s) &= q_0 \mathrm{e}^{msT} + q_1 \mathrm{e}^{(m-1)sT} + \ldots + q_m \\ r(s) &= r_0 \mathrm{e}^{msT} + r_1 \mathrm{e}^{(m-1)sT} + \ldots + r_m\,. \end{aligned} \tag{A.5}$$

Henceforth, functions of the form (A.4) will be called *rational periodic functions.* Assuming that in (A.1)

$$\zeta = z^{-1} \tag{A.6}$$

we obtain the new real rational function

$$f^o(\zeta) \stackrel{\triangle}{=} f_o(z)\,|_{z=\zeta^{-1}} \tag{A.7}$$

which admits the representation

$$f^o(\zeta) = \frac{q^o(\zeta)}{r^o(\zeta)} \tag{A.8}$$

where $q^o(\zeta)$ and $r^o(\zeta)$ are co-prime polynomials in the variable ζ. In what follows, the function (A.8) will be called *inverse* with respect to $f_o(z)$. Obviously, the inversion operation is reversible, so that

$$f_o(z) = f^o(\zeta)\,\big|_{\zeta=z^{-1}}\,. \tag{A.9}$$

Assuming that in (A.8)

$$\zeta = \mathrm{e}^{-sT} \tag{A.10}$$

we obtain the new representation of the rational periodic function

$$f(s) = \left.\frac{q^o(\zeta)}{r^o(\zeta)}\right|_{\zeta=\mathrm{e}^{-sT}} \tag{A.11}$$

where

$$\begin{aligned} q^o(\zeta) &= q_0 + q_1\zeta + \ldots + q_m\zeta^m \\ r^o(\zeta) &= r_0 + r_1\zeta + \ldots + r_m\zeta^m \end{aligned} \tag{A.12}$$

are polynomials in the variable $\zeta = \mathrm{e}^{-sT}$. Regarding the rational periodic function $f(s)$ as initial, by the substitution (A.3) we can obtain the real rational function $f^o(\zeta)$. Henceforth, the functions $f_o(z)$ and $f^o(\zeta)$ will be called the *rational representations* of the rational periodic function $f(s)$.

Example A.1 Let us have the rational periodic function

$$f(s) = \frac{5\mathrm{e}^{3sT} + 2\mathrm{e}^{2sT} + 7}{\mathrm{e}^{4sT} + 3\mathrm{e}^{sT}}\,. \tag{A.13}$$

Passing to the variable z by (A.3), we obtain the rational representation

$$f_o(z) = \frac{5z^3 + 2z^2 + 7}{z^4 + 3z}\,. \tag{A.14}$$

From (A.14) by (A.6) we find the inverse function with respect to (A.14)

$$f^o(\zeta) = \frac{5\zeta + 2\zeta^2 + 7\zeta^4}{1 + 3\zeta^3}\,. \tag{A.15}$$

From (A.15) by means of (A.10) we obtain the representation of the form (A.11)

$$f(s) = \frac{\mathrm{e}^{-sT}\left(5 + 2\mathrm{e}^{-sT} + 7\mathrm{e}^{-3sT}\right)}{1 + 3\mathrm{e}^{-3sT}}\,.$$

In this example

$$q_o(z) = 5z^3 + 2z^2 + 7\,, \quad r_o(z) = z^4 + 3z$$

and

$$q^o(\zeta) = 5\zeta + 2\zeta^2 + 7\zeta^4\,, \quad r^o(\zeta) = 1 + 3\zeta^3\,.$$

□

A.2 Causal and limited rational periodic functions

A rational periodic function $f(s)$ is called *causal*, if the finite limit exists:

$$\ell^+[f(s)] \triangleq \lim_{\operatorname{Re} s\to\infty} f(s) < \infty\,. \tag{A.16}$$

It can be proved that for a causal rational periodic function the tendency to the limits in (A.16) is uniform with respect to $\operatorname{Im} s$. From (A.1) and (A.8) taking due account of (A.3) and (A.10) it follows that

$$\ell^+[f(s)] = \lim_{z\to\infty} f_o(z) \tag{A.17}$$

$$\ell^+[f(s)] = \lim_{\zeta\to 0} f^o(\zeta)\,. \tag{A.18}$$

The following propositions follow from (A.17) and (A.18).

1. A rational periodic function $f(s)$ is causal if and only if the polynomials (A.2) satisfy the condition

$$\deg r_o(z) \geq \deg q_o(z)\,. \tag{A.19}$$

2. The rational periodic function $f(s)$ is causal if and only if the function $f^o(\zeta)$ is analytic in the origin $\zeta = 0$. Since the ratio (A.8) is assumed to be irreducible, the last condition is equivalent to

$$r^o(0) \neq 0 \tag{A.20}$$

i.e., the polynomial $r^o(\zeta)$ should not have zero roots. Henceforth, the functions $f_o(z)$ and $f^o(\zeta)$ will be called *causal*, if the condition (A.16) holds.

Example A.2 The rational periodic function (A.13) is causal, because $\ell^+[f(s)] = 0$. □

Example A.3 The rational periodic function

$$f(s) = \frac{1 - \mathrm{e}^{-sT} + 3\mathrm{e}^{-2sT}}{\mathrm{e}^{-sT} - \mathrm{e}^{-2sT}}$$

is not causal, because $\ell^+[f(s)] = \infty$. □

A causal rational periodic function $f(s)$ is called *limited*, if a finite limit exists:

$$\ell^-[f(s)] \triangleq \lim_{\operatorname{Re} s\to -\infty} f(s) < \infty\,. \tag{A.21}$$

It can be shown that for a limited rational periodic function $f(s)$ the tendency to the limit in (A.21) is uniform with respect to $\operatorname{Im} s$. As follows from the aforesaid,

$$\ell^-[f(s)] = \lim_{z\to 0} f_o(z) \tag{A.22}$$

$$\ell^-[f(s)] = \lim_{\zeta\to\infty} f^o(\zeta)\,. \tag{A.23}$$

Combining Eqs. (A.17), (A.18), (A.22), and (A.23), we obtain the following proposition: A rational periodic function $f(s)$ is limited if and only if its rational representations $f_o(z)$ and $f^o(\zeta)$ are proper and analytic in origin. Under these conditions the functions $f_o(z)$ and $f^o(\zeta)$ will also be called *limited.*

A causal rational periodic function will be called *retarded*, if it is not limited. If a rational periodic function $f(s)$ is retarded, the corresponding function $f^o(\zeta)$ takes the form

$$f^o(\zeta) = \frac{q_0 + q_1\zeta + \ldots + q_{p+\kappa}\zeta^{p+\kappa}}{r_0 + r_1\zeta + \ldots + r_p\zeta^p} \tag{A.24}$$

where q_i, r_i, $\kappa > 0$, $r_0 \neq 0$, $q_{p+\kappa} \neq 0$, and $r_p \neq 0$ are constsnts. Dividing the numerator in (A.24) by the denominator, we obtain

$$f^o(\zeta) = l_\kappa\zeta^\kappa + l_{\kappa-1}\zeta^{\kappa-1} + \ldots + l_1\zeta + f_1^o(\zeta) \tag{A.25}$$

where

$$f_1^o(\zeta) = \frac{\tilde{q}_0 + \tilde{q}_1\zeta + \ldots + \tilde{q}_p\zeta^p}{r_0 + r_1\zeta + \ldots + r_p\zeta^p} \tag{A.26}$$

is a limited function and l_i $(i = 1, \ldots, \kappa)$ are constants. Passing back to the indeterminate s in (A.25), by (A.10) we obtain

$$f(s) = l_\kappa \mathrm{e}^{-\kappa sT} + l_{\kappa-1}\mathrm{e}^{-(\kappa-1)sT} + \ldots + l_1\mathrm{e}^{-sT} + f_1(s) \tag{A.27}$$

where

$$f_1(s) = \frac{\tilde{q}_0 + \tilde{q}_1\mathrm{e}^{-sT} + \ldots + \tilde{q}_p\mathrm{e}^{-psT}}{r_0 + r_1\mathrm{e}^{-sT} + \ldots + r_p\mathrm{e}^{-psT}} \tag{A.28}$$

is a limited function. Thus, a retarded function can always be represented as the sum of a limited function and a quasi-polynomial in the variable $\zeta = \mathrm{e}^{-sT}$. In the special case when $f_1(s) = 0$, we have a quasi-polynomial

$$f(s) = l_\kappa \mathrm{e}^{-\kappa sT} + l_{\kappa-1}\mathrm{e}^{-(\kappa-1)sT} + \ldots + l_1\mathrm{e}^{-sT}\,. \tag{A.29}$$

Example A.4 Under the conditions of Example A.1 in the representation (A.15) the degree of the numerator is higher than that of the denominator. Therefore, the rational periodic function (A.13) is retarded. Dividing the numerator by the denominator in (A.15) gives

$$f^o(\zeta) = \frac{7}{3}\zeta + \frac{\frac{8}{3}\zeta + 2\zeta^2}{1 + 3\zeta^3}\,. \tag{A.30}$$

From (A.30) we find the representatioon of the form (A.27) for $f(s)$

$$f(s) = \frac{7}{3}\mathrm{e}^{-sT} + \frac{\mathrm{e}^{-sT}\left(\frac{8}{3} + 2\mathrm{e}^{-sT}\right)}{1 + 3\mathrm{e}^{-3sT}}\,.$$

□

Example A.5 The rational periodic function

$$f(s) = \frac{3e^{sT} + 2}{2e^{sT} - 1} = \frac{3 + 2e^{-sT}}{2 - e^{-sT}}$$

is limited, because in the given case

$$\ell^{+}[f(s)] = \frac{3}{2}, \quad \ell^{-}[f(s)] = -2\,.$$

In this case we have the rational representations

$$f_o(z) = \frac{3z + 2}{2z - 1}, \quad f^o(\zeta) = \frac{3 + 2\zeta}{2 - \zeta}\,. \qquad \square$$

Using the above reasoning, we can obtain the general representation for an arbitrary rational periodic function $f(s)$, that is not causal. In this case the rational representation has the form

$$f^o(\zeta) = \frac{q^o(\zeta)}{\zeta^\lambda \tilde{r}^o(\zeta)} \tag{A.31}$$

where $q^o(\zeta)$ and $\tilde{r}^o(\zeta)$ are polynomials such that $\tilde{r}^o(0) \neq 0$, and λ is a positive integer. By means of partial fraction expansion, we can represent (A.31) in the form

$$f^o(\zeta) = \frac{l_{-\lambda}}{\zeta^\lambda} + \ldots + \frac{l_{-1}}{\zeta} + \tilde{f}^o(\zeta) \tag{A.32}$$

where l_{-i} are constants and the function $\tilde{f}^o(\zeta)$ is causal. Using the representation (A.25) for $\tilde{f}^o(\zeta)$, we obtain

$$f^o(\zeta) = \frac{l_{-\lambda}}{\zeta^\lambda} + \ldots + \frac{l_{-1}}{\zeta} + \tilde{f}_1^o(\zeta) + l_\kappa \zeta^\kappa + \ldots + l_1 \zeta \tag{A.33}$$

where $\tilde{f}_1^o(\zeta)$ is a limited real rational function . Changing the variable in (A.33) for s, we obtain

$$f(s) = l_{-\lambda} e^{\lambda sT} + \ldots + l_{-1} e^{sT} + \tilde{f}_1(s) + l_1 e^{-sT} + \ldots + l_\kappa e^{-\kappa sT} \tag{A.34}$$

with a limited rational periodic function $\tilde{f}_1(s)$.

Example A.6 The rational periodic function

$$f(s) = \frac{1 + 3e^{-sT} + 2e^{-2sT}}{e^{-sT} - 2e^{-2sT}}$$

is not causal. Its representation (A.34) has the form

$$f(s) = e^{sT} + \frac{5 + 2e^{-sT}}{1 - 2e^{-sT}}\,. \qquad \square$$

A.3 Zeros and poles of rational periodic functions

Since the exponential function is unique, it follows from (A.34) that only the poles of the limited function $\tilde{f}_1(s)$ can be singular points of the function $f(s)$ in any finite region of the complex plane. Hence, the rational periodic function $f(s)$ is a meromorphic function in the indeterminate s. In this section we investigate some general properties of the zeros and poles of an arbitrary rational periodic function which are useful for the further exposition.

1.

Lemma A.1 *Let us consider a quasi-polynomial*

$$r(s) = r_0 + r_1 e^{-sT} + \ldots + r_p e^{-psT} \tag{A.35}$$

where r_i are constants and $r_p \neq 0$, and the corresponding polynomial $r^o(\zeta)$ given by

$$r^o(\zeta) = r_0 + r_1\zeta + \ldots + r_p\zeta^p\,. \tag{A.36}$$

Then, if $\tilde{s}$ is a root of the equation

$$r(s) = r_0 + r_1 e^{-sT} + \ldots + r_p e^{-psT} = 0 \tag{A.37}$$

of multiplicity ν, the number $\tilde{\zeta} = e^{-\tilde{s}T}$ is a root of the polynomial $r^o(\zeta)$ of the same multiplicity.
The inverse proposition is also valid. If $\tilde{\zeta}$ is a root of the polynomial $r^o(\zeta)$ of multiplicity ν, any number $\tilde{s}$ with $e^{-\tilde{s}T} = \tilde{\zeta}$ is also a root of Eq. (A.37) of the same multiplicity.

Proof Without loss of generality we may take $r_p = 1$. Let the polynomial $r^o(\zeta)$ have zeros $\zeta_1, \ldots, \zeta_m$ of multiplicities $\nu_1, \ldots, \nu_m$, respectively. Then,

$$r^o(\zeta) = (\zeta - \zeta_1)^{\nu_1} \cdots (\zeta - \zeta_m)^{\nu_m}\,. \tag{A.38}$$

Assuming that $\zeta = e^{-sT}$ in (A.38),

$$r(s) = \left(e^{-sT} - \zeta_1\right)^{\nu_1} \cdots \left(e^{-sT} - \zeta_m\right)^{\nu_m}\,. \tag{A.39}$$

Denote

$$r_\nu(s, a) \triangleq \left(e^{-sT} - a\right)^{\nu} \tag{A.40}$$

where $\nu > 0$ is an integer. It can be immediately checked that all roots of the equation

$$r_1(s, a) = e^{-sT} - a = 0$$

are simple, because

$$\frac{\partial r_1(s, a)}{\partial s} = -Te^{-sT} \neq 0\,.$$

Therefore, all roots of the equation

$$r_\nu(s, a) = 0$$

have multiplicity ν. Taking due account of this fact, a comparison of the expansions (A.38) and (A.39) proves the proposition of the lemma. ■

Using Lemma A.1, we can determine the general character of the location of the zeros of the function $r(s)$ (A.35) in the complex plane. Since

$$r(s) = r(s + \mathrm{j}\omega), \quad \omega = 2\pi/T$$

together with a root s the values $s_k = s + k\mathrm{j}\omega$ for any integer k are also roots of Eq.(A.37). Let us consider the expansion (A.39) and let the numbers $\tilde{s}_i$ for each i be such that

$$\tilde{s}_i = -\frac{1}{T}\ln\zeta_i .$$

Then, as follows from Lemma A.1, in any point

$$\tilde{s}_{ik} = \tilde{s}_i + k\mathrm{j}\omega \tag{A.41}$$

the function $r(s)$ has a zero of multiplicity ν. The aggregate of the numbers (A.41) gives the whole set of zeros of the function $r(s)$. Among the numbers (A.41) for a fixed i there is a number $\tilde{s}_{i0}$ such that $-\omega/2 \leq \operatorname{Im}\tilde{s}_{i0} < \omega/2$. The number $\tilde{s}_{i0}$ will be called a *primary root* of the function $r(s)$ corresponding to i. As follows from the aforesaid, if the expansion (A.39) holds, the function $r(s)$ has m primary roots $\tilde{s}_{i0}$ of multiplicities ν_i. In this case the entire set of the roots of (A.41) can be represented in the form

$$\tilde{s}_{ik} = \tilde{s}_{i0} + k\mathrm{j}\omega \tag{A.42}$$

with any integer k. Since the number of primary roots of the function $r(s)$ is finite, the entire set of roots lie on a finite number of vertical lines $\operatorname{Re}s = \operatorname{Re}\tilde{s}_{i0}$ $(i = 1, \ldots, m)$.

2. Let us have the rational periodic function

$$f(s) = \frac{q_0 + q_1\mathrm{e}^{-sT} + \ldots + q_p\mathrm{e}^{-psT}}{r_0 + r_1\mathrm{e}^{-sT} + \ldots + r_p\mathrm{e}^{-psT}} = \frac{q(s)}{r(s)} \tag{A.43}$$

and the corresponding rational representation

$$f^o(\zeta) = \frac{q_0 + q_1\zeta + \ldots + q_p\zeta^p}{r_0 + r_1\zeta + \ldots + r_p\zeta^p} = \frac{q^o(\zeta)}{r^o(\zeta)} . \tag{A.44}$$

The function (A.43) is called *reducible*, if there exists a value of the indeterminate $s = \tilde{s}$ that simultaneously

$$q(\tilde{s}) = 0, \quad r(\tilde{s}) = 0 . \tag{A.45}$$

If there are not such numbers satisfying (A.45) the function $f(s)$ will be called *irreducible*.

Lemma A.2 *The function (A.43) is reducible if and only if the real rational function (A.44) is reducible.*

Proof Let $f(s)$ be reducible, so that for $s = \tilde{s}$ Eq. (A.45) holds. Then, by Lemma A.1, the number $\tilde{\zeta} = \mathrm{e}^{-\tilde{s}T}$ is simultaneously a root of the polynomials $q^o(\zeta)$ and $r^o(\zeta)$, i.e., the ratio $f^o(\zeta)$ is reducible. Conversely, let the ratio (A.44) be reducible, i.e., its numerator and denominator have a common root $\tilde{\zeta}$. Then, by Lemma A.1, any number $\tilde{s}$ such that $\mathrm{e}^{-\tilde{s}T} = \tilde{\zeta}$ satisfies (A.45), i.e., the rational periodic function (A.43) is reducible. ∎

3. Let the rational periodic function $f(s)$ (A.43) be irreducible. Then, the set of its poles coincide with the set of the roots of its denominator $r(s)$. Let the numbers $\tilde{s}_{i0}$ $(i = 1, \ldots, m)$ be the primary roots of (A.37). Then, the set of the poles of $f(s)$ is given by (A.42). In this case the poles $\tilde{s}_{i0}$ will be called *primary*. Hence, all poles of $f(s)$ are located in the complex plane on a finite number of vertical lines passing through the primary poles $\tilde{s}_{i0}$ in a finite interval one from another (Fig. A.1).

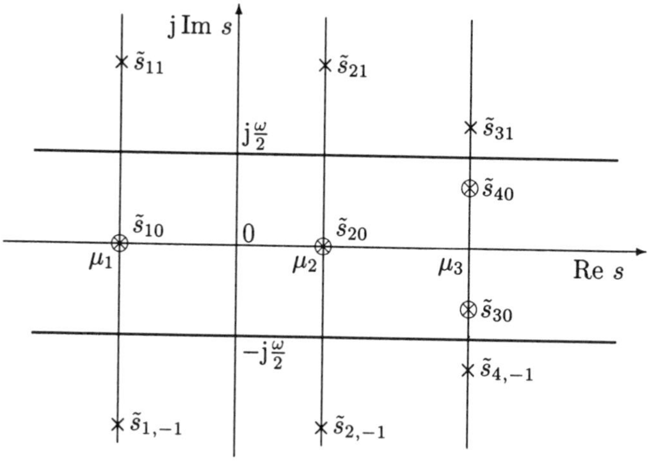

Figure A.1: Location of the poles of a rational periodic function

Denoting by μ_i $(i = 1, \ldots, r)$ the different real parts of the primary poles s_{i0} so that $\mu_i < \mu_{i+1}$, we obtain that the whole complex plane is separated into a finite number of stripes of finite width and left and right half-planes, inside which the function $f(s)$ is analytic, except for, possibly, the infinite point. Denoting $\mu_0 = -\infty$ and $\mu_{r+1} = \infty$, the intervals $S_i\,(\mu_i, \mu_{i+1})$ will be called the *regularity intervals* of the function $f(s)$. The numbers μ_i will be called the *characteristic indices* of the rational periodic function $f(s)$.

A.4 Partial fraction expansion of limited rational periodic functions

Generalizing the definition given in Section A.2, an arbitrary function $f(s)$ in the complex variable s will be called limited, if there exist the following finite limits

$$\ell^{\pm}[f(s)] = \lim_{\text{Re}\, s \to \pm\infty} f(s) \triangleq \ell^{\pm} \tag{A.46}$$

and the trend to the limit is uniform with respect to $\text{Im}\, s$, $-\omega/2 \le \text{Im}\, s \le \omega/2$. The following theorem dealing with limited functions plays a major part in the further exposition.

Theorem A.1 *Let $f(s)$ be a periodic function in s that is analytic everywhere, except for its poles. Let $f(s)$ have a purely imaginary period*

$$f(s) = f(s + \mathrm{j}\omega) \tag{A.47}$$

where ω is a real number. Assume that the limits (A.46) exist uniformly with respect to $\text{Im}\, s$ *in* $-\omega/2 \le \text{Im}\, s < \omega/2$. *Let also the function $f(s)$ have a finite number of poles $s_1, \dots, s_\rho$ with multiplicities $\nu_1, \dots, \nu_\rho$ in the stripe* $-\omega/2 \le \text{Im}\, s < \omega/2$, *which will be called the* primary stripe. *These poles will be henceforth called* primary. *Then, the following identities hold:*

$$f(s) = \frac{T}{2} \sum_{i=1}^{\rho} \operatorname*{Res}_{q=s_i} \left[f(q) \coth \frac{(s-q)T}{2} \right] + \frac{\ell^+ + \ell^-}{2} \tag{A.48}$$

$$f(s) = T \sum_{i=1}^{\rho} \operatorname*{Res}_{q=s_i} \left[f(q) \frac{1}{\mathrm{e}^{(s-q)T} - 1} \right] + \ell^+ \tag{A.49}$$

$$f(s) = T \sum_{i=1}^{\rho} \operatorname*{Res}_{q=s_i} \left[f(q) \frac{1}{1 - \mathrm{e}^{(q-s)T}} \right] + \ell^- \tag{A.50}$$

where $T = 2\pi/\omega$.

Proof First, we prove the identity (A.48), assuming that none of the primary poles s_i lie on the boundary of the primary stripe $-\omega/2 \le \text{Im}\, s < \omega/2$. Consider the contour integral

$$J(s) = \frac{1}{2\pi \mathrm{j}} \oint_{ABCD} f(q) \coth \frac{(q-s)T}{2}\, \mathrm{d}q \tag{A.51}$$

taken along the boundary of the rectangle ABCD (Fig.A.2), inside which all of the primary poles s_i are located, where the point s does not coincide with any pole. Then, the function

$$b(q) \triangleq f(q) \coth \frac{(q-s)T}{2} \tag{A.52}$$

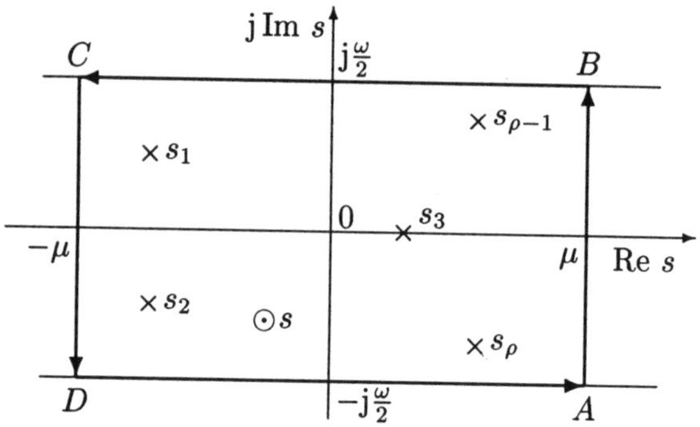

Figure A.2: Integration contour for $f(s)$

has primary poles of multiplicities ν_i at the points s_i and a simple pole $q = s$ with the residue

$$\operatorname*{Res}_{q=s} b(q) = \frac{2}{T} f(s)\,. \tag{A.53}$$

Using the residue theorem to the integral (A.51) and taking into account (A.53), we obtain

$$\begin{aligned} J(s) &= \frac{1}{2\pi \mathrm{j}}\left[\int_A^B b(q)\,\mathrm{d}q + \int_B^C b(q)\,\mathrm{d}q + \int_C^D b(q)\,\mathrm{d}q + \int_D^A b(q)\,\mathrm{d}q\right] \\ &= \sum_{i=1}^{\rho} \operatorname*{Res}_{q=s_i}\left[f(q)\coth\frac{(q-s)T}{2}\right] + \frac{2}{T} f(s)\,. \end{aligned} \tag{A.54}$$

The second and fourth integral in the square brackets annihilate due to (A.47), because the integration is performed in reverse directions. Therefore, (A.54) can be rewritten as

$$\begin{aligned} &\frac{2}{T} f(s) + \sum_{i=1}^{\rho} \operatorname*{Res}_{q=s_i}\left[f(q)\coth\frac{(q-s)T}{2}\right] = \\ &= \frac{1}{2\pi\mathrm{j}} \int\limits_{\mu-\mathrm{j}\omega/2}^{\mu+\mathrm{j}\omega/2} f(q)\coth\frac{(q-s)T}{2}\,\mathrm{d}q + \frac{1}{2\pi\mathrm{j}} \int\limits_{-\mu+\mathrm{j}\omega/2}^{-\mu-\mathrm{j}\omega/2} f(q)\coth\frac{(q-s)T}{2}\,\mathrm{d}q\,. \end{aligned} \tag{A.55}$$

The identity (A.55) holds for all sufficiently large $\mu > 0$. Since

$$\coth\frac{(q-s)T}{2} = \frac{\mathrm{e}^{qT} + \mathrm{e}^{sT}}{\mathrm{e}^{qT} - \mathrm{e}^{sT}}$$

then

$$\lim_{\operatorname{Re} q\to\infty} \coth\frac{(q-s)T}{2} = 1\,, \qquad \lim_{\operatorname{Re} q\to-\infty} \coth\frac{(q-s)T}{2} = -1\,. \tag{A.56}$$

Taking account of (A.56) and (A.46) we obtain

$$
\begin{aligned}
&\lim_{\mu\to\infty}\frac{1}{2\pi \mathrm{j}}\int_{\mu-\mathrm{j}\omega/2}^{\mu+\mathrm{j}\omega/2} f(q)\coth\frac{(q-s)T}{2}\,\mathrm{d}q=\frac{\ell^+}{T}\\
&\lim_{\mu\to\infty}\frac{1}{2\pi \mathrm{j}}\int_{-\mu-\mathrm{j}\omega/2}^{-\mu+\mathrm{j}\omega/2} f(q)\coth\frac{(q-s)T}{2}\,\mathrm{d}q=\frac{\ell^-}{T}\,.
\end{aligned}
\tag{A.57}
$$

Taking the limit in (A.55) as $\mu\to\infty$ and taking account of (A.57) yields

$$
\frac{2}{T}\,f(s)+\sum_{i=1}^{\rho}\operatorname*{Res}_{q=s_i}\left[f(q)\coth\frac{(q-s)T}{2}\right]=\frac{\ell^+ + \ell^-}{T}
$$

which is equivalent to (A.48). The formula is proved for the case when $f(s)$ is free of poles on the boundary of the primary stripe. If such poles are present, the proof remains if the integration contour is somewhat vertically displaced. Formulas (A.49) and (A.50) are proved in a similar way, taking into account that

$$
\lim_{\operatorname{Re} q\to\infty}\frac{1}{\mathrm{e}^{(q-s)T}-1}=0\,,\qquad \lim_{\operatorname{Re} q\to-\infty}\frac{1}{\mathrm{e}^{(q-s)T}-1}=-1
$$

$$
\lim_{\operatorname{Re} q\to\infty}\frac{1}{1-\mathrm{e}^{(s-q)T}}=1\,,\qquad \lim_{\operatorname{Re} q\to-\infty}\frac{1}{1-\mathrm{e}^{(s-q)T}}=0\,.
$$

■

Corollary 1 [Whittaker and Watson (1927)] If under the conditions of Theorem A.1 all the poles s_i are simple with residues f_i, we obtain the following equivalent representations

$$
f(s)=\frac{T}{2}\sum_{i=1}^{\rho} f_i\coth\frac{(s-s_i)T}{2}+\frac{\ell^+ + \ell^-}{2}
\tag{A.58}
$$

$$
f(s)=T\sum_{i=1}^{\rho}\frac{f_i}{\mathrm{e}^{(s-s_i)T}-1}+\ell^+
\tag{A.59}
$$

$$
f(s)=T\sum_{i=1}^{\rho}\frac{f_i}{1-\mathrm{e}^{(s_i-s)T}}+\ell^-\,.
\tag{A.60}
$$

Proof If s_i is a simple primary pole with residue f_i,

$$
\operatorname*{Res}_{q=s_i}\left[f(q)\coth\frac{(s-q)T}{2}\right]=f_i\coth\frac{(s-s_i)T}{2}
$$

which, taking account of (A.48), yields (A.58). Relations (A.59) and (A.60) can be proved analogously. ■

Corollary 2 Under the conditions of Theorem A.1 the function $f(s)$ is a limited rational periodic function, Rosenwasser (1994c).

Example A.7 Consider the rational periodic function

$$F(s,a) \triangleq \frac{1}{e^{-sT} - a} \tag{A.61}$$

where $a \neq 0$ is a constant. In this case

$$\ell^{+}[F(s,a)] = -\frac{1}{a}, \quad \ell^{-}[F(s,a)] = 0 \tag{A.62}$$

i.e., the function $F(s,a)$ is limited. The function $F(s,a)$ has simple poles at the points which are the roots of the equation

$$e^{-sT} - a = 0\,.$$

Let $\tilde{s}$ be a primary root of the last equation with residue

$$\operatorname*{Res}_{s=\tilde{s}} F(s,a) = -\frac{1}{Te^{-\tilde{s}T}} = -\frac{1}{Ta}\,. \tag{A.63}$$

Substituting (A.62) and (A.63) in (A.58) gives

$$F(s,a) = -\frac{1}{2a}\left[1 + \coth\frac{(s-\tilde{s})T}{2}\right]\,. \tag{A.64}$$

The correctness of Eq. (A.64) can be checked directly, because

$$\coth\frac{(s-\tilde{s})T}{2} = \frac{ae^{sT}+1}{ae^{sT}-1}\,.$$

□

A.5 Boundedness of rational periodic functions

In this section we formulate general propositions dealing with the boundedness of rational periodic functions in some regions of the complex plane. As a preliminary we consider the rational periodic function

$$\tilde{f}(s) \triangleq \coth\frac{(s-a)T}{2} = \frac{e^{sT}+e^{aT}}{e^{sT}-e^{aT}} \tag{A.65}$$

which is limited due to (A.56). Therefore, the function is analytic over the whole complex plane, except for simple poles at the points $a_k = a + kj\omega$, $\omega = 2\pi/T$ (Fig.A.3). Construct a closed circle R_k around every pole a_k, so that

$$|s - a_k| \leq \epsilon \tag{A.66}$$

where $\epsilon > 0$ is so small that no other poles a_k lie inside or on the circle R_k. Denote by C_ϵ the complex plane minus all the circles R_k (see Fig. A.3).

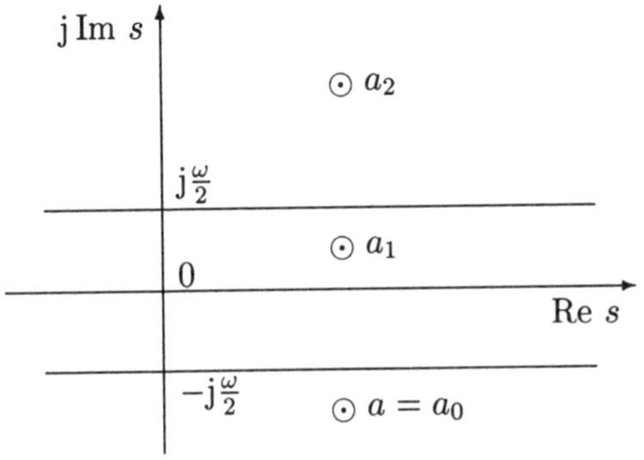

Figure A.3: Poles of $\tilde{f}(s)$

Lemma A.3 *The function (A.65) is in C_ϵ uniformly bounded, so that*

$$|\tilde{f}(s)| < K = \text{const.}, \quad s \in C_\epsilon\,. \tag{A.67}$$

The proof of Lemma A.3 is given in Priwalow (1967). ∎

The result of Lemma A.3 can be extended to limited rational periodic functions of general form. Let $f(s)$ be an irreducible limited rational periodic function with primary poles at the points $\tilde{s}_{i0}$ $(i = 1, \ldots, m)$, so that the complete set of poles of $f(s)$ is described by (A.42). Encircle each pole $\tilde{s}_{ik}$ by a closed disc R_{ik} of radius ϵ, which is so small that there are no other poles in this disc. Denote by $\tilde{C}_\epsilon$ the complex plane minus the discs R_{ik}.

Theorem A.2 *The limited rational periodic function $f(s)$ is uniformly bounded in $\tilde{C}_\epsilon$, so that*

$$|f(s)| < K = \text{const.}, \quad s \in \tilde{C}_\epsilon\,. \tag{A.68}$$

Proof Since $f(s)$ is assumed to be limited, in the representation (A.27) we have $l_j = 0\,, (j = 1, \ldots, \kappa)$. Therefore, using (A.26) gives

$$f^o(\zeta) = l_0 + \frac{\lambda_0 + \lambda_1\zeta + \ldots + \lambda_{p-1}\zeta^{p-1}}{r_0 + r_1\zeta + \ldots + r_p\zeta^p} \tag{A.69}$$

with known constants l_0, λ_i and r_i . Then, the following partial fraction expansion holds

$$f^o(\zeta) = l_0 + \sum_{i=1}^{m}\sum_{k=1}^{\nu_i} \frac{f_{ik}}{(\zeta - \zeta_i)^k} \tag{A.70}$$

where f_{ik} are known constants and there are no zero numbers among the ζ_i. Passing to the variable s in (A.70), taking account of (A.10) we find

$$f(s) = l_0 + \sum_{i=1}^{m}\sum_{k=1}^{\nu_i} \frac{f_{ik}}{(\mathrm{e}^{-sT} - \zeta_i)^k}\,.$$

Using (A.64) makes it possible to represent the last equation in the form

$$f(s) = l_0 + \sum_{i=1}^{m} \sum_{k=1}^{\nu_i} i(-1)^k \frac{f_{ik}}{(2\zeta_i)^k} \left[1 + \coth \frac{(s - \tilde{s}_{i0})T}{2}\right]^k .$$

Each of the functions in the square brackets is, by virtue of Lemma A.3, bounded in $\tilde{C}_\epsilon$. Therefore, the function $f(s)$ possesses this property as well. ■

Corollary 1 Any rational periodic function $f(s)$ is uniformly bounded in any stripe of finite width in which it is analytic.

Corollary 2 Let the rational periodic function $f(s)$ be causal and $\mu_1 \ldots, \mu_r$ be the set of its finite characteristic indices. For an arbitrarily small $\epsilon > 0$ the function $f(s)$ is uniformly bounded in any stripe $\mu_i + \epsilon \le \operatorname{Re} s \le \mu_{i+1} - \epsilon$ and in the right half-plane $\operatorname{Re} s \ge \mu_r + \epsilon$.

A.6 Calculating integrals from rational periodic functions

In this section we consider the calculation of integrals of the form

$$I(c) \triangleq \frac{T}{2\pi \mathrm{j}} \int_{c-\mathrm{j}\omega/2}^{c+\mathrm{j}\omega/2} f(s)\, \mathrm{d}s \tag{A.71}$$

where $f(s)$ is a rational periodic function with period $\mathrm{j}\omega$ and c is regarded as a parameter. Let the function $f(s)$ have m primary poles $\tilde{s}_{10}, \ldots, \tilde{s}_{m0}$ in the primary stripe $-\omega/2 \le \operatorname{Im} s < \omega/2$, so that

$$\begin{aligned} \operatorname{Re} \tilde{s}_{i0} &< c \quad (i = 1, \ldots, q) \\ \operatorname{Re} \tilde{s}_{i0} &> c \quad (i = q+1, \ldots, m) . \end{aligned} \tag{A.72}$$

Theorem A.3 *Let the following limit exist*

$$\ell^-[f(s)] = \ell^- = \lim_{\operatorname{Re} s \to -\infty} f(s) . \tag{A.73}$$

Then, the integral (A.71) is given as

$$I(c) = T \sum_{i=1}^{q} \operatorname*{Res}_{\tilde{s}_{i0}} f(s) + \ell^- . \tag{A.74}$$

If the following limit exists

$$\ell^+[f(s)] = \ell^+ = \lim_{\operatorname{Re} s \to \infty} f(s) \tag{A.75}$$

the integral (A.71) appears as

$$I(c) = -T \sum_{i=q+1}^{m} \operatorname*{Res}_{\tilde{s}_{i0}} f(s) + \ell^{+} . \tag{A.76}$$

If both limits ℓ^{+}, ℓ^{-} exist, i.e., the function $f(s)$ is limited, Eqs. (A.74) and (A.76) hold simultaneously.

Proof To prove (A.74) consider the contour integral

$$I_k \triangleq \frac{T}{2\pi \mathrm{j}} \oint_{ABCD} f(s)\,\mathrm{d}s \tag{A.77}$$

along the rectangle shown in Fig. A.4, assuming that the poles $\tilde{s}_{10}, \ldots, \tilde{s}_{q0}$ are

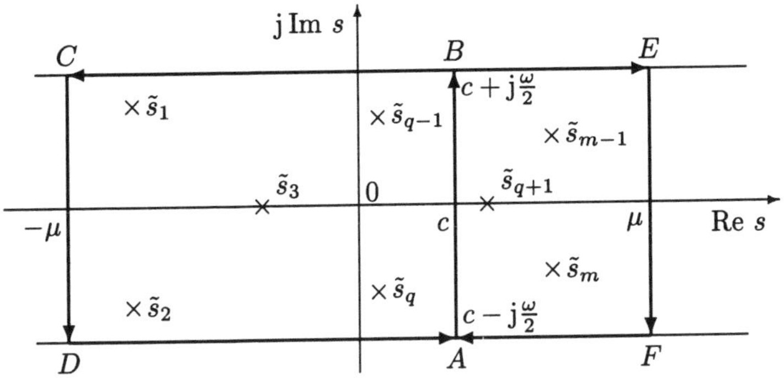

Figure A.4: Integration contour for I_k

located inside this contour. By the residue theorem,

$$I_k = T \sum_{i=1}^{q} \operatorname*{Res}_{\tilde{s}_{i0}} f(s) . \tag{A.78}$$

On the other hand,

$$I_k = \frac{T}{2\pi \mathrm{j}} \left[\int_A^B f(s)\,\mathrm{d}s + \int_B^C f(s)\,\mathrm{d}s + \int_C^D f(s)\,\mathrm{d}s + \int_D^A f(s)\,\mathrm{d}s \right] . \tag{A.79}$$

The integrals along BC and DA annihilate due to the reverse directions and the periodicity of the function $f(s)$. The integral along CD has the form

$$\frac{T}{2\pi \mathrm{j}} \int_C^D f(s)\,\mathrm{d}s = \frac{T}{2\pi \mathrm{j}} \int_{-\mu+\mathrm{j}\omega/2}^{-\mu-\mathrm{j}\omega/2} f(s)\,\mathrm{d}s \tag{A.80}$$

where μ is a sufficiently large positive number. For $\mu \to \infty$ taking account of (A.73) we have

$$\lim_{\mu \to \infty} \frac{T}{2\pi \mathrm{j}} \int_{-\mu+\mathrm{j}\omega/2}^{-\mu-\mathrm{j}\omega/2} f(s)\,\mathrm{d}s = -\ell^{-} . \tag{A.81}$$

Then, for $\mu \to \infty$ from (A.78) and (A.79) we have

$$-\ell^- + \frac{T}{2\pi \mathrm{j}} \int_{c-\mathrm{j}\omega/2}^{c+\mathrm{j}\omega/2} f(s)\,\mathrm{d}s = T \sum_{i=1}^{q} \operatorname*{Res}_{\tilde{s}_{i0}} f(s)$$

which is equivalent to (A.74). Formula (A.76) is proved similarly by contour integration along the boundary of the rectangle $ABEF$ (see Fig. A.4). ■

Remark If $f(s)$ has poles on the boundary of the primary stripe, the validity of the theorem can be proved by a vertical displacement of the integration contour.

Corollary 1 If $c > \operatorname{Re} \tilde{s}_{i0}$ $(i = 1, \ldots, m)$, i.e., the function $f(s)$ is analytic for $\operatorname{Re} s \geq c$ and there exists the limit ℓ^+, from (A.76) it follows that

$$I(c) = \ell^+ . \tag{A.82}$$

Corollary 2 If $\ell^- = 0$, Eq. (A.74) yields

$$I(c) = T \sum_{i=1}^{q} \operatorname*{Res}_{\tilde{s}_{i0}} f(s) \tag{A.83}$$

while for $\ell^+ = 0$ from (A.76) it follows that

$$I(c) = -T \sum_{i=q+1}^{m} \operatorname*{Res}_{\tilde{s}_{i0}} f(s) . \tag{A.84}$$

Thus, the residue theorem is applicable for calculating integrals of the form (A.71) only if the corresponding limits ℓ^+, ℓ^- are zero.

Example A.8 Consider the integral (A.71) with

$$f(s) = \frac{\mathrm{e}^{2sT} + 1}{(\mathrm{e}^{sT} - 0.5)(\mathrm{e}^{sT} - 2)} = \frac{1 + \mathrm{e}^{-2sT}}{(1 - 0.5\mathrm{e}^{-sT})(1 - 2\mathrm{e}^{-sT})} .$$

In this case $\ell^+ = 1$ and $\ell^- = 1$, i.e., the function $f(s)$ is limited. The function $f(s)$ has primary poles $\tilde{s}_{20} = T^{-1} \ln 2$ in the right half-plane and $\tilde{s}_{10} = T^{-1} \ln 0.5 = -\tilde{s}_{20}$ in the left half-plane. Then,

$$\operatorname*{Res}_{\tilde{s}_{10}} f(s) = -\frac{5}{3T} , \quad \operatorname*{Res}_{\tilde{s}_{20}} f(s) = \frac{5}{3T} .$$

For $c > \tilde{s}_{20}$ Eq. (A.82) gives $I(c) = \ell^+ = 1$. For $\tilde{s}_{10} < c < \tilde{s}_{20}$ the integral can be calculated by any of the formulae (A.74) or (A.76), so that $I(c) = -\frac{2}{3}$. For $c < \tilde{s}_{10}$ Eq. (A.74) gives $I(c) = \ell^- = 1$. □

Example A.9 Consider the integral

$$I = \frac{T}{2\pi \mathrm{j}} \int_{-\mathrm{j}\omega/2}^{\mathrm{j}\omega/2} \frac{\mathrm{e}^{3sT}+1}{(\mathrm{e}^{sT}-0.1)(\mathrm{e}^{sT}-0.2)}\,\mathrm{d}s\,. \tag{A.85}$$

In this case the limit ℓ^+ does not exist, while $\ell^- = 50$. Using Eq. (A.74) and the technique of Example A.8 gives $I = 0.3$. The integral (A.85) can also be computed in another way. We note that

$$\frac{\mathrm{e}^{3sT}+1}{(\mathrm{e}^{sT}-0.1)(\mathrm{e}^{sT}-0.2)} = \mathrm{e}^{sT} + \frac{0.3\mathrm{e}^{2sT}-0.02\mathrm{e}^{sT}+1}{(\mathrm{e}^{sT}-0.1)(\mathrm{e}^{sT}-0.2)}$$

so that

$$I = \frac{T}{2\pi \mathrm{j}} \int_{-\mathrm{j}\omega/2}^{\mathrm{j}\omega/2} \frac{0.3\mathrm{e}^{2sT}-0.2\mathrm{e}^{sT}+1}{(\mathrm{e}^{sT}-0.1)(\mathrm{e}^{st}-0.02)}\,\mathrm{d}s\,. \tag{A.86}$$

The integrand in (A.86) is a limited rational periodic function, all poles of which are located left of the imaginary axis. Since $\ell^+ = 0.3$, using (A.82) gives $I = 0.3$ again. □

A.7 Integration of products of rational periodic functions

In this section we investigate some properties of the integral

$$J \triangleq \frac{T}{2\pi \mathrm{j}} \int_{-\mathrm{j}\omega/2}^{\mathrm{j}\omega/2} d(s)r(s)\,\mathrm{d}s \tag{A.87}$$

where $d(s)$ and $r(s)$ are limited rational periodic functions, so that

$$d(s) = d(s+\mathrm{j}\omega)\,, \quad r(s) = r(s+\mathrm{j}\omega)\,, \quad \omega = 2\pi/T \tag{A.88}$$

and

$$d(s) = d(-s)\,, \quad r(s) = r(-s)\,. \tag{A.89}$$

Moreover, due to the assumptions on symmetry and boundedness there exist the limits

$$\ell_d \triangleq \lim_{\mathrm{Re}\, s\to\pm\infty} d(s)\,, \quad \ell_r \triangleq \lim_{\mathrm{Re}\, s\to\pm\infty} r(s)\,. \tag{A.90}$$

Theorem A.4 *Let the functions $d(s)$ and $r(s)$ have no poles on the imaginary axis or on the boundary of the primary stripe $-\omega/2 \le \mathrm{Im}\, s < \omega/2$. Let, in the half-plane $\mathrm{Re}\, s < 0$, the function $d(s)$ have simple primary poles $d_1, \dots, d_n$ and $r(s)$ have simple primary poles $r_1, \dots, r_m$, so that there are not equal numbers among them. By a_i we denote the residues of $d(s)$ at the poles d_i and by b_l the residues of $r(s)$ at r_l. Then,*

$$J = \ell_d\ell_r + T\ell_r\sum_{i=1}^{n} a_i + T\ell_d\sum_{l=1}^{m} b_l - T^2\sum_{i=1}^{n}\sum_{l=1}^{m} a_i b_l \coth\frac{(d_i+r_l)T}{2}\,. \tag{A.91}$$

Proof Denote

$$d(s)r(s) = f(s)\,. \tag{A.92}$$

Under the given assumptions the rational periodic function $f(s)$ is symmetric and limited, and, with account for (A.90),

$$\ell^+[f(s)] = \ell^-[f(s)] = \ell_d \ell_r\,. \tag{A.93}$$

In this case the function $f(s)$ has $n+m$ simple poles $d_i\,(i=1,\dots,n)$ and $r_l\,(l=1,\dots,m)$ in the half-stripe $\operatorname{Re} s < 0$, $-\omega/2 \le \operatorname{Im} s < \omega/2$. The residues at these poles are given by

$$\operatorname*{Res}_{d_i} f(s) = a_i r(d_i)\,, \quad \operatorname*{Res}_{r_l} f(s) = b_l d(r_l)\,. \tag{A.94}$$

Formula (A.74) gives

$$J = \ell_d \ell_r + T \sum_{i=1}^{n} \operatorname*{Res}_{d_i} f(s) + T \sum_{l=1}^{m} \operatorname*{Res}_{r_l} f(s)\,. \tag{A.95}$$

Taking account of (A.94), this is equivalent to

$$J = \ell_d \ell_r + T \sum_{i=1}^{n} a_i r(d_i) + T \sum_{l=1}^{m} b_l d(r_l)\,. \tag{A.96}$$

For further transformations, we note that under the given assumptions the function $d(s)$ has in the primary stripe the poles $d_1,\dots,d_n$ and also the poles $d_{n+1} = -d_1,\dots,d_{2n} = -d_n$ located in the right half-plane. Let a_i $(i = n+1,\dots,2n)$ be the residues of $d(s)$ at these poles. Then, using the partial fraction expansion, we may write

$$d(s) = \ell_d + \frac{T}{2}\sum_{i=1}^{n} a_i \coth\frac{(s-d_i)T}{2} + \frac{T}{2}\sum_{i=1}^{n} a_{i+n} \coth\frac{(s+d_i)T}{2}\,. \tag{A.97}$$

Substituting $-s$ for s, taking account of (A.89) we have

$$d(s) = \ell_d - \frac{T}{2}\sum_{i=1}^{n} a_i \coth\frac{(s+d_i)T}{2} - \frac{T}{2}\sum_{i=1}^{n} a_{i+n} \coth\frac{(s-d_i)T}{2}\,. \tag{A.98}$$

Comparing the expansions (A.97) and (A.98), we obtain $a_{i+n} = -a_i$, so that the expansion (A.97) takes the form

$$d(s) = \ell_d + \frac{T}{2}\sum_{i=1}^{n} a_i \left[\coth\frac{(s-d_i)T}{2} - \coth\frac{(s+d_i)T}{2}\right]\,. \tag{A.99}$$

Similarly, the expansion

$$r(s) = \ell_r + \frac{T}{2}\sum_{l=1}^{m} b_l \left[\coth\frac{(s-r_l)T}{2} - \coth\frac{(s+r_l)T}{2}\right] \tag{A.100}$$

can be proved. From (A.99) and (A.100) it follows that

$$\begin{aligned} d(r_l) &= \ell_d + \frac{T}{2}\sum_{i=1}^{n} a_i \left[\coth\frac{(r_l-d_i)T}{2} - \coth\frac{(r_l+d_i)T}{2}\right] \\ r(d_i) &= \ell_r + \frac{T}{2}\sum_{l=1}^{m} b_l \left[\coth\frac{(d_i-r_l)T}{2} - \coth\frac{(d_i+r_l)T}{2}\right]. \end{aligned} \tag{A.101}$$

Substituting (A.101) into (A.96), we obtain

$$\begin{aligned} J = \ell_d\ell_r + T\sum_{i=1}^{n} a_i &\left[\ell_r + \frac{T}{2}\sum_{l=1}^{m} b_l \left[\coth\frac{(d_i-r_l)T}{2} - \coth\frac{(d_i+r_l)T}{2}\right]\right] + \\ +T\sum_{l=1}^{m} b_l &\left[\ell_d + \frac{T}{2}\sum_{i=1}^{n} a_i \left[\coth\frac{(r_l-d_i)T}{2} - \coth\frac{(r_l+d_i)T}{2}\right]\right]. \end{aligned} \tag{A.102}$$

After combining the similar terms in (A.102) we obtain (A.91). ■

A.8 Calculation of parameter integrals

In the present book we systematically employ integrals of the following type for investigation of continuous-time processes in sampled-data systems

$$y(t) = \frac{T}{2\pi \mathrm{j}} \int_{c-\mathrm{j}\omega/2}^{c+\mathrm{j}\omega/2} f(s,t)\,\mathrm{d}s \tag{A.103}$$

where c is a real constant, $f(s,t)$ is the discrete Laplace transform of a known function, and the following conditions hold:

i) For any t the function $f(s,t)$ is a rational periodic function and

$$f(s,t+T) = f(s,t)\mathrm{e}^{sT}. \tag{A.104}$$

ii) The poles s_i of the function $f(s,t)$ are independent of t and meet

$$c > \operatorname{Re} s_i. \tag{A.105}$$

iii) For $0 < t = \varepsilon < T$ the function $f(s,t)$ is causal, i.e., there exists a finite limit

$$\ell^+[f(s,\varepsilon)] = \lim_{\operatorname{Re} s\to\infty} f(s,\varepsilon) \triangleq \ell_f^+(\varepsilon). \tag{A.106}$$

Consider the general procedure of calculating the integral (A.103) under the above restrictions. For simplicity we assume that $f(s,t)$ does not have any poles on the horizontal lines $\operatorname{Im} s = \pm \mathrm{j}\omega/2$, though the final result is independent of this assumption.

Since the function $f(s,\varepsilon)$ is assumed to be causal, taking account of (A.27) we obtain

$$f(s,\varepsilon) = f_0(s,\varepsilon) + \tilde{f}_1(s,\varepsilon) \tag{A.107}$$

where

$$f_0(s,\varepsilon) = q_\kappa(\varepsilon)\mathrm{e}^{-\kappa sT} + q_{\kappa-1}(\varepsilon)\mathrm{e}^{-(\kappa-1)sT} + \ldots + q_0(\varepsilon)\,. \tag{A.108}$$

In (A.108) $q_i(\varepsilon)$ are known functions, and for the function $\tilde{f}_1(\varepsilon)$ we have

$$\ell^-[\tilde{f}_1(s,\varepsilon)] = \lim_{\mathrm{Re}\, s\to-\infty} \tilde{f}_1(s,\varepsilon) = 0\,. \tag{A.109}$$

Henceforth, we use the notation

$$\begin{aligned} f_0(s,t) &\triangleq f_0(s,t-mT)\mathrm{e}^{msT}, \quad mT < t < (m+1)T \\ \tilde{f}_1(s,t) &\triangleq \tilde{f}_1(s,t-mT)\mathrm{e}^{msT}, \quad mT < t < (m+1)T\,. \end{aligned} \tag{A.110}$$

Using (A.110) gives

$$y(t) = y_0(t) + y_1(t) \tag{A.111}$$

where

$$y_0(t) = \frac{T}{2\pi \mathrm{j}} \int_{c-\mathrm{j}\omega/2}^{c+\mathrm{j}\omega/2} f_0(s,t)\,\mathrm{d}s\,, \quad y_1(t) = \frac{T}{2\pi \mathrm{j}} \int_{c-\mathrm{j}\omega/2}^{c+\mathrm{j}\omega/2} \tilde{f}_1(s,t)\,\mathrm{d}s\,. \tag{A.112}$$

From (4.19) and (4.20) it follows that the function $y_0(t)$ is finite, which corresponds to a deadbeat process.
Then,

$$y_0(t) = \begin{cases} 0 & t > (\kappa+1)T \\ q_m(t-mT) & mT < t < (m+1)T, \ (m = 0,\ldots,\kappa) \\ 0 & t < 0\,. \end{cases} \tag{A.113}$$

To calculate the function $y_1(t)$ we may use Theorem A.3 and its Corollaries. Using (A.104) and (A.109), we have

$$\ell^+[\tilde{f}_1(s,t)] = 0\,, \quad t < 0\,.$$

Then, using (A.103) and (A.82) yields

$$y_1(t) = 0\,, \quad t < 0\,. \tag{A.114}$$

For $t > 0$ Eqs. (A.104) and (A.109) give

$$\ell^-[\tilde{f}_1(s,t)] = \lim_{\mathrm{Re}\, s\to-\infty} f(s,t) = 0\,.$$

Then, using (A.83) we receive

$$y_1(t) = T\sum_i \operatorname*{Res}_{\tilde{s}_i} \tilde{f}_1(s,t)\,, \quad t > 0 \tag{A.115}$$

where the residues are taken at all the primary poles of the function $\tilde{f}_1(s,t)$ which coincide with the primary poles of the function $f(s,t)$. Indeed, as follows from (A.108) and (A.109), the function $f_0(s,t)$ is analytic with respect to s for all t. Therefore,

$$\operatorname*{Res}_{\tilde{s}_i} \tilde{f}_1(s,t) = \operatorname*{Res}_{\tilde{s}_i} f(s,t)$$

and from (A.115) we find

$$y_1(t) = \begin{cases} T\sum_i \operatorname{Res}_{\tilde{s}_i} f(s,t) & t > 0 \\ 0 & t < 0\,. \end{cases}$$

Combining the above relations, we obtain the general expression for the integral (A.103)

$$y(t) = \begin{cases} T\sum_i \operatorname*{Res}_{\tilde{s}_i} f(s,t) & t > (\kappa+1)T \\ q_m(t-mT) + T\sum_i \operatorname*{Res}_{\tilde{s}_i} f(s,t) & \begin{array}{c} mT < t < (m+1)T \\ (m = 0,\dots,\kappa) \end{array} \\ 0 & t < 0\,. \end{cases} \tag{A.116}$$

From (A.116) it follows that under the above restrictions we may use the residue theorem for calculating the integral (A.103) for sufficiently large $t > 0$.

Example A.10 Consider an integral of the form (A.103), with

$$f(s,\varepsilon) = \frac{f_0(\varepsilon) + f_1(\varepsilon)\mathrm{e}^{-sT} + f_2(\varepsilon)\mathrm{e}^{-2sT}}{1 - a\mathrm{e}^{-sT}}$$

where $|a| < 1$, $c > 0$ and (A.104) holds. Obviously, the representation (A.107) has the form

$$f(s,\varepsilon) = q_1(\varepsilon)\mathrm{e}^{-sT} + q_0(\varepsilon) + \tilde{f}_1(s,\varepsilon)$$

where

$$q_1(\varepsilon) = -\frac{f_2(\varepsilon)}{a}\,, \quad q_0(\varepsilon) = -\frac{f_1(\varepsilon)}{a} - \frac{f_2(\varepsilon)}{a^2}$$

$$\tilde{f}_1(s,\varepsilon) = \frac{f_0(\varepsilon) + \frac{f_1(\varepsilon)}{a} + \frac{f_2(\varepsilon)}{a^2}}{1 - a\mathrm{e}^{-sT}}\,.$$

In this case the function $f(s,\varepsilon)$ has a single primary pole $\tilde{s}$ with residue

$$\operatorname*{Res}_{\tilde{s}} f(s, t-mT) = \frac{1}{T}\left[f_0(t-mT) + \frac{f_1(t-mT)}{a} + \frac{f_2(t-mT)}{a^2}\right] a^m$$

$$mT < t < (m+1)T\,.$$

As a result, the general formula (A.116) yields

$$y(t) = \begin{cases} f_0(t-mT)a^m + f_1(t-mT)a^{m-1} + \\ \quad + f_2(t-mT)a^{m-2} & \begin{array}{c} mT < t < (m+1)T \\ m \geq 2 \end{array} \\ f_0(t-T)a + f_1(t-T) & T < t < 2T \\ f_0(t) & 0 < t < T \\ 0 & t < 0\,. \end{cases}$$

□

Appendix B

DirectSD – a toolbox for direct polynomial design of SD systems

A new MATLAB toolbox *DirectSD* for analysis and direct synthesis of sampled-data systems in continuous time is presented. The design methods used in the toolbox are based on the theoretical base of the book, the main part of which is the parametric transfer function concept, which makes it possible to investigate SD systems in continuous time in the frequency domain. The toolbox includes macros for analysis and direct $\mathcal{H}_2-$ and $\mathcal{H}_\infty-$optimization of single-input single-output (SISO) SD systems under deterministic and stochastic disturbances. [1]

B.1 Introduction

Due to mass introduction of digital computers into modern control systems the effective computer-aided design of optimal control laws became one of the most important problems. Being considered in continuous time, SD systems are periodically time-variable systems Rosenwasser (1973). Therefore, the theory of LTI systems is not applicable for their rigorous analysis and design. On the other hand, recent investigations de Souza and Goodwin (1984); Lennartson et al. (1989) have shown that SD systems design based on approximative discrete-time models can lead to systems that are practically irrelevant owing to intensive intersample behaviour.

There are several approaches to the investigation of SD systems in continuous time, among them are the hybrid state space method Yamamoto (1994); Hara et al. (1994), the "lifting" technique Chen and Francis (1995) and the

[1] A restricted trial version of the *DirectSD* toolbox can be downloaded from http://www-at.e-technik.uni-rostock.de/~bl

"FR-operator" method Hagiwara and Araki (1995). Nevertheless, the practical application of these methods to controller synthesis for complex systems with neutral plants and pure-delay elements is connected with a number of fundamental difficulties.
Many of these difficulties can be conquered on the basis of the parametric transfer function (PTF) methods, Rosenwasser and Lampe (1997), that make it possible to perform a direct digital controller design for continuous-time plants without a transition to approximated discrete-time models. Application examples of the PTF theory to control problems can be found in Rosenwasser et al. (1996, 1998, 1999b). For the practical realization of this approach a polynomial equations method is employed, Polyakov (1998, 1999).
Recently, MATLAB became an universal environment for scientific computations, also due to its possibility to append new toolboxes. In the present report a new toolbox ***DirectSD*** for analysis and direct design of SISO SD systems in continuous time is presented. The toolbox is based on the PTF theory and polynomial equation techniques. The size of files is about 600 Kb including demonstration programs. The toolbox has been tested on PCs and Workstations under the operating systems Windows 95, Windows NT, UNIX and Linux, using MATLAB 4.2 and 5.2.
Section B.2 shortly describes the theoretical basis for analysis and $\mathcal{H}_2-$ and $\mathcal{H}_\infty-$optimization of SD systems in continuous time using the PTF method. Section B.3 presents the basic data structures used in the toolbox. In Section B.4 the problems of $\mathcal{H}_2-$ and $\mathcal{H}_\infty-$optimization under stochastic disturbances are considered together with techniques of their solution using the toolbox. Section B.5 describes the design of $\mathcal{H}_2-$ and $\mathcal{H}_\infty-$optimal tracking systems for deterministic signals. Section B.6 presents other possibilities of the toolbox and describes the demo programs. Some nontrivial numerical examples of optimal systems design using the ***DirectSD*** toolbox are considered in Section B.7. Appendix B ends with conclusions.

B.2 Theoretical basis

Using the methods of the linear theory of digital control in continuous time, given in the present book, it is possible to associate any periodically time-variable system with an universal system characteristic called the *parametric transfer function* $W(s,t)$, which is analogous to the ordinary transfer function used for LTI systems and possesses similar properties. For example, the variance $v_y(t)$ of the output $y(t)$ of such a system under a stationary stochastic disturbance is calculated by

$$v_y(t) = \frac{1}{2\pi \mathrm{j}} \int_{-\mathrm{j}\infty}^{\mathrm{j}\infty} W(-s,t) S_x(s) W(s,t) \, \mathrm{d}s \tag{B.1}$$

where $S_x(s)$ is the spectral density of the input signal $x(t)$. As is evident from (B.1), the variance is a function of the time t. Since the PTF $W(s,t)$

is periodic with a sampling period T, the variance is also periodic with this period T. For integral evaluation of the system behaviour in continuous time the *average variance* is used:

$$\bar{v}_y = \int_0^T v_y(t)\,\mathrm{d}t\,.$$

For the investigation of SD systems under deterministic disturbances the *Laplace transform* of the continuous-time output is employed. For example,

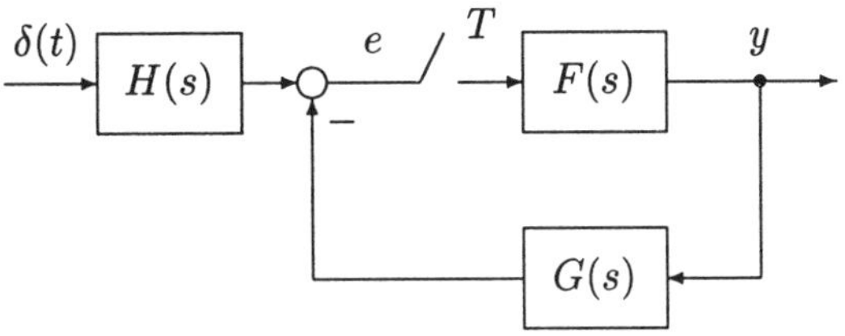

Figure B.1: Closed-loop SD system

for the simplest closed-loop system shown in Fig. B.1 the Laplace transform of the output $y(t)$ takes the form

$$Y(s) = \frac{F(s)\mathcal{D}_H(T,s,+0)}{1+\mathcal{D}_{FG}(T,s,+0)}\,.$$

where $\mathcal{D}_H(T,s,t)$ and $\mathcal{D}_{FG}(T,s,t)$ denote the discrete Laplace transforms for the functions $H(s)$ and $F(s)G(s)$, respectively. If the system under investigation must follow a reference signal $\hat{y}(t)$ having the Laplace transform $\hat{Y}(s)$, the integral quadratic tracking error in continuous time can be calculated using Parseval's formula:

$$E = \int_0^\infty [y(t)-\hat{y}(t)]^2\,\mathrm{d}t = \frac{1}{2\pi \mathrm{j}}\int_{-\mathrm{j}\infty}^{\mathrm{j}\infty} [Y(s)-\hat{Y}(s)][Y(-s)-\hat{Y}(-s)]\,\mathrm{d}s\,.$$

The above expressions make it possible to reduce problems of continuous-time optimization of SD systems to equivalent discrete-time problems. These problems can be handled by the polynomial methods of Chapter 18, (similar methods for LTI continuous-time and discrete-time systems are given in Kučera (1979); Grimble (1995)). In the case of a quadratic problem the cost functional can be written in the form

$$J_2 = \frac{1}{2\pi \mathrm{j}}\oint_\Gamma A(\zeta)\tilde{C}(\zeta)\tilde{C}(\zeta^{-1}) - B(\zeta)\tilde{C}(\zeta) - B(\zeta^{-1})\tilde{C}(\zeta^{-1}) + E(\zeta)\,\frac{\mathrm{d}\zeta}{\zeta} \to \min$$

where $\zeta = \mathrm{e}^{-sT}$; $A(\zeta)$, $B(\zeta)$ and $E(\zeta)$ are real rational functions such that $A(\zeta) = A(\zeta^{-1})$ and $E(\zeta) = E(\zeta^{-1})$; $\tilde{C}(\zeta)$ is a function that uniquely defines the controller. For robust optimization we use the following functional:

$$J_\infty = \|A(\zeta)\tilde{C}(\zeta)\tilde{C}(\zeta^{-1}) - B(\zeta)\tilde{C}(\zeta) - B(\zeta^{-1})\tilde{C}(\zeta^{-1}) + E(\zeta)\|_\infty \to \min$$

which is the $\mathcal{H}_\infty$−norm of an equivalent discrete-time system.

B.3 System description

All methods used in the *DirectSD* toolbox are based on a frequency domain philosophy for SD systems analysis and design. Therefore, it seems natural to describe all networks in terms of their transfer functions. Such an approach is important because elements located in different parts of the system exert different influence onto the design procedure and the results. If the system is described in state space, these structural peculiarities are ignored what may lead to erroneous results or unnecessary restrictions.
The *DirectSD* toolbox assumes that real-rational transfer functions are stored in a special format as rectangular matrices. Their first and second rows are the numerator and denominator of the transfer function, while the third one contains its poles.
To create a transfer function structure the macro `tfn` is used. Its parameters are the numerator and denominator of the transfer function, respectively. For instance, the transfer function $F(s) = (2s+1)/(s^2+2s+3)$ can be formed as follows:

```
F = tfn ( [2 1], [1 2 3] ).
```

Such a transfer function can be transferred to an argument in any macro in the *DirectSD* toolbox.
Spectra of stationary stochastic signals are also stored as real-rational functions. They are created using the macro `stf`. It accepts two parameters: the numerator and denominator of the spectral density which are polynomials in ω^2. For example, the spectrum $S_x(\omega) = 4/(\omega^2+4)$ can be formed as follows:

```
Sx = stf ( [4], [1 4] ).
```

Other initial data (sampling period, pure delays in continuous elements, etc.) are scalar values. The zero-order hold is used as an extrapolator in the macros of the *DirectSD* toolbox, though this is not mandatory, see Chapter 10. The toolbox includes macros for analysis and design of sufficiently general structures that incorporate most practical problems as special cases. For each scheme the structure is taken as given, which is simultaneously an advantage (it is possible to enhance the reliability of numerical methods taking into account structural peculiarities) and disadvantage (for a different block-diagram it is required to build a new macro).

B.4 Optimal control under stochastic disturbances

B.4.1 Optimal digital filtering

The system shown in Fig. B.2 is acted upon by an additive mixture of a reference signal $x(t)$ and a noise $n(t)$, both of them are described as independent stationary stochastic processes with known spectral densities. It is required to

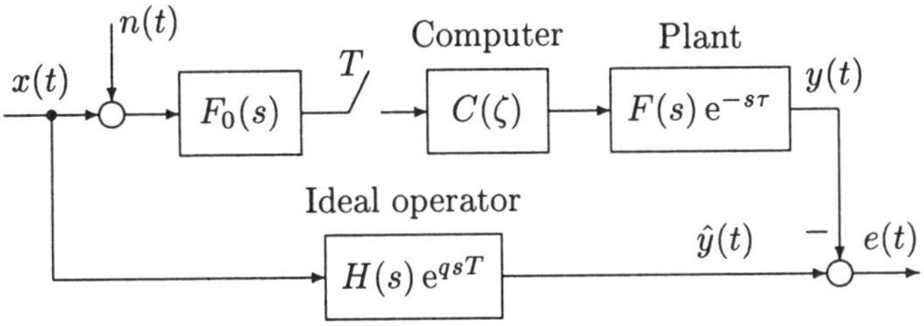

Figure B.2: Optimal digital filtering

determine the transfer function of a stable digital filter $C(\zeta)$ such that the average variance of the error $e(t)$ is minimal. The ideal operator is a real rational function $H(s)$ with signal shift by an integral number q of sampling periods. Three possible cases, namely $q < 0$, $q = 0$ and $q > 0$, give the problems of optimal smoothing, filtering and prediction, respectively.
The optimal controller is designed using the macro `optfil` as follows:

```
[C,var] = optfil ( F, F0, H, Sx, Sn, T, Delay, q );
```

where F, F_0 and H are transfer functions; S_x and S_n are spectral densities of input signals; T, *Delay* and q denote the sampling period, pure delay τ and shift of the reference signal. As a result, we obtain the transfer function of the optimal controller $C(\zeta)$ (in increasing powers of ζ) and the average error variance *var* in the optimal system.

B.4.2 $\mathcal{H}_2-$ and associated $\mathcal{H}_\infty-$optimal control in closed-loop systems

The problem was investigated in Sections 18.2 and 18.4. The initial system is shown in Fig. 18.1. It consists of a plant with transfer functions $F_1(s)$ and $F_2(s)$, actuator $H(s)$, dynamic feedback $G(s)$ and two pure delay elements with delays τ_1 and τ_2. Exogenous disturbances are described as independent stationary white noises $w(t)$ and $m(t)$ transformed by forming filters $F_w(s)$ and $F_m(s)$, respectively. Dynamic weighting functions $V_1(s)$ and $V_2(s)$ are used to form desired frequency properties of the system, Grimble (1995).
The following value, which includes the sum of average variances of the signals $z_y(t)$ and $z_u(t)$, is used for $\mathcal{H}_2-$optimization:

$$J_2 = \bar{v}_y + \bar{v}_u \,.$$

This cost function can be reduced to an integral from a quadratic form $X(\zeta)$ that is to be minimized over the set of stabilizing controllers:

$$J_2 = \frac{1}{2\pi \mathrm{j}} \oint_\Gamma X(\zeta)\, \frac{\mathrm{d}\zeta}{\zeta} \,.$$

This problem is solved by the macro `optfdb`:

```
[C, Poles] = optfdb ( F1, F2, H, G, Fw, Fm, ...
                      V1, V2, T, Delay1, Delay2 );
```

where *Delay*1 and *Delay*2 denote the delays τ_1 and τ_2, respectively, and other notations are evident. As a result, we obtain the discrete transfer function of the optimal digital controller $C(\zeta)$ and the poles of the closed-loop system in the variable *Poles*.

For robust optimization an idea due to Grimble (1995) is used. Taking this approach, the cost function for robust optimization can be chosen as

$$J_\infty = \|X(\zeta)\|_\infty \to \min$$

that equals the $\mathcal{H}_\infty$–norm of an equivalent discrete-time system. Obviously, the weight functions must be chosen in a different way compared to the quadratic optimization, Grimble (1995). The problem of $\mathcal{H}_\infty$–optimization is solved by the macro **sdhinf**:

```
[C,Poles,Lam] = sdhinf ( F1, F2, H, G, Fw, Fm, ...
                         V1, V2, T, Delay1, Delay2 );
```

In addition to the optimal controller and the poles of the optimal system the macro returns the value *Lam* defined as

$$\lambda = \|X(\zeta)\|_\infty^{1/2}$$

for the optimal system.

B.5 Design of optimal tracking systems

B.5.1 One degree of freedom systems

Figure B.3 shows the block-diagram of a digital tracking system that consists of a plant with transfer function $F(s)$, actuator $H(s)$, dynamic feedback $G(s)$, prefilter $F_0(s)$ and digital controller with discrete transfer function $C(\zeta)$. The elements $F(s)$, $H(s)$ and $F_0(s)$ may include integrators. Pure delays in the continuous elements and the computational delay are simulated by the pure delay networks $\mathrm{e}^{-s\tau_1}$ and $\mathrm{e}^{-s\tau_2}$. In addition, it is assumed that the input signal may appear not at $t = 0$, but with a delay τ_0.

The aim of the system is to follow, as close as possible, a reference signal that is defined as a result of the transformation of a delta-function $\delta(t)$ by a forming filter $R(s)$. The desired linear transformation of the input signal is given by a transfer function $Q(s)$. The desired form of the control signal $\hat{u}(t)$ is given in the form of the Laplace image $\hat{U}(s) = Q_u(s)R(s)\mathrm{e}^{-s\tau_0}$, where the transforming function $Q_u(s)$ must be chosen in a special way Polyakov (1999). In order to reach the desired frequency properties of the system dynamic weighting functions $V_1(s)$ and $V_2(s)$ are introduced as shown in Fig. B.4. The optimization criterion appears as the sum of the integral quadratic output and control errors $z_y(t)$ and $z_u(t)$:

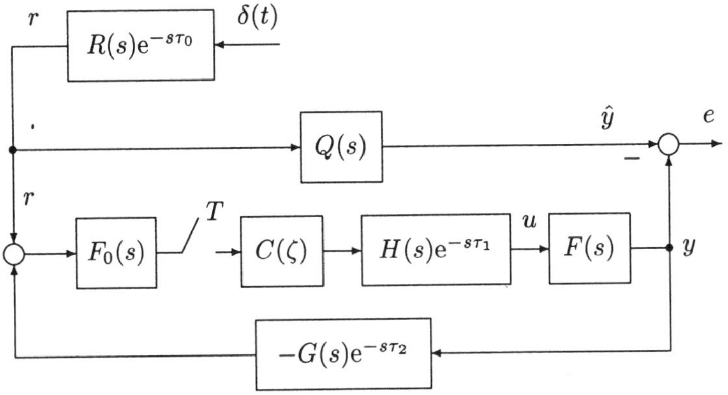

Figure B.3: Block-diagram of digital tracking system

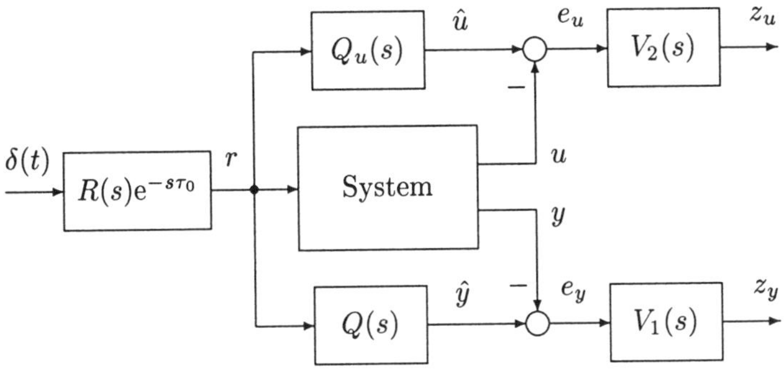

Figure B.4: Design block-diagram of tracking system

$$J_2 = \int_0^\infty z_y^2(t)\,\mathrm{d}t + \int_0^\infty z_u^2(t)\,\mathrm{d}t \;\to \min$$

This problem is solved by means of the macro `trackh2`:

```
[C,Poles] = trackh2 ( F, H, G, Fo, Q, Qu, R, ...
                      V1, V2, T, Delay0, Delay1, Delay2 );
```

which returns the transfer function of the optimal controller and the poles of the closed-loop system.

Such a system can be an object of robust optimization that is equivalent to the minimization of the $\mathcal{H}_\infty$−norm of an equivalent discrete-time system. In Polyakov (1999) it is shown that with proper choice of weighting functions such an approach allows one to obtain good system tracking capabilities and robustness. The $\mathcal{H}_\infty$−optimization problem is solved by the macro `trackinf`:

```
[C,Poles,Lam] = trackinf ( F, H, G, Fo, Q, Qu, R, ...
                           V1, V2, T, Delay0, Delay1, Delay2 );
```

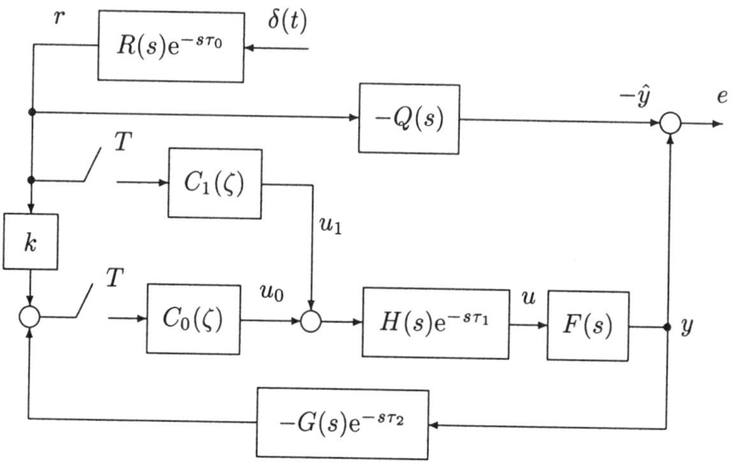

Figure B.5: System with two controllers

B.5.2 Systems with two controllers

The *DirectSD* toolbox can be used for analysis and design of SD systems with two controllers (Fig. B.5). Dependent on the value k, it can be a two-degrees of freedom system (for $k = 0$) or a $2\frac{1}{2}$ degrees-of-freedom system (for $k = 1$) Grimble (1995). The synthesis is implemented in two stages. First, we construct a stabilizing controller C_0 that ensures desired properties of the closed loop (for instance, by a quadratic or $\mathcal{H}_\infty$−criterion using the macro `trackh2` or `trackinf`, respectively). Then, by means of the macro `lq2dof` the optimal tracking controller C_1 is designed:

```
C1 = lq2dof ( F, H, G, C0, Q, Qu, R, V1, V2, T, k, ...
              Delay0, Delay1, Delay2 );
```

B.6 Other possibilities

B.6.1 Analysis of SD systems in continuous time

The *DirectSD* toolbox includes the following procedures for the analysis of open-loop and closed-loop SD systems under stochastic and deterministic disturbances:

- macro `varfil` calculates the instantaneous and average error variance for the filtering system shown in Fig. B.2:
  calculating the instantaneous variance at specified time instants

```
t = 0:0.2*T:T;
vart = varfil ( F, Fo, H, Sx, Sn, C, T, Delay, q, t );
```

calculating the average variance over the sampling period

```
varm = varfil ( F, Fo, H, Sx, Sn, C, T, Delay, q );
```

- macro **varfdb** calculates the instantaneous and average error and control variance in the closed-loop stochastic system shown in Fig. 18.1: calculating the instantaneous error and control variance at specified time instants

```
t = 0:0.2*T:T;
[vart,varut] = varfdb (F1, F2, H, G, Fw, Fm, V1, V2, ...
                       C, T, Delay1, Delay2, t );
```

calculating the average variance over the sampling period

```
[varm,varum] = varfdb (F1, F2, H, G, Fw, Fm, V1, V2, ...
                       C, T, Delay1, Delay2 );
```

- macro **errtrack** calculates the integral quadratic output and control errors in the tracking system shown in Fig. B.3:

```
[Err,ErrU] = errtrack ( F, H, G, Fo, C, Q, Qu, R, V1, V2, ...
                        T, Delay0, Delay1, Delay2 );
```

- macro **err2dof** calculates the integral quadratic output and control errors of the tracking system shown in the system with two controllers in Fig. B.5:

```
[Err,ErrU] = err2dof ( F, H, G, C0, C1, Q, Qu, R, V1, V2, ...
                       k, T, Delay0, Delay1, Delay2 );
```

B.6.2 Synthesis with a mixed $\mathcal{H}_2$–$\mathcal{H}_\infty$ criterion

Sometimes a compromise solution is required that can be obtained by including both $\mathcal{H}_2-$ and $\mathcal{H}_\infty-$terms into a single cost function. Such an approach was proposed for stochastic systems in Grimble (1995). The ***DirectSD*** toolbox realizes this idea as follows:

- design of an optimal closed-loop stabilization systems with a mixed $\mathcal{H}_2$–$\mathcal{H}_\infty$ criterion is performed using the macro **sdhinf**:

```
[C,Poles,Lam] = sdhinf ( F1, F2, H, G, Fw, Fm, V1, V2, ...
                         T, Delay1, Delay2, WeightH2 );
```

where the last input argument is a scalar weight of the quadratic term in the cost function.

- design of optimal tracking systems with a mixed $\mathcal{H}_2$–$\mathcal{H}_\infty$ criterion is performed using the macro **trackinf** in a similar way:

```
[C,Poles,Lam] = trackinf ( F, H, G, Fo, Q, Qu, R, V1, V2, ...
                           T, Delay0, Delay1, Delay2, WeightH2 );
```

B.6.3 Demonstration programs

The *DirectSD* toolbox includes the following demonstration programs:

- `demofil`: optimal digital filtration of continuous-time signals
- `demofdb`: design of $\mathcal{H}_2$–optimal closed-loop stochastic systems
- `demohinf`: design of $\mathcal{H}_\infty$–optimal closed-loop systems
- `demomix`: design of optimal closed-loop stochastic systems using a mixed $\mathcal{H}_2$–$\mathcal{H}_\infty$ criterion
- `demoeqff`: model matching problem for open-loop systems
- `demotrh2`: design of optimal tracking system
- `demotrhi`: design of $\mathcal{H}_\infty$–optimal tracking system
- `demo2dof`: design of optimal tracking system with two controllers (systems with 2 or $2\frac{1}{2}$ degrees of freedom)

Demonstration programs also include examples of using the analysis macros of the *DirectSD* toolbox.

B.7 Numerical examples

B.7.1 Design of optimal stabilizing system

Let us demonstrate the use of the *DirectSD* toolbox for the design of an optimal digital controller for the stochastic stabilizing system shown in Fig. 18.1. This structure includes the structure of Fig. 17.9. For the special values of Example 17.5, we have to choose

$$
\begin{aligned}
&F_1(s) = H(s) = G(s) = V_1(s) = 1\,, \qquad F_m(s) = V_2(s) = 0\\
&F_2(s) = \frac{s+0.5}{s^2-s}\,, \qquad F_w(s) = \frac{2}{s+2}\\
&T = 0.2\,, \quad \tau_1 = 0.093\,, \quad \tau_2 = 0\,.
\end{aligned}
$$

The macros should be called in the following order:

```
F1 = tfn ( 1, 1 );
F2 = tfn ( [1 0.5], [1 -1 0] );
H  = tfn ( 1, 1);
G  = tfn ( 1, 1);
V1 = tfn ( 1, 1);
V2 = tfn ( 0, 1);
Fw = tfn ( 2, [1 2] );
Fm = tfn ( 0, 1 );
T  = 0.2;
Delay1 = 0.093;
Delay2 = 0;
[C, Poles] = optfdb ( F1, F2, H, G, Fw, Fm, V1, V2, ...
                      T, Delay1, Delay2 );
```

As a result, we obtain the optimal controller (17.90). It was shown in Example 17.5 that this controller ensures good performance of the closed system. Furthermore this example demonstrates that the synthesis on the basis of a discrete-time model can lead to useless systems that are close to the stability boundary.

B.7.2 Redesign problem

The *DirectSD* toolbox allows one to solve the redesign problem which consists in replacing an analogue controller by a digital one in such a way that the transient in the redesigned system is as close to the original as possible. It is easy to see that this problem can be reformulated as a problem of optimal tracking system design (Fig. B.3) for the case when $Q(s)$ is equal to the transfer function of the initial continuous-time system and $R(s) = 1/s$.
Let us construct the optimal digital controller for the system shown in Fig. B.3 with the special values of the example given in Rattan (1989):

$$F_0(s) = H(s) = G(s) = V_1(s) = 1, \quad V_2(s) = 0$$
$$F(s) = \frac{10}{s(s+1)}, \quad (s) = \frac{1}{s}$$
$$T = 0.04, \qquad \tau_1 = 0.01, \quad \tau_0 = \tau_2 = 0.$$

The continuous-time controller has the form

$$C_{cont}(s) = \frac{0.416s+1}{0.139s+1}$$

so that

$$Q(s) = \frac{29.9281s + 71.9424}{s^3 + 8.1942s^2 + 37.1223s + 71.9424}.$$

The script to solve this problem with the *DirectSD* toolbox is given below:

```
F  = tfn ( 10, [1 1 0] );
H  = tfn ( 1, 1 );
G  = tfn ( 1, 1 );
Fo = tfn ( 1, 1 );
Q  = tfn ( [29.9281 71.9424], ...
           [1.0000 8.1942 37.1223 71.9424] );
Qu = tfn(0,1);
R  = tfn ( 1, [1 0] );
T  = 0.04;
Delay0 = 0;
Delay1 = 0.01;
Delay2 = 0;
V1 = tfn ( 1, 1 );
V2 = tfn ( 0, 1 );
[C,Poles] = trackh2 ( F, H, G, Fo, Q, Qu, R, V1, V2, ...
                     T, Delay0, Delay1, Delay2 );
```

As a result, we obtain

$$C_{opt}(\zeta) = \frac{4.0371 - 7.8317\zeta + 4.3600\zeta^2 - 0.8149\zeta^3 + 0.2640\zeta^4}{1 - 1.2124\zeta - 0.0909\zeta^2 + 0.3104\zeta^3 + 0.0106\zeta^4 - 0.0001\zeta^5}.$$

Let us compare this controller with those obtained using other known redesign methods. The widely known Tustin transformation gives

$$C_T(\zeta) = \frac{2.742(1 - 0.9083\zeta)}{1 - 0.7484\zeta}.$$

In Rattan (1989) the following controller was constructed:

$$C_R(\zeta) = \frac{3.704(1 - 0.8785\zeta)}{1 - 0.55\zeta}.$$

The transients for the redesigned systems with the digital controllers $C_{opt}(\zeta)$,

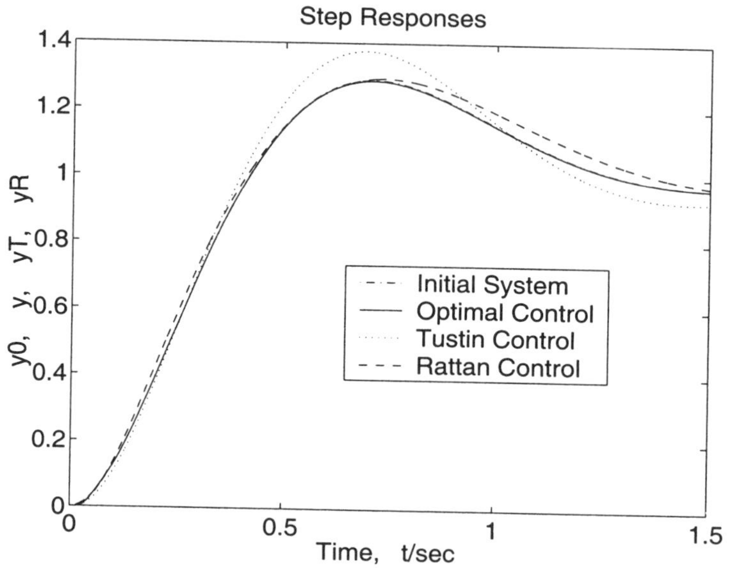

Figure B.6: Transients for redesigned systems

$C_T(\zeta)$, and $C_R(\zeta)$ are shown, together with that of the initial continuous-time system, in Fig. B.6. The controller $C_T(\zeta)$ gives the worst redesign quality (integral quadratic error $39 \cdot 10^{-4}$), controller $C_R(\zeta)$ takes the second place (error $12 \cdot 10^{-4}$), while the optimal controller $C_{opt}(\zeta)$ ensures almost complete coincidence with respect to the graphical resolution (error $0.0288 \cdot 10^{-4}$).

B.8 Conclusions

As was shown above, the ***DirectSD*** toolbox for the MATLAB environment makes it possible to solve a wide class of problems of analysis and direct design of SD systems in continuous time. It uses new theoretical ideas based on the parametric transfer function concept that is free of many difficulties and restrictions of traditional methods for analysis and design of SD systems. Though the toolbox can be used currently only for the investigation of SISO systems, a multivariable version of the toolbox is being prepared.

Bibliography

J. Ackermann. *Sampled-Data Control Systems: Analysis and Synthesis, Robust System Design.* Springer-Verlag, Berlin, 1985.

J. Ackermann. *Abtastregelung.* Springer-Verlag, Berlin, 3rd edition, 1988.

F.A. Aliev, V.B. Larin, K.I. Naumenko, and V.I. Suntsev. *Optimization of linear time-invariant control systems.* Naukova Dumka, Kiev, 1978. (in Russian).

B.D.O. Anderson. Controller design: Moving from theory to practice. *IEEE Control Systems*, 13(4):16–24, 1993.

B.D.O. Anderson and J.B. Moore. *Optimal Filtering.* Prentice-Hall, Englewood Cliffs, NJ, 1979.

M. Araki, T. Hagiwara, and Y. Ito. Frequency response of sampled-data systems II. Closed-loop considerations. In *Proc. 12th IFAC World Congr.*, volume 7, pages 293–296, Sydney, 1993.

M. Araki and Y. Ito. Frequency response of sampled-data systems I. Open-loop considerations. In *Proc. 12th IFAC World Congr.*, volume 7, pages 289–292, Sydney, 1993.

K.J. Åström. *Introduction to stochastic control theory.* Academic Press, NY, 1970.

K.J. Åström, P. Hagender, and J. Sternby. Zeros of sampled-data systems. *Automatica*, 20(4):31–38, 1984.

K.J. Åström and B. Wittenmark. *Computer controlled systems: Theory and design.* Prentice-Hall, Englewood Cliffs, NJ, 1984.

B.A. Bamieh and J.B. Pearson. A general framework for linear periodic systems with applications to $\mathcal{H}_\infty$ sampled-data control. *IEEE Trans. Autom. Contr*, AC-37(4):418–435, 1992a.

B.A. Bamieh and J.B. Pearson. The H_2 problem for sampled-data systems. *Syst. Contr. Lett*, 19(1):1–12, 1992b.

B.A. Bamieh, J.B. Pearson, B.A. Francis, and A. Tannenbaum. A lifting technique for linear periodic systems with applications to sampled-data control systems. *Syst. Contr. Lett*, 17:79–88, 1991.

Ch. Blanch. Sur les equation differentielles lineares a coefficients lentement variable. *Bull. technique de la Suisse romande*, 74:182–189, 1948.

S. Bochner. *Lectures on Fourier Integrals.* University Press, Princeton, NJ, 1959.

G.D. Brown, M.G. Grimble, and D. Biss. A simple efficient $\mathcal{H}_\infty$ controller algorithm. In *Proc. 26th IEEE Conf. Decision Contr*, Los Angeles, 1987.

S.S.L. Chang. *Synthesis of optimum control systems.* McGraw Hill, New York, Toronto, London, 1961.

T. Chen and B.A. Francis. *Optimal sampled-data control systems.* Springer-Verlag, Berlin, Heidelberg, New York, 1995.

T.A.C.M. Claasen and W.F.G. Mecklenbräuker. On stationary linear time varying systems. *IEEE Trans. Circuits and Systems*, CAS-29(2):169–184, 1982.

R.E. Crochiere and L.R. Rabiner. *Multirate digital signal processing.* Prentice-Hall, Englewood Cliffs, NJ, 1983.

J.A. Daletskii and M.G. Krein. *Stability of solutions of differential equations in Banach-space.* Nauka, Moscow, 1970. (in Russian).

C.E. de Souza and G.C. Goodwin. Intersample variance in discrete minimum variance control. *IEEE Trans. Autom. Contr*, AC-29:759–761, 1984.

G. Doetsch. *Anleitung zum praktischen Gebrauch der Laplace Transformation und z-Transformation.* Oldenbourg, München, Wien, 1967.

J.C. Doyle. Guaranteed margins for LQG regulators. *IEEE Trans. Autom. Contr*, AC-23(8):756–757, 1978.

G.E. Dullerud. *Control of uncertain sampled-data systems.* Birkhäuser, Boston, 1996.

EUROPOLY. The European network of excellence for industrial application of polynomial design methods, since 1998. Coord. M. Sebec, http://www.utia.cas.cz/europoly.

A. Feuer and G.C. Goodwin. Generalised sample and hold functions - frequency domain analysis of robustness, sensitivity and intersampling difficulties. *IEEE Trans. Autom. Contr*, AC-39(5):1042–1047, 1994.

A. Feuer and G.C. Goodwin. *Sampling in digital signal processing and control.* Birkhäuser, Boston-Basel-Berlin, 1996.

N. Fliege. *Multiraten-Signalverarbeitung.* B.G. Teubner, Stuttgart, 1993.

G.F. Franklin, J.D. Powell, and H.L. Workman. *Digital control of dynamic systems.* Addison Wesley, New York, 1990.

B. Friedland. *Advanced control system design.* Prentice-Hall, Englewood Cliffs, NJ, 1995.

F.R. Gantmacher. *The theory of matrices.* Chelsea, New York, 1959.

G.C. Goodwin and M. Salgado. Frequency domain sensitivity functions for continuous-time systems under sampled-data control. *Automatica*, 30(8): 1263–1270, 1994.

M.J. Grimble. *Robust Industrial Control.* Prentice-Hall, UK, 1995.

M.J. Grimble and V. Kučera, editors. *Polynomial methods for control systems design.* Springer, London, 1996.

M. Günther. *Kontinuierliche und zeitdiskrete Regelungen.* B.G. Teubner, Stuttgart, 1997.

T. Hagiwara and M. Araki. FR-operator approach to the $\mathcal{H}_2$-analysis and synthesis of sampled-data systems. *IEEE Trans. Autom. Contr*, AC-40(8): 1411–1421, 1995.

S.R. Hall. Comments on two methods for designing a digital equivalent to a continuous control system. *IEEE Trans. Autom. Contr*, AC-39:420–421, 1994.

S. Hara, H. Fujioka, and P.T. Kabamba. A hybrid state-space approach to sampled-data feedback control. *Linear Algebra and Its Applications*, 205-206:675–712, 1994.

S. Hara, R. Kondo, and H. Katori. Properties of zeros in digital control systems with computational time-delay. *Int. J. Control*, 49:493–511, 1989.

R. Isermann. *Digitale Regelungssysteme. Band I: Grundlagen, deterministische Regelungen. Band II: Stochastische Regelungen, Mehrgrößenregelungen Adaptive Regelungen, Anwendungen.* Springer-Verlag, Berlin, 2 edition, 1987.

M.A. Jevgrafov. *Analytic functions.* Nauka, Moscow, 1965. (in Russian).

G. Jorke, B.P. Lampe, and N. Wengel. *Arithmetische Algorithmen der Mikrorechentechnik.* Verlag Technik, Berlin, 1989.

E.I. Jury. *Sampled-data control systems.* John Wiley, New York, 1958.

P.T. Kabamba and S. Hara. Worst-case analysis and design of sampled-data control systems. *IEEE Trans. Autom. Contr*, AC-38(9):1337–1357, 1993.

T. Kaczorek. *Linear control systems*, volume II - Synthesis of multivariable systems. J. Wiley, New York, 1993.

T. Kailath. *Linear Systems.* Prentice Hall, Englewood Cliffs, NJ, 1980.

R. Kalman and J.E. Bertram. A unified approach to the theory of sampling systems. *J. Franklin Inst.*, 267:405–436, 1959.

S. Karlin. *A first course in stochastic processes.* Academic Press, New York, 1966.

V.J. Katkovnik and R.A. Polnektov. *Discrete multidimensional control.* Nauka, Moscow, 1966. (in Russian).

J.P. Keller and B.D.O. Anderson. H_∞-Optimierung abgetasteter Regelsysteme. *Automatisierungstechnik*, 40(4):114–123, 1993.

P.T. Khargonekar and N. Sivarshankar. $\mathcal{H}_2$-optimal control for sampled-data systems. *Systems & Control Letters*, 18:627–631, 1992.

V. Kučera. *Discrete Linear Control.* Academic Press, Prague, 1979.

V. Kučera. *Analysis and Design of Discrete Linear Control Systems.* Prentice Hall, London, 1991.

B.C. Kuo and D.W. Peterson. Optimal discretization of continuous-data control systems. *Automatica*, 9(1):125–129, 1973.

H. Kwakernaak. Minimax frequency domain performance and robustness optimisation of linear feedback systems. *IEEE Trans. Autom. Contr*, AC-30 (10):994–1004, 1985.

H. Kwakernaak. The polynomial approach to $\mathcal{H}_\infty$ regulation. In *Lecture Notes in Mathematics, $\mathcal{H}_\infty$ control theory*, volume 1496, pages 141–221. Springer-Verlag, London, 1990.

B.P. Lampe, G. Jorke, and N. Wengel. *Algorithmen der Mikrorechentechnik.* Verlag Technik, Berlin, 1984.

B.P. Lampe and U. Richter. Digital controller design by parametric transfer functions - comparison with other methods. In *Proc. 3. Int. Symp. Methods Models Autom. Robotics*, volume 1, pages 325–328, Miedzyzdroje, Poland, 1996.

B.P. Lampe and U. Richter. Experimental investigation of parametric frequency response. In *Proc. 4. Int. Symp. Methods Models Autom. Robotics*, pages 341–344, Miedzyzdroje, Poland, 1997.

B.P. Lampe and E.N. Rosenwasser. Design of hybrid analog-digital systems by parametric transfer functions. In *Proc. 32nd CDC*, pages 3897–3898, San Antonio, TX, 1993.

B.P. Lampe and E.N. Rosenwasser. Application of parametric frequency response to identification of sampled-data systems. In *Proc. 2. Int. Symp. Methods Models Autom. Robotics*, volume 1, pages 295–298, Miedzyzdroje, Poland, 1995.

B.P. Lampe and E.N. Rosenwasser. Best digital approximation of continuous controllers and filters in $\mathcal{H}_2$. In *Proc. 41st KoREMA*, volume 2, pages 65–69, Opatija, Croatia, 1996.

B.P. Lampe and E.N. Rosenwasser. Parametric transfer functions for sampled-data systems with time-delayed controllers. In *Proc. 36th IEEE Conf. Decision Contr*, pages 1609–1614, San Diego, CA, 1997a.

B.P. Lampe and E.N. Rosenwasser. Sampled-data systems: The L_2–induced operator norm. In *Proc. 4. Int. Symp. Methods Models Autom. Robotics*, pages 205–207, Miedzyzdroje, Poland, 1997b.

F.H. Lange. *Signale und Systeme*, volume 1–3. Verlag Technik, Berlin, 1971.

B. Lennartson and T. Söderström. Investigation of the intersample variance in sampled-data control. *Int. J. Control*, 50:1587–1602, 1990.

B. Lennartson, T. Söderström, and Sun Zeng-Qi. Intersample behavior as measured by continuous-time quadratic criteria. *Int. J. Control*, 49:2077–2083, 1989.

O. Lingärde and B. Lennartson. Frequency analysis for continuous-time systems under multirate sampled-data control. In *Proc. 13th IFAC World Congr*, volume 2a–10, 5, pages 349–354, San Francisco, USA, 1996.

A.G. Madievski and B.D.O. Anderson. A lifting technique for sampled-data controller reduction for closed-loop transfer function consideration. In *Proc. 32nd IEEE Conf. Decision Contr*, pages 2929–2930, San Antonio, TX, 1993.

S.G. Michlin. *Lectures on linear integral equations.* GIFML, Moscow, 1953. (in Russian).

T.P. Perry, G.M.H. Leung, and B.A. Francis. Performance analysis of sampled-data control systems. *Automatica*, 27(4):699–704, 1991.

U. Petersohn, H. Unger, and Wardenga W. Beschreibung von Multirate-Systemen mittels Matrixkalkül. *AEÜ*, 48(1):34–41, 1994.

J.P. Petrov. *Design of optimal control systems under incompletely known input disturbances.* University press, Leningrad, 1987. (in Russian).

C.L. Phillips and H.T. Nagle. *Digital control system analysis and design.* Prentice-Hall, Englewood Cliffs, NJ, 3rd edition, 1995.

K.Y. Polyakov. An algorithm for synthesis of digital control systems on the basis of the parametric transfer function method. *Izd. Russ. Akad. Nauk*, 3: 32–39, 1998. (in Russian).

K.Y. Polyakov. Polynomial design of optimal sampled-data tracking systems. *Automatika i Telemechanika*, 1999. (to appear, in Russian).

K.Y. Polyakov, E.N. Rosenwasser, and B.P. Lampe. Robust sampled-data control systems design on basis of parametric transfer function method. In *Proc. Int. Conf. on Computer Methods for Control Systems*, pages 15–21, Szczecin, Poland, December 1997.

K.Y. Polyakov, E.N. Rosenwasser, and B.P. Lampe. Direct optimal design of digital control tracking system with neutral plants. In *Proc. 5. Int. Symp. Methods Models Autom. Robotics*, volume 2, pages 433–436, Miedzyzdroje, Poland, 1998.

K.Y. Polyakov, E.N. Rosenwasser, and B.P. Lampe. Associated $\mathcal{H}_\infty$−problem for sampled-data systems. In *Proc. Conf. Adaptive Control*, pages –, St. Petersburg, Russia, Sept. 1999a.

K.Y. Polyakov, E.N. Rosenwasser, and B.P. Lampe. DirectSD - a toolbox for direct design of sampled-data systems. In *Proc. IEEE Intern. Symp. CACSD'99*, pages 357–362, Kohala Coast, Island of Hawai'i, Hawai'i, USA, 1999b.

I.I. Priwalow. *Einführung in die Funktionentheorie.* 3. Aufl., B.G. Teubner, Leipzig, 1967.

J.R. Ragazzini and G.F. Franklin. *Sampled-data control systems.* McGraw-Hill, New York, 1958.

J.R. Ragazzini and L.A. Zadeh. The analysis of sampled-data systems. *AIEE Trans.*, 71:225–234, 1952.

K.S. Rattan. Digitalization of existing control systems. *IEEE Trans. Autom. Contr*, AC-29:282–285, 1984.

K.S. Rattan. Compensating for computational delay in digital equivalent of continuous control systems. *IEEE Trans. Autom. Contr*, AC-34:895–899, 1989.

E.N. Rosenwasser. *Vibrations of nonlinear systems, the method of integral equations.* Techn. Information Service, Springfield, VA, 1970.

E.N. Rosenwasser. *Periodically nonstationary control systems.* Nauka, Moscow, 1973. (in Russian).

E.N. Rosenwasser. *Lyapunov indices in linear control theory.* Nauka, Moscow, 1977. (in Russian).

E.N. Rosenwasser. Discrete stabilisation of linear continuous-time plants. I. Algebraic theory of complete linear plants. *Automation and Remote Control*, 55(7, Part 1):974–987, 1994b.

E.N. Rosenwasser. Discrete stabilisation of linear continuous-time plants. II. Construction of the set of stabilising programs. *Automation and Remote Control*, 55(8, Part 1):1148–1160, 1994a.

E.N. Rosenwasser. *Linear theory of digital control in continuous time*. Nauka, Moscow, 1994c. (in Russian).

E.N. Rosenwasser. Mathematical description and analysis of multivariable sampled-data systems in continuous-time - I. Parametric transfer functions and weight functions of multivariable sampled-data systems. *Automation and Remote Control*, 56(4, Part 1):526–540, April 1995a.

E.N. Rosenwasser. Minimization of a class of functionals with applications to sampled-data systems theory. *Dokl. Russ. Akad. Nauk*, 342(3):326–329, 1995b. (in Russian).

E.N. Rosenwasser. Synthesis of multivariable linear systems with a given characteristic polynomial. *Automation and Remote Control*, 57(8):1091–1107, 1996a.

E.N. Rosenwasser. Transfer functions and impulse response of multivariable continuous-digital systems. *Dokl. Russ. Akad. Nauk*, 346(5):606–609, 1996b. (in Russian).

E.N. Rosenwasser and B.P. Lampe. *Digitale Regelung in kontinuierlicher Zeit - Analyse und Entwurf im Frequenzbereich*. B.G. Teubner, Stuttgart, 1997.

E.N. Rosenwasser and K.Y. Polyakov. Optimal SISO digital control system design under uncertain disturbances. Technical Report A-440, State Marine Tech. Univ., St. Petersburg, Russia, 1996. (in Russian).

E.N. Rosenwasser, K.Y. Polyakov, and B.P. Lampe. Entwurf optimaler Kursregler mit Hilfe von Parametrischen Übertragungsfunktionen. *Automatisierungstechnik*, 44(10):487–495, 1996.

E.N. Rosenwasser, K.Y. Polyakov, and B.P. Lampe. Frequency domain method for $\mathcal{H}_2$–optimization of time-delayed sampled-data systems. *Automatica*, 33 (7):1387–1392, 1997.

E.N. Rosenwasser, K.Y. Polyakov, and B.P. Lampe. Optimal discrete filtering for time-delayed systems with respect to mean-square continuous-time error criterion. *Int. J. Adapt. Control Signal Process.*, 12:389–406, 1998.

E.N. Rosenwasser, K.Y. Polyakov, and B.P. Lampe. Application of Laplace transformation for digital redesign of continuous control systems. *IEEE Trans. Automat. Contr*, 4(4):883–886, April 1999a.

E.N. Rosenwasser, K.Y. Polyakov, and B.P. Lampe. Comments on "A technique for optimal digital redesign of analog controllers". *IEEE Trans. Control Systems Technology*, 7(5):633–635, September 1999b.

M. Saeki. Method of solving a polynomial equation for an $\mathcal{H}_\infty$ optimal control problem. *IEEE Trans. Autom. Contr*, AC-34:166–168, 1989.

M. Sågfors. *Optimal Sampled-Data and Multirate Control.* PhD thesis, Faculty of Chemical Engineering, Åbo Akademi University, Finland, 1998.

L. Schwartz. *Methodes mathematiques pour les sciences physiques.* Hermann 115, Paris VI, Boul. Saint-Germain, 1961.

H. Schwarz. *Optimale Regelung und Filterung - Zeitdiskrete Regelungssysteme.* Akademie-Verlag, Berlin, 1981.

L.S. Shieh, B.B. Decrocq, and J.L. Zhang. Optimal digital redesign of cascaded analogue controllers. *Optimal Control Appl. Methods*, 12:205–219, 1991.

L.S. Shieh, J.L. Zhang, and J.W. Sunkel. A new approach to the digital redesign of continuous-time controllers. *Control Theory Adv. Techn.*, 8:37–57, 1992.

V.B. Sommer, B.P. Lampe, and E.N. Rosenwasser. Experimental investigations of analog-digital control systems by frequency methods. *Automation and Remote Control*, 55(Part 2):912–920, 1994.

E.D. Sontag. *Mathematical control theory – deterministic finite dimensional systems.* Springer-Verlag, New York, 1998.

A. Spring and H. Unger. Schnelles Identifikationsverfahren für rekursive Multirate-Systeme. *FREQUENZ*, 48(3–4):72–78, 1994.

D.S. Stearns. *Digitale Verarbeitung analoger Signale.* R. Oldenbourg Verlag, München, 1988.

R.F. Stengel. *Stochastic optimal control. Theory and application.* J. Wiley & Sons, Inc., New York, 1986.

I.Z. Stokalo. Generalisation of symbolic method principal formula onto linear differential equations with variable coefficients. *Dokl. Akad. Nauk SSR*, 42: 9–10, 1945. (in Russian).

E.C. Titchmarsh. *The theory of functions.* Oxford, 1932.

H.T. Toivonen. Sampled-data control of continuous-time systems with an $\mathcal{H}_\infty$-optimality criterion. *Automatica*, 28(1):45–54, 1992.

H.T. Toivonen. Worst-case sampling for sampled-data H_∞ design. In *Proc. 32nd IEEE Conf. Decision Contr*, pages 337–342, San Antonio, TX, 1993.

J. Tou. *Digital and Sampled-Data Control Systems.* McGraw-Hill, New York, 1959.

H.L. Trentelmann and A.A. Stoorvogel. Sampled-data and discrete-time $\mathcal{H}_2$–optimal control. In *Proc. 32nd Conf. Dec. Contr.*, pages 331–336, San Antonio, TX, 1993.

H. Unger, U. Petersohn, and S. Lindow. Zur Beschreibung hybrider Multiraten-Systeme mittels Matrixkalküls. *FREQUENZ*, 1997. (einger.).

K.G. Valejew. Application of Laplace transform for analysis of linear systems. In *Proc. Intern. Conf. on Nonlin. Oscill.*, volume I, pages 126–132, Kiev, 1970. (in Russian).

B. van der Pol and H. Bremmer. *Operational calculus based on the two-sided Laplace integral.* University Press, Cambridge, 1959.

L.N. Volgin. *Optimal discrete control of dynamic systems.* Nauka, Moscow, 1986. (in Russian).

S. Volovodov, B.P. Lampe, and E.N. Rosenwasser. Application of method of integral equations for analysis of complex periodic behaviors in Chua's circuits. In *Int. Conf. Control of Oscillations and Chaos*, St. Petersburg, Russia, August 1997.

S.K. Volovodov, E.N. Rosenwasser, and Smolnikov A.V. Device for determination of frequency response of sampled-data control systems. *Russian patent*, (71620993), 1991. Disclosure and registration.

A. Weinmann. *Uncertain models and robust control.* Springer-Verlag, Wien, 1991.

E.T. Whittaker and G.N. Watson. *A course of modern analysis.* University Press, Cambridge, 4 edition, 1927.

R.A. Yackel, B.C. Kuo, and G. Singh. Digital redesign of continuous systems by matching of states at multiple sampling periods. *Automatica*, 10:105–111, 1974.

Y. Yamamoto. A function space approach to sampled-data systems and tracking problems. *IEEE Trans. Autom. Contr*, AC-39(4):703–713, 1994.

Y. Yamamoto and P. Khargonekar. Frequency response of sampled-data systems. *IEEE Trans. Autom. Contr*, AC-41(2):161–176, 1996.

D.C. Youla, H.A. Jabr, and J.J. Bongiorno (Jr.). Modern Wiener-Hopf design of optimal controllers. Part II - The multivariable case. *IEEE Trans. Autom. Contr*, AC-21(3):319–338, 1976.

L.A. Zadeh. Circuit analysis of linear varying-parameter networks. *J. Appl. Phys.*, 21(6):1171–1177, 1950.

L.A. Zadeh. On stability of linear varying-parameter systems. *J. Appl. Phys.*, 22(4):202–204, 1951.

K. Zhou and J.C. Doyle. *Essentials of robust control.* Prentice-Hall Intern., Upper Saddle River, NJ, 1998.

K. Zhou, J.C. Doyle, and K. Glover. *Robust and optimal control.* Prentice-Hall, Englewood Cliffs, NJ, 1996.

J.Z. Zypkin. *Sampling systems theory.* Pergamon Press, New York, 1964.

Index